Fundamentals of
MECHANICAL DESIGN

Fundamentals of
MECHANICAL DESIGN

RICHARD M. PHELAN
Professor of Mechanical Engineering, Cornell University

International Student Edition

McGraw-Hill Book Company, Inc.

New York   San Francisco
Toronto   London   *Second Edition*

Kōgakusha Company, Ltd.

Tokyo

FUNDAMENTALS OF MECHANICAL DESIGN

INTERNATIONAL STUDENT EDITION

Exclusive rights by Kōgakusha Co., Ltd. for manufacture and export from Japan. This book cannot be re-exported from the country to which it is consigned by Kōgakusha Co., Ltd. or by McGraw-Hill Book Company Inc. or any of its subsidiaries.

I

TOSHO PRINTING CO., LTD., TOKYO, JAPAN

To Olive

PREFACE

Most teachers of engineering subjects begin their careers while graduate students or immediately upon receiving an advanced degree. Relatively few enter the teaching profession after many years in industry, and most of those who do come from narrow fields of intense specialization. In general, regardless of background, each finds much new material to learn, or possibly to understand for the first time, and, naturally, he is content to cover the same material in pretty much the same way until something happens to disrupt the routine. As the teacher gains experience in his field, he begins questioning the whys and wherefores of what he has been doing over the past years and reaches some conclusions about the relative merits of different approaches and the relative values of topics. Although in the normal process of organizing his courses he has been able to make minor changes in coverage and emphasis, major changes can be made only by switching to another text or by writing his own.

The author would undoubtedly still be reasonably content with the traditional pattern of courses and the normal, slow rate of change had he not been confronted in 1955 with the necessity of organizing a new course which was to cover as much of mechanical design as possible in a total allotment of four semester-credit hours. The only feasible approach was to integrate material from the traditional sequence of mechanisms or kinematics, dynamics of machinery, and design of machine elements. The absence of a suitable text led to the writing of notes, which in turn led to the development of the first edition of *Fundamentals of Mechanical Design*.

Now, on the basis of six years' experience teaching this course along with other courses in the traditional sequence, the author has come to the conclusion that the integrated treatment of mechanisms, kinematics, and design of machine elements is, *by far*, the most logical and efficient way of teaching mechanical design.

It is logical because the student sees design as a whole—the end product is always in sight, and the means to the end are kept in proper perspective rather than becoming ends in themselves. For example, it has always seemed incongruous to discuss displacements of mechanisms in one course, the resulting velocities and accelerations in a second course, and finally, in a third course, the problems of strength, rigidity, wear, etc., that must be considered when converting a line into a usable, three-dimensional component.

The obvious reason for increased efficiency is a decrease in repetition; but of even greater importance is the fact that, when considered from the broad viewpoint, much of the traditional coverage is seen to be trivial *to all but the specialist,* and can be eliminated from the required course without appreciable loss.   In fact, eliminating the trivial and the resulting overemphasis of minute details will result in a better understanding of the fundamentals and increased interest in the subject of design.

The role of the mechanical engineer has been, is, and will continue to be changing rapidly.   Thus, the relative importance of topics is constantly changing, and time must be found for introducing new material. All areas within mechanical engineering are faced with this problem, but the greatest need in the design area is to find time for extending the breadth and depth of dynamics of machinery, particularly in vibrations and feedback control systems.   In this age of automation, and in view of the unlimited future possibilities, it is difficult to conceive of graduating designers, or even mechanical engineers in general, who have no understanding of the dynamics of feedback control systems.

It also seems fairly obvious that the total time allotted to the design area is not going to increase; in fact, it may even decrease.   Something must give, and something is going to give.   The question is whether the giving is to be in the direction of maintaining, or even increasing, the strength of the design sequence or in the direction of eliminating mechanical design from the required areas of study for mechanical engineers. The author believes that when design is eliminated, mechanical engineering is also eliminated.   Mechanical engineers have legitimate interests in many areas, such as thermodynamics, fluid mechanics, heat transfer, materials processing, that, although vital to mechanical engineering, are still not uniquely mechanical engineering.   However, with very few exceptions, mechanical engineers are the only engineers with any appreciable depth of education in the area of mechanical design.

The author is also convinced that a "last ditch" battle to maintain the status quo is neither practical nor desirable.   Rather, the increasing diversity of interests and the pressure of time for more mathematics, physics, etc., should be recognized, and the required sequence in design should be tailored to meet the needs of all mechanical engineers, while still providing a good background for those who wish to specialize in the area of design.

In the author's opinion, the *minimal* requirements can be met in a sequence of three courses. In terms of semester-credit hours, the sequence is:

1. A four-credit course, with three recitations and one computing period each week, that covers material from courses presently labeled mechanisms, kinematics, design of machine elements, etc.

2. A four-credit course, also with three recitations and one computing period each week, that covers material on dynamics of rigid bodies, dynamics of elastic bodies (vibrations), and dynamics of control systems.

3. A two-credit course, with two design periods each week, in which the complete design of simple machines would be considered—requiring the integration of material from other areas as well as the extension in depth of several topics previously covered in courses 1 and 2.

An additional two credits applied to the first course would be a major step in upgrading the program from minimal to adequate. The author would use them to expand the first course into a two-semester sequence consisting of two recitations plus one computing period each week. In any case, the program outlined above should be supplemented by advanced elective courses in kinematics, dynamics, vibrations, control systems, and design of machine elements and machines for those students who desire to specialize in the area of design.

This book is an attempt to provide the text for the six-credit first course and some additional material in the area of vibrations for use by those students who would not continue with the course in dynamics of machinery.

Although somewhat oversimplified, the author's philosophy may be expressed by defining mechanical design as the marriage of statics, dynamics, strength of materials, and properties of materials *with reality*. One cannot hope to do more than purely routine design without a good background in mechanics and materials, as well as in manufacturing processes; but *equally important*, one cannot do more than insignificant design with a background solely in these courses. The main justification for design courses lies in the fact that a designer must work in a real world with real materials and the accompanying factors of wear, corrosion, unknown loads, finite life, the effect of surface finish and environment on endurance strength, etc., that defy mathematical exactitude. In this text, major emphasis is given to those topics not covered, or not covered adequately, in mechanics and materials.

It should be pointed out that, although mathematics through differential equations will be encountered, no attempt has been made to use mathematics of a level higher than required in a given situation. It is the author's intention to use whatever method will assist the student in gaining a better understanding of and a deeper insight into the physical behavior of mechanical components and how to design them to meet the specifications established for a given situation. In line with this, velocity and acceleration analyses of mechanisms are treated almost entirely by use of graphical vector methods. Although many readers will have previously studied analytical vector methods, it is felt that, in the limited time available, a much better understanding is gained by seeing on paper each vector in its correct relationship to the others. The possibilities and advantages of analytical solutions in combination with high-speed digital computers in handling repetitive solutions of all types are noted, and references are given where appropriate. One important result of the advent of the digital computer is that there is no longer justification for giving time and space to certain special techniques, such as the Ritterhaus construction for finding accelerations of slider-crank mechanisms.

There are no bibliographies at the ends of the chapters, and in general, most references are to books and not papers in technical journals. The reasons are simply that the reader will benefit most by having one or, at most, a few well-chosen references given at specific points and that whenever there is an up-to-date book available, the reader will find the perspective given by the wide range of source material included in a book of more use than just a list of the original papers.

The reader will also observe that no tables of rated capacities, i.e., catalogue data, are included for some elements, while rather extensive data of commercial significance are given for others. Certain off-the-shelf items, such as V belts, chains, and ball and roller bearings, either are available in such a wide range of types and constructions or are used under such widely varying conditions of lubrication, temperature, etc., that a designer would be foolish to make his selection on the basis of tables in a textbook. Therefore, the discussion of these elements is limited to those fundamentals of operation and design that will assist the designer in making the proper choice from the many possibilities and, also, acquaint him with the peculiar characteristics and problems of each so that he may most efficiently and intelligently utilize the information presented by the manufacturers in their catalogues.

Although no single book, or for that matter any number of books, can possibly provide all the theory plus the very latest on new materials and design data required in critical applications, it is possible to design many elements, such as clutches, brakes, flat leather belts, spur gears, and journal bearings, on the basis of a textbook discussion and be certain that the design not only will work but will be reasonably efficient and economical. Many of the concepts related to these elements carry over into areas that do not lend themselves to an adequate treatment in such limited space. Thus, for an introduction to mechanical design, it seems logical (1) to emphasize those topics that can be treated in a textbook in a fairly rational and thorough manner, (2) to point out the extensions to more complicated designs, and (3) to recommend sources of information for use by the reader as the need for further study develops.

Those familiar with the first edition will find many small changes and a number of major changes and additions, including: (1) the addition of a derivation of equations for transferring the acceleration between points that, although coincident, have relative motion (the Coriolis component of acceleration); (2) an increased emphasis on continuous third derivatives in base curves for cam design; (3) the addition of an introduction to the concept of reliability and its relation to mechanical design, to give some measure of perspective with respect to the present state of knowledge and to the possibilities—as well as limitations—for the future; (4) the addition of a section on designing when elastic deformations are significant, with major emphasis on the application of graphical-numerical methods in determining slopes and deflections of stepped shafts; (5) an extensive revision of the chapter on brakes in the interest of emphasizing the lever-mounted shoe brake; (6) an entirely new treatment of the design of spur gears, which is based on the newest AGMA information and for the first time provides a reasonably rational approach in terms of allowable stresses, dynamic effects, overloads, and reliability considerations; (7) an extensive revision of the discussion on journal bearings, including, for the first time in any book, a method, based upon experimental results, for designing journal bearings for dynamic loads when the significant support is due to the squeeze-film; (8) a revision of the discussion of the load capacity of rolling-contact bearings, with increased emphasis on the statistical reliability aspects, and the extension to the determination, in terms of cumulative damage, of the basic dynamic load capacity required under any specified conditions of life and load, including loads that vary with time; and (9) the addition of a section on the application of Rayleigh's method for determining the lowest natural frequency for lateral vibrations or whirling of a shaft with many masses.

As with the first edition, the examples are presented as design problems, and the assumptions, choices, and estimates necessary in engineering applications are introduced in context. Many of the examples actually extend the coverage of the basic text.

The problems in the back of the book are numerous and cover a wide range of complexity. In most cases, the problems are set up with the broadest practical specifications, and the student must make many of the choices and assumptions that he will be faced with in his professional career. Although some complications are introduced because there will be many different satisfactory answers to these problems—unless the instructor prefers to suggest specific values—it is the author's firm belief that this important facet of mechanical design cannot be overemphasized.

The author is indebted to many of his colleagues, both in industry and in education, for their thought-provoking comments and suggestions. In the development of this edition, particular thanks are due to Professors A. H. Burr and J. F. Booker of Cornell University, E. S. Ault of Purdue University, P. H. Black of the University of Ohio, J. E. Shigley of the University of Michigan, and J. P. Vidosic of the Georgia Institute of Technology.

Richard M. Phelan

CONTENTS

CHAPTER I

INTRODUCTORY DEFINITIONS

A *machine* is a combination of rigid or resistant bodies having definite motions and capable of performing useful work. The term *mechanism* is applied to the physical arrangement that provides for the definite motion of the parts of a machine.

Kinematics of machinery describes the study of relative motions without regard to the effects of mass and force upon the motion.

Dynamics of machinery is generally considered to include not only the external forces but also the effects of balanced and unbalanced forces due to the mass and accelerations of the parts.

Since by definition the important concept of a mechanism is that of constraint so that the parts move with definite relative motions, or *constrained* motion, it is desirable to have a special term to describe the joint that permits this relative motion. The joint is called a *pair*, and the geometrical forms making up the joint are called *pairing elements*. The terms *lower* pairs and *higher* pairs are often used to distinguish between joints which permit surface contact and those which permit only point or line contact.

A *link* is a rigid body having elements of two or more pairs through which it is connected to other bodies. A *simple* link has only two pairing elements, whereas a *compound* link has more than two pairing elements.

The term *chain* is given to combinations of links. A chain, as in Fig. 1-1a, which permits no relative motion is called a *locked* chain or *structure*. A chain, as in Fig. 1-1b, in which with one link fixed every point in every other link must move in a definite path is called a *constrained* or *kinematic* chain with *one degree of freedom*.

One degree of freedom means that specifying one parameter, such as the angle θ in Fig. 1-1b, will completely define the instantaneous position of every point in the chain relative to all other points. Unless otherwise stated, all mechanisms will be assumed to have one degree of freedom. The term *unconstrained* chain is given to the type of chain in Fig. 1-1c, in which, with one link fixed, the points in the other links are not constrained to move in fixed paths.

It should be noted that, although the chain in Fig. 1-1c is unconstrained

1

with one degree of freedom, it becomes a kinematic chain with two degrees of freedom when two parameters, such as the angles θ and ϕ, are specified.

A *mechanism* is a kinematic chain with one link fixed. Thus a kinematic chain with N links may be used to give N mechanisms by making each link in sequence the fixed link. These N mechanisms are said to be *inversions* of the chain. The concept of inversion is very useful and will be discussed in more detail in later chapters.

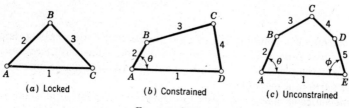

(a) Locked   (b) Constrained   (c) Unconstrained

FIG. 1-1. Chains.

A body is considered to have *plane motion* when all its points move parallel to a fixed plane. Plane motion may be either rotation, translation, or a combination of rotation and translation.

A body has *plane rotation* if its plane motion is such that each point in it remains a constant distance from a fixed axis perpendicular to the plane of motion.

A body has *plane translation* if its plane motion is such that a line connecting two points on a plane parallel to the plane of motion remains parallel to all previous positions. Plane translation is *rectilinear* if all

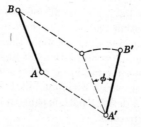

FIG. 1-2

points move in parallel straight-line paths and is *curvilinear* if all points move in identical curved paths parallel to the plane of motion. *Unless otherwise specified, the term translation is understood to mean rectilinear translation.*

Any plane motion may be shown to be the equivalent of a rotation plus a translation. Figure 1-2 shows two positions, AB and $A'B'$, of a link and one way in which the motion may be resolved into a combination of rotary and translatory motion.

Helical motion describes the motion of a body having rotation combined with translation along the axis of rotation. The nut on the screw in Fig. 1-3 is an example of a body having helical motion.

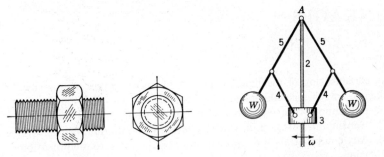

FIG. 1-3   FIG. 1-4. Centrifugal flyball governor.

A body has *spherical motion* if each point in it remains at a constant distance from a fixed point. All points in link 5 of the centrifugal flyball governor in Fig. 1-4 have spherical motion.

A *cycle* of motion is completed when the links of a mechanism have passed through all possible positions and have returned to their starting positions. The *period* is the time interval required for a complete cycle. A *phase* shows the simultaneous relative positions of the links at any instant.

CHAPTER 2

LINKAGES

A classification of mechanisms that is of major importance is that of linkages. Such mechanisms are used to convert the following:
1. Continuous rotation into continuous rotation
2. Continuous rotation into reciprocation or oscillation, or the reverse
3. Any combination of reciprocation or oscillation into reciprocation or oscillation

An infinite variety of motions may be obtained by varying proportions of the links and by combining several simple linkages into a compound mechanism.

In general, the designer is interested in determining the dimensions required to convert a given input into a desired output. This aspect of design is commonly called *synthesis* to distinguish it from the *analysis* of the input-output relationship of existing mechanisms. Although much work has been done and is in progress in the synthesis of linkages, the limited number of variables associated with a practical number of links imposes somewhat severe restrictions on the problems that can be solved.

Synthesis is design, and it will be utilized wherever possible, e.g., in the design of cams and gear trains. However, the pressures of time and space require placing major emphasis here on analysis. This chapter will be concerned with the characteristics and applications of several of the more fundamental and commonly used linkages, and the reader is referred to the books by Hall,[1] Hinkle,[2] and Faires[3] for more detailed information on synthesis and to the set of books edited by Jones[4] and Horton[5] for detailed descriptions of many ingenious mechanisms.

[1] A. S. Hall, Jr., "Kinematics and Linkage Design," Prentice-Hall, Inc., Englewood Cliffs, N.J., 1961.
[2] R. T. Hinkle, "Kinematics of Machines," 2d ed., Prentice-Hall, Inc., Englewood Cliffs, N.J., 1960.
[3] V. M. Faires, "Kinematics," McGraw-Hill Book Company, Inc., New York, 1959.
[4] F. D. Jones (ed.), "Ingenious Mechanisms for Designers and Inventors," vols. I and II, The Industrial Press, New York, 1930, 1936.
[5] H. L. Horton (ed.), "Ingenious Mechanisms for Designers and Inventors," vol. III, The Industrial Press, New York, 1951.

4

This chapter will be concerned with several of the more fundamental and commonly used linkages.

2-1. Four-bar Linkages. The simplest complete kinematic chain has four links, as shown in Fig. 2-1a and b.   In all figures of four-bar mechanisms, except for the discussion in Sec. 2-11 on inversion, 1 will be the fixed link, 2 the driver, 3 the connecting link, and 4 the follower.

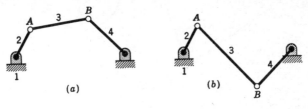

Fig. 2-1

2-2. Oscillation of Driver and Follower.   Figure 2-2 shows the case where the proportions are such that both the driver and follower can only oscillate.   The four limiting phases are shown dotted.   Point A moves through the arc $A'A''A'''A''''$ and back to A', while B moves through

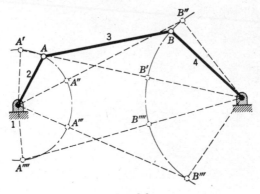

Fig. 2-2

the arc $B'B''B'''B''''$ and back to B'.   It should be evident that, when A is at A' or A'''' and links 3 and 4 are collinear, there is no tendency for the rotation of link 2 to cause rotation of link 4.   These phases are known as *dead-center* phases.   Therefore, the arc of motion of A must be smaller than $A'A''''$ if operation is to be smooth.   The dead-center phases for the case with link 4 as the driver are those when A is at A'' and A'''.

2-3. Rotation of Driver and Oscillation of Follower.   Figure 2-3a and b shows two four-bar linkages that permit the driver to rotate completely

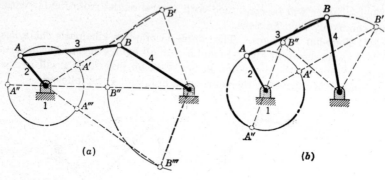

(a)   (b)

FIG. 2-3

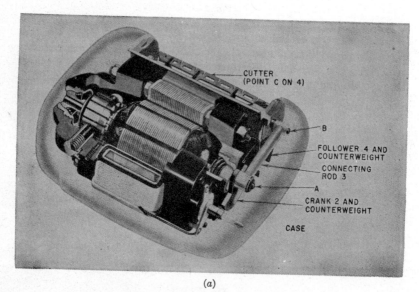

CUTTER
(POINT C ON 4)

B

FOLLOWER 4 AND
COUNTERWEIGHT

CONNECTING
ROD 3

A

CRANK 2 AND
COUNTERWEIGHT

CASE

(a)

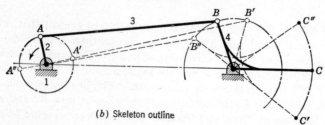

(b) Skeleton outline

FIG. 2-4. The Sunbeam Shavemaster. (Courtesy Sunbeam Corporation.)

while the follower oscillates. The important difference between the two mechanisms is that, in Fig. 2-3a, A'' and B'' are simultaneous dead-point positions of A and B. This means that the motion of the follower is not completely constrained, in that, as A moves through A'', B may move from B'' toward either B' or B'''.

Figure 2-4 shows this linkage as applied to the Sunbeam Shavemaster electric shaver. The crank 2 is directly connected to the universal motor shaft and rotates at a uniform speed of approximately 8,600 rpm. The cutter C is attached to the follower 4, as shown in the skeleton outline in Fig. 2-4b, and makes one complete cycle of oscillation per revolution of the motor. The counterweights, on the crank and follower in Fig. 2-4a, are utilized in minimizing vibration by counterbalancing the inertia forces and couples resulting from the accelerations of the several parts.

The book by Hrones and Nelson[1] may be of considerable use to the designer who is concerned with converting rotation into oscillation; it contains the displacement paths for over 7,000 variations of this simple four-bar linkage.

2-4. Rotation of Driver and Follower. The requirement for continuous rotation of the driver and follower is that there must be no dead points. Figure 2-5 shows a mechanism which meets this requirement. The descriptive name *drag-link* mechanism has been given to this particular linkage.

In many instances, the useful part of a machine's operation is a motion in only one direction. Therefore, the maximum useful output is given when the major part of the period is spent during the working stroke and a minimum time is wasted in returning the mechanism to its starting position. The drag link is an important member of the group of *quick-return* mechanisms which are used to provide this characteristic motion.

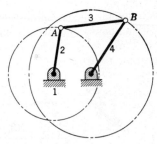

FIG. 2-5. Drag-link mechanism.

There are numerous other quick-return mechanisms, and several will be discussed in this and later chapters. The drag link is the only one of the group that has no sliding pairs in the basic quick-return part of the mechanism. Figure 2-6 shows the basic drag-link mechanism and the additional links required to transform uniform rotation of the crank 2 into a quick-return reciprocating motion of the ram 6. The crank rotates through the angle α as the ram moves on its working stroke to the left and through the angle β as the ram returns to its starting position. The

[1] J. A. Hrones and G. L. Nelson, "Analysis of the Four-bar Linkage: Its Application to the Synthesis of Mechanisms," Technology Press, M.I.T., Cambridge, Mass., and John Wiley & Sons, Inc., New York, 1951.

ratio of α to β, or the ratio of the time for the working stroke to the time for the return stroke, is known as the *time ratio*.

Some electric locomotives use a *side-rod drive* to transmit power from the motor or from one set of wheels to the other wheels in much the same manner as for the ordinary steam locomotive. Figure 2-7 is a skeleton diagram illustrating how the wheels 2 and 4 and side rod 3 are the equivalent of a four-bar linkage. When A and B are at A' and B', respectively, neither wheel can drive the other. The practical solution of this problem

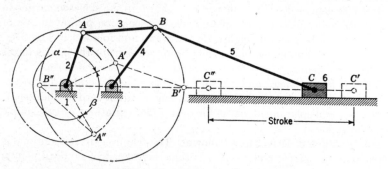

FIG. 2-6. Drag-link quick-return mechanism.

is to mount a side rod on the wheels on the other side of the locomotive, with a phase angle, usually 90° (at A'' and B''), between the driving links or cranks so that when one side rod is on dead center the other is in a position to drive.

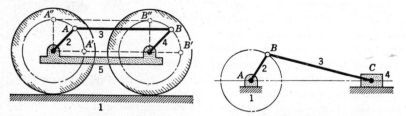

FIG. 2-7. Locomotive side-rod drive.   FIG. 2-8. Slider-crank mechanism.

2-5. Slider Crank. If the follower of a four-bar linkage were made infinitely long, a point on the axis of the pair joining the connecting link to the follower would reciprocate in a straight line. In a practical application, the follower becomes a block or piston constrained to move in a straight line by guides or by a cylinder. This special case of the four-bar linkage is the common *slider-crank* mechanism shown in Fig. 2-8. The crankshaft, connecting rod, piston, and cylinder block of an internal-combustion engine, as shown in Figs. 2-9 and 2-10, make up the most familiar slider-crank mechanism in use today.

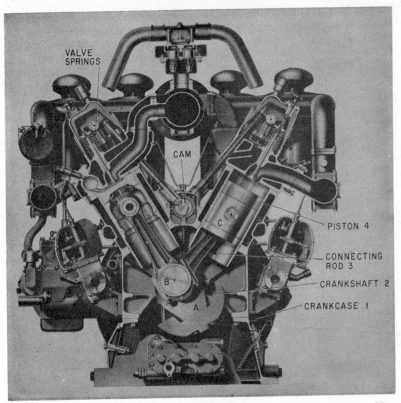

VALVE
SPRINGS

CAM

C

B

A

PISTON 4

CONNECTING
ROD 3

CRANKSHAFT 2

CRANKCASE 1

Fɪɢ. 2-9. Transverse section of a V-12 diesel engine. (*Courtesy Waukesha Motor Company.*)

2-6. Eccentric and Rod. Quite often a reciprocating motion is required, but the required stoke is comparatively short in relation to the driving-shaft dimensions. In such a case, it may not be practical to use the usual form of throw on the crankshaft. However, with only a slight change in appearance, the basic slider-crank mechanism becomes the kinematically equivalent *eccentric-and-rod* mechanism shown in Fig. 2-11. The operation by which the eccentric-and-rod mechanism is derived from the slider-crank mechanism is called *expansion of pairs*. In this example, the pair at *B* in Fig. 2-8 has been expanded so that it includes all of link 2 and the

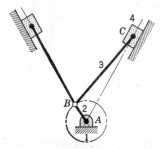

Fɪɢ. 2-10. Skeleton outline of the basic slider-crank mechanism for the diesel engine in Fig. 2-9.

pair at A. This principle may be used to derive many kinematically identical mechanisms with quite different outward appearances.

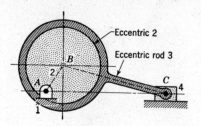

FIG. 2-11. Eccentric-and-rod mechanism.

2-7. Offset Slider Crank. If the crank is rotating at a uniform rate, the slider of an ordinary slider-crank mechanism requires the same time for a stroke in either direction. However, if the path of the slider is offset so that its line of motion does not pass through the center of the crank circle, a simple quick-return mechanism is created. In Fig. 2-12 the time ratio is α/β.

2-8. Connecting Rod of Infinite Length. By expanding the pair at C of Fig. 2-8 to include the pair at B, the variation in Fig. 2-13 is the result.

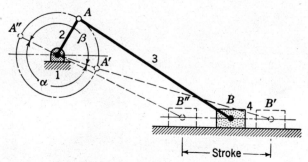

FIG. 2-12. Offset-slider-crank mechanism.

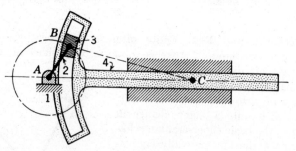

FIG. 2-13

If the slot in Fig. 2-13 is then made straight, as shown in Fig. 2-14, the effect is the same as making the connecting rod infinitely long. This mechanism is called the *Scotch yoke* and has been used on engines, pumps, etc., where compactness is important. A typical arrangement with a steam engine driving a pump is shown in Fig. 2-15.   The flywheel is carried on shaft A.   A somewhat more recent use of this mechanism is as the driving unit, or forcing function, for vibrating or shake tables.   The important factor is that, since the reciprocating rod 4 moves as the projection on the diameter of a point moving on the circumference of a circle, the uniform angular rotation of the crank is transformed into simple harmonic reciprocation of the rod.   Because only the fundamental vibration is present and the higher harmonics do not cloud the picture, simple harmonic motion is useful in calibrating vibration-measuring instruments and in vibration tests of parts or assemblies, e.g., electronic equipment.

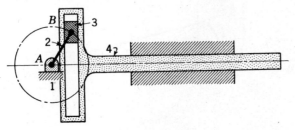

FIG. 2-14. Scotch-yoke mechanism.

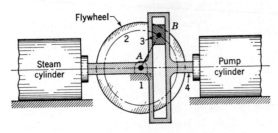

FIG. 2-15

The important factor which limits the usefulness of this mechanism to relatively small machines operating under light loads is that the sliding block (link 3 in Fig. 2-14) must transmit large forces to link 4 by sliding contact.   As will be discussed in Chap. 18, it is much more difficult to provide an adequate film of lubricating oil for separating sliding parts than for turning pairs.

2-9. Connecting Rod of Finite Length.   Since the ordinary slider-crank mechanism does not have an infinitely long connecting rod, it will not give simple harmonic motion to the slider.   The expression *angularity of the connecting rod* is often used to describe the cause of the deviation from simple harmonic motion.   In Fig. 2-16, R is the length of the crank, L is the length of the connecting rod, and x is the displacement of the slider to the left of the extreme right-hand position, i.e., top dead center. Then,

$$\begin{aligned}
x &= R + L - R \cos \theta - L \cos \phi \\
&= R(1 - \cos \theta) + L(1 - \cos \phi) \\
&= R(1 - \cos \theta) + L(1 - \sqrt{1 - \sin^2 \phi}) \\
&= R(1 - \cos \theta) + L\left[1 - \sqrt{1 - \left(\frac{R}{L}\right)^2 \sin^2 \theta}\right]
\end{aligned} \tag{2-1}$$

This is the exact solution, but it is usually more convenient to simplify the expression by expanding the quantity under the radical in accordance

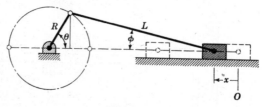

FIG. 2-16

with the binomial theorem and to keep only those terms of the series that are significant. If this is done, then

$$\begin{aligned}
x = R(1 - \cos \theta) + R\Bigg[\frac{1}{2}\frac{R}{L} \sin^2 \theta + \frac{1}{2 \times 4}\left(\frac{R}{L}\right)^3 \sin^4 \theta \\
+ \frac{1 \times 3}{2 \times 4 \times 6}\left(\frac{R}{L}\right)^5 \sin^6 \theta + \cdots\Bigg]
\end{aligned} \tag{2-2}$$

Since $x = R(1 - \cos \theta)$ would be the equation for the case of the infinite connecting rod, the remaining terms are due to the angularity of the connecting rod. Also, it is important to note that, as the ratio R/L becomes smaller, the contribution of the higher-order harmonics rapidly becomes less prominent. Extensive use is made of this equation, after it has been differentiated twice with respect to time to get the acceleration of the slider, in the study of the degree of balance of the forces and couples due to the inertia of the reciprocating parts in internal-combustion engines.[1] For this particular application, where R/L is generally less than 1/4, all terms beyond the $\sin^2 \theta$ term may be neglected.

2-10. Toggle Mechanisms. The *toggle mechanism* shown in Fig. 2-17 is used in applications such as rock crushers, presses, riveting machines,

---

[1] C. W. Ham, E. J. Crane, and W. L. Rogers, "Mechanics of Machinery," 4th ed., chap. 12, McGraw-Hill Book Company, Inc., New York, 1958.

H. H. Mabie and F. W. Ocvirk, "Mechanisms and Dynamics of Machinery," chap. 12, John Wiley & Sons, Inc., New York, 1957.

clutches, vise-grip pliers, etc., where a large force is required to act through a short distance. As the slider-crank mechanism (links 4, 5, and 6) approaches dead center, there is a rapid rise in the ratio of the useful force Q to the input force P. If links 4 and 5 are equal in length

$$\frac{Q}{P} = \frac{\cos \alpha}{2 \sin \alpha} = \frac{1}{2 \tan \alpha}$$

It follows that, if a motor delivering a relatively constant torque is being used to drive the machine, links 2 and 3 should be collinear when the slider-crank mechanism approaches dead center (α approaches zero). When a toggle mechanism is used to apply a steady force, as in the engagement of a clutch, it is often made self-locking in the engaged position by permitting B to go slightly below dead center (or *over center*), where it is held against a stop by the force Q.

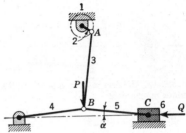

Fig. 2-17. Toggle mechanism.

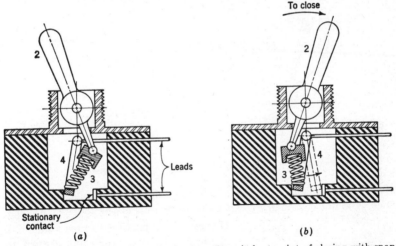

Fig. 2-18. Toggle switch. (a) Switch open; (b) switch at point of closing with snap action.

The toggle principle is also effectively used in switches, circuit breakers, and other mechanisms where a snap action is required. Figure 2-18a shows a toggle switch with the contacts open. In Fig. 2-18b, the handle 2 has been pushed to the right, and the compression spring 3 has been forced over center so that the spring force on link 4 is suddenly reversed to act to the right and link 4 snaps to the position (dotted) in which the

circuit is closed. Subsequent motion of the handle to the left will reverse
the procedure and open the contacts with a similar snap action.

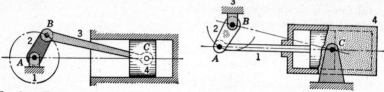

FIG. 2-19. The first inversion of the slider-crank mechanism.

FIG. 2-20. The second inversion of the slider-crank mechanism.

2-11. Inversions. In Chap. 1 it was stated that a mechanism is a
kinematic chain with one link fixed and that a chain with N links would

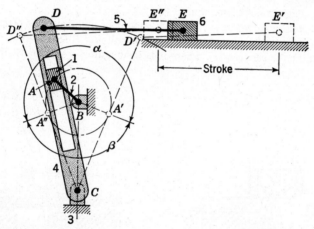

FIG. 2-21. The crank-shaper quick-return mechanism.

give N mechanisms. Since all four-bar linkages are similar in appearance, it is not customary to speak of inversions of the four-bar linkage.

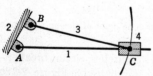

FIG. 2-22. The third inversion of
the slider-crank mechanism.

However, it is possible to take a four-link
chain and, by fixing each link in turn, to
have all the motions discussed in Secs. 2-2
to 2-4.

The slider-crank mechanism is more
readily identified, and it is both easy and
useful to recognize its inversions as such.

1. The first inversion (Fig. 2-19) is the
ordinary slider-crank mechanism, and common applications are in steam
and internal-combustion engines with the piston 4 driving and in pumps
with the crank 2 driving.

2. In the second inversion, the connecting rod 3 is fixed, and link 4 now oscillates. Some steam engines have been built using an oscillating cylinder, as shown in Fig. 2-20, because this permits the use of simple ports instead of a complicated valve mechanism to control the steam. Toy

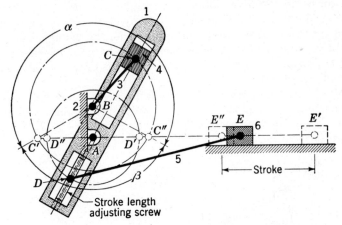

FIG. 2-23. The Whitworth quick-return mechanism.

steam engines are usually constructed in this manner. A more practical use of this inversion is the quick-return part of the *crank-shaper* mechanism shown in Fig. 2-21, where uniform rotation of the crank 2 is converted into a slow working stroke and a quick return of the ram 6.

3. Fixing the crank 2 gives the third inversion in Fig. 2-22. The *Whitworth quick-return* mechanism in Fig. 2-23, in which link 3 is the driver rotating at a uniform angular velocity, is used in shapers, other machine tools, and textile machinery. Figure 2-24 is a rotary engine

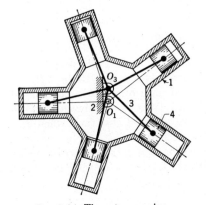

FIG. 2-24. The rotary engine.

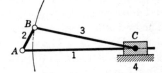

FIG. 2-25. The fourth inversion of the slider-crank mechanism.

which was used extensively during World War I before the radial engine was highly developed. The crank 2 was fixed to the frame, and the propeller was attached to the rotating crankcase 1.

4. Fixing the slider 4 gives the fourth inversion in Fig. 2-25. One of the few examples of its use is the hand-operated well pump shown in Fig. 2-26.

2-12. Straight-line Mechanisms. Obviously, the slider-crank mechanism can be used to give motion in a straight line, but quite often the friction and structural difficulties accompanying the use of guides make it highly desirable to substitute turning pairs for the sliding pairs. Only a few of the large number of mechanisms which have been devised to give either approximate or exact straight-line motions by use of turning pairs will be discussed here.

At the time James Watt was building his early steam engines there were no means available for machining crosshead guides that would accurately constrain the piston to move in a straight line. The ingenious mechanism developed to circumvent this difficulty is the *Watt straight-line*

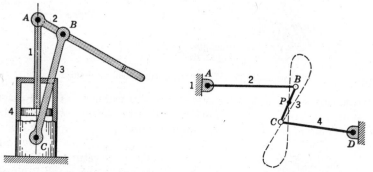

FIG. 2-26. Hand-operated pump.   FIG. 2-27. Watt straight-line mechanism.

mechanism. If, in Fig. 2-27, the segments BP and CP are proportioned inversely as the lengths of the adjacent arms, that is, $BP/CP = CD/AB$, the point P will follow a figure-8 type of path of which a portion will very closely approximate a straight line. To get a maximum length of straight-line motion it is necessary to locate A and D so that, when links 2 and 4 are parallel, they will be perpendicular to link 3.

The best-known exact straight-line mechanism is the *Peaucellier* mechanism in Fig. 2-28. The requirements which must be met if P is to move in a straight line are as follows:[1]

Link 2 must equal AB.

Links 3 and 4 are equal.

Links 5, 6, 7, and 8 are equal.

2-13. Pantograph. When it is desired to duplicate some motion exactly but to a reduced or enlarged scale, the copying device used is

[1] Ham, Crane, and Rogers, *op. cit.*, pp. 36–37.

some form of a *pantograph*.   This mechanism consists basically of a four-bar linkage in the form of a parallelogram.   If one corner of the parallelogram is fixed and the points p and P are located on a line passing through the fixed pivot, as in Fig. 2-29, it can be shown by use of similar triangles that p and P will follow similar paths and that the motion of p will be exactly Ap/AP or BC/BP times the motion of P.

Pantographs are used for reducing or enlarging drawings and maps; to guide cutting tools, cutting torches, etc., in the duplicating of complicated shapes such as cams, turbine blades, etc.; and to permit more convenient recording of small or large motions.

In the experimental study of the performance of steam engines, air compressors, etc., it is important to know the pressure in the cylinder as a function of the position of the piston.   A typical indicator designed for

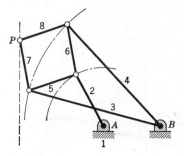

Fig. 2-28.   Peaucellier straight-line mechanism.

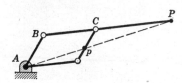

Fig. 2-29. Pantograph.

use at speeds up to 400 rpm is shown in Fig. 2-30.   The pressure being recorded moves the piston against the force of the calibrated spring.   In order to keep the change in volume small and to keep the inertia forces due to accelerating the parts small, the piston moves only a short distance, and a very light-weight linkage is used to make the stylus P move in a magnified duplicate motion of the piston.   A comparison of Fig. 2-30*b* with Fig. 2-29 will show that the indicator mechanism is a modification that closely approximates the basic pantograph linkage.   The design approach would be to choose dimensions for links 3, 5, and 6 and the locations of point A and the centerline of the piston travel so that

$$\frac{pC}{AB} = \frac{PC}{PB}$$

and links 3 and 5 are parallel when A, p, and P lie on a straight line. Now, if P is constrained to move in a straight line parallel to the piston motion p while keeping points P, p, and A on a straight line, a point on

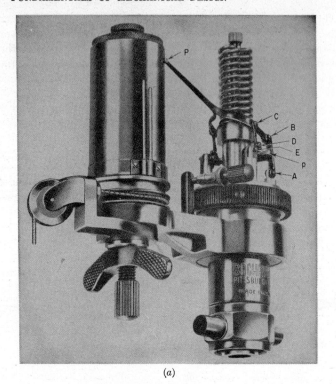

(a)

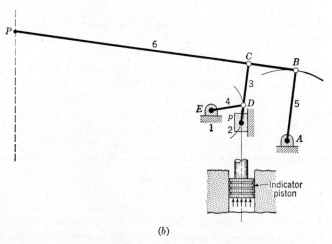

(b)

Fig. 2-30. (a) Bacharach type 2-CP engine indicator; (b) skeleton outline of modified pantograph linkage. (*Courtesy Bacharach Industrial Instrument Company.*)

link 3 can be found that moves on a curved path that very closely approximates the arc of a circle.   Then, if this point (D on 3) is connected to the center of curvature of its path (E) by a link (4), the artificial constraint on the motion of P may be removed, and the linkage will now give P a motion similar to that of p but multiplied by the ratio PB/CB.   In the actual mechanism, the dimensions and ratios of dimensions that would be given by the procedure outlined above are altered somewhat to give the best possible approximation over the widest range of travel.

When the indicator is used, a sheet of special paper is fastened to the drum.   The drum oscillation is made proportional to the engine piston travel by connecting the cord to a linkage, often another pantograph, that gives it a reduced motion that is otherwise similar to that of the engine piston.

2-14. Parallel Linkages.   A class of four-bar linkages with somewhat limited use consists of those linkages which appear in the form of a parallelogram.   These are known as *parallel linkages*, and the simplest one is the *parallel rule* (Fig. 2-31) which has been largely replaced by drafting machines.

Fig. 2-31. Parallel rule.

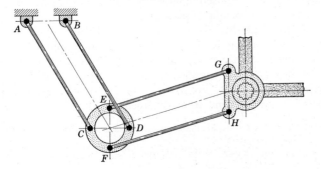

Fig. 2-32. Drafting machine.

The early *drafting machines* were similar to Fig. 2-32, in which the ring $CDEF$ couples the two parallelograms $ABCD$ and $EFGH$ so that the angle set on the head will be maintained at all positions on the drawing.   The major disadvantage is that in extreme positions, where the parallelograms become very narrow, any lost motion or play in the pin connections adversely affects the accuracy of the drawing.

Modern drafting machines (Fig. 2-33) still rely upon the basic parallelogram but use steel bands wrapped around circular disks. These bands remain a uniform distance apart and are kept under enough tension so that there is no lost motion and no slipping.

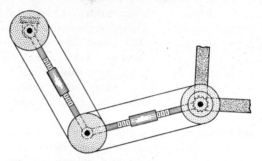

FIG. 2-33. Drafting machine.

2-15. Computing Mechanisms. Since the definition of a mechanism states that all points move with definite relative motion, it follows that for every mathematical equation there will be a mechanism or combination of mechanisms that will solve the equation. However, except for some of the simpler cases, it is practically impossible to discover the proper mechanism. The literature[1] contains numerous examples of mechanisms that can be used in computers or controlling devices. Some of the types of operations readily performed by linkages, gear trains (Chap. 17), and variable-speed drives (Chap. 17) are addition, subtraction, integration, multiplication, division, determination of linear values for trigonometric functions, the taking of square roots, solution of quadratic equations, and combinations of these operations. It should also be pointed out that cams (Chap. 5) can often be used to express relations for which linkages have yet to be discovered.

[1] A. Svoboda, "Computing Mechanisms and Linkages," McGraw-Hill Book Company, Inc., New York, 1948.
  J. S. Beggs, "Mechanism," chap. 9, McGraw-Hill Book Company, Inc., New York, 1955.
  J. H. Billings, "Applied Kinematics," 3d ed., chap. 14, D. Van Nostrand Company, Inc., Princeton, N.J., 1953.
  Hinkle, *op. cit.*
  Mabie and Ocvirk, *op. cit.*, chap. 8,

CHAPTER 3

MOTION IN MACHINES

The previous chapters have discussed some of the general types of motion and the ways in which the four-bar linkage may be altered to give motions fitting certain specifications within fairly broad limits.

This and following chapters will be given to a more thorough study of displacement, velocity, and acceleration.

Those readers who become involved in detailed computations for many mechanisms or variations of a mechanism, particularly if the motion occurs in space, i.e., in three dimensions, rather than in a plane, will find it worthwhile to consider the analytical vector methods[1] because they reduce most problems to almost routine bookkeeping procedures and are, therefore, ideal for solution by use of high-speed digital computers. However, the classical graphical methods will receive particular attention here because they bring one closer to an understanding of the fundamentals involved and also permit a more rapid solution in the great majority of design situations.

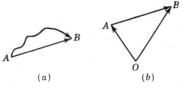

(a)   (b)

FIG. 3-1

3-1. Displacement. The *displacement* of a point is the change in position of the point and is a vector quantity in that it has both magnitude and direction. In Fig. 3-1a, a point moves along the path shown from A to B. The displacement of the point during this interval is represented by the line **AB**.

If it is more desirable to define the positions of a point along its path in terms of a displacement from some fixed position, i.e., relative to another point, then Fig. 3-1b is another way of presenting the same motion as in Fig. 3-1a. The vector **OA** is the initial displacement of P, the vector **OB**

---

[1] J. E. Shigley, "Kinematic Analysis of Mechanisms," McGraw-Hill Book Company, Inc., New York, 1959.

J. S. Beggs, "Mechanism," McGraw-Hill Book Company, Inc., New York, 1955.

R. T. Hinkle, "Kinematics of Machines," 2d ed , Prentice-Hall, Inc., Englewood Cliffs, N.J., 1960.

is the final displacement, and

$$OB = OA + AB \qquad (3\text{-}1)$$

3-2. Linear Speed and Velocity of a Point. *Speed* is a scalar quantity, i.e., has magnitude only, and is determined by dividing the distance traveled by the time interval for the motion. *Velocity* is a vector quantity, i.e., has both magnitude and direction, and is defined as the rate of change of position with respect to time or simply the time rate of displacement.

In Fig. 3-2 a point moves along the path shown from A to B in the time interval Δt. ΔS is the actual distance traveled, and ΔL is the displacement. The average speed is $\Delta S/\Delta t$, and the average velocity has a magnitude of $\Delta L/\Delta t$ and is in the direction of ΔL. In the limit, as Δt approaches zero, ΔS and ΔL become equal. The instantaneous speed dS/dt equals the magnitude of the instantaneous velocity dL/dt, and the direction of the instantaneous velocity is tangent to the path of motion.

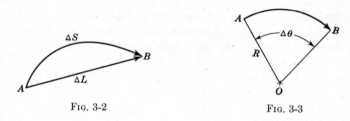

FIG. 3-2   FIG. 3-3

In almost all cases, the designer is interested in the instantaneous velocity, and therefore *the use of the term velocity without qualification will denote the instantaneous value.*

3-3. Angular Velocity. *Angular velocity* is a vector quantity having both magnitude and direction and is defined as the time rate of angular displacement of a line. In Fig. 3-3 a point has moved along the arc of a circle from A to B in the time interval Δt. During this time the radius vector R has turned through the angle $\Delta\theta$. In the limit, as Δt approaches zero, the instantaneous angular velocity becomes $\omega = d\theta/dt$ in a clockwise (cw) direction.

3-4. Linear Acceleration of a Point. *Acceleration* is a vector quantity and is defined as the rate of change of velocity with respect to time.

In Fig. 3-4a the vectors representing the velocity of P at two points on the curvilinear path are labeled $\mathbf{V}_1$ and $\mathbf{V}_2$. The radii of curvature of the path at P_1 and P_2 are R_1 and R_2, respectively. The change in velocity as shown in Fig. 3-4b is

$$\Delta\mathbf{V} = \mathbf{V}_2 - \mathbf{V}_1 \qquad (3\text{-}2)$$

$\Delta \mathbf{V}$ may also be expressed as

$$\Delta \mathbf{V} = \mathbf{AC} + \mathbf{CB} \tag{3-3}$$

The vector $\mathbf{CB}$ represents the change in magnitude of the velocity, and the vector $\mathbf{AC}$ represents the change in velocity due to its change in direction. Therefore,

$$\Delta \mathbf{V} = 2V_1 \sin \frac{\Delta \theta}{2} \nleftrightarrow (V_2 - V_1) \tag{3-4}$$

and the average acceleration becomes

$$\mathbf{A}_{av} = \frac{\Delta \mathbf{V}}{\Delta t} = \frac{2V_1 \sin (\Delta \theta / 2)}{\Delta t} \nleftrightarrow \frac{V_2 - V_1}{- \Delta t} \tag{3-5}$$

In the limit, as Δt approaches zero, the component $[2V_1 \sin (\Delta \theta/2)]/\Delta t$ is perpendicular to $\mathbf{V}_1$ and $\mathbf{V}_2$, or normal to the path; the component

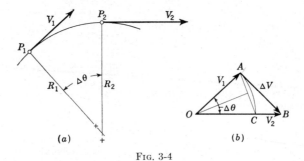

Fig. 3-4

$(V_2 - V_1)/\Delta t$ is in the direction of $\mathbf{V}_1$ and $\mathbf{V}_2$, or tangent to the path; and R_1 and R_2 become the instantaneous radius of curvature of the path R. The equation becomes

$$\mathbf{A} = V \frac{d\theta}{dt} \nleftrightarrow \frac{dV}{dt} \tag{3-6}$$

Thus, it is seen that *the instantaneous acceleration of a point is composed of a normal and a tangential component.* The total acceleration may be expressed as

$$\mathbf{A} = V \frac{d\theta}{dt} \nleftrightarrow \frac{dV}{dt} = \mathbf{A}^n + \mathbf{A}^t \tag{3-7}$$

where $$A = \sqrt{(A^n)^2 + (A^t)^2}$$

It is important to note that the normal component ($\mathbf{AC}$) is always directed *toward the center of curvature,* whereas the tangential component

(CB) may be in the same or the opposite direction as the velocity, depending on whether its magnitude is increasing or decreasing.

3-5. Angular Acceleration. The angular acceleration, or the time rate of change of angular velocity, of a line is

$$\alpha = \frac{d^2\theta}{dt^2} = \frac{d\omega}{dt} \tag{3-8}$$

The magnitude of the velocity of a point moving on a curved path is

$$V = R\omega \tag{3-9}$$

where R is the radius of curvature of the path and ω is the angular velocity of the radius of curvature. After rearranging terms and differentiating with respect to time, Eq. (3-9) becomes

$$\frac{dV}{dt} = \frac{d(R\omega)}{dt} = R\frac{d\omega}{dt} + \omega\frac{dR}{dt} \tag{3-10}$$

or in terms taken from Eqs. (3-7) and (3-8)

$$A^t = R\alpha + \omega\frac{dR}{dt} \tag{3-11}$$

In many cases of interest the radius of curvature is a constant and Eq. (3-11) becomes

$$A^t = R\alpha \tag{3-12}$$

It is often convenient to be able to express the normal acceleration in terms of angular velocity or vice versa. The magnitude of the normal acceleration can be expressed as

$$A^n = V\frac{d\theta}{dt} = R\omega^2 = \frac{V^2}{R} \tag{3-13}$$

Although it has not been emphasized, it is worthwhile at this point to note that the preceding discussion has been concerned with the acceleration of a point moving along a path on a stationary member. In this case the velocity and accelerations are considered to be *absolute*.

3-6. Relative Motion. The term *relative motion* is used when expressing the displacement, velocity, or acceleration of one point or member in terms of the motion of another point or member. As noted immediately above, motion relative to the particular member that is fixed or stationary is called absolute.

The velocity of B relative to A is that vector which must be added to the velocity vector of A to give the velocity vector for B as a resultant.

This may be written as

$$V_B = V_A + V_{B/A} \qquad (3\text{-}14)$$

or

$$V_{B/A} = V_B - V_A \qquad (3\text{-}15)$$

and the vector diagram is shown in Fig. 3-5a.

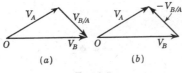

(a)   (b)

FIG. 3-5

The vector equation may be manipulated in the same manner as an ordinary algebraic equation. The information presented in Fig. 3-5a may be expressed in several ways by rearranging the vector equation:

$$V_B = V_A + V_{B/A}$$
$$V_A = V_B - V_{B/A} \qquad (3\text{-}16)$$
$$V_B = V_A - V_{A/B}$$

As these examples point out, $V_{A/B}$ is equal in magnitude but opposite in direction to $V_{B/A}$, or

$$V_{A/B} = -V_{B/A} \qquad (3\text{-}17)$$

This principle is sometimes useful in that it permits a vector subtraction to be written as an addition. Figure 3-5b shows how the vector solution of the equation

$$V_A = V_B - V_{B/A} \qquad (3\text{-}18)$$

may be simplified by rewriting it as

$$V_A = V_B + (-V_{B/A}) \qquad (3\text{-}19)$$

Relative angular motions may be handled in a similar manner. However, since angular motions must be either clockwise (cw) or counterclockwise (ccw), it is not necessary to draw the vectors because the directions can be kept straight by use of plus and minus signs. The usual convention is to use plus for counterclockwise and minus for clockwise motions.

3-7. Transmission of Motion and Force. The function of a mechanism is to provide a definite motion, and the function of a machine is to combine motion and force to do useful work. The manners in which motion and force may be transmitted from one member to another have so much in common that there is no need to consider them separately.

Both motion and force may be transmitted from one member to another in one of the following ways:

1. Use of intermediate rigid connectors such as links, rollers, etc. (Fig. 3-6a)

2. Use of intermediate flexible connectors such as belts, ropes, chains, springs, fluids, etc. (Fig. 3-6b)

3. Direct contact, as in gears, friction wheels, cams, etc. (Fig. 3-6c)

4. Use of magnetic fields and fluids

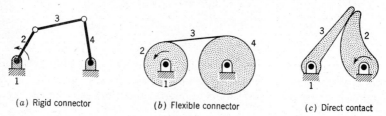

(a) Rigid connector   (b) Flexible connector   (c) Direct contact

FIG. 3-6. Transmission of motion and force.

3-8. Line of Transmission. Each method of transmitting motion and force has its own well-defined characteristics and limitations which, in general, are related to the line along which the motion or force is effectively transmitted from the driver to the follower. This line is called the *line of transmission*.

The important consideration here is that usually only the component of motion or force lying along the line of transmission contributes to the usefulness of the mechanism or machine. Because of this, the line of transmission provides a useful tool in determining the ratio of the angular velocities of the driver and the follower and the ratio of the torques acting on the shafts of the driver and the follower.

The components of motion and force *not* acting along the line of transmission are nevertheless present and must be considered in the design of a machine because of their effect on the forces acting on shafts, bearings, and other members, as well as on the dissipation or wasting of energy due to the presence of friction and motion.

3-9. Intermediate Rigid Connectors. Figure 3-7 shows a typical four-link mechanism transmitting motion from the driver 2 to the follower 4. The pin-connected link 3 can transmit motion or force only in tension or compression, and therefore the line of transmission must be along link 3.

If the driver 2 has an angular velocity of ω_2 in a clockwise direction, A will have an absolute, i.e., relative to the fixed link or ground, velocity represented by the vector $\mathbf{AM_2}$ which is equal in magnitude to $O_2A\omega_2$. The useful component of this velocity is that labeled $\mathbf{AS_2}$ along link 3. Since link 3 is rigid, B must have the same component ($\mathbf{BS_4} = \mathbf{AS_2}$) of velocity. However, link 4 is constrained to rotate about O_4, and the absolute velocity of B must be perpendicular to link 4 and of such a magnitude that it has a component $\mathbf{BS_4}$ in the direction along link 3.   The required velocity of B is represented by the vector $\mathbf{BM_4}$.

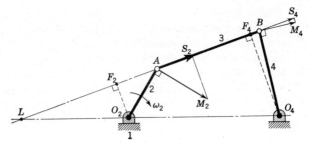

FIG. 3-7

The angular velocity of any body is expressed as

$$\omega = \frac{V}{R} \tag{3-20}$$

where ω = angular velocity
  V = absolute velocity of any point in body
  R = radius from center of rotation to point
This gives

$$\omega_2 = \frac{AM_2}{O_2A} \quad \text{and} \quad \omega_4 = \frac{BM_4}{O_4B} \tag{3-21}$$

and
$$\frac{\omega_2}{\omega_4} = \frac{AM_2}{O_2A}\frac{O_4B}{BM_4} \tag{3-22}$$

Lines O_2F_2 and O_4F_4 pass through the centers of rotation O_2 and O_4, respectively, and are perpendicular to the line of transmission. The triangles O_2AF_2 and AM_2S_2 are similar because the corresponding sides are perpendicular. The same is true for triangles O_4BF_4 and BM_4S_4. The triangles LO_2F_2 and LO_4F_4 are also similar.

After making the proper substitutions, Eq. (3-22) will give the following:

$$\frac{\omega_2}{\omega_4} = \frac{O_4F_4}{O_2F_2} \tag{3-23}$$

and
$$\frac{\omega_2}{\omega_4} = \frac{O_4L}{O_2L} \tag{3-24}$$

These same relationships for the angular-velocity ratio may be as readily derived for any of the methods of transmitting motion or force by mechanical means. The relation expressed by these equations is sometimes called the *angular-velocity theorem* and may be stated as follows: *The angular velocities of driver and follower are inversely as the length of the perpendiculars from the centers of rotation to the line of transmission, or, more usefully, inversely as the segments into which the line of centers is cut by the line of transmission.*

The reader should recognize that the vector $\mathbf{AM}_2$ in Fig. 3-7 could have just as readily been a force instead of a velocity, and the components would have been determined in a somewhat similar manner. There is no particular advantage in carrying the derivation as far as finding an equation for the torque ratio. It should be sufficient to point out that, if losses are neglected, the power transmitted must be the same for the driver and the follower. Therefore, *the torque ratio is the inverse of the angular-velocity ratio.* This useful relationship comes up quite regularly in designing machines.

3-10. Intermediate Flexible Connectors. This class of connectors is characterized by the fact that motion and force may be transmitted along the connector only in the direction which loads the connector in tension.

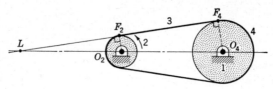

FIG. 3-8

Figure 3-8 shows the case of a pulley 2 driving another pulley 4 by means of the flexible flat belt 3. Belts, chains, etc., will be discussed more thoroughly in later chapters, but at this time it is sufficient to note that, for rotation of the driving pulley 2 as shown, the top belt will be in tension and will be the line of transmission. In this particular case, it is useful to note that the perpendiculars from the center of rotation to the line of transmission are equal to the radii of the cylinders plus one-half the thickness of the belt. If the relatively small belt thickness (Sec. 14-2) is neglected, the angular-velocity ratio then becomes

$$\frac{\omega_2}{\omega_4} = \frac{R_4}{R_2} \tag{3-25}$$

and the velocity of the belt is constant and equal to

$$\omega_2 R_2 \quad \text{or} \quad \omega_4 R_4 \tag{3-26}$$

3-11. Direct Contact. When motion or force, with friction neglected, is transmitted from or to a surface, it must be done in a direction perpendicular or normal to the surface. Therefore, the line of transmission for members in direct contact must lie along the common normal, as shown in Fig. 3-9. Either definition of the angular-velocity theorem may be used, but generally the definition in terms of the segments into which the line of centers is cut by the line of transmission (or common normal) is the more convenient.

Direct-contact mechanisms are compact in that only three links are required to transfer motion from the driver to the follower. However, this is done at the expense of limiting the direction of rotation in which the follower is positively moved by the driver and by introducing sliding pairs instead of turning pairs.

The first consideration is important in that, even with the driver rotating in only one direction, some force, usually supplied by gravity or by a spring, must keep the follower in contact at all times. This problem will be considered in more detail in Chap. 5 on cams.

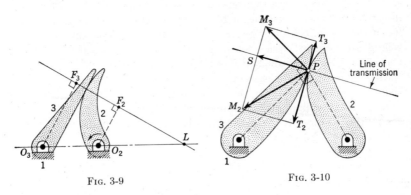

FIG. 3-9   FIG. 3-10

The latter consideration is important in that the product of the transmitted force and the coefficient of friction for the sliding contact results in a frictional force tangent to the contact surfaces. The combination of a frictional force and sliding both dissipates energy and causes wear of the contacting surfaces. Thus, it is often desirable to know something about the velocity with which one member slides on the other in relation to the design and operation of cams and gears. In Fig. 3-10 the vector $\mathbf{PM}_2$ is the absolute velocity of the contact point P as a point in 2. This vector has two components: $\mathbf{PS}$ normal to the contact surfaces and $\mathbf{PT}_2$ tangent to the contact surfaces. If it is assumed that the two members remain in contact with negligible deformation of the surfaces, the absolute velocity of P as a point in 3 must have the same component of velocity along the common normal as it does as a point in 2. This requirement is met if the absolute velocity of P in link 3 is $\mathbf{PM}_3$ and the tangential component of this velocity becomes $\mathbf{PT}_3$.

The velocity of sliding is the vector difference of the tangential components of the velocities of the points of contact. Since these tangential components lie on the same line, + and − may be assigned to directions from P, and the algebraic difference and the vector difference become the same. It is worthwhile to note that the velocity of sliding is also equal to the vector difference of the absolute velocities of the contact points.

3-12. Constant Angular-velocity Ratio. Several conclusions concerning the transmission of motion from one link to another are important enough to be discussed separately. One of these conclusions is that, for one link to drive another with a constant angular-velocity ratio, the line of transmission must always intersect the line of centers at the same point. This is the basic principle underlying the selection of curves suitable for the profiles of the teeth on gears.

3-13. Pure Rolling. Pure rolling implies the complete absence of sliding; therefore, the absolute velocities of the points of contact in a direct-contact mechanism must be equal, and the point of contact must lie on the line of centers. Some of the families of curves that give pure rolling are tangent circles rotating about their centers; pairs of equal ellipses, each rotating about one of its foci with a center distance equal to the common major axis; pairs of similar logarithmic spirals rotating about their foci; pairs of equal parabolas; and pairs of equal hyperbolas.

It should be apparent that rolling circles are the only curves that fulfill the conditions for both a constant angular-velocity ratio and pure rolling.

VELOCITY AND ACCELERATION ANALYSIS

Quite often in the design of mechanisms or machines, it is necessary to know more about the velocities and accelerations than can easily be found by use of the material in Chap. 3. Consequently, it is desirable to continue the development of the basic principles to the point where more or less well-defined methods or techniques are available. These methods of velocity and acceleration analysis will be discussed in detail in this chapter.

4-1. Analytical Determination of Velocities and Accelerations. A simple method for determining the velocity and the acceleration of a point is to write the equation for the displacement of the point as a function of time and differentiate it with respect to time, either once or twice, as the case may be. The difficulty is that, except for the slider-crank mechanism and some of its variations, it is practically impossible to write the necessary equation, and the designer must resort to some form of vector analysis.

The slider-crank mechanism has had such widespread use in high-speed internal-combustion engines that the series terms (Sec. 2-9) are available in tables for different ratios of connecting-rod length to crank radius.[1]

4-2. Determination of Linear Velocities by Composition and Resolution. In Fig. 3-7, Sec. 3-9, the velocity of A as a point in link 3 was resolved into a component along the link and a component perpendicular to the link, the former being due to translation of the link and the latter to rotation. Since the link is considered to be rigid, all points in the link must have the same translation component of velocity, whereas the rotational component will vary from point to point in the link. The velocity of B was found by composing a vector which is the result of the translation component, the known direction of the rotational component, and the known direction of the velocity of B.

[1] L. C. Lichty, "Internal Combustion Engines," 6th ed., McGraw-Hill Book Company, Inc., New York, 1951.

This procedure is relatively simple to apply to those points which have known paths of motion. However, when applied to other points on a mechanism, the procedure is considerably more complicated, and other methods are usually more convenient.

4-3. Instant Centers. One of the most convenient means of determining the velocity of any point in a mechanism is based upon the fact that at any instant the motion of one body relative to another body is such that there is a point in the plane of motion where the two bodies have no relative motion. This point may or may not be on the actual links; in fact, it may even be at infinity. Since there is no relative velocity between the points, it is obvious that at this point both bodies must have the same absolute velocity.

Other points in the bodies do not have zero relative velocities, and, since the links are rigid, the relative velocities must be due to relative rotation about the common point. This point actually represents the instantaneous axis of rotation of one link relative to the other and is known as the *instant center, instantaneous center, velocity pole,* or *centro.*

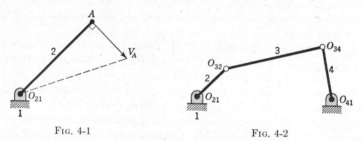

Fig. 4-1   Fig. 4-2

In order to use the instant centers in velocity analysis, it is first necessary to locate them. In Fig. 4-1 it is obvious that link 2 must at all times rotate relative to the fixed link 1 about the point O_{21}. The notation O_{21} means the instant center of rotation of link 2 relative to link 1. In this particular case, O_{21} has no motion and is known as a *fixed center.* For any point in link 2, the direction of the velocity will be perpendicular to and the magnitude of the velocity will be proportional to its radius from the center O_{21}.

In Fig. 4-2 the fixed centers O_{21} and O_{41} are located as above. The instant center of rotation of link 3 relative to link 2 must be the point O_{32} at which the links are mechanically connected. It should be apparent that O_{32} could just as well have been considered as the instant center of link 2 relative to link 3 and noted as O_{23}. The instant center O_{34} or O_{43} is located in the same manner. Because the center O_{32} is a result of a mechanical connection between the links and consequently must always be at the same place on the links, it is sometimes called a *permanent center.*

It is always easy to determine the locations of fixed and permanent centers by inspection. However, the usefulness of this method lies in the fact that links with complicated motions, such as link 3 in Fig. 4-2, not only have instant centers relative to the members to which they are actually connected but also have instant centers relative to all other members. The instant center O_{31} is of particular importance in that it is the center of rotation of link 3 with respect to the fixed link 1. As already discussed, the velocity of any point in a link is proportional to the distance from the point to the center of the link's rotation relative to the ground. Therefore, once the velocity of any point in link 3 and the location of the instant center O_{31} are known, the veloc-ities of all other points in link 3 may be found by simply setting up similar tri-angles which will give the desired pro-portional relationships.

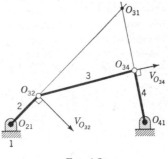

The immediate problem is to locate the center O_{31}. The procedure for this is illustrated in Fig. 4-3. The centers O_{32} and O_{34} must have velocities perpen-dicular to links 2 and 4, respectively. Therefore, if O_{32} and O_{34} are considered as points in link 3, the instant center for the rotation of link 3 relative to the

Fig. 4-3

ground must lie on lines which are perpendicular to the velocities of O_{32} and O_{34}; that is, extensions of the lines O_{21}-O_{32} and O_{41}-O_{34} intersect at the center O_{31}.

By utilizing the principle of inversion and fixing link 2 instead of link 1, the reasoning above may be applied to locate the instant center O_{42} which will be at the intersection of the extended lines O_{21}-O_{41} and O_{32}-O_{34}. This is illustrated in Fig. 4-4.

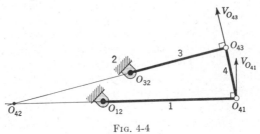

Fig. 4-4

4-4. Kennedy's Theorem. The reader may not have observed the pattern that exists, but it is important and is known as *Kennedy's theorem* or *the law of three centers*. The statement of the theorem is as follows:

If any three bodies 1, 2, and 3 have plane motion, their instant centers O_{12}, O_{13}, and O_{23} must lie on a straight line.

This simple statement provides a ready basis for locating all the instant centers in any mechanism.

4-5. Locating Instant Centers.   Every link in a mechanism has an instant center with respect to every other link.   Therefore, the number of instant centers n, for any mechanism containing N links, may be calculated from

$$n = \frac{N(N - 1)}{2} \tag{4-1}$$

After the number of instant centers is known, it is desirable to use some system which lists or records all the centers.   The easily recognized centers are noted and checked off, and Kennedy's theorem is then applied in locating the remaining instant centers.

4-6. Slider-crank Mechanism.   The slider-crank mechanism in Fig. 4-5 has four links and six instant centers.   The only new concept here is

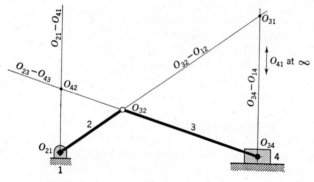

FIG. 4-5

that, since link 4 slides in a straight line on link 1, the radius of curvature of the relative path is infinitely long.   Thus, the instant center O_{41} is at infinity and is so noted in the figure.   It is important to visualize that all lines parallel to the line labeled O_{41} at ∞ will also pass through the instant center O_{41}.   The instant centers O_{12}, O_{23}, and O_{34} are readily located.   With practice, it is easy to recognize the names of the two remaining instant centers and how to use the known centers in locating them.   But this becomes almost impossible in a more complicated case, and the usual practice is to make use of a *star diagram*, as in Fig. 4-6, both to keep track of the known centers and to point out the way to locate the remaining ones.

The method is as follows: First, draw a circle and space evenly around its circumference the numbers of the links.   Every instant center is represented by a straight line connecting two numbers; for example, O_{12} is represented by the line connecting 1 and 2. Thus, lines representing O_{12}, O_{23}, O_{34}, and O_{14} may immediately be drawn in. At the same time, it is apparent that O_{31} and O_{24} are missing. Now, as previously shown, the instant center O_{31} will lie at the intersection of the line through the instant centers O_{32} and O_{12} with the line through the instant centers O_{34} and O_{14}. Each set of instant centers falling on a line is represented by a triangle in the star diagram. If a dotted line is drawn indicating the unknown center O_{31}, it is evident that the two lines, O_{32}-O_{12} and O_{34}-O_{41}, that will pass through O_{31} can be drawn, because the dotted line completes the triangles representing the sets of instant centers O_{31}, O_{32}, O_{21} and O_{31}, O_{34}, O_{41}.

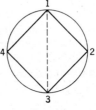

Fig. 4-6

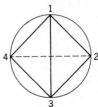

Fig. 4-7

A similar analysis (Fig. 4-7) will show that the instant center O_{24} will lie at the intersection of the lines O_{21}-O_{41} and O_{23}-O_{43}.

When the mechanism contains more than four links, there will be more than two lines passing through each instant center.

4-7. Direct-contact Mechanisms. The cam, follower, and frame in Fig. 4-8 make up a three-link direct-contact mechanism which has three instant centers. The instant centers O_{21} and O_{31} can be located by inspection. According to Kennedy's theorem, the remaining center O_{32} must lie on the line O_{31}-O_{21}, or the line of centers. But, since there are no other instant centers, Kennedy's theorem cannot be used to find another line which will intersect the line O_{31}-O_{21} at the instant center O_{32}.

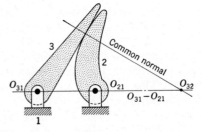

Fig. 4-8

The necessary additional information may be obtained by considering that the instant center O_{32} will be the point about which links 2 and 3 have relative motion and that, as shown in Sec. 3-11, the only relative motion possible between the points of contact must be along the tangent to the surfaces, or perpendicular to the common normal. Therefore, since the relative motion of points in link 2 and link 3 must be due to rotation about the instant center O_{32}, the instant center O_{32} must lie on the common normal, and its intersection with the line O_{31}-O_{21} will locate the center O_{32}.

4-8. Determination of Linear Velocities by Use of Instant Centers. A velocity analysis of any mechanism, no matter how complicated, may be made as a series of simple steps based entirely upon two properties of the instant center.

1. *The absolute velocity of any point in a link will have a magnitude proportional to and a direction perpendicular to the radius of the point from the instant center of the link's rotation relative to the fixed link or ground.*

2. *The instant center is a point in common with the two links and will act as if there were instantaneously an actual pin connection between the two links at that point.*

The first property (1) provides a convenient basis for determining the velocity of any point in a given link after the velocity of any one point in the link is known. In Fig. 4-9 the velocity of A is known and is represented by a vector drawn to a convenient scale. If A is considered as a point in link 3, the instant center O_{31} is the center of rotation of link 3 relative to the ground. The center O_{31} is often called the *pivot point* for link 3, for reasons which will soon be apparent. Since the velocity of any point in link 3 is proportional to its radius from O_{31}, a gauge line passing through O_{31} and the tip of the vector $\mathbf{V}_A$ establishes the correct proportions for velocity vectors for all points in link 3. The velocity of B is found by first swinging an arc, with O_{31} as the pivot point, from B to B'. Then a vector is drawn from B' perpendicular to the line O_{31}-A, and the intersection of this perpendicular and the gage line determines the proper length for the vector representing the velocity of B. This vector is then drawn from B perpendicular to the radius to O_{31}.

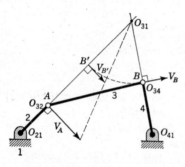

Fig. 4-9

The second property (2) must be utilized in transferring a velocity from one link to another. In other words, the instant center will have a velocity which is common to two points in different links even though the links are rotating about different centers. The instant center, when used in this manner, is known as a *transfer point*. It is most important to understand that this transfer point must be the instant center whose notation contains the numbers of the two links concerned. For example, if in Fig. 4-10 it is desired to determine the velocity of B as a point in link 4 directly from the velocity of A as a point in link 2, it is first necessary to determine the velocity of the point in common to the two links—the transfer point O_{24}. The velocity of O_{24} is determined by drawing the gage line for link 2 through O_{21} and the tip of the vector $\mathbf{V}_{A'}$. A new gage line for link 4 is drawn through O_{41} and the tip of the vector for the velocity of O_{24}. The radius $O_{41}\text{-}B$ is then pivoted about O_{41}, and the remaining procedure should be clear from study of the preceding example.

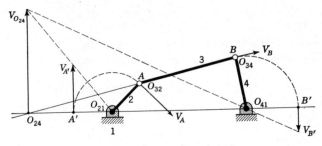

FIG. 4-10

It has probably occurred to the reader that the velocity of B can more quickly be found by considering both A and B to be points in link 3, avoiding the need for a transfer point. This is true for this simple example, but if A and B are separated by several links, considerable time and effort may be saved by going directly through the transfer point for the two links concerned.

4-9. Parallel-line Construction. The only special technique for using instant centers in determining linear velocities that is important enough to include in this brief discussion is the *parallel-line construction*. This method is based on similar triangles and is particularly useful when a necessary pivot point is located off the paper, as is the case for link 3 in Fig. 4-11. In this example, the velocity of A is known, and the velocity of C is to be determined. The procedure is to swing the length of the vector $\mathbf{V}_A$ around so that it falls on the line $O_{21}\text{-}O_{23}\text{-}O_{31}$ at a. Next a line is drawn parallel to AB so that it

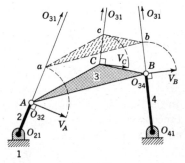

FIG. 4-11

intersects the line $O_{41}\text{-}O_{34}\text{-}O_{31}$ at b. Then lines parallel to BC and AC are drawn from b and a, respectively. The intersection of these lines is labeled c. The triangles ABC and abc are similar, and the extension of the line Cc will pass through the pivot point O_{31}. Therefore, $Cc/CO_{31} = Aa/AO_{31}$, and the velocity vector $\mathbf{V}_C$ will have a length equal to, and a direction perpendicular to, the line Cc. Similarly, the velocity vector $\mathbf{V}_B$ will have a length equal to, and a direction perpendicular to, the line Bb.

4-10. Determination of Angular Velocities by Use of Instant Centers. The angular velocity of a link can be found by dividing the velocity of any point in the link by the radius to the pivot point. In Fig. 4-9,

$$\omega_3 = \frac{V}{R} = \frac{V_A}{O_{31}A} = \frac{V_B}{O_{31}B} \qquad \text{ccw} \qquad (4\text{-}2)$$

This method obviously cannot be applied when the instant center is too far off the paper, as in Fig. 4-11, and a different approach must then be used.

4-11. Method of Relative Velocities. Probably the most useful method for making a velocity analysis of a mechanism is that based upon the fundamental relationship between the velocity of one point in a link relative to any other point in the link.

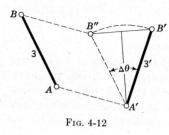

FIG. 4-12

In Chap. 1 the statement was made that any plane motion may be shown to be the equivalent of a rotation plus a translation. This is illustrated in Fig. 4-12, where A has moved to A' and B to B' in the time interval Δt. The dotted lines show how the motion of B may be considered to be the sum of a translation with A and a rotation about A. The vector equation for the average velocity of B is

$$\mathbf{V}_{B,\text{av}} = \frac{BB''}{\Delta t} \nleftrightarrow \frac{B''B'}{\Delta t} \qquad (4\text{-}3)$$

but

$$\frac{BB''}{\Delta t} = \frac{AA'}{\Delta t} = \mathbf{V}_{A,\text{av}} \qquad (4\text{-}4)$$

and

$$\frac{B''B'}{\Delta t} = \frac{2AB \sin (\Delta\theta/2)}{\Delta t} = \mathbf{V}_{B/A,\text{av}} \qquad (4\text{-}5)$$

In the limit, as Δt approaches zero, the average velocities become instantaneous velocities and $[2 \sin (\Delta\theta/2)]/\Delta t$ becomes $d\theta/dt$ or ω. The vector equation then becomes

$$\mathbf{V}_B = \mathbf{V}_A + \mathbf{V}_{B/A} \qquad (4\text{-}6)$$

The important concept here is that *the velocity of any point in a rigid body relative to any other point in the same body is a vector quantity equal in magnitude to the product of the angular velocity of the body and the distance between the two points and is directed perpendicular to the line connecting the points.*

Conversely, the angular velocity of any link will have a magnitude equal to the velocity of any point in the body relative to any other point in the same body divided by the distance between the two points, and its direction can be determined by inspection.

4-12. The Velocity Polygon or Velocity Vector Diagram. A vector equation such as

$$\mathbf{V}_B = \mathbf{V}_A + \mathbf{V}_{B/A} \tag{4-7}$$

is perfectly general. Thus, if all vectors representing absolute velocities are drawn from a common pole, the velocity of any point relative to any other point may be readily determined.

In Fig. 4-13a the velocity of A is known and is drawn to scale as the vector $\mathbf{V}_A$. It is desired to find the velocity of B. The vector $\mathbf{V}_A$ is drawn from the pole O in Fig. 4-13b. All that is known about the remaining terms is that $\mathbf{V}_B$ must be perpendicular to link 4 and that $\mathbf{V}_{B/A}$ must be perpendicular to the line AB. Since $\mathbf{V}_B$ is an absolute velocity, a line

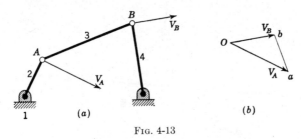

Fig. 4-13

perpendicular to link 4 is drawn through the pole O. By definition, the vector $\mathbf{V}_{B/A}$ must pass through the head of both vectors $\mathbf{V}_A$ and $\mathbf{V}_B$. Therefore, a line perpendicular to the line AB is drawn through the head of $\mathbf{V}_A$. The resulting triangle gives the vector solution of the equation and is known as a *velocity polygon* or a *velocity vector diagram*.

Arrows will be shown on only those vectors representing absolute velocities. The reason for not showing arrows on the relative-velocity vectors is that $\mathbf{V}_{B/A}$ and $\mathbf{V}_{A/B}$ are equal in magnitude but opposite in direction and the sense may be determined by inspection as follows: The notation $\mathbf{V}_{B/A}$ means the velocity of B relative to A, or the vector which must be added to $\mathbf{V}_A$ to give $\mathbf{V}_B$. In Fig. 4-13b, it can be seen that this means that $\mathbf{V}_{B/A}$ has the direction from a to b. Conversely, $\mathbf{V}_{A/B}$ has the direction from b to a. In other words, the relative velocity is directed *away* from the small letter representing the point of reference.

4-13. The Velocity Image. Figure 4-14 shows a linkage with a known velocity $\mathbf{V}_A$. It is desired to find the velocity of points B and C and the angular velocity of link 3. The velocity polygon is drawn in exactly the same manner as in the preceding section.   The magnitude of the angular velocity of link 3 may be found by dividing the relative velocity of any two points in link 3 by the distance between them.   Thus,

$$\omega_3 = \frac{V_{A/C}}{AC} = \frac{V_{C/A}}{AC} = \frac{V_{B/A}}{AB} = \frac{V_{A/B}}{AB} = \frac{V_{C/B}}{CB} = \frac{V_{B/C}}{CB} \tag{4-8}$$

The direction of the angular velocity can be determined by noting the direction of the velocity of any point in the link relative to another point in the same link; e.g., the velocity of A relative to B is seen to be from b to a in the velocity polygon.   Therefore, link 3 is rotating in a counterclockwise direction.

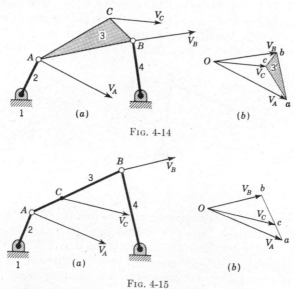

FIG. 4-14

FIG. 4-15

It should be noted that, since the lines ba, bc, and ac in the velocity polygon were drawn perpendicular to lines BA, BC, and AC, respectively, in the configuration diagram, the triangles abc and ABC are similar, with the triangle abc being rotated 90°.   The *triangle abc is called the velocity image of link* 3.   Even a link which consists of a single line has a velocity image; e.g., the velocity image of link 2 is the line Oa in the velocity polygon.   This principle is very useful in that the velocity of any point on a link may be found by using the simple proportional relationship between the link and its velocity image; that is, if C in Fig. 4-15 is one-third the distance from A to B, then c will be one-third the distance from a to b, and the vector Oc is the absolute velocity of C.

4-14. The Velocity Polygon for Direct-contact Mechanisms. The direction of the relative velocity between points in direct contact has been discussed in Secs. 3-11 and 4-7. A fact fundamental to the drawing of the velocity polygons for these mechanisms is that the velocity of the contact point on one link relative to the coincident contact point on the other link must be directed along the tangent to the contact surfaces.

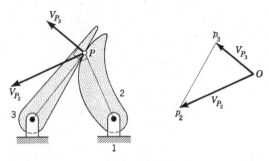

FIG. 4-16

In Fig. 4-16 the angular velocity of link 2 is known. The linear velocity of P as a point on 2 is calculated and laid off from the pole point O of the velocity polygon. Lines in the direction of the velocity of P as a point on 3 and in the direction of the velocity of P on 3 relative to P on 2 are drawn from O and p_2, respectively. The intersection of these lines locates p_3 on the velocity polygon.

4-15. The Acceleration Polygon or Acceleration Vector Diagram. There are not as many methods or techniques available for making an acceleration analysis of a mechanism as there are for velocity analysis. Composition and resolution have little or no meaning or usefulness when applied to accelerations. The reason that instant centers can be used in the study of velocities is that actually this method is one of relative velocities with reference to points that have zero instantaneous absolute velocity. As previously shown, these reference points, i.e., the instant center of the rotation of the link relative to the fixed link, may be located without resort to any knowledge other than the geometry of the figure. Locating the instant center of acceleration of a link is not so simple, because its position depends upon the values of angular velocity and angular acceleration as well as upon the geometry of the mechanism. Therefore, the instant center of acceleration has little practical use in an acceleration analysis, and its academic value does not warrant its inclusion in this book.

Except for the straightforward, analytical solution of the slider-crank mechanism, the practical methods for acceleration analysis of mechanisms in general are the graphical and analytical vector solutions and the graphical differentiation of velocity-time graphs. Of these, the most useful method is the graphical solution of the vector equations involving relative accelerations. The procedure is quite similar to that used in the relative-velocity analysis and is based on the simple vector equation

$$\mathbf{A}_B = \mathbf{A}_A + \mathbf{A}_{B/A} \qquad (4\text{-}9)$$

It was pointed out in Sec. 3-4 that the acceleration of a point is the vector sum of two components, normal and tangential. Figure 4-17 shows a link 2 with angular velocity ω and angular acceleration α in the directions indicated. The normal component $\mathbf{A}_B^n$ is equal to $R\omega^2$ and is directed toward the center of curvature, while the tangential component $\mathbf{A}_B^t$ is equal to $R\alpha$ and is directed in the same sense as α. The total acceleration of B is the vector sum of $\mathbf{A}_B^n$ and $\mathbf{A}_B^t$.

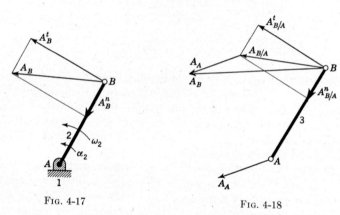

FIG. 4-17

FIG. 4-18

It should be clear that this example is a special case of relative acceleration, in that the equation

$$\mathbf{A}_B = \mathbf{A}_A + \mathbf{A}_{B/A} \qquad (4\text{-}10)$$

where $\mathbf{A}_A = 0$ and $\mathbf{A}_{B/A} = \mathbf{A}_{B/A}^n + \mathbf{A}_{B/A}^t$, gives the correct solution.

The more usual case is that in which A also has a known acceleration, as in Fig. 4-18. However, since A is still the center of curvature of the path of B relative to A, the equation

$$\begin{aligned} \mathbf{A}_B &= \mathbf{A}_A + \mathbf{A}_{B/A} \\ &= \mathbf{A}_A + \mathbf{A}_{B/A}^n + \mathbf{A}_{B/A}^t \end{aligned} \qquad (4\text{-}11)$$

is still correct.

Two important points that must be thoroughly understood are the following:

1. *The normal component of acceleration of a point relative to any other point in the same rigid body is a function of only the angular velocity of the link and the distance between the points and is always directed toward the reference point.*

2. *The tangential component of acceleration of a point relative to any other point in the same rigid body is a function of the angular acceleration of the link and the distance between the two points, and it will be directed perpendicular to the line connecting the two points with the same sense as the angular acceleration.*

Example 4-1.   The angular velocity and angular acceleration of link 2 of the mechanism in Fig. 4-19a are, respectively, 10 radians/sec, ccw, and 50 radians/sec², ccw.   It is desired to draw the complete acceleration polygon and to determine the angular accelerations of links 3 and 4.

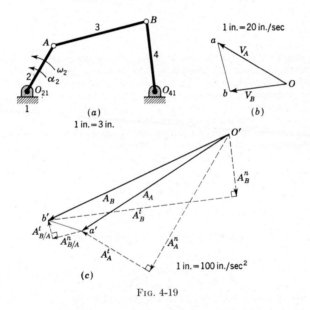

FIG. 4-19

The velocity of A is calculated, and the velocity polygon in Fig. 4-19b is drawn to scale:

$$V_A = R\omega = 1.8 \times 10 = 18 \text{ in./sec}$$

The normal and tangential components of acceleration are calculated and are added vectorially to give the vector $\mathbf{A}_A$, which is drawn from the pole O', as was done in the case of the velocity polygon:

$$A_A^n = R\omega^2 = 1.8 \times (10)^2 = 180 \text{ in./sec}^2$$
$$A_A^t = R\alpha = 1.8 \times 50 = 90 \text{ in./sec}^2$$

B is a point both in link 3 and in link 4, and the equations

$$A_B = A_A + A_{B/A}$$
$$= A_A + A_{B/A}^n + A_{B/A}^t$$

and

$$A_B = A_B^n + A_B^t$$

may be written. It is not possible to calculate either tangential component ($A_{B/A}^t$ or A_B^t), because nothing is known about the magnitudes of the angular accelerations of links 3 and 4. It is known that the direction of $A_{B/A}^t$ will be perpendicular to AB and the direction of A_B^t will be perpendicular to $O_{41}B$.

The normal components, $A_{B/A}^n$ and A_B^n, can be completely determined. $A_{B/A}^n$ will be in the direction from B to A, and its magnitude is $AB(\omega_3)^2$, or, more conveniently, $(V_{B/A})^2/AB$. A_B^n will be in the direction from B to O_{41}, and its magnitude is $(V_B)^2/O_{41}B$. From the velocity polygon, $V_{B/A} = 10.7$ in./sec, and $V_B = 13$ in./sec.

$$A_{B/A}^n = \frac{(V_{B/A})^2}{BA} = \frac{10.7^2}{3.3} = 34.7 \text{ in./sec}^2$$

$$A_B^n = \frac{(V_B)^2}{BO_{41}} = \frac{13^2}{2.4} = 70.4 \text{ in./sec}^2$$

The vector A_B^n is drawn from the pole O' in Fig. 4-19c, and a line in the direction of A_B^t is drawn through its tip. The vector $A_{B/A}^n$ is drawn from a', and a line in the direction of $A_{B/A}^t$ is drawn through its tip. The intersection of the two lines drawn in the directions of the tangential components locates b'. This point is the only one that satisfies the two equations for the acceleration of B. The vector $O'b'$ is the acceleration of B, and the vector a' to b' is the acceleration of B relative to A.

The arrows on the tangential components must be at the ends that indicate addition to the normal components of acceleration. The angular acceleration of link 3 is found from

$$\alpha_3 = \frac{A^t}{R} = \frac{A_{B/A}^t}{AB}$$

and its direction will have the same sense as the tangential acceleration about A, that is, counterclockwise.

Similarly, the angular acceleration of link 4 is found from

$$\alpha_4 = \frac{A_B^t}{O_{41}B}$$

and its direction is counterclockwise. From the acceleration polygon, $A_B = 230$ in./sec^2, $A_{B/A}^t = 20$ in./sec^2, and $A_B^t = 219$ in./sec^2.

$$\alpha_3 = \frac{A_{B/A}^t}{AB} = \frac{20}{3.3} = 6.1 \text{ radians/sec}^2, \text{ ccw}$$

$$\alpha_4 = \frac{A_B^t}{O_{41}B} = \frac{219}{2.4} = 91.3 \text{ radians/sec}^2, \text{ ccw}$$

4-16. The Acceleration Image. Considering that the acceleration of one point in a link relative to another point in the same link may be expressed as

$$\mathbf{A}_{B/A} = \mathbf{A}^n_{B/A} + \mathbf{A}^t_{B/A} = \sqrt{(A^n_{B/A})^2 + (A^t_{B/A})^2}$$

and that

$$A^n_{B/A} = AB\omega^2$$

and

$$A^t_{B/A} = AB\alpha$$

will lead to the conclusion

$$A_{B/A} = AB\sqrt{\omega^4 + \alpha^2} \qquad (4\text{-}12)$$

Thus, since ω and α are properties of the entire link, it is evident that the acceleration of any point relative to another point in the same link is proportional to the distance between the points. Therefore, each link will have an acceleration image as well as a velocity image, and simple ratios can be used in determining the acceleration of specific points.

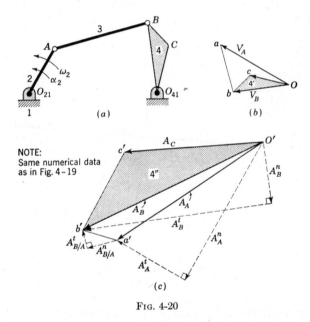

Fig. 4-20

This principle may be used also in constructing the acceleration polygon, but care must be taken to ensure that the acceleration image of a link is not flipped over; i.e., if ABC are in a clockwise direction on the configuration diagram, $a'b'c'$ must likewise be in a clockwise direction on the acceleration polygon. In Fig. 4-20 the mechanism and data are the same as in Fig. 4-19 except that link 4 is now shown as a triangle.

4-17. Relative Acceleration of a Pair of Coincident Points—the Coriolis Component. It is important to understand that the two preceding sections have been concerned with the acceleration of one point relative to another point on the same link where the radius of curvature of the relative path is a constant value and with the transfer of accelerations from one link to another through a *point common to both links at all times,* i.e., a pin joint, or *between nonrotating sliding members.* These conditions are not met when there is sliding contact between rotating members, such as between links 3 and 4 of the crank-shaper mechanism in Fig. 4-21.

In this situation the acceleration of points on links 2 and 3 can be calculated as before and the problem becomes the determination of the acceleration of a point on link 4. The simplest approach utilizes a point on link 4 which, at this instant, is coincident with a point whose acceleration has already been determined. Here, the logical point to use is A_4, the point on link 4 that is coincident with A on links 2 and 3.

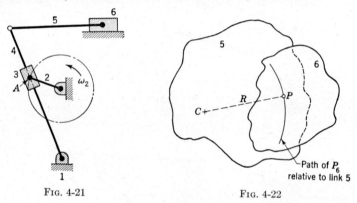

Fig. 4-21   Fig. 4-22

As in Sec. 3-4, the most significant factor in handling this type of problem will be the curvature of the path. However, the path of interest is no longer that of the absolute motion; instead, it will be *the path relative to a moving,* i.e., rotating, *link.*

Figure 4-22 shows two links of a mechanism. Let us assume that a complete velocity analysis has been made and that the acceleration analysis has progressed through pin connections up through link 5. There is no point that is common to links 5 and 6 at all times, and we wish to transfer the acceleration from P_5 to its coincident point P_6.

The acceleration of P_6 can be expressed as

$$\mathbf{A}_{P_6} = \mathbf{A}_{P_5} + \mathbf{A}_{P_6/P_5} \tag{4-13}$$

where $\mathbf{A}_{P_5}$ has already been determined. Now, recalling that all links and points in a mechanism move with definite relative motions—irrespective of which link is fixed—we see that the path of P_6 relative to link 5 would become the absolute path of P_6 if the mechanism were to be inverted by fixing link 5. Therefore, the motion of P_6 about the center of curvature of its path relative to link 5 is analogous to that of point P about its center of curvature on the stationary member in Fig. 3-4.

From this point of view, we can write

$$\mathbf{A}_{P_6/P_5} = \mathbf{A}_{P_6/C} + \mathbf{A}_{C/P_5} \tag{4-14}$$

The components are

$$\mathbf{A}_{P_6/C} = \mathbf{A}^n_{P_6/C} + \mathbf{A}^t_{P_6/C}$$

$$= R\omega_R{}^2 \nrightarrow \frac{d(R\omega_R)}{dt} \tag{4-15}$$

and

$$\mathbf{A}_{C/P_5} = \mathbf{A}^n_{C/P_5} + \mathbf{A}^t_{C/P_5}$$

$$= CP\omega_5{}^2 \nrightarrow CP\,\frac{d\omega_5}{dt} \tag{4-16}$$

where $\omega_R =$ angular velocity of the radius of curvature R.

The tangential terms have been left as derivatives because, as previously discussed, the only information generally available is that their directions must be tangent to the path of motion.

Equations (4-15) and (4-16) might be used directly when $\omega_R \neq \omega_5$, but the application to practical problems will be simplified considerably by putting them in different forms.

Since both Eqs. (4-15) and (4-16) contain normal and tangential components, it is convenient to consider Eq. (4-14) as

$$\mathbf{A}_{P_6/P_5} = \mathbf{A}^n_{P_6/P_5} + \mathbf{A}^t_{P_6/P_5} \tag{4-17}$$

where

$$\mathbf{A}^n_{P_6/P_5} = R\omega_R{}^2 \nrightarrow CP\,\omega_5{}^2 = R\omega_R{}^2 \nrightarrow R\omega_5{}^2 \tag{4-18}$$

and

$$\mathbf{A}^t_{P_6/P_5} = \frac{d(R\omega_R)}{dt} \nrightarrow CP\,\frac{d\omega_5}{dt} \tag{4-19}$$

It should be noted that the direction of $\mathbf{A}^n_{P_6/P_5}$ is dependent upon the relative magnitudes of ω_R and ω_5; that is, if $\omega_R > \omega_5$, the direction is from P to C, and if $\omega_R < \omega_5$, the direction is from C to P.

Since only the general sense of direction of the tangential components is known, it is convenient to lump them together, using a capital script $\mathcal{C}$, that is, $\mathcal{C}^t_{P_6/P_5}$ instead of $\mathbf{A}^t_{P_6/P_5}$, to emphasize that this component *is not the ordinary tangential component* found when going from one point to another point on the same rigid member.

The problem of the direction of the normal acceleration can be eliminated by designating the direction from P to C as positive. Doing so permits rewriting Eq. (4-18) as

$$\mathbf{A}^n_{P_6/P_5} = R(\omega_R{}^2 - \omega_5{}^2) \tag{4-20}$$

where R now has magnitude only. Adding and subtracting $2R\omega_R\omega_5 + 2R\omega_5{}^2$ and rearranging gives

$$\begin{aligned}
\mathbf{A}^n_{P_6/P_5} &= R(\omega_R{}^2 - 2\omega_R\omega_5 + \omega_5{}^2) + 2R\omega_R\omega_5 - 2R\omega_5{}^2 \\
&= R(\omega_R - \omega_5)^2 + 2R(\omega_R - \omega_5)\,\omega_5 \tag{4-21}
\end{aligned}$$

In previous sections it was found to be more convenient to express the normal components in terms of linear velocities because this information was available in the velocity polygon. This is still true here, but the fact that Eq. (4-21) is indeterminate when $R = \infty$, at which time $\omega_R - \omega_5 = 0$, makes it not only desirable but necessary in most problems to rewrite the equation as

$$\mathbf{A}^n_{P_6/P_5} = \frac{(V_{P_6/P_5})^2}{R} \mathrel{+\mkern-10mu+} 2V_{P_6/P_5}\omega_5 \tag{4-22}$$

The first term has the appearance of a normal component of acceleration and has the direction from P to C. However, this is only part of the story, and to emphasize this, again a capital script $\mathcal{C}$ is used, that is, $\mathcal{C}^n_{P_6/P_5}$. The second term is called the *Coriolis* component of acceleration, and its magnitude can be seen to equal *two times the product of the relative velocity of the coincident points and the angular velocity of the reference member.* Its direction may be either positive, from P to C, or negative, from C to P, depending upon whether $\omega_5(\omega_R - \omega_5)$ is positive or negative, respectively. Although this can be determined in some cases without much trouble, it is generally more convenient to use the following rule: *The Coriolis component will be directed in the sense of the relative velocity vector after it has been rotated 90° in the direction of the angular velocity of the reference link.*

The preceding development results in the following equation for the relative acceleration of coincident points:

$$\mathbf{A}_{P_6/P_5} = \mathcal{C}^n_{P_6/P_5} + \mathcal{C}^t_{P_6/P_5} \mathrel{+\mkern-10mu+} 2V_{P_6/P_5}\,\omega_5 \tag{4-23}$$

Example 4-2. Link 2 of the crank-shaper mechanism in Fig. 4-23 is rotating cw at a uniform speed of 50 rpm. It is desired to determine the acceleration of the ram 6 in feet per second per second.

The velocity of A as a point on links 2 and 3 is calculated, and the velocity polygon is drawn to scale in Fig. 4-23b:

$$V_{A_{2,3}} = O_2A\omega_2 = \tfrac{8}{12} \times 2\pi \times \tfrac{50}{60} = 3.49 \text{ fps}$$

Since the tangential acceleration of $A_{2,3}$ is zero, the total acceleration is the normal acceleration, which is calculated and drawn from the pole O' in Fig. 4-23c:

$$A_{A_{2,3}} = A^n_{A_{2,3}} = \frac{(V_{A_{2,3}})^2}{O_2A} = \frac{3.49^2}{\tfrac{8}{12}} = 18.28 \text{ ft/sec}^2$$

It now becomes necessary to determine the acceleration of a point on link 4. Since link 4 slides in link 3 and both rotate, the Coriolis component must be included and one logical approach would appear to be to use the equation

$$\begin{aligned}
\mathbf{A}_{A_4} &= \mathbf{A}_{A_2} + \mathbf{A}_{A_4/A_2} \\
&= \mathbf{A}_{A_2} + \mathcal{C}^n_{A_4/A_2} + \mathcal{C}^t_{A_4/A_2} \mathrel{+\mkern-10mu+} 2V_{A_4/A_2}\omega_2 \tag{a}
\end{aligned}$$

However, this is not convenient or desirable, because to solve

$$G^n_{A_4/A_2} = \frac{(V_{A_4/A_2})^2}{R}$$

requires determining the radius of curvature of the path of A_4 on link 2. This not only would be difficult and time consuming but would also result in a relatively

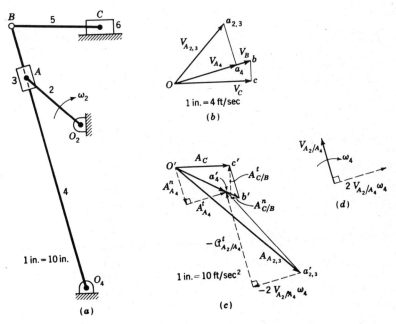

FIG. 4-23

inaccurate answer. The simplest way to avoid this difficulty would be to find $\mathbf{A}_{A_4/A_2}$, since the radius of curvature of A_4 on link 3 is ∞. However, for illustrative purposes, let us continue to work with links 2 and 4 by rewriting Eq. (a) as

$$\mathbf{A}_{A_4} = \mathbf{A}_{A_2} + (-G^n_{A_2/A_4}) + (-G^t_{A_2/A_4}) + (-2V_{A_2/A_4}\omega_4) \tag{b}$$

R is now the radius of curvature of the path of A_2 on link 4, which is infinite. Thus,

$$G^n_{A_2/A_4} = \frac{(V_{A_2/A_4})^2}{R} = \frac{(1.93)^2}{\infty} = 0$$

$G^t_{A_2/A_4}$ cannot be calculated, but it will be directed along the relative path, i.e., parallel to O_4A.
Then

$$2V_{A_2/A_4}\omega_4 = 2V_{A_2/A_4}\frac{V_{A_4}}{O_4A}$$

$$= 2 \times 1.93 \times \frac{2.90}{24.1/12} = 5.57 \text{ ft/sec}^2$$

Using the rule given above, the direction of the Coriolis component is determined as shown in Fig. 4-23d. The vector $-2V_{A_2/A_4}\omega_4$ is drawn from $a_{2,3}$, and a line in the direction of $-\mathfrak{C}^t_{A_2/A_4}$ is drawn from the head of $-2V_{A_2/A_4}\omega_4$.

Since the magnitude of the tangential acceleration cannot be calculated, it now becomes necessary to consider the absolute acceleration of A_4, which can be written as

$$\mathbf{A}_{A_4} = \mathbf{A}^n_{A_4} + \mathbf{A}^t_{A_4}$$

where

$$A^n_{A_4} = \frac{(V_{A_4})^2}{O_4A} = \frac{2.90^2}{24.1/12} = 4.18 \text{ ft/sec}^2$$

$\mathbf{A}^n_{A_4}$ is drawn from O' in the direction from A to O_4. A line in the direction of $\mathbf{A}^t_{A_4}$, that is, perpendicular to O_4A, is then drawn from the head of $\mathbf{A}^n_{A_4}$.

The intersection of the lines labeled $-\mathfrak{C}^t_{A_2/A_4}$ and $\mathbf{A}^t_{A_4}$ locates a_4.

The acceleration of B is now found by drawing the acceleration image of link 4 in accordance with the relationship

$$\frac{O'b'}{O'a_4'} = \frac{O_4B}{O_4A}$$

Now that the acceleration of B is known, the determination of the acceleration of C can be made in a straightforward manner, as in Example 4-1. The equations of interest are

$$\mathbf{A}_C = \mathbf{A}_B + \mathbf{A}_{C/B}$$
$$= \mathbf{A}_B + \mathbf{A}^n_{C/B} + \mathbf{A}^t_{C/B}$$

and

$$\mathbf{A}_C = \mathbf{A}^n_C + \mathbf{A}^t_C$$

where

$$A^n_{C/B} = \frac{(V_{C/B})^2}{CB} = \frac{0.90^2}{10/12} = 0.97 \text{ ft/sec}^2$$

and

$$A^n_C = 0$$

$\mathbf{A}^n_{C/B}$ is drawn from b' in the direction from C to B, and a line in the direction of $\mathbf{A}^t_{C/B}$ is drawn through the tip of $\mathbf{A}^n_{C/B}$. The intersection of the line through O' in the direction of $\mathbf{A}^t_C$ and the line labeled $\mathbf{A}^t_{C/B}$ determines the location of c' and thus the acceleration of C, which is found to be directed to the right with a magnitude of 6.20 ft/sec^2.

4-18. Graphical Differentiation. The use of graphical differentiation in velocity and acceleration analysis has too often been given less emphasis than it deserves. The method does have one major limitation in that it can be applied only to the changes in *magnitude* of the vector quantities of both linear and angular displacement, velocity, and acceleration.

One of the important uses of graphical differentiation is the application to displacement-time or velocity-time data taken from an oscillograph record strip or from high-speed motion pictures of a mechanism, projectile, etc.

The magnitude of the velocity at any instant is

$$V = \frac{dS}{dt} \tag{4-24}$$

and the magnitude of the acceleration at any instant is

$$A = \frac{dV}{dt} = \frac{d^2S}{dt^2} \tag{4-25}$$

It is obvious that the acceleration-time curve can be determined either by using one of the standard methods of velocity analysis for plotting the velocity-time curve and then differentiating this curve only once or by graphically differentiating the displacement-time curve to get the velocity-time curve and then differentiating the resulting velocity-time curve to get the acceleration-time curve. The former method, where applicable, would be expected to give more accurate results in that it involves determining the tangent to a curve only once instead of twice.

Figure 4-24 shows a portion of a displacement-time curve. The first step in the procedure for finding the velocity at time T_1 is to draw a tangent to the curve at P. The accuracy of the results will depend largely upon the accuracy of the tangent. One method often used is to place a small pocket mirror or reflecting bar perpendicular to the curve

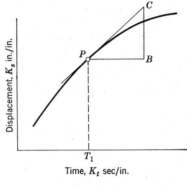

FIG. 4-24

so that the reflected portion of the curve appears to be continuous with the actual curve. The plane of the mirror is the normal to the curve, and the tangent is then drawn perpendicular to the normal. However, with a little practice, reasonable accuracy in drawing the tangent may be obtained without use of a mirror. After the tangent is drawn, PB is drawn with some convenient length parallel to the time axis. A line drawn from B parallel to the displacement axis intersects the tangent at C. The magnitude of the velocity at this instant is

$$V = \frac{dS}{dt} = \frac{(BC)K_s}{(PB)K_t} \tag{4-26}$$

where K_s and K_t are the scales for displacement and time, respectively, on the graph.

If the entire velocity-time curve is desired, it is convenient to use the same distance PB for all points. This makes the velocity at any instant equal to

$$BC \frac{K_s}{(PB)K_t} \tag{4-27}$$

or the distances BC may be directly plotted on the same time axis as the displacements, with the scale for the velocity curve becoming

$$K_v = \frac{K_s}{(PB)K_t} \quad \text{in./sec/in.} \tag{4-28}$$

This same reasoning is applied when differentiating the velocity-time curve to get the acceleration-time curve.

4-19. Closure.   It is hoped that the reader appreciates the fact that in many situations, where it may not be necessary, it may be worthwhile to combine several of the methods of this chapter in making a motion analysis of a mechanism.

It is also important to appreciate that only a few basic rules and applications have been given here.   There are many extensions of principles, special techniques, etc., worth studying and understanding if one must spend considerable time working with problems of displacement, velocity, and acceleration related to mechanisms.   In particular, the reader should not lose sight of the possibilities offered by the combination of analytical vector methods and high-speed digital computers.

CHAPTER 5

CAMS

With very few exceptions, the most convenient means for imparting a specified motion to a member is by means of a cam-and-follower mechanism. Not only can the motion be completely specified, but the resulting physical configuration is both rugged and compact. Some of the basic considerations underlying the transmission of motion and force through a cam and follower have already been discussed, and some of the fundamental considerations of strength, wear, and durability will be discussed in a later chapter on the particular type of cam and follower known as gears.

Thus, this chapter will be limited to a discussion of general classifications of cams, the most desirable motion characteristics, and the procedure and techniques for designing cams which will operate with a follower in the desired manner.

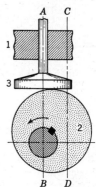

FIG. 5-1. Disk cam with radial flat-face follower.

5-1. Disk Cams. The majority of cams are in the form of shaped disks which either rotate or oscillate and give either a reciprocating or oscillatory motion to a follower. There are innumerable variations, but several of the more common types of *disk cams* are shown in Figs. 5-1 to 5-7.

Figure 5-1 shows a disk cam 2 driving a *flat-face follower* 3. In this case, the follower reciprocates with a motion that is independent of the location of the follower-stem axis relative to the axis of rotation of the cam; i.e., the axis of the follower stem may be through CD as well as through AB without affecting the motion. The force analysis of the mechanism, particularly the couple acting on the guide, will be appreciably affected by the displacement or offset of the axes. In general, reciprocating flat-face followers are designed with intersecting axes to minimize the guide couple and thus reduce friction and wear.

A common modification of the disk cam with a reciprocating flat-face follower is achieved by offsetting the axis of the follower stem in the direc-

53

tion of the axis of the cam, as in Fig. 5-2, so that the friction force rotates the follower, which more evenly distributes the wear over the face of the follower.   The term *mushroom* follower is applied to this particular circular follower face.   The followers used to open and close the valves in automobile engines are generally this type.

If the force being transmitted is large and the sliding contact between the cam and follower surfaces introduces too much friction wear and wasted energy, it is usually desirable to replace the flat-face follower with a *roller* follower, as shown in Fig. 5-3.[1]   As will be discussed in a later

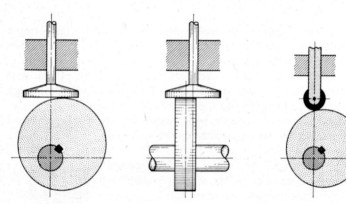

FIG. 5-2. Disk cam with mushroom follower.   FIG. 5-3. Disk cam with radial roller follower.

chapter on bearings, standard as well as special-purpose bearings (Fig. 18-53) are commercially available for this application.

It is worthwhile to point out here that, for a given cam profile, the motion of a roller follower is dependent upon the relationship of the follower-stem axis to the cam rotation axis, i.e., whether a radial or an offset roller follower is used.

One obvious disadvantage of the cams just discussed is that there is not positive contact between the cam and the follower.   If the outward motion of the follower is upward and the speed of rotation of the cam is slow, the dead weight of the follower may be sufficient to maintain contact between the cam and the follower at all times.   However, if the rotation speed of the cam is so high that the acceleration of the follower on its downward stroke must be greater than that due to gravity, an additional force must be present to maintain contact at all times.   If the cam and the follower are permitted to separate, there will be inaccurate

[1] See Fig. 2-9 for the application of a disk cam with a radial roller follower in the valve mechanism of a heavy-duty diesel engine.

timing, and the impact between the surfaces when contact is remade may adversely affect the life of the mechanism. Springs, e.g., the valve springs in the diesel engine in Fig. 2-9, are ordinarily used in this situation even though cams, such as in Figs. 5-4 and 5-5, may be designed to ensure positive return of the follower.

Figure 5-4 shows a *positive-return disk* cam with a flat-face follower. The design of such cams will be considered in Sec. 5-13, but it can be pointed out at this time that, since the distance between the faces must be constant at all times, the cam cannot be designed to give different motion characteristics during the outward and the return strokes.

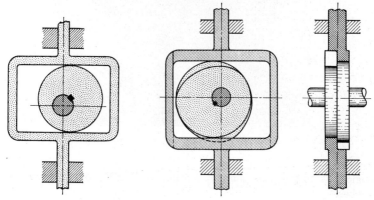

FIG. 5-4. Positive-return disk cam.   FIG. 5-5. Double-disk positive-return cam.

The reader should note that if the cam is circular in shape, as shown in the figure, the resulting mechanism is a variation of the Scotch yoke and the cam rotating with a uniform angular velocity will impart simple harmonic motion to the follower.

In order to achieve different motions on the outward and the return strokes with a positive-return disk cam, it is necessary to use two disks so designed that each disk drives in one direction. An example of a *double-disk positive-return* cam is shown in Fig. 5-5.

It should also be obvious that manufacturing tolerances and wear will cause *backlash* or clearance between the unloaded face of the follower and the cam. Consequently, there will be undesirable noise and vibration due to the impact occurring each time the transmitted force reverses direction and the cam contacts a different face of the follower. This problem is more acute for high-speed operation and is one of the reasons why high-speed cams generally operate with spring-loaded followers.

Naturally, two rollers with a fixed center distance could be substituted for the flat faces of the follower. The same problems of complexity in manufacturing, lack of control over motion characteristics of both outward and return strokes, and backlash exist also for this case.

A positive-return cam with a roller follower is shown in Fig. 5-6. Here the follower roller rides in a groove machined in the face of the disk. A *face* cam has the same problem of backlash as the positive-return cams just discussed. It has the advantage that it can be designed with a

FIG. 5-6. Face cam with oscillating roller follower. (*Courtesy The Torrington Company.*)

single disk to give independent motions during the outward and return strokes.

A type of cam which is related to the face cam is that shown in Fig. 5-7. This cam is known as an *adjustable-plate* cam and consists of a disk to

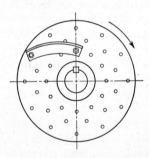

FIG. 5-7. Adjustable-plate cam.

which curved pieces may be bolted to the face in any desirable position. It is limited to slow speeds because of the lack of control over accelerations, etc., but within its field it is one of the most practical means of providing adjustable time and displacement characteristics for automatic machinery designed to be used for more than one specific operation or set of operations.

5-2. Cylindrical Cams. Sometimes the design of a machine requires the axis of rotation of a cam to be parallel to the desired motion of the follower. Here it is generally more convenient to use a *cylindrical* cam than to complicate the train of motion by adding gears, etc., to permit the use of a disk cam.

Cylindrical cams can be designed to transform rotation of the cylinder into reciprocation of the follower parallel to the axis of the cam, but the usual case is that illustrated in Fig. 5-8.   Here a conical roller running in a groove machined into the surface of the cylinder imparts a positive oscillation of the follower.   Automatic lathes and similar machine tools often use a cylindrical cam that is similar in principle to the adjustable-plate cam, in that different shapes may be bolted to the cylinder to permit more flexibility in the use of the machine.   This particular type of cylindrical cam is called a *drum* or *barrel* cam.

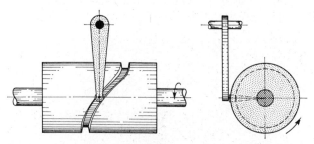

FIG. 5-8. Cylindrical cam.

5-3. Three-dimensional Cams.   The cams and followers discussed in the previous two sections are limited to use when the follower motion is a function of only one variable.   In many situations, particularly in the areas of shell ballistics, automatic navigation, and control of guided missiles, it is necessary to have an output signal that is the function of two variables.   If the functions are relatively complex, the best solution is probably a *three-dimensional* cam.

Figure 5-9 shows a three-dimensional cam with an oscillating follower. Both input signals are in the form of rotary motions; one rotates the cam, and the other rotates a translation or lead screw which moves the plane of oscillation of the follower along the axis of the cam.   Therefore, the resulting angular position of the follower is a function of both input signals.   Potentiometers are utilized in converting the displacements into electrical signals.

The cam itself may be considered to be made up of an infinite series of disk cams of infinitesimal thickness.   It should be noted that neither a flat-face nor a roller follower may be used because of the double curvature of the cam surface.   A spherical contact point, as shown, is usually most satisfactory.

5-4. Other Classifications of Cams.   The majority of cams are of either the disk or the cylindrical type.   However, there are numerous other cams which, though more or less related to the disk and cylindrical cams, have had sufficient use to justify special names such as *stamp-mill* cam, *translation* cam, *conical* cam, *spherical* cam, etc.

These and other special cams are discussed in detail in many textbooks and handbooks.[1]

5-5. Base-curve Selection.   Usually a cam must be manufactured according to dimensions specified on a drawing.   These dimensions are most readily expressed as functions of a distance, either a radius or the depth of cut from a reference plane, and an angle.   Thus, the basic item in any cam design is a *displacement-cam angle graph* for the motion of the follower.   In most cases, the cam will rotate at a uniform speed and the cam angle will be directly proportional to time.   Even though it may not

FIG. 5-9. Three-dimensional cam.   (*Courtesy Sperry Gyroscope Company.*)

always be correct, the term *displacement-time* is commonly used as being synonymous with displacement-cam angle.   Unless otherwise stated, every cam discussed here will be assumed to rotate at a uniform speed. The curve on the graph is known as a *base curve.*

In many design problems the follower is to move outward during a certain part of a revolution of the cam, to dwell for a part of a revolution of the cam, and then to return to the initial position.   This type of motion is often referred to as *rise-dwell-return* or, simply, R-D-R.   If the speed is low, almost any displacement-cam angle curve will be satis-

[1] H. A. Rothbart, "Cams," John Wiley & Sons, Inc., New York, 1956.
F. D. Jones (ed.), "Ingenious Mechanisms for Designers and Inventors," vols. I and II, The Industrial Press, New York, 1930, 1936.
H. L. Horton (ed.), "Ingenious Mechanisms for Designers and Inventors," vol. III, The Industrial Press, New York, 1951.

factory.   Though it is seldom used, the simplest base curve is the *straight-line*, or *constant-velocity*, base curve, shown with its accompanying velocity and acceleration diagrams in Fig. 5-10.   It will be noted that theoretically there will be an infinite acceleration for an infinitesimal time interval whenever there is a change in velocity.   This infinite acceleration would require an infinite force.   Obviously, this cannot exist in a practical case.   At low speeds, the elasticity of the parts, i.e., cam surfaces, shafts, belts, etc., limits the force involved to a reasonable value.   However, at higher speeds, the forces may become too large for satisfactory operation.   Thus, in all cases it is desirable, and in most cases imperative, to consider the accelerations given by the base curve.   In fact, the present trend in the design of high-speed cam mechanisms is to go at least one step further and consider the rate of change of acceleration with respect to time.   This third derivative of displacement with respect to time describes the rate of change of force and has been descriptively

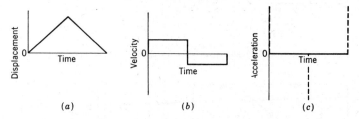

Fig. 5-10. Motion-time characteristics for the constant-velocity base curve.

called *jerk*.   Jerk is considered to be responsible for many service problems encountered in relation to vibration, noise, and wear.   At the present time, little can be done quantitatively with jerk.   It is a difficult matter to calculate the dynamic behavior of a train of elastic parts under the best of conditions.   It becomes even more difficult, if not impossible, if an attempt is made to include the effects of errors in the cam profile. Jerk may be important whenever there is a discontinuity in the acceleration-time curve.   It will be particularly important when the discontinuity consists of an instantaneous change (step) in acceleration which gives an infinite jerk.

There are any number of ways by which the straight-line base curve can be modified to give better acceleration characteristics.   Since such curves are arbitrary and possess no particular advantages, they will not be considered here.

Generally speaking, base curves are either trigonometric or polynomial functions.   If jerk is to be finite, one must look for acceleration functions that are continuous, i.e., without sudden or step changes.   The simplest trigonometric function is the sine acceleration which results in the *cycloidal* base curve whose motion-time characteristics are shown in Fig. 5-11.   The simplest polynomial is the triangular-acceleration curve which is, in effect, a straight-line approximation of a sine and results in a series of cubics, as shown in Fig. 5-12.   Many other curves have been used and

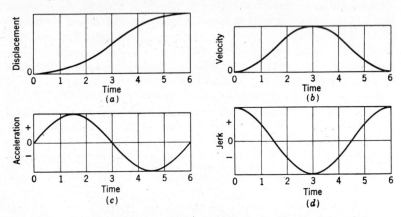

FIG. 5-11. Motion-time characteristics for the sine-acceleration base curve.

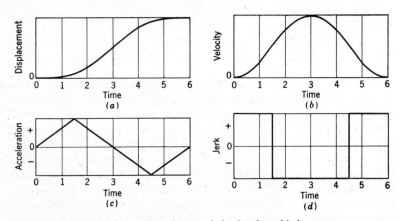

FIG. 5-12. Motion-time characteristics for the cubic base curve.

are recommended for high-speed cams.[1]   However, for present purposes it will be more profitable to limit this discussion to three commonly used curves that can be completely studied by simple analytical and graphical means.   These curves are the *uniform-acceleration*, or *parabolic*, and the

[1] Rothbart, *op. cit.*
    M. Kloomok and R. V. Muffley, Cam Design: With Emphasis on Dynamic Effects, *Prod. Eng.*, vol. 26, pp. 186ff., February, 1955.

harmonic base curves for low- and moderate-speed cams and the *cycloidal* base curve for high-speed cams.

5-6. Uniform-acceleration Base Curve.   The displacement, velocity, and acceleration-time diagrams for the rise motion with a uniform-acceleration base curve are shown in Fig. 5-13.   The basic equations for the first half of the rise are

$$\text{Acceleration } A = \text{const} \tag{5-1}$$
$$\text{Velocity } V = At \tag{5-2}$$
$$\text{Displacement } S = \tfrac{1}{2}At^2 \tag{5-3}$$

where t is measured from $t = 0$.

Since the displacement increases as the square of the elapsed time, the simple graphical construction illustrated in Fig. 5-13a may be used to construct any displacement-time curve for uniformly accelerated motion. For example, in the figure the displacement at time 3 (when the initial acceleration changes) is divided into the same number (three) of equal

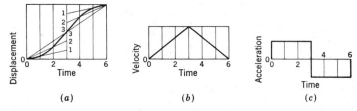

Fig. 5-13. Motion-time characteristics for the uniform-acceleration base curve.

parts as the time intervals.   Then, when straight lines are drawn from these points of division through the origin, it can be seen that they intersect the corresponding time ordinates at heights proportional to the squares of the number of time intervals; i.e., at time 1 the displacement is $\tfrac{1}{3} \times \tfrac{1}{3} = \tfrac{1}{9}$ that at time 3, and at time 2 the displacement is $\tfrac{2}{3} \times \tfrac{2}{3} = \tfrac{4}{9}$ that at time 3.

The equations for the second half of the rise become more complicated because the initial velocity is not zero.   The simplest way to construct this part of the displacement-time diagram is to start from the top, where the velocity is zero, and work back down in the same manner as for the first half of the curve.

If the rise is to be a combination of constant-acceleration and constant-velocity curves, the fact that the average velocity over the period of uniform acceleration from rest is equal to one-half the final velocity greatly simplifies the location of the points of transition from one type of motion to another, because it permits a solution by graphical means rather than by simultaneous equations.   In the interest of keeping the motion as smooth as possible, it is desirable to have the adjacent parts of the displacement-time curve tangent at the point of transition. This in turn means that there should be no sudden changes in velocity at these transition points. Therefore a point moving with a uniform velocity during a time interval equal to the sum of one-half the acceleration interval, the uniform velocity interval, and one-half the deceleration interval will have the same total displacement and will pass through the same transition points as the desired combination of curves. This method is illustrated in Fig. 5-14, where it is used to locate the transition points B and C when the rise of the follower is to take place in a total of 12 time intervals consisting of four time intervals with constant acceleration, three time intervals with constant velocity, and five time intervals with constant deceleration. The previously described graphical construction is then used in drawing the uniform-acceleration parts of the curve.

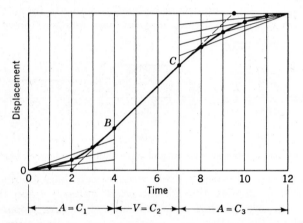

FIG. 5-14. Graphical construction of the uniform-acceleration base curve.

This use of the average velocity can be readily expanded to permit quick graphical construction of the curves for any situation involving uniformly accelerated motion. It may also be of considerable assistance in solving for the numerical values of displacement, velocity, and acceleration.

As already noted, this base curve gives tangent displacement curves, and the acceleration may be easily calculated. However, the fact that there are instantaneous changes in acceleration, or infinite jerks, limits its use to moderate and low speeds. Its major advantage is that it gives the lowest possible accelerations and, therefore, the lowest possible inertia forces.

5-7. Harmonic Base Curve. The harmonic base curve is so called because it gives simple harmonic motion to the follower. Since simple harmonic motion can be represented by the projection on the diameter of a point moving at a uniform speed on the circumference of a circle, the simple graphical construction shown in Fig. 5-15a may be used to construct the displacement-time diagram.   It is important to note that the semicircles are divided into the same number of segments as there are time intervals for the rise and return strokes.   Examination of the acceleration-time diagram in Fig. 5-15c will show that the harmonic base curve may also result in an instantaneous change in acceleration, or infinite jerk, of the follower.   It should be noted that, if there are no dwell

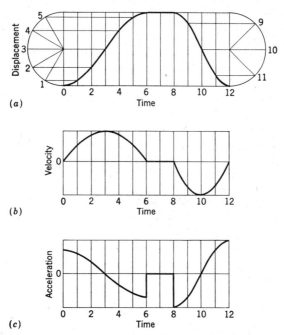

FIG. 5-15. Motion-time characteristics for the harmonic base curve.

periods and if the rise and return strokes have the same time intervals, the acceleration-time curve becomes continuous and jerk is not a problem.

The main, though relatively unimportant, disadvantages of the harmonic base curve when compared with the uniform-acceleration base curve are that (1) for the same length of stroke and time interval, the harmonic base curve results in higher acceleration and (2) the harmonic base curve does not lend itself to an easy solution if any part of the motion is to be at constant velocity.

5-8. Sine-acceleration or Cycloidal Base Curve.   Starting with the sine-acceleration curve in Fig. 5-11c, integrating twice, and solving for the constants of integration results in the following equation for the displacement as a function of time:

$$S = L\left(\frac{t}{T} - \frac{1}{2\pi}\sin 2\pi\,\frac{t}{T}\right) \tag{5-4}$$

where L = rise or lift of the follower and T = time interval for the rise. As a function of cam angle, Eq. (5-4) becomes

$$S = L\left(\frac{\theta}{\beta} - \frac{1}{2\pi}\sin 2\pi\,\frac{\theta}{\beta}\right) \tag{5-5}$$

where β = angle through which the cam rotates during the rise of the follower.

Figure 5-16 illustrates the graphical procedure for drawing the displacement-time diagram. First, a circle with a circumference equal to

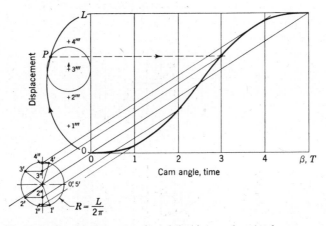

FIG. 5-16. Graphical construction of the sine-acceleration base curve.

the rise is drawn with its center on a straight line passing through the initial and final points on the curve. The circle is then divided into the same number of segments as the time interval, with the zero point located on the horizontal line through the center of the circle. The points on the circle are then projected horizontally to intersect the vertical line through the center of the circle. Lines drawn through these points of intersection and parallel to the line through the initial and final points of the curve intersect the ordinates at the displacements for the respective times.

5-9. Cam-profile Layout. Once the displacement-time diagram has been specified, one can solve for velocities and accelerations, and most cams may be manufactured without further information. However, the designer must also consider such factors as interference or space considerations, wear or strength problems, and whether or not the cam will actually give the displacement-time characteristics used in its design. These, and most other problems encountered in cam design, are related to the size and shape of the cam and follower, and it becomes necessary to consider the cam profile in some detail, either analytically or graphically. If extreme accuracy is necessary, one should consider the use of special analytical methods.[1] As before, the analytical methods are particularly useful in conjunction with high-speed digital computers when many variations are to be studied. However, for most purposes, the information required by the designer can readily be determined with sufficient accuracy by using the more general graphical means. Not only is the graphical approach more general but in a limited time should contribute more to a basic understanding of principles under which a cam and follower must operate. Therefore, graphical methods only will be considered in this book.

The layout of a cam profile always seems a complicated procedure. This is generally because of the necessity of using figures which show several steps at the same time. Actually, every cam design is based upon the application of only three basic ideas, as follows:

1. *The displacement-time diagram is always given for a point, called the reference point, on the follower which may be used to draw the follower in its correct position at any time.*

2. *The cam profile will be tangent to the follower at all times and will thus be determined by the envelope of the follower profiles drawn in successive positions.*

3. *The step outlined in item (2) is most readily accomplished by making use of the principle of inversion; i.e., for the purpose of drawing the envelope of follower profiles, it is necessary to invert the mechanism by holding the cam fixed and rotating the frame in the direction opposite to that specified for the cam.*

The above three points are general and provide an adequate guide for designing all types of cams; however, certain techniques and a little ingenuity can simplify the solution. Several examples are given below to illustrate the application of these ideas. It should be remembered that there are usually a number of equally good ways to solve each problem,

[1] Rothbart, *op. cit.*

W. B. Carver and B. E. Quinn, An Analytical Method of Cam Design, *Trans. ASME.* vol. 67, pp. 523–526, 1945.

S. L. Linderoth, Jr., Calculating Cam Profiles, *Machine Design*, vol. 23, pp. 115–119, July, 1951.

M. Kloomok and R. V. Muffley, Plate Cam Design: Pressure Angle Analysis, *Prod. Eng.*, vol. 26, pp. 155ff., May, 1955; and Plate Cam Design: Radius of Curvature, *Prod. Eng.*, vol. 26, pp. 186ff., September, 1955.

and the reader will find it profitable to give some thought to alternative approaches.

5-10. Disk Cam with Reciprocating Flat-face Follower. The displacement-time diagram and the cam which, rotating clockwise, will give the required motion to the flat-face follower are shown in Fig. 5-17. The first step in drawing the cam profile is to locate the positions (1″, 2″, 3″, etc.) of the follower on the centerline (1′, 2′, 3′, etc.) corresponding to the time intervals represented by the angles on the cam. Note that the

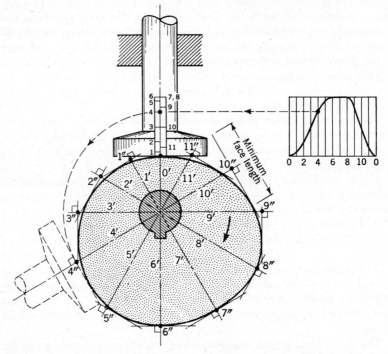

FIG. 5-17. Cam-profile layout for disk cam with reciprocating flat-face follower.

angular positions on the cam are located in the direction (ccw) opposite to the rotation of the cam. The manner in which all positions of the follower face are located is illustrated completely for position 4. Since the face of the follower will have risen to 4 when the cam rotates through the angle which will put the radial line 4′ in the top vertical position, an arc with the radius from the cam axis to 4 is swung counterclockwise to intersect the line 4′ at 4″. The follower face always makes the same angle with a radial line through the cam axis, and it is now drawn perpendicular to the radial line 4′ through the point 4″. This same procedure is followed for all positions. The cam is then drawn tangent to all the follower faces in their inverted positions. It should be apparent that the accuracy of the cam profile depends upon the number of positions used. This also is true in manufacturing a cam, since the usual method is to rotate the cam in small increments and to feed a milling cutter or grinding wheel so that the cutting surface corresponds exactly to the follower face. Even with rotational increments of only 1°, the resulting cam surface will consist of a series of plane surfaces, and a craftsman must hand-file the cam so that its surface is smooth. Additional cams can then be made by use of the master cam as a template on a duplicating machine.

The layout of the cam also gives important information about the required length of the follower face. Since the follower face must be at least long enough to include its point of tangency with the cam surface, a simple check with dividers at each position will quickly locate the critical one. The minimum useful length is shown in the figure at position 10''. However, since it is undesirable to have contact occurring at the end of the follower face, the usual practice is to make the length somewhat greater than that determined from the drawing. The amount by which the length of the follower face is increased should vary with the size of the cam, the number of positions used in the graphical construction, and the accuracy of the drawing. The designer must use his own judgment, but $\frac{1}{16}$ in. may be sufficient for small cams, whereas $\frac{1}{4}$ in. or more may be required on large cams. A common value is $\frac{1}{8}$ in.

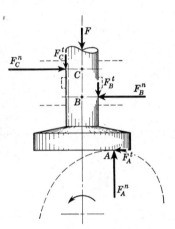

FIG. 5-18. Free-body diagram of a reciprocating flat-face follower.

As discussed in Sec. 3-11, the resultant force between sliding surfaces will be composed of a transmitted force normal to the surfaces and a friction force tangent to the surfaces. In Fig. 5-18 the useful force $\mathbf{F}$ acts along the follower axis. If friction were negligible, $\mathbf{F}_A^n$ would equal $\mathbf{F}$, and the resulting couple would be balanced by the couple due to equal and opposite forces $\mathbf{F}_C^n$ and $\mathbf{F}_B^n$. However, since friction is present, each normal force is accompanied by a friction force, and the free-body diagram of the follower will be as shown in the figure. It should be noted that friction results in the transmitted force $\mathbf{F}_A^n$, between the cam and follower, being considerably larger than the useful force $\mathbf{F}$. Thus, the required cam torque and the rate of wear will be greater than anticipated if the effect of friction is neglected. The magnitudes of the forces can be calculated after the conditions are fully specified by using Newton's laws of motion, which state that the summation of forces in any direction and the summation of moments or couples about any point must equal zero. At this time, it is sufficient to recognize that these forces and couples exist and should be considered in the design of an actual cam.

As a point of interest, the reader should use the concept of instant centers to prove to himself that, for a given displacement-time diagram,

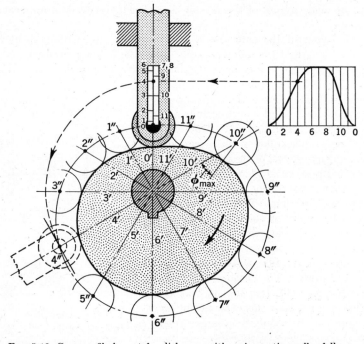

FIG. 5-19. Cam-profile layout for disk cam with reciprocating roller follower.

the displacement of the point of contact from the follower axis is independent of the cam size. This means that varying the size of the cam will have no effect upon either the required minimum face length or the force analysis for a flat-face follower driven by a disk cam.

5-11. Disk Cam with Radial Roller Follower. The approach to the design of a disk cam with any roller follower is practically the same as described above for the case of a flat-face follower. The main consideration is that the cam surface is no longer drawn tangent to a series of straight lines but rather to a series of circles. Therefore, the displacement-time diagram must apply to the center of the roller and must be used to locate the roller in each of its inverted positions, as illustrated in Fig. 5-19.   The centerline passing through the centers of the roller circles defines what is known as the *pitch surface* of the cam.

In the preceding section, it was pointed out that the force analysis of a disk cam with a flat-face follower was independent of the size of the cam. Such is not the case when a roller follower is used.

As already observed, in general, the point of contact of the roller and the cam will not lie on the follower axis.   This is true for both radial and offset roller followers.

Since the transmitted force will act perpendicular to the contact surfaces, it will lie on a line through the contact point and the center of the roller.   The angle between this line of transmission and the line of motion of the follower is called the *pressure angle* ϕ.   For the cam and follower in Fig. 5-19 the maximum pressure angle occurs, as indicated, at 10''.   Figure 5-20 shows the forces involved and the manner in which the pressure angle resolves the transmitted force at the center of the roller into a useful component along the follower axis and a side-pressure component perpendicular to the follower axis.   The latter component must be balanced by a force and a couple provided by the guide.   Obviously, the force and couple acting on the guide may lead to problems related to friction and wear and thus are undesirable.   Figure

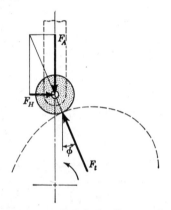

FIG. 5-20. Free-body diagram of the roller of a reciprocating roller follower.

5-21 illustrates that, for a given displacement-time diagram, the larger the cam the smaller the pressure angle and therefore the smaller the force and couple on the guide.   The easiest way to visualize this is to consider that, for an infinitely large cam, the profile would be circular and the pressure angle would be 0°.   It is usually considered good practice to limit the pressure angle to 30°, but if rolling-contact bearings are applied to slides (see Figs. 8-25 and 18-54) as well as to pivots, it is possible to use pressure angles as high as 50° without difficulty.   Large cams and large rollers also result in lower contact stresses and thus permit the use of cheaper materials or material treatments.

5-12. Disk Cams with Oscillating Followers.   The only new idea required in the design of a disk cam with an oscillating flat-face or a roller follower is that the displacement-time diagram applies to a point moving on the arc of a circle.   Therefore the displacement-time diagram is most conveniently constructed and used when based on angular displacement.

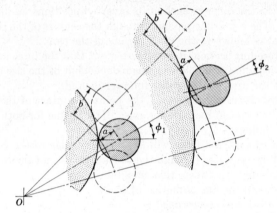

Fig. 5-21. The effect of cam size on pressure-angle magnitude.

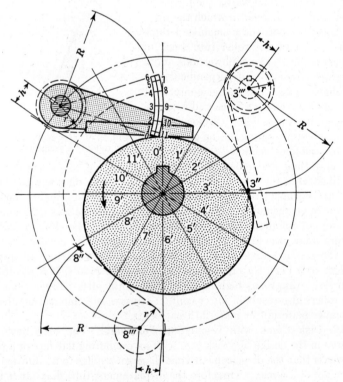

Fig. 5-22. Cam-profile layout for disk cam with an oscillating flat-face follower.

The construction of a typical disk cam with oscillating flat-face follower is shown in Fig. 5-22. It should be noted that the inverted follower faces are drawn through points transferred from the displacement scale tangent to the circle, with radius r, drawn at each inverted position of the center of oscillation of the follower.

5-13. Positive-return Disk Cams. As mentioned earlier in this chapter, it is possible to specify the displacement-time characteristics for

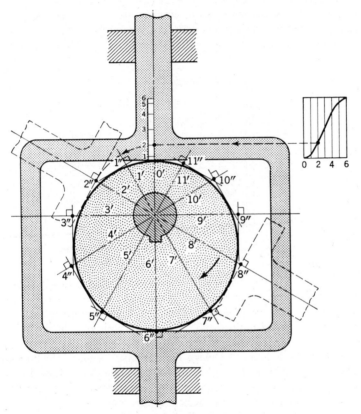

FIG. 5-23. Cam-profile layout for a single-disk positive-return cam.

either direction of motion of a yoke follower operating with a single-disk cam, but it is not possible to do so for both directions. Figure 5-23, which illustrates the construction of the cam, shows that, since the same half of the cam drives the follower during both the outward and return strokes, the two strokes must each take 180° of cam rotation, must have the same displacement-time curves, and must be in opposite directions.

Since the two follower faces must always be the same distance apart, the name *constant-breadth* cam is used for the positive-return single-disk cam with reciprocating yoke follower.

If two rollers are substituted for the flat faces, the distance from one roller center to the other must remain constant; such cams are called *constant-diameter* cams.

A particular group of constant-breadth cams known as *positive-return single-disk circular-arc* cams are often useful because of the relative ease in generating surfaces made up of circular arcs. An example of a cam made up of arcs of two different radii is given in Fig. 5-24. The requirement to give constant breadth is that the centers O, A, and B must lie at

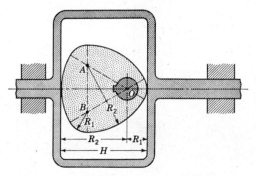

FIG. 5-24. Single-disk circular-arc positive-return cam.

apexes of an equilateral triangle. Inspection of the figure will show that this makes the breadth H constant and equal to $R_1 + R_2$ and the total displacement equal to $R_2 - R_1$. Both the outward and return strokes of the follower would require 120° of cam rotation, and there would be a dwell of 60° of cam rotation at the end of each stroke.

5-14. General Kinematic Design Limitations. Some of the limitations in relation to face length, size, etc., have been discussed in specific cases. However, there are several more or less general and obvious limitations that must be considered when designing any cam.

Size of Disk Cam with Flat-face Follower. When a disk cam is designed to operate with a flat-face follower, the procedure, as already developed, is to draw the cam profile tangent to a series of inverted follower faces. However, if the cam is too small, the inverted follower faces may lie so that it is impossible to draw a surface tangent to all faces. This is illustrated in Fig. 5-25a, where the follower face drawn through 2″ does not contact the cam. The situation is corrected by making the cam larger. In Fig. 5-25b the same motion is used as in Fig. 5-25a, and the size of the cam has been increased so that the follower face at $2''$ can now contact the cam.

Size of Cam and Roller with Roller Follower. The problems related to size are much the same for both flat-face and roller followers in that a continuous cam profile must be drawn tangent to a series of inverted followers. Figure 5-26a, in which the roller follower at $3''$ cannot contact the cam, illustrates the case of a too-small cam. The same motion can be properly given if the cam is made larger, as in Fig. 5-26b.

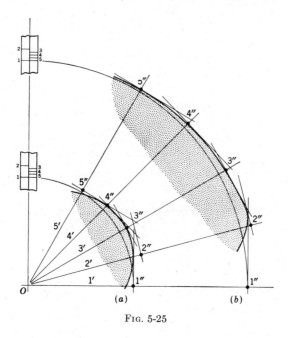

Fig. 5-25

If it is impractical or impossible to use a larger cam, it is sometimes possible to correct the situation by using a smaller-diameter roller follower. This method cannot be used to make any major corrections and, in general, is not recommended because of the introduction of additional problems related to the strength, rigidity, wear, etc., of the parts.

5-15. Closure. The brief discussion in this chapter has been limited to the presentation of some considerations fundamental in selecting a suitable base curve and in applying the principle of inversion to lay out the profile of the cam. It should be appreciated that the actual machining of a cam for use with a flat-face follower requires only the displacement-time (or angle) curve, as discussed in Sec. 5-10. If a roller follower is to be used, the roller diameter becomes an additional consideration.

The cam-profile layout is useful largely as a simple means of checking over-all space requirements, the minimum length for a flat-face follower, and the maximum pressure angle for a roller follower and to be certain that the cam is large enough so that the follower motion will actually be as specified.

The reader must not forget that this discussion has been concerned largely with the kinematics of *ideal* cams with absolutely accurate profiles

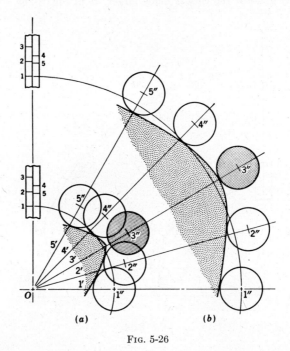

(a)   (b)

FIG. 5-26

and with infinitely rigid members. The fact that neither of these conditions exists in reality becomes more important as speeds, and thus accelerations, increase. Under extreme conditions, the actual follower motion may differ considerably from that predicted by consideration of the base curve alone, and the net result may be incorrect, noisy operation and a short service life. Although at the present time the understanding of the area of cam dynamics is still more qualitative than quantitative, except for some highly idealized and special situations, the review of the literature provided by Rothbart[1] will be of interest and use in the design of cams for critical applications.

[1] Rothbart, *op. cit.*

CHAPTER 6

FUNDAMENTALS OF DIMENSION
DETERMINATION

Many factors enter into the specification of dimensions for parts of a machine; in general, they are related to motion, strength, and manufacturing considerations.

As discussed in preceding chapters, the motion characteristics of a mechanism may be thoroughly studied by considering the mechanism in terms of simple drawings or skeleton outlines. However, when the mechanism becomes the basis for a machine, the links are known as *parts* or *machine elements* and must be considered as three-dimensional bodies. Usually, one, or, in the case of cams, two, of the dimensions will be determined in the motion analysis, and the primary condition to be met by the remaining dimensions is that of adequate strength to support the forces acting on the part.

In many cases, a limitation of the material or the manufacturing process will be the final criterion for some dimensions. This is particularly true when the part is being made as a casting or a weldment (Chap. 7) or when heat-treatment is involved.

One of the most difficult yet fascinating problems encountered in machine design is that there is almost never a unique answer. In fact, there are often several solutions that are equally good. It is also true that most of the practical considerations do not lend themselves to expression as exact theoretical equations, and consequently exact solutions of design problems are seldom found.

This chapter will introduce some of the considerations that apply in a general way to the design of many machine elements. At the same time, the information gained from previous courses in materials and mechanics of materials will be reviewed and correlated with the practical considerations to establish a solid basis for designing any machine member.

6-1. Strength and Failure. A primary consideration in designing any part is the necessity of making it strong enough so that it will not fail in service. Most people, regrettably including engineers, commonly associate the word *failure* with the actual breaking of a part. While there is no question that a broken part has failed, a designer must be concerned with a broader interpretation and understanding of what really determines whether a part has failed or not. *A designer will consider a part to have failed when it is no longer capable of performing its required function in a satisfactory manner.*

This definition includes many types of failure in addition to actual breaking. For example, failure may be related to the effects of wear on the correct operation of mating parts, the effects of wear on the noise generated by a machine, the effects of both elastic and plastic deformations of machine members upon the performance of other related parts of the machine as well as of the deformed part, and the effects of creep of materials under stress at elevated temperatures.

The designer, in considering all the possible modes of failure, must base his calculations upon one of the following:

1. Allowable stress
2. Rigidity
3. Stability
4. Wear

The dimensions determined by calculation must often be modified by a fifth consideration of the requirements for making the part by a particular manufacturing process. In almost every case, the calculations will give the minimum permissible dimensions, and any further increase will generally make the part stronger, more rigid, or more stable.

These five bases for dimension specification will be considered in detail in the remainder of this chapter and in later chapters where problems associated more closely with particular machine elements will be studied.

6-2. Service Life and Reliability. Reliability has long been a major concern of the designer of mechanical equipment. However, the term had been used only infrequently until the advent of the space age when missile failures due to lack of reliability of components became a matter of world-wide concern.

Actually, reliability, as the term is now most commonly used, is a statistical problem in specifying the probability that a part or a system (machine) will perform satisfactorily for a specified length of time or number of cycles of operation. As far as mechanical design is concerned, two broad and rather obvious classifications may be set up in terms of desired service life: (1) a finite life and (2) an indefinite or infinite life.

In addition to those components where fracture is the mode of failure, the first category also includes those for which creep and corrosion are major considerations, e.g., turbine blades, rocket nozzles, etc., and those for which wear is a major design consideration, e.g., clutches, brakes, belts, translation screws, gears, bearings, etc. Discussions of creep and corrosion are beyond the scope of this book, but wear will be encountered many times, because, wherever motion exists, it will be a problem. Unfortunately, there are practically no statistical data available in relation to wear and service life, and, as will be evident, the designer is forced to rely on the experience of others and use so-called *allowable* values of design data that have been found to give a reasonable life. Where failure by fracture is the major consideration, the significance of the classifications of finite and indefinite life can be approached in a more rigorous, quantitative manner.

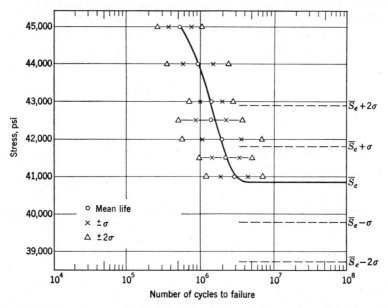

FIG. 6-1. Endurance strength data from bending tests on annealed SAE 1050 steel specimens. (*From "Symposium on Fatigue with Emphasis on Statistical Approach—II," p. 35, American Society for Testing Materials, Philadelphia, 1952.*)

Figure 6-1 shows the results of an analysis by Epremian and Mehl[1] of *endurance strength* data from bending-fatigue tests made on annealed SAE 1050 steel specimens. A thorough discussion of statistics is not possible here, but for the immediate purpose of gaining some insight into the possibilities and limitations of designing from the reliability point of view, it will be sufficient to note that the *mean*[2] is simply the average,

[1] "Symposium on Fatigue with Emphasis on Statistical Approach—II," pp. 25–54, American Society for Testing Materials, Philadelphia, 1952.

[2] In this particular application the mean is the log mean, i.e., the antilog of the arithmetic mean of the log of the number of cycles N, because the distribution has been found to be normal with respect to log N rather than N itself. Similarly, the standard deviation is also based on log N.

that σ is the *standard deviation*, and that for a normal distribution, the range of $\pm\sigma$ about the mean includes 68 per cent, $\pm2\sigma$ includes 95 per cent, and $\pm3\sigma$ includes 99.7 per cent of the observations. Although, even here, the data cover only a small range of values of stress and life, it can be seen that in the region to the left of the relatively sharp break in the curve there may be a considerable variation in the number of cycles to failure at a given stress level or a considerable variation in stress level for a given number of cycles to failure. For example, at 42,000 psi, there is a 95 per cent probability that the fatigue life will be between 5.6×10^5 and 6.8×10^6 cycles, the latter being over ten times the former.

The percentage variation in expected life becomes even greater as the stress level increases and the mean life decreases. Therefore, where weight is important and where the required life is short, both being particularly important in missiles, the designer must work with relatively high stresses and is forced to consider the statistical aspects of failure. Actually, at the present time, little of this can be applied to general mechanical design for a number of reasons: (1) There are very meager data available, practically none, for even the most common engineering materials; (2) there is no information available relating the statistics of the failure of fatigue specimens to the statistics of the failure of actual machine parts; (3) the designer seldom has available any statistical information about the loading of the part or machine; and (4) the relationship of the loading statistics to failure of a part is still not understood in any general, applicable terms.

The direct result of this gap in knowledge is that, in general, one cannot design actual parts or machines to meet a reliability specification, such as 98 per cent are to have a service life of 5,000 cycles of operation. However, after the designer has done the best he can with the information available, life tests of a sufficient number of parts or machines will provide the information for calculating the reliability, and extensive redesigning and testing will usually be required before the original specifications can be met. This procedure is time consuming and expensive and is ordinarily justified only where weight and size are extremely critical, as in the design of missiles and, to a somewhat lesser extent, aircraft.

For a few specific elements, e.g., springs, wire rope, and gears, which will be discussed in later chapters, some information is available in relation to failure by fracturing which will permit the designer to consider finite life in at least a fairly rational, if not statistically rigorous, manner. The only group of elements for which complete statistics are available is that known as rolling-contact bearings, i.e., ball and roller bearings. The running of the large number of tests required to furnish sufficient data for a statistical analysis has been both feasible and economically justified, because in practical use these bearings have finite service lives, the number used is so great, and the selection of materials and subsequent heat-treatment and processing during manufacturing are carefully controlled.

The *endurance limit* s_e is usually defined as the limiting value of reversed stress below which the material can withstand an infinite number of stress cycles.   In practice, infinity is difficult to reach, and the number of cycles is usually 10^7.   As shown in Fig. 6-1, the break in the curve for the mean endurance limit s_e occurs at about 3×10^6 cycles for this particular material.   For the horizontal part of the curve the number of cycles no longer has any significance and the standard deviation must be applied to the stress level itself.

From a practical viewpoint, it soon becomes apparent that the great majority of machines designed for use in industry or in the home should be designed—in so far as possible—on the basis of an infinite life.   For example, the shaft of a 1,750-rpm electric motor will have completed 3×10^6 cycles in less than 29 hr.   Therefore, *unless otherwise noted*, the discussion in this and following chapters will be concerned with designing for infinite life.

6-3. Factor of Safety.   The term *factor of safety* is one of the most misunderstood and misused expressions in the language of machine design. There are numerous ways in which factors of safety have been defined, but, in general, the variations correspond closely to one of the two equations

$$\text{f.s.} = \frac{\text{damaging stress}}{\text{design stress}} \tag{6-1}$$

and

$$\text{f.s.} = \frac{\text{maximum safe load}}{\text{normal service load}} \tag{6-2}$$

Equation (6-1), in terms of stress, is the more common definition, and most machine parts are designed on the basis of a limiting or allowable stress.   It is very important to recognize that the rigidity and elastic stability of parts are usually not direct functions of stress, and therefore, in these cases, it is necessary to consider the factor of safety in terms of loads, as in Eq. (6-2).   There should not be much difficulty in keeping the above limitations in mind, but, if in doubt, remember that it is always correct to use the ratio of loads for determining factors of safety.

Since these equations result in numerical answers, the common pitfall lies in interpreting the answer as correct in absolute magnitude.   As will be brought out in relation to choosing a factor of safety upon which to base design calculations, there are a number of factors which may result in a considerable difference between the calculated and the true values. Nevertheless, the concept of a factor of safety, or design factor, as it is sometimes known, is very useful and practically necessary in designing machines.

6-4. Factor of Safety with Static Loading. Static loading does not mean that the load carried by the part is literally static and never varies in any manner; it does mean that the load varies relatively few times and that the variations occur slowly enough so that neither fatigue nor impact need be considered.

Before a factor of safety can be selected and applied to a design problem, it is necessary to consider the behavior of the chosen material under the specified operating conditions. For the case of static loading, the basic material property is that of ductility, and it is desirable to classify a material as either ductile or brittle.

Unfortunately, there is no absolute dividing line between the two classes. The usual engineering definition is that ductile materials can sustain over 5 per cent elongation before rupture. Most steels, aluminums, brasses, etc., are thus considered to be ductile. The main brittle material used in machine parts is cast iron.

If a part made of a ductile material is subjected to a load sufficiently large, the yield strength of the material will be exceeded, and the part will be permanently deformed. This permanent deformation will generally result in unsatisfactory functioning, and therefore the part will have failed. It should also be noted that the deformation will often result in the transfer of part or all of the load to another member and that in any case the part still has appreciable load-carrying capacity before the ultimate strength is reached and it breaks. Another important attribute of a ductile material under static loading is that the high stresses accompanying stress concentration at sharp corners, holes, etc., will be relieved by local yielding at the region of high stresses. Thus, if the minor dimensional changes due to the local yielding can be tolerated, stress concentrations do not need to be considered when designing a part of a ductile material to be used under steady-load conditions.

Therefore, the design of any part to be made from a ductile material and to operate under essentially static loads will be based upon the yield strength, and the factor of safety f.s. can be defined as

$$\text{f.s.} = \frac{s_y}{s} \tag{6-3}$$

and the *allowable stress* s_a is

$$s_a = \frac{s_y}{\text{f.s.}} \tag{6-4}$$

or

$$s_a = s \tag{6-5}$$

where s_y = yield strength, psi, and s = nominal calculated or design stress, psi.

The actual selection of a factor of safety to be used in designing any machine element must be based on a number of considerations, as will be discussed in Secs. 6-14 to 6-18.

A part made from a brittle material would give no warning in the form of yielding before breaking, nor would there be any local yielding in the regions of high stresses due to stress concentrations. Stress concentrations or stress raisers will be discussed in detail in a later section; at this time it will be sufficient to point out that the stress-concentration factor K can be defined as

$$K = \frac{\text{actual stress}}{\text{nominal stress}} \tag{6-6}$$

where the nominal stress is the value calculated using the ordinary equations of strength of materials and the actual stress is the higher real value present.

Consideration of the behavior of a brittle material under static loading leads to a definition of factor of safety for this case of

$$\text{f.s.} = \frac{s_u}{K_t s} \tag{6-7}$$

and the allowable stress is

$$s_a = \frac{s_u}{\text{f.s.}} \tag{6-8}$$

or

$$s_a = K_t s \tag{6-9}$$

where s_u = ultimate strength, psi, and K_t = static stress-concentration factor.

It should be noted that the factor of safety for a design with a brittle material under static loads is based upon the ultimate strength of the material, while the factor of safety for a ductile material under similar conditions is based on the yield strength. Thus, if the same value for factor of safety is used for both materials, the factor of safety *in terms of failure by fracturing* is considerably higher for the ductile material. To compensate for this difference, the usual practice with brittle materials and static loading is to use a factor of safety twice that selected for a ductile material under the same conditions.

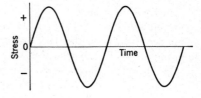

FIG. 6-2. Stress-time diagram for reversed-stress loading.

6-5. Factor of Safety with Fatigue Loading. *The elementary concept of fatigue loading is that based upon a complete reversal of stress during each cycle.* A stress-time diagram for the reversed-stress case is shown in Fig. 6-2. As discussed in Sec. 6-2, the material property associated with tests made with this loading is known as the endurance strength when the stress can be carried for only a specified finite number of cycles before breaking and as the endurance limit s_e when it can be carried an indefinite, or infinite, number of cycles without breaking. Local yielding will no longer be effective in relieving the peak stresses in regions of stress concentration, as for ductile materials under static loads, because repeated reversals of the direction of plastic deformation will rapidly break down the material and a crack will soon appear. The crack is, in itself, a source of severe stress concentration, and failure by breaking will occur in a relatively few additional cycles. Consequently, the factor of safety for both ductile and brittle materials must include the effects of stress concentration and can be defined as

$$\text{f.s.} = \frac{s_e}{K_f s} \tag{6-10}$$

and the allowable stress is

$$s_a = \frac{s_e}{\text{f.s.}} \tag{6-11}$$

or

$$s_a = K_f s \tag{6-12}$$

where s_e = endurance limit, psi, and K_f = fatigue stress-concentration factor.

Although the S-N diagrams for some materials, notably many alloys of aluminum, magnesium, and copper and most plastics, do not indicate a sharp break or knee, as shown in Fig. 6-1, practically all members of the most important family of engineering materials, the ferrous alloys, do have a true endurance limit. For steels, the endurance limit has been found to be closely correlated with the ultimate strength s_u. Values of s_e lie within the range of 0.45 to $0.60s_u$ with an upper limit of about 100,000 psi, and the relation $s_e = 0.5s_u$ is commonly used in design. For materials other than steel, no such simple relationship exists and the reader should refer to the extensive tables compiled from many sources by Grover, Gordon, and Jackson.[1]

The endurance limit is ordinarily given as the reversed stress which a polished specimen will carry without failure for 10^7 cycles. The two main problems presented here are that (1) the smoothness of the surface has a pronounced effect on the endurance limit, and few machine parts, except in the aircraft industry, are made with polished surfaces and (2) many machine parts have a load-time curve which, though varying, does not indicate a completely reversed stress.

Values of endurance limits for polished surfaces are not of much direct use in designing machine parts because of the expense involved in secur-

[1] H. J. Grover, S. A. Gordon, and L. R. Jackson, "Fatigue of Metals and Structures," pp. 274–388, NAVWEPS 00-25-534, Government Printing Office, Washington, D.C., 1960.

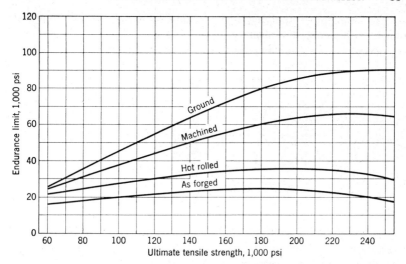

FIG. 6-3. Endurance limit of steel as related to ultimate strength and surface condition. (*From G. C. Noll and C. Lipson, Allowable Working Stresses, Proc. SESA, vol. 3 no. 2, 1946.*)

ing such surfaces. Thus, tables or curves, as in Fig. 6-3, relating the expected endurance limit of the finished part to the quality of the surface must be utilized in practical problems.

The above discussion concerning the endurance limit has been limited to the case of reversed normal stress, i.e., tensile and compressive stresses. When the need arises for using the mechanical properties of *steel* under shear loading, the appropriate values of shear yield and shear endurance limit may be calculated using the relations

$$s_{sy} = \tfrac{1}{2} s_y \qquad\qquad (6\text{-}13)$$

and
$$s_{se} = \tfrac{1}{2} s_e \qquad\qquad (6\text{-}14)$$

Fortunately, many machines and machine parts are not required to operate under fatigue loads in contact with water or other corrosive substances. There are meager data available, but tests have shown that the presence of clean water may decrease the endurance limit to less than one-half the usual value, and the effect of salt water on the endurance limit is even more pronounced.[1]

[1] The following books are recommended as excellent sources of information related to corrosion, fatigue, and many other topics of interest to the designer:

Grover, Gordon, and Jackson, *op. cit.*

O. J. Horger (ed.), "Metals Engineering Design," McGraw-Hill Book Company, Inc., New York, 1953.

George Sines and J. L. Waisman (eds.), "Metal Fatigue," McGraw-Hill Book Company, Inc., New York, 1959.

The problem of applying an endurance limit determined from tests made with completely reversed loads to the case shown in Fig. 6-4 is of importance. In this case, there is a cyclic variation in load, but the

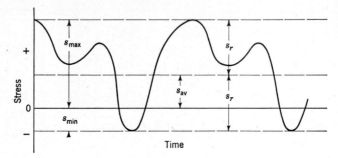

FIG. 6-4. Stress-time diagram for repeated stress loading.

variation s_r is above and below a mean or average value s_{av}. In other words, the loading consists of a reversed component

$$s_r = \frac{s_{max} - s_{min}}{2} \tag{6-15}$$

superposed upon an average stress

$$s_{av} = \frac{s_{max} + s_{min}}{2} \tag{6-16}$$

It should be noted that this average stress is not a true average in the sense of being the area under a curve divided by the length.

The failure points from fatigue tests made with different steels and combinations of average and reversed stresses are plotted in Fig. 6-5 as functions of s_r and s_{av}. The most significant observation is that, in general, the failure point is little related to the average stress when it is compressive but is very much a function of the average stress when it is tensile. In practice, this means that fatigue failures are rare when the average stress is compressive or negative; therefore, the greater emphasis must be given to the combination of a reversed stress and a steady tensile stress. There are several ways in which problems involving this combination of stresses may be solved. The method presented here, as given by Soderberg,[1] is the most straightforward and best adapted for design purposes.

Since the failure line in Fig. 6-5 is not explicitly defined and since ductile materials are considered to have failed if the part yields, it is more practical to use a straight line connecting s_e and s_y as the failure line.

[1] C. R. Soderberg, Factor of Safety and Working Stress, *Trans. ASME*, vol. 52, APM 52-2, 1930.

C. R. Soderberg, Working Stresses, *Trans. ASME*, vol. 57, pp. A106–108, 1935.

This modification ignores the effects of strain hardening and results in conservative values.

As previously discussed, the appropriate stress-concentration factor must be applied to both ductile and brittle materials under fatigue loads,

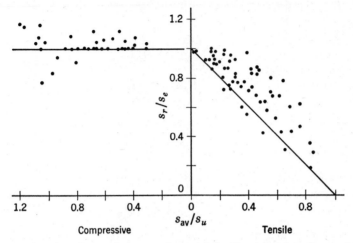

FIG. 6-5. Test data for notch-free specimens of ductile materials under combined steady and reversed axial stress. (*After J. O. Smith, The Effect of Range of Stress on the Fatigue Strength of Metals, Univ. Illinois Eng. Expt. Sta. Bull. 334, 1942.*)

whereas stress concentration may be neglected for the case of ductile materials under steady loads. This reasoning, when applied to the case of varying loads, results in using stress-concentration factors with only the reversed component of stress for ductile materials and with both components for brittle materials.

A typical Soderberg triangle is shown in Fig. 6-6. Here the failure line

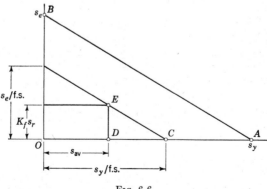

FIG. 6-6

connects s_e and s_y. After a value for factor of safety is chosen which relates design stresses to failure stresses, a line is drawn from $s_e/\text{f.s.}$ to $s_y/\text{f.s.}$ This line then is the locus of all combinations of steady and reversed stresses that correspond to the chosen factor of safety.

If E is the design point for a ductile material, OD is s_{av} and DE is $K_f s_r$. From the similar triangles CDE and AOB,

$$\frac{DE}{OC - OD} = \frac{OB}{OA} \tag{6-17}$$

or, in terms of stress,

$$\frac{K_f s_r}{s_y/\text{f.s.} - s_{av}} = \frac{s_e}{s_y} \tag{6-18}$$

Equation (6-18) may be rewritten as

$$\frac{1}{\text{f.s.}} = \frac{K_f s_r}{s_e} + \frac{s_{av}}{s_y} \tag{6-19}$$

Solving Eq. (6-19) for the factor of safety gives

$$\text{f.s.} = \frac{s_y}{s_{av} + (s_y/s_e)K_f s_r} \tag{6-20}$$

An interesting and useful physical interpretation can be given to this equation (6-20) by noting that the denominator is the distance OC. Thus, the factor s_y/s_e may be considered to weight the reversed-stress component $K_f s_r$ before it is added to the steady-stress component s_{av} to form a sum which is divided into the yield point s_y.

6-6. Application of Soderberg's Equations. The equations developed above are useful in analyzing an existing part with known dimensions but are not directly useful in designing a part when the problem is to determine the dimensions. In this section, the basic method is applied to the use of a ductile material in a number of the most common loading conditions.

Axial Loading

$$s = \frac{P}{A} \tag{6-21}$$

where P = load, lb, and A = cross-section area, in.^{2}. Substituting this relation into Eq. (6-20) gives

$$\text{f.s.} = \frac{A s_y}{P_{av} + (s_y/s_e)K_f P_r} \tag{6-22}$$

or

$$A = \frac{P_{av} + (s_y/s_e)K_f P_r}{s_y/\text{f.s.}} \tag{6-23}$$

Simple Bending

$$s = \frac{Mc}{I} \tag{6-24}$$

where M = bending moment, lb-in.

    c = distance from neutral axis to outermost fiber, in.

    I^* = moment of inertia of the area, in.[4]

Equation (6-20) becomes

$$\text{f.s.} = \frac{s_y I/c}{M_{av} + (s_y/s_e)K_f M_r} \tag{6-25}$$

and
$$\frac{I}{c} = \frac{M_{av} + (s_y/s_e)K_f M_r}{s_y/\text{f.s.}} \tag{6-26}$$

The term I/c is called the *section modulus* and is often given the symbol Z.

Simple Torsion of Circular Shafts

$$s_s = \frac{Tc}{J} \tag{6-27}$$

where T = torque, lb-in.

    c = radius to outermost fiber, in.

    J = polar moment of inertia of cross-section area, in.[4]

Equation (6-20) becomes

$$\text{f.s.} = \frac{s_{sy} J/c}{T_{av} + (s_{sy}/s_{se})K_{fs} T_r} \tag{6-28}$$

and
$$\frac{J}{c} = \frac{T_{av} + (s_{sy}/s_{se})K_{fs} T_r}{s_{sy}/\text{f.s.}} \tag{6-29}$$

where K_{fs} = torsion-fatigue stress-concentration factor.

For the usual case in which the shaft is made of a ductile steel, Eqs. (6-28) and (6-29) may be rewritten in terms of s_y and s_e as

$$\text{f.s.} = \frac{(s_y/2)J/c}{T_{av} + (s_y/s_e)K_{fs} T_r} \tag{6-30}$$

and
$$\frac{J}{c} = \frac{T_{av} + (s_y/s_e)K_{fs} T_r}{s_y/2\text{f.s.}} \tag{6-31}$$

Combined Bending and Torsion of Circular Shafts. A common machine element is a shaft used to transmit torque. Usually a gear, pulley, sprocket, or similar device is attached to the shaft. This arrangement results in a bending moment as well as a torque acting on the shaft, and the relationship of combined stresses to fatigue failures must be considered. The theory of failure under biaxial combined stresses that agrees most closely with experimental evidence is that known as the *maximum-distortion-energy* theory (also known as the *von Mises-Huber-Hencky* theory).[1] This theory considers failure to occur when the distortion energy due to the components of stress equals the distortion energy at failure under a uniaxial stress. The agreement with experimental data

---

* Refer to Appendix B for properties of some of the more common sections.

[1] J. Marin and J. A. Sauer, "Strength of Materials," 2d ed., chap. 9, The Macmillan Company, New York, 1954.

    Horger, *op. cit.*, pp. 317–327.

is excellent for ductile materials and fair, although on the safe side, for brittle materials. Unfortunately, the maximum-distortion-energy theory of failure cannot be used directly in design; it can be used only to analyze a part with given dimensions and loads and is, therefore, limited to trial-and-error solutions or to checking the final design.

For design of parts to be made from ductile materials the most useful theory of failure under combined stresses is that known as the *maximum-shear-stress* theory of failure. For most purposes the difference between the maximum-shear-stress and the maximum-distortion-energy theories is inconsequential. Actually, the maximum-shear-stress theory is conservative, i.e., on the safe side, and its use will result in the part having an actual factor of safety that is slightly higher than the value used in the design equation.

When the maximum-shear-stress theory of failure is applied to the combination of normal and shear stresses acting on a cross section of the shaft, the critical or maximum shear stress becomes

$$s_{s,\max} = \sqrt{\left(\frac{s}{2}\right)^2 + s_s{}^2} \qquad (6\text{-}32)$$

and, by definition, the factor of safety for static loads becomes

$$\text{f.s.} = \frac{s_{sy}}{s_{s,\max}} = \frac{s_{sy}}{\sqrt{(s/2)^2 + s_s{}^2}} \qquad (6\text{-}33)$$

Application of the concept of adding the average or steady component and the weighted reversed component, as discussed in Sec. 6-5 for the case of cyclic loads, leads to

$$\left(\frac{s}{2}\right)^2 = \tfrac{1}{4}s^2 = \tfrac{1}{4}[s_{\text{av}} + (s_y/s_e)K_f s_r]^2 \qquad (6\text{-}34)$$

and

$$s_s{}^2 = [s_{s,\text{av}} + (s_{sy}/s_{se})K_{fs}s_{sr}]^2 \qquad (6\text{-}35)$$

Substitution of Eqs. (6-34) and (6-35) into Eq. (6-33) gives

$$\text{f.s.} = \frac{s_{sy}}{\sqrt{\tfrac{1}{4}[s_{\text{av}} + (s_y/s_e)K_f s_r]^2 + [s_{s,\text{av}} + (s_{sy}/s_{se})K_{fs}s_{sr}]^2}} \qquad (6\text{-}36)$$

Since $s_{\text{av}} = M_{\text{av}}c/I$, $s_r = M_r c/I$, $s_{s,\text{av}} = T_{\text{av}}c/J$, $s_{sr} = T_r c/J$, $I = \tfrac{1}{2}J$, and $c/I = 2c/J$, Eq. (6-36) may be rewritten as

$$\text{f.s.} = \frac{s_{sy}J/c}{\sqrt{[M_{\text{av}} + (s_y/s_e)K_f M_r]^2 + [T_{\text{av}} + (s_{sy}/s_{se})K_{fs}T_r]^2}} \qquad (6\text{-}37)$$

For design purposes, the equation

$$\frac{J}{c} = \frac{(\pi/32)d^4 - (\pi/32)d_i^4}{d/2}$$

for hollow shafts is written in a more convenient form as

$$\frac{J}{c} = \frac{\pi d^3}{16}(1 - C^4) \tag{6-38}$$

where $C = d_i/d$.

Substituting (6-38) into Eq. (6-37) and rearranging gives

$$(1 - C^4)\frac{\pi d^3}{16} = \frac{f.s.}{s_{sy}}\sqrt{\left(M_{av} + \frac{s_y}{s_e}K_f M_r\right)^2 + \left(T_{av} + \frac{s_{sy}}{s_{se}}K_{fs}T_r\right)^2} \tag{6-39}$$

The majority of rotating shafts carry a steady torque, and the loads remain fixed in space in both direction and magnitude. Thus, during each revolution every fiber on the surface of the shaft undergoes a complete reversal of stress due to the bending moment. Therefore, for the usual case when $M_{av} = 0$, $M_r = M$, $T_{av} = T$, and $T_r = 0$, Eq. (6-39) may be simplified as

$$\frac{\pi d^3}{16} = \frac{f.s.}{s_{sy}(1 - C^4)}\sqrt{\left(\frac{s_y}{s_e}K_f M\right)^2 + T^2} \tag{6-40}$$

For steels, where $s_{sy} = 0.5s_y$, Eq. (6-40) may be written in terms of s_y as

$$\frac{\pi d^3}{32} = \frac{f.s.}{s_y(1 - C^4)}\sqrt{\left(\frac{s_y}{s_e}K_f M\right)^2 + T^2} \tag{6-41}$$

The left-hand member of these equations has been left in terms of $\pi d^3/16$ and $\pi d^3/32$ because tables of these quantities for circular shafts are often available (Appendix C). In any case, the equations may be readily solved for d^3. Most shafts are solid, but in situations where weight and reliability are of great importance, hollow shafts with a ratio of $d_i/d = 0.6$ are often used. The weight decreases more rapidly than the strength because the material near the center is not highly stressed and carries only a relatively small part of the total bending and torque loads. The center portion contributes even less to the rigidity of the shaft. The reliability of the material is increased by using hollow shafts. Very large shafts are usually machined from a forged billet, and boring out is specified to remove inclusions, holes, etc., left in the center of the billet, the last region to solidify upon cooling. A hollow shaft also permits more uniform heat-treatment and simplifies inspection of the finished part.

The shaft may be made hollow by boring, forging, or using cold-drawn seamless tubing. Unless the seamless steel tubing can be purchased so close to the required final dimensions that very little machining needs to be done, the high material cost may make it less practical than boring a hole in a solid piece.

6-7. Stress Concentration. The term *stress concentration* is applied to any situation that results in the actual stress being higher than that calculated by means of the usual equations from elementary strength of materials. The term *stress raiser* is often applied to any condition giving rise to stress concentration.

In general, stress concentration results wherever there are changes in shape, such as shoulders, holes, notches, or keyways, and where there is an interference fit between a hub or bearing race and a shaft. A device or concept that is useful in assisting the designer to visualize the presence of stress concentration and how it may be mitigated is that of *stress-flow lines*. This is illustrated in Fig. 6-7. The tensile stress will be uniformly

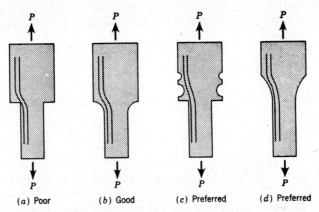

FIG. 6-7. Reduction of stress concentration in tension members.

distributed over the cross section except in the vicinity of the change in width.

As shown in Fig. 6-7a, the stress lines tend to bunch up and cut very close to the sharp reentrant corner. The remaining examples in Fig. 6-7 show methods that may be applied to improve the situation by giving more equally spaced flow lines and by decreasing their slope. In general, the more gradual the transition from one shape to another, the lower the increase in stress.

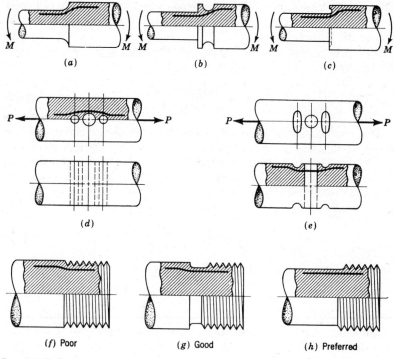

Fɪɢ. 6-8. Reduction of stress concentration in shafts and other cylindrical members.

Figure 6-8 shows several ways in which shafts and other cylindrical members with shoulders, threads, and holes may be improved.   These are self-explanatory, but it is worthwhile to note that Figs. 6-8*b* and *c* are particularly useful when it is not practical to use a large-radius fillet, as is usually the case for ball- and roller-bearing mountings.   Figure 6-9

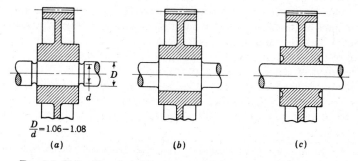

Fɪɢ. 6-9. Reduction of stress concentration in press-fitted assemblies.

illustrates several ways in which the stress-concentration effects of a press fit may be decreased by making a more gradual transition from the rigid hub to the more flexible shaft.

6-8. Stress-concentration Factors.   As discussed in Sec. 6-4, the stress-concentration factor is defined as

$$K = \frac{\text{actual stress}}{\text{nominal stress}} \tag{6-42}$$

The nominal stress is usually calculated for the minimum section involved, but this is not always the case; hence the designer should be careful to note the basis upon which any curve or table is constructed.

The actual stress can be calculated for relatively few cases.   One fundamental analytical solution is the Kirsch solution for an elliptical hole in an infinite plate.[1]   The value of the maximum stress is

$$s_{\max} = \left(1 + 2\frac{a}{b}\right)s \tag{6-43}$$

or the theoretical stress-concentration factor

$$K_t = \frac{s_{\max}}{s} = \left(1 + 2\frac{a}{b}\right) \tag{6-44}$$

When a/b is large, the ellipse approaches a crack transverse to the load, and the value of K_t becomes very large.   When a/b is small, the ellipse approaches a longitudinal slit, and the increase in stress is small.   When the hole is circular, $a/b = 1$, and the maximum stress is three times the nominal value.   The approximate stress distributions for several cases are shown in Fig. 6-10.

Sections 6-4 and 6-5 considered the factor of safety in terms of a theoretical stress-concentration factor K_t when a brittle material is statically loaded and a fatigue stress-concentration factor K_f when the material is subjected to a dynamic load.   In general, the two factors are not equal. The theoretical factor is a function of only the geometry, i.e., relative proportions of the part, while the fatigue factor is not only related to the geometry but also dependent upon the material, the treatment or processing of the material, and the absolute dimensions of the part.

The effect of the size of the part is illustrated in Fig. 6-11, in which a comparison is made of the theoretical and fatigue factors for three series of specimens with different degrees of stress concentration.   Each series is made up of geometrically similar normalized nickel-molybdenum-steel specimens.

[1] S. Timoshenko and J. N. Goodier, "Theory of Elasticity," 2d ed., pp. 78–85, McGraw-Hill Book Company, Inc., New York, 1951.

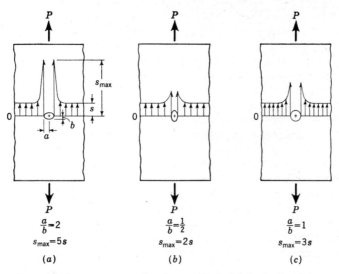

Fig. 6-10. Stress concentration due to elliptical holes in flat plates.

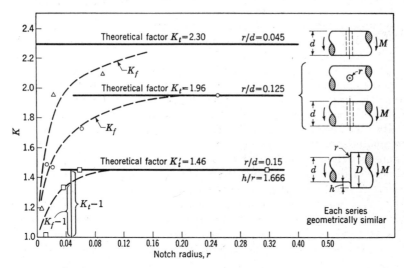

Fig. 6-11. Comparison of theoretical and experimental stress-concentration factors for normalized nickle-molybdenum steel. (*From R. E. Peterson, Relation between Stress Analysis and Fatigue of Metals, Proc. SESA, vol. 11, no. 2, pp. 199–206, 1954.*)

The two major observations are that (1) for small parts, thus small notch radii, K_f may be considerably less than K_t and (2) for large parts and large notches, K_f will become approximately equal to K_t, although the scatter of data may indicate some values above as well as below K_t.[1]

The first consideration points out the desirability of using fatigue values when known and explains why a scratch with a theoretically infinite stress-concentration factor does not automatically mean failure of the part. The second consideration points out that, from a practical viewpoint, the designer may safely use the theoretical value when better information is not available.

Since almost all design problems in which stress concentration is a factor are concerned with dynamic loads, it is apparent that the designer is more interested in values of K_f than he is in values of K_t. Unfortunately, it is not a simple matter to determine values of K_f, and design data are very meager.

The fatigue stress-concentration factor is determined from experimental tests as

$$K_f = \frac{\text{endurance limit without stress concentration}}{\text{endurance limit with stress concentration}} \qquad (6\text{-}45)$$

Considering that each point requires two complete endurance-limit tests, involving a large number of specimens, it can be seen that the determination of values of K_f for even one combination of geometrical shape, material, and material treatment is a time-consuming and expensive task. When all the possible variations are considered, it becomes obvious that obtaining complete data from fatigue tests is an almost impossible task.

While analytical solutions for theoretical stress-concentration factors are not too plentiful, many of the techniques of experimental stress analysis,[2] such as photoelasticity, brittle models, rubber models, membrane and other analogies, short gage-length extensometers, and strain gages, have been effectively used in obtaining theoretical factors. The method of photoelasticity has been most widely used, in verifying analytical solutions as well as in finding values of K_t for design use.

One of the most interesting theories for explaining the decrease in the effect of stress concentration on small parts under fatigue loading, or the *size effect*, is based on statistics. Briefly, the statistical theory of failure considers that failure starts with a particularly weak crystal or grain in the material. Since larger parts have more material in the region of high stress, the probability of this material containing one of these weak spots

[1] Sines and Waisman, *op. cit.*, p. 293.
[2] M. Hetenyi, "Handbook of Experimental Stress Analysis," John Wiley & Sons, Inc., New York, 1949.

is higher, and thus the larger part is more likely to fail under otherwise identical conditions.

For design purposes it would be desirable to have a means for determining values of K_f when only values of K_t are known. Considerable progress has been made in this direction, but, at the present time, the information is not complete enough to apply to every situation.[1]

As shown for the lower curves in Fig. 6-11, the ordinate for the fatigue factor is $K_f - 1$ and for the theoretical factor is $K_t - 1$. The ratio of

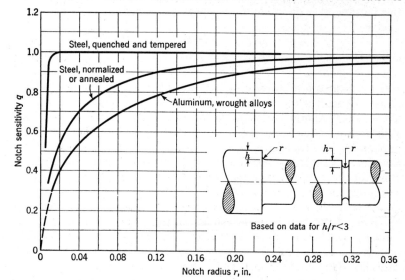

Fig. 6-12. Average notch sensitivity. ("*Metals Handbook,*" 1954 *Supplement, p.* 102, *Metal Progr., July,* 15, 1954.)

the fatigue-factor ordinate to the theoretical-factor ordinate is a measure of the sensitivity of the specimen to the presence of a notch and is known as the *notch-sensitivity factor q.*

Thus,

$$q = \frac{K_f - 1}{K_t - 1} \tag{6-46}$$

or

$$q = \frac{K_{fs} - 1}{K_{ts} - 1} \tag{6-47}$$

The notch-sensitivity factor is a function of the size of the notch, the material, and the treatment of the material. Very few curves are available, but Fig. 6-12 shows the variation of q with notch radius for quenched

[1] The book "Stress Concentration Design Factors" by R. E. Peterson, John Wiley & Sons, Inc., New York, 1953, is recommended as the outstanding single source of design information relative to stress concentration and fatigue.

and tempered steel, normalized or annealed steel, and wrought aluminum alloys.[1] Even though the results are from fatigue tests on specimens with $h/r < 3$, there is reason to believe that the data are also useful when h/r is considerably greater than 3.* It should be noted that, for nearly all practical purposes, $q = 1$ for the quenched and tempered steels.

The major usefulness of the notch-sensitivity factor is that it permits determining the fatigue factor when the theoretical factor is known and the appropriate curve of q as a function of notch radius is available. For this purpose, it is desirable to rearrange terms in Eqs. (6-46) and (6-47) to give

$$K_f = 1 + q(K_t - 1) \tag{6-48}$$

and

$$K_{fs} = 1 + q(K_{ts} - 1) \tag{6-49}$$

A common use of stress-concentration factors is in the designing of shafts. Almost every shaft has shoulders, grooves, keyways, and interference fits. Another common use is in the design of tension members, such as connecting rods, which carry varying loads. The function of most shafts is to transmit a torque from one member to another. Sometimes it is feasible to forge integrally the shaft and gears, pulleys, etc. However, it is usually more economical to make or purchase the parts separately and then assemble them. The torque is usually transmitted to and from the shaft by means of one or more keys, but in some cases the bore of the hub is made enough smaller than the shaft that the torque may be transmitted by the friction between the interfering or press-fitted members. Even when keys are used, a light press fit is desirable to eliminate rocking and wobbling and possible serious trouble due to misalignment. Both the keyway and the interference fits are unavoidable stress raisers.

There is little information available for guidance when two or more causes of stress concentrations occur at the same location. In most cases the resultant K is neither the sum nor the product but is somewhat higher than the maximum single value. The best procedure is to avoid overlapping of stress concentrations.

The information concerning the stress concentration at the end of a keyway in a shaft is somewhat inadequate. Tests have shown that the shape of the end of the keyway is of major importance. The two most common shapes are shown in Fig. 6-13. As would be expected, the sled-runner keyway a gives lower stresses due to bending loads than does the profiled keyway b. The values of K_f also depend upon the material and its treatment. It is suggested that $K_f = 1.6$ for sled-runner keyways

---

[1] For an extensive treatment of notch sensitivity see Sines and Waisman, *op. cit.*, chap. 13.

* Personal communication from R. E. Peterson, April, 1955.

and $K_f = 2.0$ for profiled keyways are reasonable values to be used with nominal bending stress calculated for the full cross section.

Stress-concentration factors for press-fitted members have been given as high as 3.9 for the case of a roller-bearing inner race pressed on a shaft. Additional laboratory tests have indicated that with press fits even the strongest steels have seldom operated satisfactorily with a nominal stress greater than 20,000 psi.

The effects of a press fit may be greatly reduced by the methods previously illustrated in Fig. 6-9 and by properly cold-rolling the fitted portion of the shaft. Horger and Maulbetsch[1] have shown that, by using small-diameter rollers and large forces, the surface of the shaft can be plastically deformed in such a manner that the compressive residual

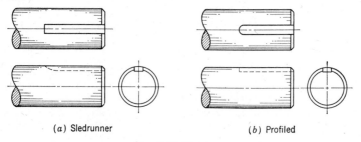

(a) Sledrunner   (b) Profiled

FIG. 6-13. Keyways.

stresses and improved mechanical properties introduced in the critical region will make the press-fitted region almost as strong as the remainder of the shaft. In the absence of better information, it is recommended that a value of $K_f = 2.5$ be used with press-fitted assemblies under bending loads.

The stress concentration introduced by shoulders and grooves is much better documented than the cases of keyways and press-fitted members. Figures 6-14 to 6-19 give curves which should cover most situations within the scope of this book. The curves are self-explanatory, but it should be noted that the value of K_t should be corrected by the use of Eqs. (6-48) and (6-49) and then applied to the nominal stress on the minimum cross section.

Screw threads are another important source of stress concentration and will be considered in Secs. 8-2 and 8-6.

Curved members such as crane hooks, press frames, and the wire in coil springs are not exactly sources of stress concentration as usually understood, but the curvature does result in the stress on the inside of the curved member being higher than that given by the elementary the-

[1] O. J. Horger and J. L. Maulbetsch, Increasing the Fatigue Strength of Press-fitted Axle Assemblies by Surface Rolling, *Trans. ASME*, vol. 58, p. A91, 1936.

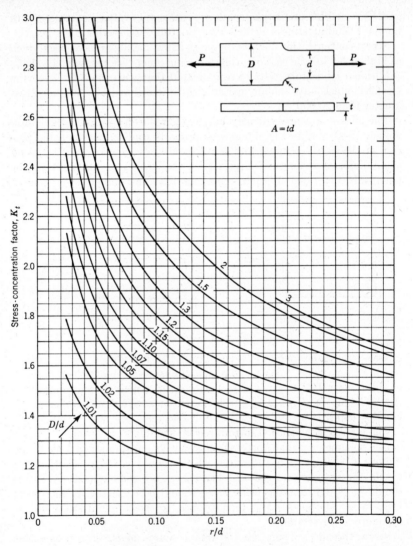

FIG. 6-14. Theoretical stress-concentration factor K_t for a flat bar with a shoulder fillet in tension. (*Reprinted with permission from R. E. Peterson, "Stress Concentration Design Factors," John Wiley & Sons, Inc., New York, 1953.*)

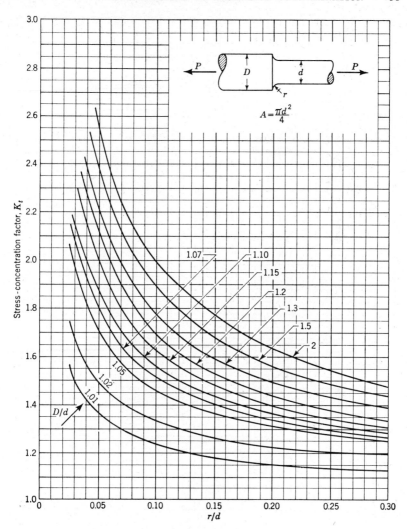

FIG. 6-15. Theoretical stress-concentration factor K_t for a shaft with a shoulder fillet in tension. (*Reprinted with permission from R. E. Peterson, "Stress Concentration Design Factors," John Wiley & Sons, Inc., New York, 1953.*)

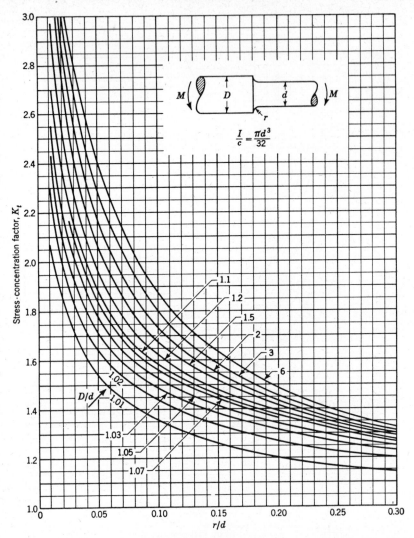

FIG. 6-16. Theoretical stress-concentration factor K_t for a shaft with a shoulder fillet in bending. (*Reprinted with permission from R. E. Peterson, "Stress Concentration Design Factors," John Wiley & Sons, Inc., New York, 1953.*)

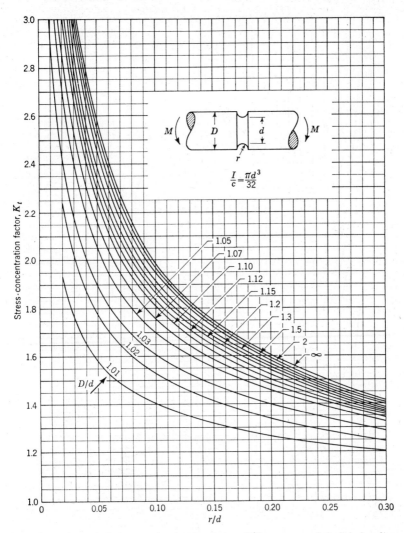

FIG. 6-17. Theoretical stress-concentration factor K_t for a grooved shaft in bending. (*Reprinted with permission from R. E. Peterson, "Stress Concentration Design Factors," John Wiley & Sons, Inc., New York, 1953.*)

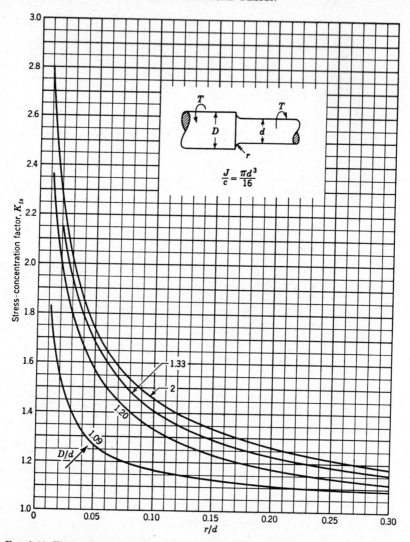

FIG. 6-18. Theoretical stress-concentration factor K_{ts} for a shaft with a shoulder fillet in torsion. (*Reprinted with permission from R. E. Peterson, "Stress Concentration Design Factors," John Wiley & Sons, Inc., New York, 1953.*)

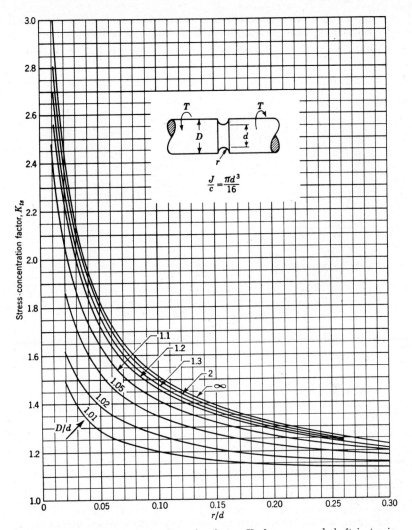

FIG. 6-19. Theoretical stress-concentration factor K_{ts} for a grooved shaft in torsion. (*Reprinted with permission from R. E. Peterson, "Stress Concentration Design Factors," John Wiley & Sons, Inc., New York, 1953.*)

ory of bending.   The general case of curved bars in bending is covered by Timoshenko,[1] and the special case of springs will be discussed in detail in Chap. 9.   For many applications, the curves in Fig. 6-20 will be sufficient.

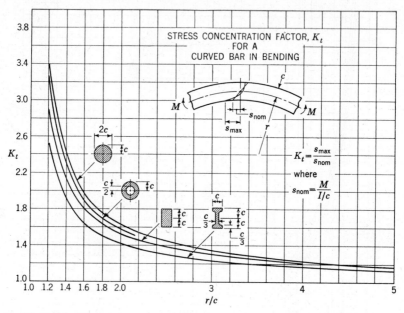

Fig. 6-20. Theoretical stress-concentration factor K_t for a curved bar in bending. *(Reprinted with permission from R. E. Peterson, "Stress Concentration Design Factors," John Wiley & Sons, Inc., New York, 1953.)*

Example 6-1.   A centrifugal blower rotating at 600 rpm is to have a capacity of 25,000 cfm.   A belt drive is to be used to connect the blower to a 20-hp 1,750-rpm electric motor.   The blower shaft is to be machined from hot-rolled 1020 steel, and the only part of the shaft under significant stress is the overhung part on which the pulley is mounted.   The belt forces resolve into a torque of 2,100 lb-in. and a force of 550 lb on the shaft.   Figure 6-21 shows the locations of the bearings, the steps in the shaft, and the plane in which the resultant belt force and torque act.   The ratio of the journal diameter to the overhung-shaft diameter is to be approximately 1.2, that is, $D/d \cong 1.2$, and the fillet is to be a full fillet, that is, $r = (D - d)/2 \cong 0.1d$.

We are asked to specify the dimensions of d, D, and r for the shaft to have a factor of safety of 3.

Solution.   Since the load is essentially static, the shaft will carry a steady torque plus a reversed bending moment, and the design equation is Eq. (6-41):

$$\frac{\pi d^3}{32} = \frac{\text{f.s.}}{s_y(1 - C^4)} \sqrt{\left(\frac{s_y}{s_e} K_f M\right)^2 + T^2}$$

[1] S. Timoshenko, "Strength of Materials," pt. I, 3d ed., chap. 12, D. Van Nostrand Company, Inc., Princeton, N.J., 1955.

and   f.s. = 3

  s_y = 43,000 psi, from Appendix D

  $C = 0$

  s_e = 26,000 psi, from Fig. 6-3 for s_u = 65,000, and machined surface

  K_f and M depend upon section

  T = 2,100 lb-in.

The two points at which failure may occur are at the end of the keyway and at the shoulder fillet. The critical section will be the one with the larger product K_fM. Since the notch-sensitivity factor q is dependent upon the unknown absolute dimensions of the notch and since the curves for notch-sensitivity factor (Fig. 6-12) are not applicable to keyways, the product K_tM will be the basis for comparing the two sections.

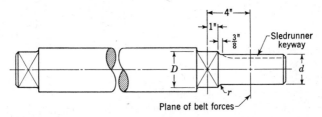

Fig. 6-21

At the end of the keyway, K_tM = 1.6 × 550 × 2.625 = 2,310, and at the shoulder fillet, K_tM = 1.62 × 550 × 3 = 2,670, where K_t = 1.62 from Fig. 6-16 for D/d = 1.2 and r/d = 0.1. Therefore, considering the shoulder fillet as the critical section,

$$\frac{\pi d^3}{32} = \frac{3}{43,000(1-0)} \sqrt{\left(\frac{43,000}{26,000} \times 1.62 \times 550 \times 3\right)^2 + (2,100)^2} = 0.341$$

From Appendix C, to the next higher sixteenth of an inch, d = $1\frac{9}{16}$ in. (from $\pi d^3/32$ = 0.341, d = 1.515 in.). Using d = $1\frac{9}{16}$ in. (1.5625), we find that

$$r = 0.1d = 0.156 \text{ in. } (\tfrac{5}{32})$$
$$D = 1.2d = 1.875 \text{ in. } (1\tfrac{7}{8})$$

Now that r is known, it is possible to determine K_f by use of the notch-sensitivity factor (from Fig. 6-12) and Eq. (6-48). Thus,

$$K_f = 1 + q(K_t - 1) = 1 + 0.93(1.62 - 1) = 1.58$$

If desired, a new value of d may be calculated, using K_f = 1.58, but in view of the closeness of the magnitudes of K_f and K_t and the fact that any error introduced by using K_t instead of K_f will be on the safe side, the recalculation may logically be omitted.

Therefore, the dimensions to be specified are

$$d = 1\tfrac{9}{16} \text{ in.}$$
$$r = \tfrac{5}{32} \text{ in.}$$
$$D = 1\tfrac{7}{8} \text{ in.}$$

It should be noted that, if it is more convenient to use d = $1\frac{1}{2}$ in., the factor of safety is decreased to 2.92, which is close enough to the originally specified value of 3 for most purposes.

6-9. Elastic Deformation.   A relatively large elastic deformation, or deflection, is a desired attribute of one general class of machine elements known as springs, which will be discussed in Chap. 9.   With the exception of special situations, such as bolts that must carry repeated or impact loads (Sec. 8-6), it is usually desirable to keep the deformation as small as possible in order to minimize difficulties arising from misalignments and, possibly, interference between mating parts.

The elastic deformation of a bar with a uniform cross section, made from a material with a linear stress-strain curve and carrying a tension load, can be calculated directly by using the definition of the modulus of elasticity E, which is

$$E = \frac{s}{\epsilon} = \frac{P/A}{\delta/L} \qquad \text{psi} \tag{6-50}$$

where s = stress, psi

ϵ = strain, in./in.

P = load, lb

A = cross-sectional area, in.[2]

δ = deformation (elongation) when load is applied, in.

L = length over which the deformation takes place, in.

In an analysis problem, Eq. (6-50) would be solved for δ, whereas, in a design situation, the deformation would be specified and the problem would become the determination of the dimensions of the cross section or the length of the rod.

Equation (6-50) can also be used for rods under compression loads provided they are "short" and stability (Secs. 6-10 to 6-13) need not be considered.

For the angular deformation, or twist, under pure torsion of a bar with a uniform circular cross section and made from a material with a linear shear-stress–shear-strain curve, we have from elementary strength of materials

$$\theta = \frac{TL}{JG} \tag{6-51}$$

where θ = angle of twist, radians

T = torque, lb-in.

L = length over which twisting occurs, in.

J = polar moment of inertia of the cross section, in.[4]

G = shear modulus of elasticity, psi

Noncircular bars are used infrequently to carry torque in mechanical designs; the reader may refer to the extensive list of equations given by Roark[1] for further information.

The elastic, and even plastic, deformation between bodies in direct

[1] R. J. Roark, "Formulas for Stress and Strain," 3d ed., chap. 9, McGraw-Hill Book Company, Inc., New York, 1954.

contact, e.g., cams, gear teeth, ball and roller bearings, etc., is essential in that it permits the transmission of finite forces by producing a finite area of contact in place of the kinematic point or line contact between the members.   Some effects of this contact deformation will be considered in relation to gear teeth (Secs. 16-8 to 16-13) and rolling-contact bearings (Sec. 18-21), but for information related to calculating deformations the reader must refer to the literature.[1]

The deflection of a shaft is often a serious problem.   Troubles may arise from (1) angular misalignment between the shaft and the bearing bore, between the teeth of mating gears, etc.; (2) interference between a rotating member and the frame or between rotating members; (3) deflections large enough to impair the accuracy of machining operations; and (4) vibrations of the shaft (Secs. 19-5 to 19-6).

If the shaft, or beam for that matter, has a uniform section and is loaded in a relatively simple manner, the equations in Appendix A or given by Roark[2] for many more conditions can be used to solve directly for the dimensions required to limit the slope or deflection to any specified value.   Unfortunately, those shafts for which slopes and deflections will be significant practically always have a number of steps in diameter and the simple equations no longer apply.

When confronted with a stepped shaft and the assumption of an "equivalent" uniform shaft will not give sufficient accuracy in calculations, the designer is forced to fall back upon the basic relationship

$$\frac{1}{R} = \frac{M}{EI} \qquad (6\text{-}52)$$

and its equivalent, for the small deflections encountered in machines,

$$\frac{d^2y}{dx^2} = \frac{M}{EI} \qquad (6\text{-}53)$$

where R = radius of curvature of the deflected shaft, in.
    y = deflection, in.
    x = distance along the shaft, in.

For a stepped shaft both M and I are functions of x, neither of which can normally be expressed in a single function that is good for all values of x.   Thus, direct double integration requires considering the shaft as a series of shafts with uniform sections.   The matching of boundary conditions, when solving for the constants of integration, becomes a tedious and time-consuming task.   In general this type of problem is best handled

[1] S. Timoshenko and J. N. Goodier, "Theory of Elasticity," 2d ed., pp. 372–382, McGraw-Hill Book Company, Inc., New York, 1951.
    Roark, *op. cit.*, chap. 13.
[2] Roark, *op. cit.*, chap. 8.

by the use of numerical or graphical techniques or combinations of them in performing the integrations.

The use of numerical methods with high-speed computers offers many advantages, primarily in relation to accuracy of calculation and the rapidity with which the effect of changes in dimensions, i.e., in I, can be studied. The latter is particularly important here because this type of problem can no longer be solved directly for dimensions as a design problem. Rather, the final design must be selected from trial-and-error (or iterative) analyses of shafts with assumed dimensions. It should be noted that, in general, the dimensions required for adequate strength will have previously been determined, and the problem now is to find out if these dimensions are also adequate to meet the rigidity requirements and, if not, to find dimensions that will.

Since the primary purpose of this discussion is the presentation of ideas, only a simplified solution, combining graphical and numerical methods, will be presented here in Example 6-2. Almost every calculus text includes a coverage of more sophisticated numerical methods, and numerous machine design and vibrations texts present purely graphical techniques.

Example 6-2.   The critical dimensions of the shaft for a centrifugal blower were determined on the basis of stress only in Example 6-1.   The questions now facing us are (1) can ordinary bearings be used or will the slope of the shaft be so large that self-aligning bearings (Chap. 18) must be specified and (2) what should be the minimum clearance between the outside diameter (OD) of the impeller and the inside diameter (ID) of the housing.

Solution.   Figure 6-22 shows the information available at this stage of the over-all design.   As usual in mechanical design problems, the information is not complete,

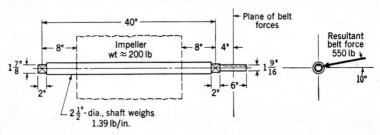

Fig. 6-22

and we must make some assumptions and/or approximations in order to get an answer.

1. The weights of the impeller and shaft act in the vertical direction, and the resultant belt force acts on a line slightly (10°) above the horizontal direction. We cannot combine these forces and then solve for the slopes and deflections but must consider each plane separately and then add the deflections vectorially to get the resultant.   This will not be too difficult in terms of magnitudes at each point along

the shaft, but the resulting deflection curve lies in a warped plane, and determining the slopes becomes a somewhat more difficult problem. Fortunately, we can usually determine the effect of the slope by considering each plane separately.

2 The weight of the impeller is not concentrated at a point but is spread over a finite length of shaft in some not very well defined manner. For convenience and to be on the safe side, we shall assume that the impeller weight is distributed uniformly over the 24-in. length of shaft.

3. The impeller hub will be a press fit on the shaft and will contribute to the rigidity of the shaft. However, since the magnitude of this contribution is unknown and difficult to estimate, we shall be conservative and neglect it.

4. The bearing reactions will not act at a point and the bearings will not be true "simple supports" with respect to moments unless they are self-aligning. Again, neither effect can be considered in a rigorous manner, and to be conservative we shall consider the bearings to act as simple supports.

Vertical plane. Figure 6-23a shows the loading that will be considered to be significant in the vertical plane. The shaft weight is relatively small, and the effect of all but the central $2\frac{1}{2}$-in.-diameter portion will be neglected. Using the information in Fig. 6-23a, the moment at any point can be calculated. For convenience the shaft length is divided into a number of divisions, and within each division the functions, i.e., moment, slope, and deflection, will be approximated by straight lines. At the expense of increased time and effort, an increase in accuracy can be obtained by using more divisions.

The equation for the moments on sections 0 through 2 is

$$M = R_L x - w_s(x - 1)\frac{(x - 1)}{2}$$

For example,

$$M_1 = 126.4 \times 4 - 1.39(4 - 1)\frac{(4 - 1)}{2} = 499 \text{ lb-in.}$$

Similarly, the equation for the moments on sections 2 through 8 is

$$M = R_L x - w_s(x - 1)\frac{(x - 1)}{2} - w_i(x - 8)\frac{(x - 8)}{2}$$

and, for example,

$$M_4 = 126.4 \times 16 - 1.39(16 - 1)\frac{(16 - 1)}{2} - 8.33(16 - 8)\frac{(16 - 8)}{2}$$
$$= 1,602 \text{ lb-in.}$$

Since the shaft and loading are symmetrical, it is not actually necessary to calculate the moments for sections 6, 7, 8, 9, 10 because they must equal those at 4, 3, 2, 1, and 0, respectively.

From Eq. (6-53), we can write the equation for the slope as

$$\frac{dy}{dx} = \int \frac{M}{EI} dx + C_1 \tag{6-54}$$

Since E is a constant for the whole shaft, it could be taken outside the integral, but it is simpler to leave it inside and to work with the M/EI diagram.

For the $1\frac{7}{8}$-in.-diameter parts of the shaft

$$I = \frac{\pi d^4}{64} = \frac{\pi 1.875^4}{64} = 0.607 \text{ in.}^4$$

and for the $2\frac{1}{2}$-in.-diameter part of the shaft

$$I = 1.918 \text{ in.}^4$$

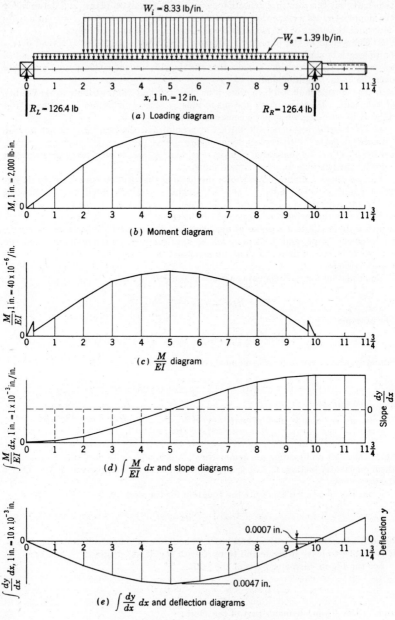

FIG. 6-23. Vertical plane.

Figure 6-23c is a plot of the values of M/EI calculated on the assumption that $E = 30 \times 10^6$ psi. It should be noted that, wherever there is a change in diameter, there is a discontinuity in the curve. For example, at section $\frac{1}{4}(-)$, $M/EI = 6.92 \times 10^{-6}$/in. and, at $\frac{1}{4}(+)$, $M/EI = 2.19 \times 10^{-6}$/in. It can also be seen at this time that the shaft could be assumed to have a uniform diameter of $2\frac{1}{2}$ in. without introducing appreciable error.

The area under the M/EI curve between 0 and 1 will be the definite integral or the change in slope from 0 to 1, the area under the curve between 0 and 2 will be the change in slope from 0 to 2, etc. The areas can be determined by counting squares, by using a planimeter, or, more simply—since we are using a straight-line approximation to the actual curve—by using the equation for the area of a trapezoid

$$A = \tfrac{1}{2}b(h_1 + h_2) \tag{6-55}$$

where b = base distance
h_1 = height at left side
h_2 = height at right side
For 0 to 1, that is, 0 to $\frac{1}{4}(-)$ plus $\frac{1}{4}(+)$ to 1,

$$\frac{dy}{dx}\bigg|_{x=1} - \frac{dy}{dx}\bigg|_{x=0} = \int_0^1 \frac{M}{EI}\,dx$$

$$= \frac{1}{2}\left(\frac{1}{4}\,4\right)(0 + 6.92)10^{-6} + \frac{1}{2}\left(\frac{3}{4}\,4\right)(2.19 + 8.69)10^{-6}$$

$$= 3.5 \times 10^{-6} + 16.3 \times 10^{-6} = 19.8 \times 10^{-6} \text{ in./in.}$$

For 0 to 2

$$\frac{dy}{dx}\bigg|_{x=2} - \frac{dy}{dx}\bigg|_{x=0} = \int_0^2 \frac{M}{EI}\,dx = \int_0^1 \frac{M}{EI}\,dx + \int_1^2 \frac{M}{EI}\,dx$$

$$= 19.8 \times 10^{-6} + \tfrac{1}{2}(4)(8.69 + 17.0)10^{-6}$$

$$= 19.8 \times 10^{-6} + 51.4 \times 10^{-6} \doteq 71.2 \times 10^{-6} \text{ in./in.}$$

and so on, until section 10 is reached.

The plot of the summation of the areas under the M/EI curve is shown in Fig. 6-23d with the scale on the left-hand side of the diagram. We are actually interested in the indefinite integral, Eq. (6-54), because we want to know the absolute values of the slope rather than just the change occurring as we move along the shaft. This requires knowing the slope at one point on the shaft, usually the point of zero slope. In general, as will be illustrated when we consider the plane of the belt forces, this point is not known until the deflection curve has been drawn. However, by symmetry, we know that the zero slope occurs at section 5, and therefore, a horizontal line intersecting the curve at 5 is the zero line for slope measurements, as shown by the scale on the right-hand side of the diagram.

The area between the slope curve and the zero line now becomes the deflection; i.e.,

$$y = \int \frac{dy}{dx}\,dx + C_2 \tag{6-56}$$

However, as above, the integration must be performed as a summation of definite integrals and the constant determined later.

Using Eq. (6-55) we find that for 0 to 1

$$y_1 - y_0 = \int_0^1 \frac{dy}{dx}\,dx = \frac{1}{2}\,4(-375 - 355)10^{-6} = -1.46 \times 10^{-3} \text{ in.}$$

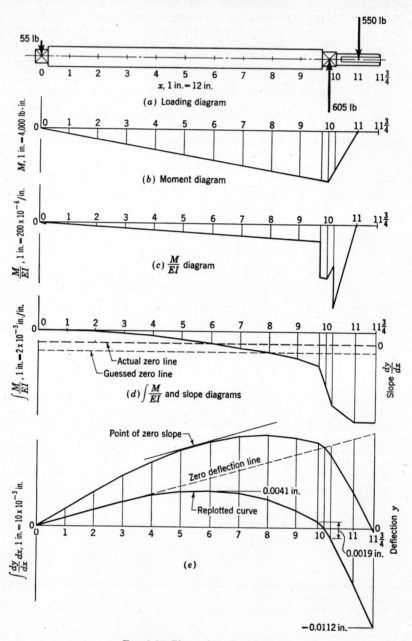

Fig. 6-24. Plane of the belt forces.

for 0 to 2

$$y_2 - y_0 = \int_0^2 \frac{dy}{dx}\, dx = \int_0^1 \frac{dy}{dx}\, dx + \int_1^2 \frac{dy}{dx}\, dx$$
$$= -1.46 \times 10^{-3} + \tfrac{1}{2}4(-355 - 304)10^{-3}$$
$$= -1.46 \times 10^{-3} - 1.32 \times 10^{-3} = -2.78 \times 10^{-3} \text{ in.}$$

and so forth, until the end of the shaft is reached.

The plot of the summation of the areas under the dy/dx curve is shown in Fig. 6-23e with the scale on the left-hand side of the diagram. Assuming the frame, or bearing supports, to be rigid, the straight line for zero deflection must intersect the curve at sections 0 and 10, the bearing locations. Since this line is horizontal, the summation curve automatically becomes the deflection curve with the scale on the right-hand side of the diagram.

Plane of the belt forces. Figure 6-24 shows the diagrams for the plane of the belt forces. The calculations were performed in exactly the same manner as for the vertical plane. It should be noted that, since we could not determine by looking at the shaft where the point of zero slope would be located, it was desirable, although not absolutely necessary, to guess at its location in order to simplify the next integration. The error in the location of the point of zero slope introduces an extraneous constant value of slope, which results in an error in the deflection curve that increases, either positively or negatively, linearly with the distance along the shaft. The zero deflection line can be located by drawing a straight line that intersects the deflection curve at points with known zero deflection, i.e., at 0 and 10 in this example. The actual deflections are then the vertical distances between the summation curve and the zero deflection line. For convenience in analyzing the results, these distances have been replotted in Fig. 6-24e with a horizontal zero line.

Conclusions. Although at this stage in the design we do not know whether a plain journal bearing with a diameter of $1\tfrac{7}{8}$ in. and a length of 2 in. can carry the load satisfactorily (see Example 18-1), we can say that a change of deflection of 0.0019 in. over the 2-in. length in the plane of the belt forces is too great and some type of self-alignment must be provided for the right-hand bearing.

For the left-hand bearing, the slope in the vertical plane is 0.00037 in./in. and in the plane of the belt forces 0.00028 in./in. Considering only the larger, we find that the change in deflection over the bearing length will be 0.00074 in. This might be satisfactory provided the bearing material would permit an appreciable amount of wear (Sec. 18-14) without failure, but good design indicates the use of a self-aligning bearing on the left end, also.

The clearance required for the impeller will be approximately the vector sum of the deflections at section 5. As shown in Fig. 6-25, the resultant is about 0.006 in. For many machines this deflection would be excessive, but the clearance between the impeller and the casing in a blower of this type and size would normally be

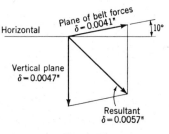

Fig. 6-25

in the order of inches, not thousandths of an inch, and thus, this deflection is negligible.

We were not asked to determine the slope and deflection at section 11 because neither will have any appreciable effect on the operation of a flat-belt drive (Chap. 14). However, if gears were to be used, and if the same forces were present, it would be a different story. The deflection, in itself, would not be significant, although it would increase the backlash somewhat (Sec. 16-3). But the slope of 0.0017 in./in. would result in a severely nonuniform load distribution across the teeth and would be intolerable. This would require redesigning the shaft on the basis of rigidity rather than stress.

6-10. Factors of Safety for Compression Members—Stability. Compression members, such as columns, beams under combined transverse and axial loads, thin plates subject to edgewise loads, and cylindrical shells subjected to external pressures, may suddenly collapse or buckle without warning as the load or pressure is slowly increased. This sudden change from satisfactory performance to disaster is considered to be a problem in stability, and the adjectives *elastic* and *plastic* are often used to describe more fully the type of behavior expected.

Since the problem of buckling of machine parts is related largely to members that may be classed as columns, e.g., connecting rods, translation or power screws (Chap. 10), and struts or other supporting links, the discussion in the following sections will be limited to this topic, and the reader is referred to the extensive literature[1] for design information on the stability of other members.

As noted in Sec. 6-3, failure due to instability is not a direct function of stress, and the factor of safety for a column must therefore be defined in terms of the working load and the *critical load* P_{cr} under which buckling occurs. Thus

$$\text{f.s.} = \frac{P_{cr}}{P} \qquad (6\text{-}57)$$

Unfortunately, there is no single equation that may be applied to all column-design problems. In fact, there are many column equations which are used in special situations—often specified and required by building codes, etc. However, consideration of only three, the Euler formula, the J. B. Johnson[2] formula for ductile materials, and a straight-line formula for a brittle material (cast iron), will be adequate for most purposes.

6-11. Long Columns. The Euler formula for the critical load on a column is

$$P_{cr} = \frac{C\pi^2 EA}{(L/k)^2} \qquad \text{lb} \qquad (6\text{-}58)$$

[1] For example, S. Timoshenko, "Theory of Elastic Stability," McGraw-Hill Book Company, Inc., New York, 1936.

Roark, *op. cit.*

API-ASME Code for Unfired Pressure Vessels, American Society of Mechanical Engineers, New York, 1951.

Rules for Construction of Unfired Pressure Vessels, ASME Boiler Code Section VIII, American Society of Mechanical Engineers, New York, 1959.

[2] Initials are used to differentiate between this and the T. H. Johnson column formula.

where C = end-fixity coefficient

E = modulus of elasticity, psi

A = area of cross section, in.^{2}

L = unsupported length of column, in.

$k = \sqrt{I/A}$ = radius of gyration of cross section, in.

It can be seen that, for a column with a given cross-section area, the critical load is a function of the *end-fixity coefficient* C and the *slenderness ratio* L/k.

Figure 6-26 shows columns with a number of different end conditions and gives the value of C for each case. The most common situation is

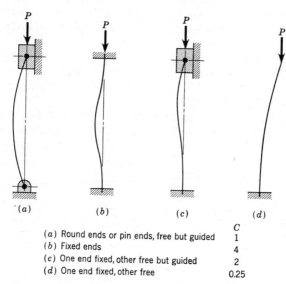

				C
(a) Round ends or pin ends, free but guided				1
(b) Fixed ends				4
(c) One end fixed, other free but guided				2
(d) One end fixed, other free				0.25

FIG. 6-26. Column-end conditions and end-fixity coefficients.

that in Fig. 6-26a, where the column ends are pin-connected and guided, as for the connecting rods in Figs. 2-4a and 2-9. Because of the flexibility of the supporting members, it is doubtful whether a column in a machine ever has ideally fixed ends, as in Fig. 6-26b and c, even when welded in place.

Values of $C = 2$ for Fig. 6-26b and $C = 1$ or $1\frac{1}{2}$ for Fig. 6-26c may be more realistic. In any case, it is worthwhile to consider the rigidity of the supporting members and to adjust the value of C accordingly.

The end conditions for almost all columns will vary with the plane of reference. For example, the column in Fig. 6-26a is pin-connected in the view shown but may well be considered to have fixed ends when viewed in the plane perpendicular to the paper.   When the column is cylindrical or square, there is little difficulty in selecting the critical plane, because it depends upon only the end conditions.   However, the majority of columns used in machines have different radii of gyration as well as different degrees of end fixity in perpendicular planes, e.g., the connecting rods referred to above.   From Eq. (6-58) it can be seen that the critical plane will be that having the least value of the product Ck^2, or, if more convenient, CI.

A serious limitation of the Euler equation is that it assumes perfect elasticity and consequently cannot be used when the stress exceeds the yield point for ductile materials.   For example, solving Eq. (6-58) for the nominal stress P_{cr}/A when the material is steel, $C = 1$, and $L/k = 10$, we find that the stress would be 2,960,000 psi, which is not a very realistic value for present-day materials.

There is no rational equation available for use when the column is so short that the Euler formula no longer applies, and the designer must rely upon empirical equations that have been found to give reasonable agreement with experimental results.

6-12. Short Columns of Ductile Materials.   The J. B. Johnson formula for the critical load for columns made of ductile materials is

$$P_{cr} = A s_y \left[1 - \frac{s_y(L/k)^2}{4C\pi^2 E} \right] \qquad \text{lb} \qquad (6\text{-}59)$$

Figure 6-27 is a plot showing the relationship of the nominal stress P_{cr}/A to the slenderness ratio L/k as given by the Johnson and Euler equations for a column made from steel with a yield strength of 30,000 psi.

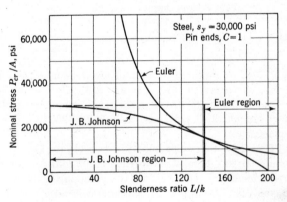

FIG. 6-27

Since the J. B. Johnson formula curve is a parabola, cutting the ordinate at s_y and the abscissa at $L/k = \sqrt{4C\pi^2E/s_y}$, the graphical construction in Sec. 5-6 may be utilized in drawing the curve for any value of yield strength.

Equating Eqs. (6-58) and (6-59) and solving for L/k gives

$$\frac{L}{k} = \sqrt{\frac{2C\pi^2E}{s_y}} \qquad (6\text{-}60)$$

at the point of tangency of the two curves. It should be noted that L/k at the point of tangency is the division point between the application of the formulas; the J. B. Johnson formula is used when $L/k < \sqrt{2C\pi^2E/s_y}$, and the Euler equation is used when $L/k > \sqrt{2C\pi^2E/s_y}$.

The majority of ductile columns used in machines fall in the J. B. Johnson region. Therefore, unless it is evident that the member will be long and slender, the J. B. Johnson formula [Eq. (6-59)] should be used as a first trial. The ratio L/k from the first trial should then be compared with the value of L/k given by Eq. (6-60) to check the assumption.

6-13. Short Columns of Brittle Materials. Experience indicates that a straight line from $P_{cr}/A = s_u$ tangent to the Euler curve is in fair agreement with results from tests on columns made from brittle materials. The straight-line formula for the critical load is

$$P_{cr} = A\left(s_u - C_1\frac{L}{k}\right) \qquad \text{lb} \qquad (6\text{-}61)$$

The end-fixity coefficient C_1 is a function of the material as well as the type of ends. For cast iron, the usual design value for ultimate strength is 34,000 psi.*

Values of C_1 and the corresponding maximum ratios of L/k are given in Table 6-1.

TABLE 6-1. END-FIXITY COEFFICIENTS FOR CAST-IRON COLUMNS

End conditions	C_1	Maximum L/k
Round......................	175	90
Fixed......................	88	160
One fixed, one round.........	116	115

* This is less than may be found from tests on small specimens but is in agreement with results from tests on full-size columns.

Example 6-3.   A trunnion-mounted hydraulic cylinder is being designed to exert a force of 2,500 lb through a working stroke of 24 in.   Figure 6-28 shows the mounting details and dimensions when the piston rod is extended its full working stroke.

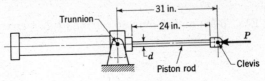

FIG. 6-28

We are asked to specify the diameter of a ground and polished cold-drawn 1020 steel rod that will have a factor of safety of $2\frac{1}{2}$.

Solution.   The piston rod is a relatively long and slender member that is subject to buckling under a compression load.   It is likely that the Euler formula will be applicable, and it should be used for the first trial.

The unsupported length is actually less than the distance between the cylinder trunnion and the rod clevis pin.   But, since the additional support given by the cylinder and clevis is indeterminate, the safest design procedure will be to consider the piston rod as a pin-ended column with an unsupported length of 31 in.

The Euler formula [Eq. (6-58)] is

$$P_{cr} = \frac{C\pi^2 E A}{(L/k)^2}$$

and $P_{cr} = $ f.s. $\times P = 2.5 \times 2,500 = 6,250$ lb, from Eq. (6-57)

$C = 1$, from Fig. 6-26 for pin ends, free but guided

$E = 30,000,000$ psi

$A = \pi d^2/4$, in.^{2}

$L = 31$ in.

$k = \sqrt{I/A} = d/4$ in.

Thus

$$6,250 = \frac{1 \times \pi^2 \times 30,000,000 \times \pi d^2/4}{(31 \times 4/d)^2}$$

Solving for d, we find

$$d = 0.802 \text{ or } 1\frac{3}{16} \text{ in.}$$

We must now compare the value of L/k for the $1\frac{3}{16}$-in.-diameter rod with that from Eq. (6-60) for the point of transition from the J. B. Johnson to the Euler formula. For the $1\frac{3}{16}$-in.-diameter rod

$$k = \frac{d}{4} = \frac{1\frac{3}{16}}{4} = 0.203 \text{ in.}$$

and

$$\frac{L}{k} = \frac{31}{0.203} = 152.6$$

From Eq. (6-70),

$$\frac{L}{k} = \sqrt{\frac{2C\pi^2 E}{s_y}} = \sqrt{\frac{2 \times 1 \times \pi^2 \times 30,000,000}{66,000}} = 94.7$$

where $s_y = 66,000$ psi, from Appendix D.   Therefore, since $152.6 > 94.7$, the initial assumption of the Euler formula was correct, and the piston-rod diameter will be specified as $1\frac{3}{16}$ in.

6-14. Selection of Factors of Safety.   It probably has already occurred to the reader that, while the previous sections in this chapter illustrate how a design may be made or analyzed in a rational manner, there is little point in considering everything in great detail unless the factor of safety itself has the proper significance.

Thus, the choice of an appropriate factor of safety is one of the most important decisions the designer must make.   Since the penalty for choosing too small a factor of safety is obvious, the tendency is to make sure that the design is safe by using an arbitrarily large value and over-designing the part.   In many instances, where only one or very few parts are to be made, overdesigning may well prove to be the most economical as well as the safest solution.   For large-scale production, however, the increased material and manufacturing costs associated with overdesigned parts result in a favorable competitive position for the manufacturer who can design and build machines which are sufficiently strong, but not too strong.

As will be evident, the cost involved in the design, research, and development necessary to give the lightest possible machine will be too great in most situations to justify the selection of a low factor of safety. An exception is in the aircraft industry, where the necessity for the light-est possible construction justifies the extra expense.   In any case, the designer need not feel lost in this matter as there are several ways in which a design value of factor of safety may be selected.   These several methods fall within the classifications discussed in the next four sections.

6-15. Legislated Safety Factors.   In many instances, particularly where danger to human life is present, the designer has little or no choice except to follow the exact design procedures outlined in a code which has been adopted by one of the engineering societies, trade associations, or a unit of civil government.   These codes may not actually mention the term *factor of safety* but, instead, may specify an allowable or design stress.   The allowable stress generally applies to only the specific case and includes a true factor of safety and other considerations, such as the effect of environment.   The codes usually also specify the equation to be used in determining the required dimension.   Considerable care should be taken when trying to adapt the allowable stress values given in a code to any situation other than that for which the code was written.

Some of the codes sponsored by the American Society of Mechanical Engineers are the Code for Pressure Piping; the Safety Code for Con-veyors, Cableways and Related Equipment; the Low Pressure Heating Boiler Code; the Miniature Boiler Code; the Power Boiler Code; and the Unfired Pressure Vessel Code.   Most of these codes have been adopted by state legislatures and government agencies and have thus achieved legal status.

There are other legal codes in relation to buildings, bridges, elevators, etc., that are too numerous to discuss here.

Many societies and trade associations, such as the American Gear Manufacturers' Association (AGMA), the American Standards Association (ASA), the American Petroleum Institute (API), the Society of Automotive Engineers (SAE), and the National Electrical Manufacturers' Association (NEMA), have adopted standards and codes which provide the particular industry with reliable, as well as standard, values of stress and procedures for use in design.

6-16. Published Factors of Safety. The literature contains many papers and articles related to successful designs in all fields of engineering. Many of the authors have become noted authorities in their specialties, and their recommendations have been adopted by and repeated in many handbooks. The "Mechanical Engineers' Handbook"[1] and the "Design and Production" volume of "Kent's Mechanical Engineers' Handbook"[2] are particularly useful in the area of mechanical design.

The major problem in using recommendations from handbooks is that the assumptions and limitations applying to the original work often do not appear in the condensed version in the handbook. Thus a critical analysis of handbook data is necessary to prevent an unwarranted extrapolation which may lead to an unsatisfactory design.

6-17. Factors of Safety from Company Experience. Most smaller companies and departments of large companies are concerned with the design of relatively few types of machines. A direct result of this concentration of attention has been the accumulation of a large amount of information about design details for both successful and unsuccessful designs. Analysis of this information has led to what is practically an internal code, in that many companies have set up standards books which list the allowable stresses and design procedures to be followed. This method works well as long as each step in extrapolation to new conditions is not large. *A particular weakness of this method is that an allowable stress is often used in designing a part whose satisfactory use depends upon its rigidity and not upon the magnitude of a stress.* Such a procedure may simplify the calculations involved in designing a new machine, but it may also lead to a poor solution when applied to a situation no longer comparable to the original design. It should be unnecessary to add that a company would be foolish if it did not make use of its experience. However, it is always good engineering to investigate each new situation somewhat beyond the point of blindly following a list of values.

6-18. Independent Selection of Factors of Safety. The previous three sections have listed various sources of factors of safety. In general, how-

[1] L. S. Marks (ed.), revised by T. Baumeister, "Mechanical Engineers' Handbook," 6th ed., McGraw-Hill Book Company, Inc., New York, 1958.

[2] C. Carmichael (ed.), "Kent's Mechanical Engineers' Handbook," "Design and Production" volume, 12th ed., John Wiley & Sons, Inc., New York, 1950.

ever, one is concerned with an application that is not covered by these sources and must then rely upon one's own judgment in selecting the factor of safety.   Some of the general considerations in choosing a factor of safety are listed below.[1]

1. The degree of uncertainty of the magnitude and kind of applied load
2. The degree of reliability of the material
3. The extent to which assumptions must be made in the analysis for nominal stress
4. The extent to which localized stress may be developed
5. The kind of environment to which the machine may be subjected
6. The extent to which human life and property may be endangered by failure of the machine
7. The uncertainty as to the exact cause of failure
8. The extent to which properties of the material may be altered during service
9. The extent to which initial or residual stresses may be set up during fabrication
10. The uncertainty as to the appropriateness of using the material properties measured on a test specimen of one size in designing a part of another size
11. The reliability required of the machine
12. The price class of the machine

As a guide for selection of factors of safety, Vidosic has given the following examples of numerical values of factor of safety based upon the yield strength[2] and endurance limit of the material:

1. f.s. = 1.25 to 1.5 for exceptionally reliable materials used under controllable conditions and subjected to loads and stresses that can be determined with certainty.   Used almost invariably where low weight is a particularly important consideration.

2. f.s. = 1.5 to 2 for well-known materials under reasonably constant environmental conditions, subjected to loads and stresses that can be determined readily.

3. f.s. = 2 to 2.5 for average materials operated in ordinary environments and subjected to loads and stresses that can be determined.

4. f.s. = 3 to 4 for untried materials used under average conditions of environment, load, and stress.

5. f.s. = 3 to 4 should also be used with better-known materials that are to be employed in uncertain environments or subjected to uncertain stresses.

---

[1] Except for a few additions and deletions, due to J. P. Vidosic, Design Stress Factors, *J. Eng. Educ.*, vol. 55, pp. 653–658, May, 1948.

[2] It should be recalled that it was recommended in Sec. 6-4 that the factor of safety for brittle materials under static loading be twice that selected for ductile materials under the same conditions.

CHAPTER 7

CASTINGS AND WELDMENTS

The fundamental approach to the determination of the dimensions required to ensure a satisfactory service life of a part was discussed in Chap. 6. At that time it was pointed out that manufacturing processes often introduce additional considerations that may in themselves be the limiting factors for some dimensions. Actually, it is not possible to separate the method of manufacturing from design, whether it is the use of powder metallurgy to manufacture a part without any subsequent machining operations or the use of an ordinary drill press to drill a hole.

An extensive discussion of the relation of the many different processes to design is beyond the scope of this book, and this chapter will be limited to a brief discussion of the casting and welding processes. These two processes have been selected because of their widespread use in making a single part as well as in mass production and because many of the principles apply equally well to other processes.

There are many good references[1] on manufacturing processes as such, but the books by Bolz[2] are of particular interest to the machine designer because of the emphasis placed on matching the design and manufacturing process to achieve the most economical production.

7-1. Casting Processes. The casting process consists of introducing a molten metal into the cavity of a mold, where it is permitted to solidify and retain the shape of the mold cavity. Because it can produce a definite shape without extensive machining, casting is used in manufacturing such widely different products as rugged machine bases made from cast iron with only modest mechanical properties to gas-turbine blades made from steels with high alloy content for use under high stresses at elevated temperatures. In the latter case, the properties of the material

---

[1] M. L. Begeman, "Manufacturing Processes," 4th ed., John Wiley & Sons, Inc., New York, 1957.

G. S. Schaller, "Engineering Manufacturing Methods," 2d ed., McGraw-Hill Book Company, Inc., New York, 1959.

[2] R. W. Bolz, "Production Processes—Their Influence on Design," vols. I and II, Penton Publishing Company, Cleveland, 1949 and 1952.

and the complicated shape make machining or grinding difficult, and the high strength at elevated temperatures makes forging impractical.

There are numerous variations, but the casting processes are most easily classified on the basis of molding method in one of the following groups:

1. Sand
2. Centrifugal
3. Permanent mold
4. Plaster mold
5. Investment
6. Die

A discussion of the advantages and disadvantages of each method would be too lengthy for inclusion here, but, in general, the factors influencing the choice of casting method are (1) the material to be cast, (2) the size of the part, (3) the shape of the part, (4) the rate of production, (5) the surface required, and (6) the dimensional accuracy required. Since the use of sand molds may be considered to be the basic method and since this method is often used for making single parts as well as in mass production, it is worthwhile to consider the interrelation of design and the sand-casting process.

7-2. Casting Design Considerations.[1]   Many of the problems arising in the manufacture of castings are the results of the following:

1. The necessity of using patterns
2. The properties of the material being cast
3. The properties of the mold

In most cases, the patternmaker and foundryman can devise means by which practically any desired shape can be cast. However, a part properly designed to be made specifically as a casting will be much cheaper to cast and will result in a sounder, stronger finished part.

7-3. Effect of Pattern-design Requirements.   Most molds are made in two parts that are joined on a plane surface, i.e., with a straight *parting line*, as shown in Fig. 7-1. After the molding sand has been rammed into place in each half of the mold, the pattern is withdrawn. The important consideration at this point is that the pattern must be designed so that it can be removed without disturbing the impression. If all surfaces are designed so that the width of the pattern decreases with distance away from the parting line, pattern removal will be simple. This condition is satisfied by allowing *draft*, or a slope of $\frac{1}{16}$ in./ft or 1° on each side. If the function of the part requires extra metal for drilling and tapping to permit the attachment of another part or if projections or *bosses* are

---

[1] Two good sources of detailed information about casting processes, materials, and design requirements are "Cast Metals Handbook," 4th ed., The American Foundrymen's Association, Des Plaines, Ill., 1957, and C. W. Briggs (ed.), "Steel Castings Handbook," 3d ed., Steel Founders' Society of America, Cleveland, 1960.

desired, as in Fig. 7-2a, to permit economical machining of a seat for a bolt head or nut, the impression must be made in the sand by a *loose piece* that can be removed after the main section of the pattern has been withdrawn. These additional operations can often be eliminated by a slight modification of the part, as indicated in Fig. 7-2b.

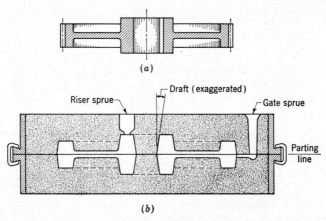

(a)

(b)

FIG. 7-1. Casting design. (a) Machine drawing of finished gear; (b) section through mold showing cavity left by pattern for casting of gear blank.

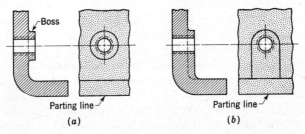

(a)   (b)

FIG. 7-2. Design b permits casting boss without using a pattern with loose piece.

The internal shape of the casting and holes through the wall are provided for by using separately made sand shapes called *cores*. Cores should be avoided wherever possible because of the added expense and the complications raised by the necessity of supporting the cores in the mold and in providing for removal of the core after solidification of the casting. The automobile engine block is one of the most complicated castings made in mass production. The extensive use of coring of the cooling-water passages through the block introduces an additional problem in that the core holes through the walls must be plugged so that water will not leak out.   This is usually accomplished by boring out the hole and pressing in a drawn sheet-metal core plug, as shown in Fig. 7-3.

7-4. Effect of Material Properties.   The material properties having the most influence on casting design, other than strength properties as considered in Chap. 6, are (1) the fluidity of the molten metal, (2) the liquid shrinkage of the metal prior to solidification, (3) the contraction during solidification, and (4) the solid contraction when cooling to room temperature.

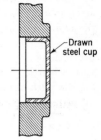

Fig. 7-3. Use of a drawn-steel cup to plug a core hole.

The fluidity of the molten metal is important in that it largely determines the minimum wall thickness that can be cast.   Nominal values of minimum wall thicknesses are given in Table 7-1 for some of the materials commonly used in sand casting.   These values cannot be used in every situation; e.g., the relatively low fluidity of steel requires the use of thicker sections in larger castings, and some high-strength cast irons require greater thickness to prevent the formation of hard spots or white cast iron at corners, edges, and other places subjected to rapid cooling.

TABLE 7-1. MINIMUM THICKNESSES FOR SAND CASTINGS

Metal	Thickness, in.
Aluminum alloys	$\frac{3}{16}$
Copper alloys	$\frac{3}{32}$
Cast irons: gray, malleable, and white	$\frac{1}{8}$
Magnesium alloys	$\frac{5}{32}$
Steel	$\frac{1}{4}$

The solidification of the molten metal normally occurs first in the thinnest section and last in the thickest section.   Since the shrinkage during solidification is appreciable, about 3 per cent for steel, a sound casting, i.e., one without *shrinkage cavities*, is produced only when the solidification occurs progressively from the extremities to the point of supply of molten metal.   In many cases the foundryman must use extra risers to supply additional metal in critical regions, or he must use chills to cool a heavy section more rapidly than normally.   It should be apparent that the designer plays a critical role in determining whether the foundryman's problem will be simple or difficult—whether the castings will be sound and cheap or possibly defective as well as expensive.   Figure 7-4 shows a number of ways by which the designer can minimize the problem due to shrinkage.

Another important consideration in casting design is that the metal not only shrinks or contracts during solidification but it will contract even more, about 7 per cent for steel, in cooling down to room temperature after solidifying. This contraction during cooling is important because the resistance offered by the mold may result in cracking of the casting. The problem is particularly critical when a long part with flanges or other projections perpendicular to its axis of maximum shrinkage is being cast with a brittle material such as cast iron. The large contraction and low hot strength of most materials often combine to cause cracks, or *hot tears*, in the region that is the last to reach room temperature, particularly if the

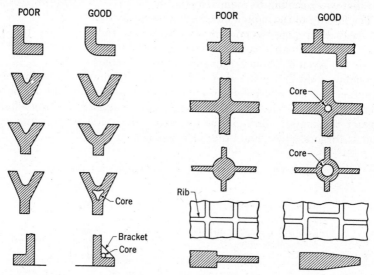

FIG. 7-4. Methods for minimizing shrinkage cavities in unfed junctions. (*Courtesy Steel Founders' Society of America.*)

hot spot is also a region of stress concentration because of the presence of a notch or other abrupt change of section.

An almost impossible casting situation can often be avoided by welding two or more castings together. This is known as *cast-weld* construction.

Progressive solidification of the metal also means that room temperature will be reached progressively. If the part makes up a closed system, the differential solidification and contraction will result in the part being subjected to internal or *residual* stresses after removal from the mold. The residual stress will add to the service or load stress and thus should be considered in relation to the strength of the part, as discussed in Chap. 6. Residual stresses cannot be neglected if subsequent machining is to be done with any degree of accuracy, because the removal of a layer of stressed material releases a force equal to the product of the stress and the section area of the removed metal. In many cases, the metal is removed from a plane surface located away from the neutral axis of the part, and a moment is also released. The force in itself will cause a uniform change in length, while the moment causes the part to bend. Consequently, when the machining is completed, the surface will be curved rather than plane, as desired. In the past, iron castings were stored for about six months, often in the shop yard where temperature variations were pronounced, but present practice for all metals is to use some form of heat-treatment to give stress relief.

In many instances, the substitution of relatively flexible curved members that will give during contraction, in place of straight members that are axially rigid, will eliminate breaking during cooling and will leave only a negligible residual stress. For example, the wheel with straight spokes in Fig. 7-5a and c is difficult to cast because the spokes and outer

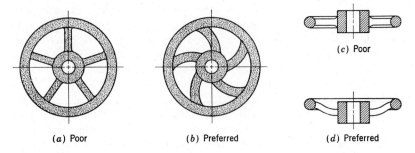

(a) Poor   (b) Preferred   (c) Poor
                                          (d) Preferred

FIG. 7-5. The use of flexible curved members to minimize the effect of contraction during cooling.

rim will cool before the hub. Then, when the hub cools, the spokes are stressed in tension because the rim resists the inward motion required if the spokes are to remain connected to the contracting hub. The juncture between the spokes and hubs is likely to fail during the cooling, before there is a chance to use another heat-treatment or annealing process to release the residual stresses. If the spokes are curved in one or both planes, as shown in Fig. 7-5b and d, the wheel can be cast without difficulty.

One of the problems arising from the contraction of the casting during cooling is that the loss in dimensions must be compensated for by making the pattern larger than the part. This is largely a problem for the foundry, but the designer must consider the effect of contraction on the accuracy with which desired dimensions can be maintained. The actual shrinkage during cooling will depend upon the shape and size of the casting, as well as on the material being cast. While exact values cannot be specified, Table 7-2 contains average values of *shrinkage allowance* for some common materials. For example, if the desired diameter of a steel cylinder is 24 in. the diameter of the pattern would be 24³⁄₁₆ in. Pattern-makers do not usually convert each dimension mathematically but use a special rule, known as a *shrinkage rule*, that is so constructed that the conversion is automatic; i.e., for steel, a 24-in. shrinkage rule would actu-

TABLE 7-2. AVERAGE PATTERN SHRINKAGE ALLOWANCES

Metal	Shrinkage, in./ft
Aluminum alloys	³⁄₃₂
Copper alloys	³⁄₁₆
Cast irons, gray	⅛
Malleable	⅛
White	³⁄₁₆
Magnesium alloys	⅛
Steel	³⁄₃₂

ally be 24³⁄₁₆ in. long. As with other casting variations, the accuracy with which specified dimensions can be cast is dependent upon the size, shape, and material of the casting. However, in general, the dimensional variation is related to the degree of contraction, and a tolerance of one-half the shrinkage allowance is adequate. Thus, on the basis of Table 7-2, the tolerance for an aluminum alloy casting would be ±³⁄₆₄ in./ft of length.

Many castings are used without finishing beyond cutting off the gate and riser sprues and other unwanted projections and cleaning the casting to remove the sand and scale. However, most castings require one or more machining operations to give closer control over dimensions. In this case, sufficient extra metal, or *finish allowance*, must be cast, as shown in Fig. 7-6, to permit machining to required dimensions. The finish

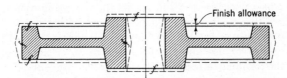
Finish allowance

FIG. 7-6. Cast gear blank from Fig. 7-1 after machining in preparation for cutting keyway and gear teeth.

allowance will depend upon the tolerance on dimensions, the character-istics of the cast surfaces of the material, and the location of the surface in the mold; that is, an extra allowance must be made for finishing surfaces which are in the top part of the mold, because impurities, such as loose sand, will rise to the top during pouring. Long surfaces require greater machining allowances to ensure cleaning up minor surface variations and to correct for warpage. In most cases, satisfactory surfaces will be obtained if a minimum of $\frac{1}{8}$ in. is removed from gray-iron and steel castings and $\frac{1}{16}$ in. from malleable-iron and nonferrous castings.

The finish allowance must also be considered in relation to the maintenance of uniform wall thickness during casting. It often happens that the finished-part drawing makes it appear that casting will be a simple matter when actually, after the finish allowances are added, the change in wall thickness will be so severe that casting will be quite difficult.

7-5. Welding. Welding is the permanent joining of two or more parts by the fusion of a common molten region. It is extensively used in fabrication (often as a replacement for bolted or riveted joints), in the repair of broken parts, in the building up of worn surfaces, and in the repair of defective castings. While there are many variations, the two broad classifications of welding processes are (1) those that use heat alone and (2) those that use a combination of heat and pressure in effecting the permanent joining of two or more pieces.

7-6. Welding Processes Requiring Heat Alone. When heat alone is used, the parts to be joined are held in position while molten metal is supplied to the joint. The molten metal may come from the parts themselves, i.e., *parent metal*, but in most cases additional or *filler metal* (normally with the composition of the parent metal) is supplied in molten form. The joint surfaces become plastic or even molten, because of the heat from the molten filler metal or other source. Thus, when the molten region solidifies or fuses, the several parts become a single piece. The descriptive term *fusion welding* is often given to this process. The three major subdivisions of thermit welding, gas welding, and electric-arc welding are based on the method of heat generation.

In *thermit welding* a mixture of iron oxide and aluminum, called *thermit*, is ignited, and the iron oxide is reduced to molten iron. The molten iron is poured into a mold constructed around the joint and fuses with the parts to be welded.

The process is often used when iron and steel parts are too large to be welded conveniently by other means. A major advantage of thermit welding is that all parts of the weld section are molten at the same time and the weld cools almost uniformly. This results in a minimum problem with residual stresses, and an additional stress-relief process is generally not required. It is particularly useful in joining together parts of large castings or forgings that are too complicated to manufacture in one piece and in the repair of parts that are more than 3 in. thick or difficult to stress-relieve, such as heavy machinery frames, locomotive frames, and ship rudder posts.

Gas welding and *electric-arc welding* are quite similar in application, and either method may be used as a manual or automatic operation. Gas welding commonly uses an oxyacetylene or a hydrogen flame to provide the heat for fusion. The main differences between gas and arc welding are that in gas welding the heating rate is slower and the operator has more control over both the heating and cooling rates. The result is that gas welding can be used on thinner materials and the joint can be preheated or postheated to decrease the severity of cooling strains and the resulting residual stresses.

The welding process most useful in building machines, and consequently of major interest to the machine designer, is the *metal-arc* process in which the filler metal is supplied by a metal welding electrode. The arc between the electrode and the part being welded transfers molten metal with sufficient force so that vertical or even overhead joints may be readily welded. This process has such utility that it will be considered in some detail in subsequent sections.

Other types of electric-arc welding that are widely used in production are the carbon, atomic-hydrogen, and inert-gas arc-welding processes.

7-7. Welding Processes Requiring Heat and Pressure. The first welds were made by hammering together parts which had been raised to the proper temperature in a furnace or forge. This process is known as *forge welding* and is little used at the present time. However, the principle of applying heat and pressure, either sequentially or simultaneously, is widely utilized in the processes known as *spot, seam, projection, upset,* and *flash welding.*

7-8. Metallic-arc-welding General Design Considerations. The major advantage of arc welding as a manufacturing process is the flexibility afforded by being able to combine a number of simple shapes into a complicated structure. The components may be individual castings or forgings but are more often pieces cut from rolled shapes such as plates, bars, strips, I beams, channels, angles, Ts, pipe, tubing, etc. The examples given in Fig. 7-7 and the gear case in Fig. 16-24 show only a few of the many possibilities, and the reader should refer to the literature[1] for additional suggestions and information.

Welds are usually classified as *bead, fillet, plug, slot,* or *groove* welds on the basis of the shape of the edges being joined together or the appearance of the deposited weld metal. The term *butt* weld is a general term describing groove welds when the parts and the weld metal lie in a plane.

[1] "Procedure Handbook of Arc Welding Design and Practice," 11th ed., Lincoln Electric Company, Cleveland, 1959.

Robert S. Green (ed.), "Design for Welding," James F. Lincoln Arc Welding Foundation, Cleveland, 1948.

H. D. Churchill and J. B. Austin, "Weld Design," Prentice-Hall, Inc., Englewood Cliffs, N.J., 1949.

R. E. Kinkead, "Practical Design for Arc Welding," vols. I, II, and III, The Hobart Brothers Company, Troy, Ohio, 1943, 1944, and 1945.

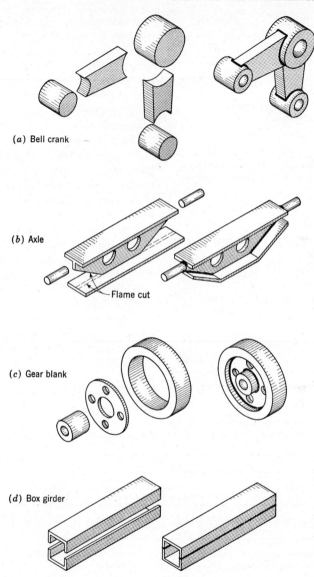

(*a*) Bell crank

(*b*) Axle

Flame cut

(*c*) Gear blank

(*d*) Box girder

Fig. 7-7. Weldments fabricated from standard rolled shapes.

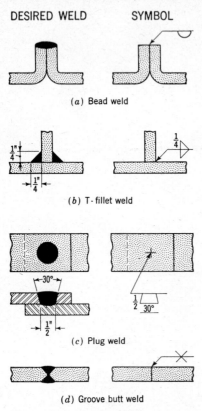

DESIRED WELD SYMBOL

(a) Bead weld

(b) T-fillet weld

(c) Plug weld

(d) Groove butt weld

FIG. 7-8. Types of welds and their designation by use of weld symbols.

The examples in Fig. 7-8 are designated in accordance with the standards of the American Welding Society (AWS),[1] as given in Figs. 7-9 to 7-11. These symbols provide a convenient means of specifying detailed instructions in a minimum space.

The contraction of the weld fusion zone, when cooling to room temperature, introduces problems similar to those due to contraction of a casting. However, since the weld fusion zone is relatively small and the temperature gradients are much higher, the problems of warpage and residual stress are more severe in the case of the weld.

While it is not practical to prescribe a single procedure that would minimize warpage in every case, some of the principles involved are general and should be considered. Probably the most used principle is that of balancing forces by use of intermittent welds (as shown in Fig. 7-12a); back-step welds (Fig. 7-12b); or alternate passes on opposite sides of the neutral axis (Fig. 7-12c).

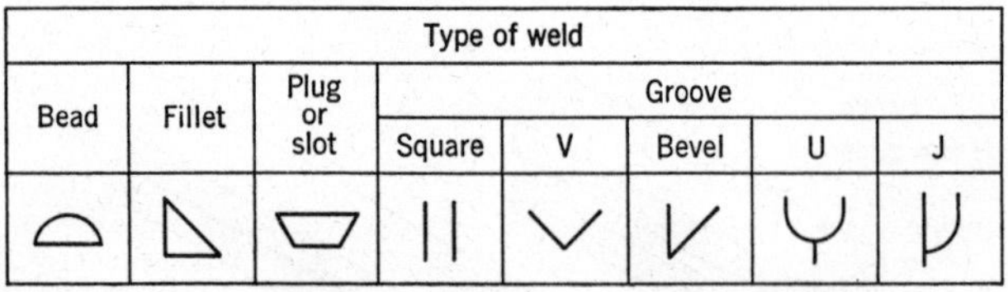

Type of weld							
Bead	Fillet	Plug or slot	Groove				
			Square	V	Bevel	U	J
⌒	△	▽	‖	⌄	⟍	⋃	⋃

FIG. 7-9. Basic arc- and gas-weld symbols. (*From Standard Welding Symbols, A 2.0–47, The American Welding Society, New York, 1947.*)

With sufficient experience or after experimentation, the parts may often be located initially out of position so that they will be in correct alignment after the weld cools. The use of hammer peening of the weld during cooling acts to expand the weld, thus counteracting the normal contraction, and helps to relieve the residual stresses.

[1] Standard Welding Symbols, A 2.0–47, American Welding Society, New York, 1947.

Weld all around	Field weld	Contour	
		Flush	Convex
○	●	—	⌒

FIG. 7-10. Supplementary symbols. (*From Standard Welding Symbols, A 2.0–47, The American Welding Society, New York, 1947.*)

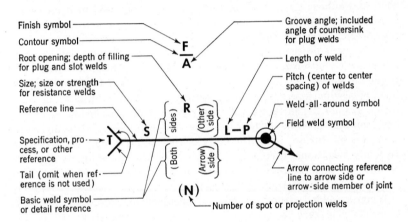

FIG. 7-11. Standard location of elements of a welding symbol. (*From Standard Welding Symbols, A 2.0–47, The American Welding Society, New York, 1947.*)

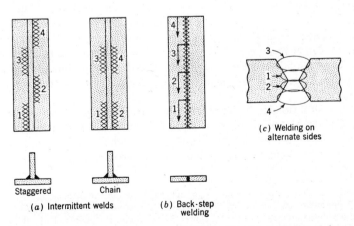

FIG. 7-12. Methods for minimizing welding distortion due to shrinkage forces.

Rugged jigs and fixtures that clamp the parts securely in place prevent most warpage by restraining the parts. In this case, the weld must deform plastically during cooling.

The problem of residual stresses cannot be completely separated from that of warpage, because residual stresses are locked in whenever there is any restraint of the contracting weld. This restraint is always present, within the material itself even when no restraint is offered by the connected members of a structure. *Preheating*, i.e., heating of the region adjacent to the joint before welding, and *postheating*, i.e., heating of the weld and adjacent region after welding, are widely used to decrease the temperature gradient across the weld area and consequently to decrease the severity of residual stress due to the differential contraction.

Subsequent processes that stress the weld and adjacent metal beyond the yield strength often permit the relaxation of the residual stresses to a negligible value. If the cold-working process is that of stretching, the relief of residual stress will occur practically uniformly through the thickness. If only the surface is plastically deformed, as by being upset during peening, the residual tensile stress may be changed to compressive stress on the surface while the body of the material remains in tension. In many situations, this is not only satisfactory but is highly desirable because, as was discussed in Chap. 6, fatigue failures are usually related to a varying tensile stress on the surface.

When a welded structure must be machined to close tolerances, it is generally necessary to use a thermal stress-relief treatment[1] to prevent additional warpage with the removal of each layer of stressed material.

7-9. Strength of Arc Welds. The design of welded structures, such as buildings, ships, and bridges, and of welded pressure vessels must usually correspond to one of the codes or standards established by the AWS, ASME, AISC, government agencies, insurance companies, and other groups.[2] These codes generally include specifications covering the selection of materials, allowable or design stresses for different types of joints, preparation or forming of parts before welding, destructive testing of weld specimens, nondestructive radiographic testing of the finished welds by means of X rays or gamma rays, and stress-relieving procedures where applicable.

Most situations in which the designer may want to utilize welding are not covered by the codes, and the problem becomes one of relating weld dimensions and material properties to the load that must be carried by the weld. In general, the principles discussed in Chap. 6 are applied to

[1] For example, for steel, the structure is heated slowly to about 1200°F, held for at least 1 hr/in. of thickness, and then allowed to cool in the furnace.

[2] A complete list cannot be included here, but some of the codes most used by machine designers are the ASME Power Boiler Code, the ASME Unfired Pressure Vessel Code, and the API-ASME Code for Unfired Pressure Vessels.

the design of welded joints. However, several factors peculiar to welds that require additional study are the impossibility of calculating the exact stress distribution across a fillet weld, the stress concentration due to the shape and surface roughness of the weld bead, the possible decrease in mechanical properties due to the weld's absorbing oxygen and nitrogen from the air while cooling from the molten state to room temperature, and the problem of securing perfect welds every time. The most practical method for considering these and other variables is that, based upon experience, presented by Jennings.[1] The following discussion is based mainly upon Jennings' work.

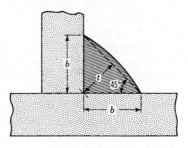

(a) Fillet weld

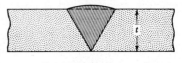

(b) Butt weld

Fig. 7-13. Throat dimensions t of welds.

Since welds usually fail at the minimum section, it is logical to base design calculations upon the minimum or *throat* area of the weld, even though the exact stress distribution may not be known. Figure 7-13 shows the location of the throat area for butt and fillet welds. It can be seen that, since the fillet-weld size is specified by the leg dimension (Fig. 7-8b), the throat thickness t, for the usual case with equal legs, may be calculated as

$$t = b \sin 45° = 0.707b \qquad (7\text{-}1)$$

The design procedure is to determine the load acting on the individual welds and then to use the appropriate allowable stress from Table 7-3

TABLE 7-3. WELD DESIGN STRESSES, PSI*

Type of weld and stress	Unshielded welding		Shielded welding	
	Static loads	Reversed loads	Static loads	Reversed loads
Butt welds:				
Tension..........	13,000	5,000	16,000	8,000
Compression......	15,000	5,000	18,000	8,000
Shear............	8,000	3,000	10,000	5,000
Fillet welds........	11,300	3,000	14,000	5,000

* From C. H. Jennings, Welding Design, *Trans. ASME*, vol. 58, p. 497, 1936.

in calculating the dimensions of the weld. These allowable stresses are for the usual case in which low-carbon (0.15 per cent C or less) welding

[1] C. H. Jennings, Welding Design, *Trans. ASME*, vol. 58, p. 497, 1936.

rods or electrodes are used on low-carbon steels (0.30 per cent C or less). Unless extra care, such as preheating the weld area, is taken, welds of steels with higher carbon content are likely to be brittle, because of the effective quenching action due to the rapid transfer of heat from the weld into the relatively large mass of cool base metal.

Unshielded welds are seldom used and are not recommended. The values have been included only to illustrate the marked decrease in mechanical properties occurring when the molten steel of the weld is permitted to absorb oxygen and nitrogen from the air. Shielding is accomplished in different ways for different methods of welding. Most manual metallic-arc welds are made with coated electrodes. The coating provides two important functions: (1) It gives off a large quantity of inert gas that completely shields the arc from the surrounding atmosphere and (2) it forms a slag that floats on the molten metal, thereby shielding the weld during cooling.

The allowable stresses in Table 7-3 include factors of safety of about 3, and the reversed-load stresses include the effects of surface roughness common to all welds. However, the additional stress concentrations due to the type and shape of the weld must be included separately whenever the welds are subject to cyclic or dynamic loads. There are few data available, but the factors presented in Table 7-4 should be used where applicable. These values may also be used in many cases to indicate the most efficient type of joint.

TABLE 7-4. STRESS-CONCENTRATION FACTORS FOR WELDS*

Type of weld	Stress-concentration factor
Reinforced butt weld	1.2
Toe of transverse fillet weld	1.5
End of parallel fillet weld	2.7
T butt joint with sharp corners	2.0

* From C. H. Jennings, Welding Design, *Trans. ASME*, vol. 58, p. 497, 1936.

When the weld must carry a repeated load, Eq. (6-23) will be the design equation, and in terms of the allowable stresses in Table 7-3, it becomes

$$A = \frac{P_{av} + (s_a/s_{ar})K_f P_r}{s_a} \tag{7-2}$$

where s_a = allowable stress for static loads and s_{ar} = allowable stress for reversed loads.

Figure 7-14 utilizes the concept of stress-flow lines, as discussed in Chap. 6, to compare the stress-concentration characteristics of a reinforced butt weld and a flush butt weld.   The reinforced butt weld may be slightly stronger under static loads, but the disruption of the straight-line flow of stress due to the excess metal introduces stress concentration ($K_f = 1.2$, from Table 7-4) and results in a lower strength under cyclic or dynamic loads.

Figure 7-15 shows three ways in which a strap or strip of steel may be welded to a larger plate.   The high value of the stress-concentration factor at the ends of the parallel fillet welds makes the joint in Fig. 7-15a undesirable when fatigue or impact loads are present. The single transverse fillet weld in Fig. 7-15b is also undesirable, even under static load, because the load acts to bend the plates so that the unwelded side of the joint tends to open up and the weld is subjected to stresses that are high and difficult to estimate.   The best design is the double transverse fillet-weld lap joint in Fig. 7-15c.   The situation is still not

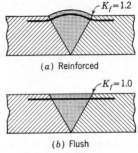

(a) Reinforced

(b) Flush

FIG. 7-14. Stress-flow lines and stress concentration in butt welds.

perfect because the toes or ends of the weld are stress raisers and the load will not be divided equally between the two welds unless the parts of the two pieces between the welds have identical elastic characteristics.   One of the common cases in which the load is equally divided between the two fillet welds is that where the two pieces are of the same material and have the same thickness and width.

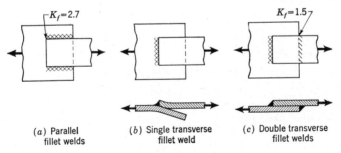

(a) Parallel fillet welds   (b) Single transverse fillet weld   (c) Double transverse fillet welds

FIG. 7-15. Lap joints.

Steel is considerably more expensive, on a per-pound basis, when it is in the form of deposited weld metal than when in its more usual forms. Consequently, it is desirable to use rolled, bent, or otherwise preformed shapes wherever possible.   It can be seen that the throat area of fillet welds increases linearly and the volume or weight increases as the square of the weld size. Therefore, if the designer has a choice, it is usually more economical to increase the strength of a weld by increasing its length rather than by increasing its size. At the same time, experience has shown that, even though the strength requirements can be met with a very small fillet weld, it is advisable to relate minimum weld size to the plate thickness, as indicated in Table 7-5.

TABLE 7-5. MINIMUM FILLET-WELD SIZES*

Plate thickness, in.	Minimum weld size, in.
$\frac{1}{8}$– $\frac{3}{16}$	$\frac{1}{8}$
$\frac{1}{4}$– $\frac{5}{16}$	$\frac{3}{16}$
$\frac{3}{8}$– $\frac{5}{8}$	$\frac{1}{4}$
$\frac{3}{4}$–1	$\frac{3}{8}$
$1\frac{1}{8}$–$1\frac{3}{8}$	$\frac{1}{2}$
$1\frac{1}{2}$ or more	$\frac{3}{4}$

* From C. H. Jennings, Welding Design, *Trans. ASME*, vol. 58, p. 497, 1936.

Whenever the length of weld is short, it is desirable to include about $\frac{1}{2}$ in. extra to allow for the decreased size due to starting and stopping of the weld bead.

Example 7-1. Determine the length of welds required to transmit a reversed load of 10,000 lb between $\frac{1}{2}$-in.-thick plates when the plates are to be joined by use of (a) parallel fillet welds, as in Fig. 7-15a, and (b) transverse fillet welds, as in Fig. 7-15c.

Solution. (a) *Parallel fillet welds*

$$P = \frac{s_a}{K_f} A$$

and $P = 10,000$ lb
    $s_a = 5,000$ psi   (Table 7-3)
    $K_f = 2.7$ (Table 7-4)
    $A = 2 \times 0.707 \times \frac{1}{2} \times L$ (two welds)
       $= 0.707L$ in.^{2}

Thus $10,000 = \dfrac{5,000}{2.7} 0.707L$

from which $L = 7.64$ or $7\frac{3}{4}$ in.

  (b) *Transverse fillet welds*

$$P = \frac{s_a}{K_f} A$$

and $P = 10,000$ lb
    $s_a = 5,000$ psi   (Table 7-3)
    $K_f = 1.5$ (Table 7-4)
    $A = 2 \times 0.707 \times \frac{1}{2} \times L$
       $= 0.707L$ in.^{2}

Thus $10,000 = \dfrac{5,000}{1.5} 0.707L$

from which $L = 4.23$ or $4\frac{1}{4}$ in.

Example 7-2. The bracket in Fig. 7-16 is designed to carry a dead weight of 3,000 lb. What size of fillet welds are required at the top and bottom of the bracket?

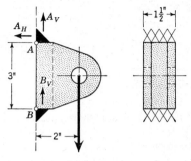

Fig. 7-16

Solution. Assume that forces act through points A and B and that the weld at A carries both a vertical force of $\frac{1}{2}W$ and a horizontal force A_H, while the weld at B carries only a vertical force of $\frac{1}{2}W$.

Vertical forces

$$\Sigma F_V = 0$$
$$A_V = B_V = \frac{1}{2}W = \frac{1}{2} \times 3,000 = 1,500 \text{ lb}$$

Horizontal force at A

$$\Sigma M_B = 0$$
$$A_H \times 3 - W \times 2 = 0$$
$$A_H = \frac{2W}{3} = \frac{2 \times 3,000}{3} = 2,000 \text{ lb}$$

Resultant force at A

$$P_A = \sqrt{(A_V)^2 + (A_H)^2} = \sqrt{(1,500)^2 + (2,000)^2} = 2,500 \text{ lb}$$

Weld at A

$$P = s_a A$$

and $P = 2,500$ lb
$s_a = 14,000$ psi   (shielded welding, static load, Table 7-3)
$A = 0.707bL$
$\quad = 0.707b \times 1$ (weld is short; $\frac{1}{2}$ in. deducted for starting and stopping)

Thus $\qquad\qquad 2,500 = 14,000 \times 0.707b$
from which $\qquad\qquad b = 0.252$ or $\frac{1}{4}$ in.

Weld at B

$$P = s_a A$$

and $P = 1,500$ lb and s_a and A are the same as above.

Thus $\qquad\qquad 1,500 = 14,000 \times 0.707b$
from which $\qquad\qquad b = 0.152$ or $\frac{3}{16}$ in.

NOTE: A somewhat more refined solution would take into account the fact that the reactions act through the welds, not through points A and B. A simple calculation will show that this would add more complications than justified, particularly in view of the necessity of making other assumptions and approximations and since the solution in the example is conservative, i.e., on the safe side.

CHAPTER 8

MECHANICAL FASTENERS

This chapter will be devoted to a survey of several groups of machine elements that, while relatively small in size, are of great importance in machine design. In general, fasteners such as screws, rivets, pins, snap rings, and keys are used to locate one member with respect to another, to transmit force from one member to another, and often to provide for easy assembly and disassembly.

8-1. Screw Fastenings. The mechanical fastening most often used in machines is the screw fastening.[1] The numerous variations in appearance and construction are generally related to the specialized functions for which the fastenings are designed, but the operation of every screw fastening depends upon the principle of the inclined plane. The inclined plane is in the form of a helix, and a torque applied to the screw results in a force or motion in the axial direction. The dimensions of the helix and thread section are important factors in relation to interchangeability, strength, operational characteristics, and ease of manufacturing. Figure 8-1 shows a typical member with right-hand threads. The hand of a thread may most easily be determined by noting the direction of rotation of a point following the helix and moving axially away from the observer. If the rotation is clockwise (cw), the thread is right-hand, and if the rotation is counterclockwise (ccw), the thread is left-hand.

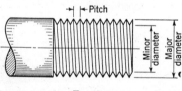

Fig. 8-1

The *major diameter* is the largest diameter and determines the nominal size; i.e., a ½-in. bolt has a major diameter of ½ in. The *minor diameter* is the smallest diameter of the thread. The term *root diameter* is often used for the minor diameter of an externally threaded member. The *pitch p* is the distance from one point on a screw thread to the corresponding point on the next thread and

$$p = \frac{1}{\text{number of threads per inch}} \quad \text{in.} \qquad (8\text{-}1)$$

---

[1] Screws used primarily to transmit motion and power will be discussed in Chap. 10.

The *lead* is the distance a threaded part moves axially, with respect to a fixed mating part, in one complete rotation. Lead is relatively unimportant when the major function is to fasten one member to another, but it is very important when power is being transmitted by use of translation screws or worm-gear units. On a single-thread screw, the pitch and lead are identical; on a double-thread screw, the lead is twice the pitch, etc.

8-2. Standard Threads. Probably the simplest form of screw thread is the V thread in Fig. 8-2a. The sharp crest and root make this thread undesirable in that it is difficult to produce, particularly in brittle mate-

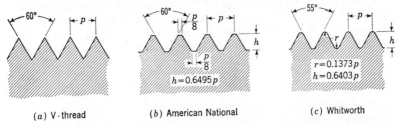

| (a) V-thread | (b) American National | (c) Whitworth |

Fig. 8-2. Screw-thread forms.

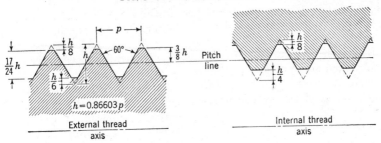

Fig. 8-3. Unified standard screw threads.

rials, and it is subject to severe stress concentration ($K \cong 3.0$). A more practical type, which has long been used in this country, is the *American National* thread in Fig. 8-2b. The flat crest and root are a considerable improvement over the V thread, but the sharp corners at the root of the thread are still bad for fatigue loads ($K \cong 2.5$).

The *Whitworth* thread, developed in Great Britain, has rounded crests and roots. This thread, shown in Fig. 8-2c, is comparatively good from a fatigue viewpoint, with $K \cong 2.0$.

The complications arising during World War II as a result of the non-interchangeability between American and English screw threads finally led to the adoption in 1948 by the United States, Canada, and Great Britain of a new standard, known as the *Unified* thread, shown in Fig. 8-3. A comparison of the thread forms will show that the Whitworth and the Unified threads are not compatible, whereas, except for a slight degree of interference at the minor diameter of an internal American National and an external Unified thread, the American National and the Unified threads are compatible and may be used interchangeably. The use of a rounded root is optional for unified external threads. After sufficient time has elapsed for replacement of tools and gages, only the Unified thread will be standard. However, for some time to come, many of the American National threads will be accepted as standard.

Table 8-1 is a compilation of some of the more useful dimensions and areas for the Unified and the American National threads. It should be noted that numbers designate the sizes of threads with diameters less than $\frac{1}{4}$ in. The American Society of Mechanical Engineers, in conjunction with the American Standards Association, has published a bulletin[1] which gives complete information about sizes, pitches, tolerances, and allowances.

The coarse-thread series (UNC and NC) is recommended for general use where conditions are favorable to quick and easy assembly.

The fine-thread series (UNF and NF) is recommended for use in automotive and aircraft work and where special conditions require a fine thread. The American National fine-thread series (NF) is based on the SAE regular screw-thread series and is still commonly known as the SAE thread.

The extra-fine-thread series (NEF) is intended for special uses where thin-walled material is to be threaded and where a maximum number of threads per inch is desired. Additional threads, such as the constant-pitch series, selected combinations, and fine threads for thin-wall tubing, are provided for use in special applications where none of the other standard thread series is suitable. The 8-thread (8 UN), 12-thread (12 UN), and 16-thread (16 UN) are the preferred constant-pitch series.

A discussion of the tolerances and allowances made necessary by the impossibility of manufacturing every part exactly the same is beyond the scope of this book. It will be sufficient for our purposes to point out that the Unified thread is divided into classes 1A, 2A, and 3A for external threads and classes 1B, 2B, and 3B for internal threads, with the tightness of fit increasing with the numbers. Classes 2A and 2B are usually specified. Before the adoption of the Unified thread standard, American National threads were divided into classes 1 through 4 for both external and internal threads. During the present transition period, only classes 2 and 3 are recognized as standard for American National threads, with class 2 being satisfactory for most purposes.

In order to simplify the designation of size, thread series, and class of

[1] Unified Screw Threads, ASA B1.1-1960, American Society of Mechanical Engineers, New York, 1960.

TABLE 8-1. UNIFIED AND AMERICAN SCREW THREADS*

Size	Major diameter, in.	Coarse-thread series			Fine-thread series			Extra-fine-thread series		
		Threads per inch	Minor† diameter, in.	Stress‡ area, in.²	Threads per inch	Minor† diameter, in.	Stress‡ area, in.²	Threads per inch	Minor† diameter, in.	Stress‡ area, in.²
0§	0.0600	....	......	......	80	0.0447	0.0018			
1§	0.0730	64	0.0538	0.0026	72	0.0560	0.0028			
2§	0.0860	56	0.0641	0.0037	64	0.0668	0.0039			
3§	0.0990	48	0.0734	0.0049	56	0.0771	0.0052			
4§	0.1120	40	0.0813	0.0060	48	0.0864	0.0066			
5§	0.1250	40	0.0943	0.0080	44	0.0971	0.0083			
6§	0.1380	32	0.0997	0.0091	40	0.1073	0.0102			
8§	0.1640	32	0.1257	0.0139	36	0.1299	0.0147			
10§	0.1900	24	0.1389	0.0175	32	0.1517	0.0200			
12§	0.2160	24	0.1649	0.0240	28	0.1722	0.0258	32	0.1777	0.0270
¼	0.2500	20	0.1887	0.0318	28	0.2062	0.0364	32	0.2117	0.0379
⁵⁄₁₆	0.3125	18	0.2443	0.0524	24	0.2614	0.0580	32	0.2742	0.0625
⅜	0.3750	16	0.2983	0.0775	24	0.3239	0.0878	32	0.3367	0.0932
⁷⁄₁₆	0.4375	14	0.3499	0.1063	20	0.3762	0.1187	28	0.3937	0.1274
½§	0.5000	13	0.4056	0.1419						
½¶	0.5000	12	0.3978	0.1374	20	0.4387	0.1599	28	0.4562	0.170
⁹⁄₁₆	0.5625	12	0.4603	0.1821	18	0.4943	0.203	24	0.5114	0.214
⅝	0.6250	11	0.5135	0.226	18	0.5568	0.256	24	0.5739	0.268
¾	0.7500	10	0.6273	0.334	16	0.6733	0.373	20	0.6887	0.386
⅞	0.8750	9	0.7387	0.462	14	0.7874	0.509	20	0.8137	0.536
1	1.0000	8	0.8466	0.606	12	0.8978	0.663	20	0.9387	0.711
1⅛	1.1250	7	0.9497	0.763	12	1.0228	0.856	18	1.0568	0.901
1¼	1.2500	7	1.0747	0.969	12	1.1478	1.073	18	1.1818	1.123
1⅜	1.3750	6	1.1705	1.155	12	1.2728	1.315	18	1.3068	1.370
1½	1.5000	6	1.2955	1.405	12	1.3978	1.581	18	1.4318	1.64
1¾	1.7500	5	1.5046	1.90						
2	2.0000	4½	1.7274	2.50						
2¼	2.2500	4½	1.9774	3.25						
2½	2.5000	4	2.1933	4.00						
2¾	2.7500	4	2.4433	4.93						
3	3.0000	4	2.6933	5.97						

* Extracted from American Standard Unified Threads, ASA B1.1-1960, with permission of the publisher, The American Society of Mechanical Engineers, New York.

† Basic minor diameter of external thread $= D - 1.22687/n$, where D is the major diameter and n is the number of threads per inch.

‡ Stress area is the assumed area of an externally threaded part, for computing tensile strength.

§ American National only.

¶ Unified only.

fit, an abbreviated notation is used on drawings, tools, and specifications; e.g., a Unified coarse-thread series, ¾-in.-diameter external thread, 10 threads per inch, class 2A fit, is designated as 3/4-10 UNC-2A. Right-hand threads are assumed unless specified as left-hand (LH).

A particular fastening situation is that of joining a pipe to a tank or another pipe so that a fluid can be transferred under pressure without leakage. The Unified and the American National threads previously discussed do not provide for effective sealing. As shown in Fig. 8-4, there

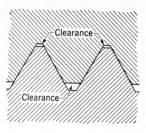

FIG. 8-4

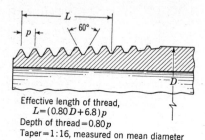

Effective length of thread,
$L = (0.80D + 6.8)p$
Depth of thread $= 0.80p$
Taper $= 1:16$, measured on mean diameter

FIG. 8-5. American Standard taper pipe threads.

are helical passages in the clearance space between crests and roots of the threads. The thread developed to give a fluid-tight seal is the *American Standard taper pipe thread* in Fig. 8-5. When it is sufficiently tightened, the interference of the tapered pipe threads gives a metal-to-metal seal. Table 8-2 lists some of the important dimensions of standard pipe

TABLE 8-2. AMERICAN STANDARD TAPER PIPE THREADS*

Nominal size, in.	Outside diameter (D), in.	Threads per inch	Nominal size, in.	Outside, diameter (D), in.	Threads per inch
⅛	0.405	27	2	2.375	11½
¼	0.540	18	2½	2.875	8
⅜	0.675	18	3	3.500	8
½	0.840	14	3½	4.000	8
¾	1.050	14	4	4.500	8
1	1.315	11½	5	5.563	8
1¼	1.660	11½	6	6.625	8
1½	1.900	11½	8	8.625	8

* Extracted from American Standard Pipe Threads (Except Dryseal), ASA B2.1-1960, with permission of the publisher, The American Society of Mechanical Engineers, New York.

and pipe threads. It is important to note that the nominal size and outside diameter may be considerably different.

8-3. Commercial Screw Fasteners. The American Standards Association has set up additional standards for screw fasteners so that the designer needs only to consider the requirements of his design in relation to stock commercial items and does not need to consider from whom the fasteners will be purchased. Almost all the handbooks contain detailed information for standard and patented fasteners.

Bolts are the basic screw fasteners and are generally used with through holes and the appropriate nuts. The bolt head may be either square or hexagon and may be unfinished, semifinished, or finished. Semifinished and finished bolts and nuts have a machined contact surface, or washer face, under the head of the bolt and on one side of the nut. The term *finished* does not imply that all surfaces are machined but refers only to a higher quality of manufacture and closer tolerances than does the term

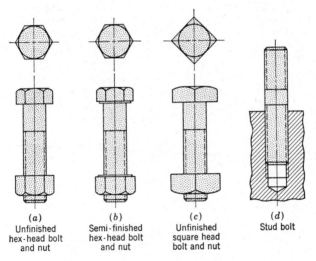

(a)
Unfinished
hex-head bolt
and nut

(b)
Semi-finished
hex-head bolt
and nut

(c)
Unfinished
square head
bolt and nut

(d)
Stud bolt

Fig. 8-6. Bolts.

semifinished. Examples of bolts and nuts are shown in Fig. 8-6, and the standard dimensions of regular bolt heads and nuts are given in Table 8-3. The threads in Fig. 8-6 correspond to the ASA simplified thread symbols.[1] As a general rule, a through hole is to be preferred to a tapped hole. A tapped hole is expensive, and the threads are easily damaged if even occasional removal is necessary. Threads are most easily damaged in soft or brittle materials such as aluminum and cast iron. If frequent removal of a bolt from a tapped hole is required, it is recommended that an insert be installed in the hole. Inserts are also extensively used in renovating tapped holes in which the threads have become worn.

[1] American Standard Drafting Manual, Screw Threads, ASA Y14.6-1957, American Society of Mechanical Engineers, New York.

TABLE 8-3. DIMENSIONS OF AMERICAN STANDARD REGULAR HEXAGON† BOLT HEADS AND NUTS*

Nominal size‡	Bolt heads			Nuts		
	Width across flats, maximum	Height, maximum		Width across flats, maximum	Height, maximum	
		Unfinished	Semifinished and finished		Unfinished	Semifinished and finished
¼	0.4375	0.188	0.163	0.4375	§	0.226§
⁵⁄₁₆	0.5000	0.235	0.211	0.5000		0.273§
⅜	0.5625	0.268	0.243	0.5625		0.337§
⁷⁄₁₆	0.6250	0.316	0.291	0.6875		0.385§
½	0.7500	0.364	0.323	0.7500		0.448§
⁹⁄₁₆	0.8125	.....	0.371	0.8750		0.496§
⅝	0.9375	0.444	0.403	0.9375		0.559§
¾	1.1250	0.524	0.483	1.1250	0.680	0.665
⅞	1.3125	0.604	0.563	1.3125	0.792	0.776
1	1.5000	0.700	0.627	1.5000	0.903	0.887
1⅛	1.6875	0.780	0.718	1.6875	1.030	0.999
1¼	1.8750	0.876	0.813	1.8750	1.126	1.094
1⅜	2.0625	0.940	0.878	2.0625	1.237	1.206
1½	2.2500	1.036	0.974	2.2500	1.348	1.317
1⅝	......	.....	.....	2.4375	.....	1.429¶
1¾	2.6250	1.196	1.134	2.6250	.....	1.540
1⅞	......	.....	.....	2.8125	.....	1.651¶
2	3.0000	1.388	1.263	3.0000	.....	1.763
2¼	3.3750	1.548	1.423	3.3750	.....	1.970
2½	3.7500	1.708	1.583	3.7500	.....	2.193
2¾	4.1250	1.869	1.744	4.1250	.....	2.415
3	4.5000	2.060	1.935	4.5000	.....	2.638

* Extracted from American Standard Square and Hexagon Bolts and Nuts, ASA B18.2-1960, with permission of the publisher, The American Society of Mechanical Engineers, New York.

† Square-head bolt and nut dimensions are, except for a few sizes, the same as for the unfinished hexagon bolts and nuts.

‡ All dimensions in inches.

§ ¼–⅝ unfinished and semifinished nuts meeting present standard specifications are not recommended for new designs.

¶ Not listed for finished nuts.

The *Heli-Coil* inserts in Fig. 8-7 consist of diamond-shaped stainless-steel or phosphor-bronze wire coiled so that it resembles a spring.   Figure 8-7*a* shows an insert being screwed into a threaded hole and an insert in position and ready for use.   A special tap is required for threading the hole to accommodate the insert.   Figure 8-7*b* shows the *Heli-Coil Screw-lock* insert which utilizes the locking effect of the friction between a deformed coil and the screw to prevent loosening.

Stud bolts (Fig. 8-6*d*) are designed to be used where a through hole is impossible.   The stud should be held fast in the tapped hole by the tightness of the threads, not by jamming at the imperfect threads or at the bottom of the hole.

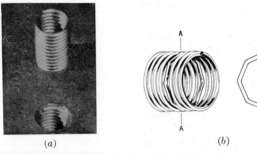

(*a*)   (*b*)

FIG. 8-7. Heli-Coil screw-thread inserts.   (*a*) Regular insert; (*b*) screw-lock insert. (*Courtesy Heli-Coil Corporation.*)

Cap screws and *machine* screws are similar in appearance, though there is a slight technical difference in that a cap screw has a chamfered end, whereas a machine screw has a square end.   The various slotted heads in Fig. 8-8 are generally used on machine screws, whereas the hexagon and socket heads are used on cap screws.   The term *cap screw* is often used to indicate a small-size bolt, even though the standards actually specify somewhat different dimensions for the heads.

Shoulder screws (Fig. 8-9) are very useful standard screw fasteners. These screws are usually made from heat-treated alloy steels with accurately ground shoulder diameters and make ideal pivot pins for linkages, rollers, etc.

Set screws are short screws used to locate parts on shafts and to transmit relatively small forces between a shaft and a hub, e.g., small motors and knobs on rheostats.   Some set screws are made with square heads, but the majority have either slotted or socket heads.   Some of the styles of points are shown in Fig. 8-10.

Carriage bolts (Fig. 8-11) are characterized by the square, ribbed, or finned neck and are used to fasten wood parts to other parts made of either wood or metal.

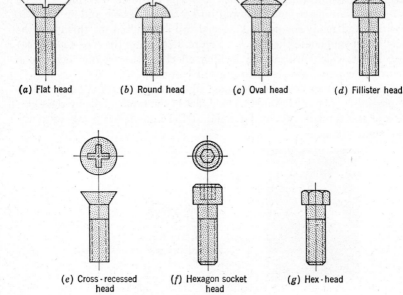

(a) Flat head   (b) Round head   (c) Oval head   (d) Fillister head

(e) Cross - recessed head   (f) Hexagon socket head   (g) Hex - head

FIG. 8-8. Cap screws and machine screws.

FIG. 8-9. Unbrako shoulder screw.   (*Courtesy Standard Pressed Steel Company.*)

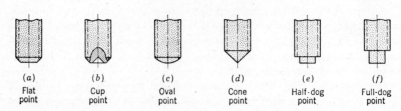

(a) Flat point   (b) Cup point   (c) Oval point   (d) Cone point   (e) Half-dog point   (f) Full-dog point

FIG. 8-10. Set-screw points.

Figure 8-12 shows some of the numerous specialized types of bolts and screws which have limited use in the design of a machine.

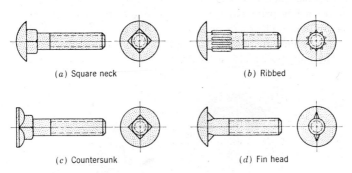

(a) Square neck   (b) Ribbed

(c) Countersunk   (d) Fin head

FIG. 8-11. Carriage bolts.

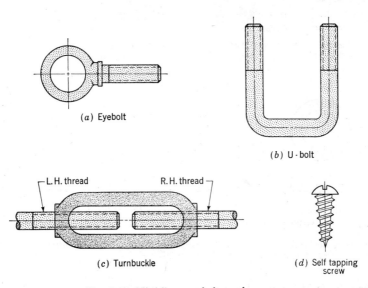

(a) Eyebolt

(b) U-bolt

(c) Turnbuckle   (d) Self tapping screw

FIG. 8-12. Miscellaneous bolts and screws.

8-4. Locking Devices. When a bolted joint is subjected to cyclic loading, there is a tendency for the nut to back off and let the joint loosen. This is particularly critical in the presence of vibration. Special threads, such as the Dardelet and the Lok-thread, have been developed to prevent this loosening by providing tapered portions of the threads which, in the final tightening, greatly increase the friction between the mating threads by a wedging action. These threads are seldom used, because of the need

where d is the bolt diameter in inches.   This equation gives forces which are larger than required in most cases, but it does show why small bolts are easily broken during assembly.   To eliminate some of the need for extra care on the part of the mechanic, it is desirable to specify ⅜-in. or larger bolts whenever practical.

8-6. Strength of Screw Fasteners under Fatigue Loading.   The transfer of a force from the bolt head through the shank and threads to the nut is complicated by the necessity of reversing the direction of the stress-flow lines and by the presence of stress concentrations, as illustrated in Fig. 8-14a.   Fatigue failures of bolts are distributed approximately with 15

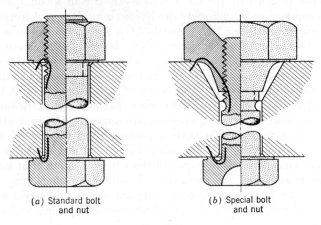

(*a*) Standard bolt and nut   (*b*) Special bolt and nut

FIG. 8-14

per cent under the head, 20 per cent at the end of the thread, and 65 per cent in the thread at the nut face.[1]   There is little information available in terms of stress-concentration factors which can be used in design (Sec. 8-2), though values as high as 3.85[*] have been found for heat-treated SAE 2320 steel bolts.   If the situation is such that the approach to bolted joints discussed later in this section cannot be utilized and it is necessary for the bolt to carry the full cyclic load, the ideas presented in Fig. 8-14b may be used to alleviate the trouble.   It should be noted that the basic approach in each case is to distribute the load more evenly or to reduce the stress concentration by making a more gradual change from one section to another.

[1] R. E. Peterson, "Stress Concentration Design Factors," p. 111, John Wiley & Sons, Inc., New York, 1953, after L. Martinaglie, Schraubenverbindungen, *Schweiz. Bauztg.*, vol. 110, p. 107, 1942.

[*] H. F. Moore and P. E. Henwood, Strength of Screw Threads under Repeated Tension, *Univ. Illinois Eng. Expt. Sta. Bull.* 264, 1934.

Probably the most useful approach to using bolted joints where the parts being bolted together are subjected to fatigue loading is based upon the relative flexibilities of the bolt and the bolted materials.   Figure 8-15 shows two methods of sealing the joint between a cylinder and cylinder head.   In one case (Fig. 8-15a), a highly compressible (soft) gasket is placed between the surfaces, and in the other (Fig. 8-15b), the O-ring

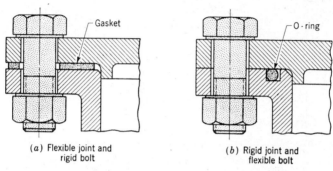

(a) Flexible joint and rigid bolt

(b) Rigid joint and flexible bolt

FIG. 8-15. Joints with different relative flexibilities of bolt and bolted materials.

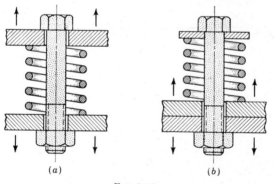

(a)   (b)

FIG. 8-16

seal permits direct contact between the relatively rigid surfaces.   Figure 8-16 shows analogous cases in which the spring in Fig. 8-16a represents the gasket in Fig. 8-15a and the spring plus the bolt in Fig. 8-16b represents the bolt in Fig. 8-15b.   It can be seen that in Fig. 8-16a the bolt must carry the initial load compressing the spring plus whatever external load is applied to the top plate.   A similar analysis of Fig. 8-16b will show that the bolt will be subjected to only the initial constant spring load until the external load applied to the top plate exceeds the initial spring force and the plates start to separate.   In other words, until the plates separate, there is no additional elongation of the bolt and therefore no additional load.   In reality, both the bolt and the joint have some flexibility, and the actual case for a joint with a relatively flexible bolt, such as in Fig. 8-15b, would be similar to that illustrated in Fig. 8-17.   Here the elongation of the bolt, δ_{ib}, is ten times the compression of the joint,

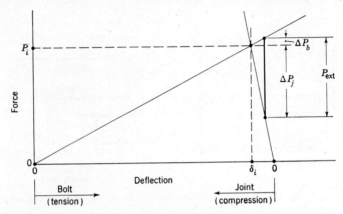

FIG. 8-17. Load-deflection curves for a joint-to-bolt stiffness ratio of 10:1.

δ_{ij}, under the initial load P_i.   Defining the *spring rate* as

$$k = \frac{P_i}{\delta_i} \qquad (8\text{-}3)$$

we find for the bolt and the joint, respectively,

$$k_b = \frac{P_i}{\delta_{ib}} \qquad (8\text{-}4)$$

and

$$k_j = \frac{P_i}{\delta_{ij}} \qquad (8\text{-}5)$$

When an external load P_{ext} is applied, the bolt elongation increases and the joint compression decreases.   By similar triangles, it can be shown that

$$\frac{\Delta P_b}{P_{ext}} = \frac{\delta_{ij}}{\delta_{ib} + \delta_{ij}} \qquad (8\text{-}6)$$

Substituting for δ_{ib} and δ_{ij} from Eqs. (8-4) and (8-5), respectively, and simplifying, we find that

$$\Delta P_b = \frac{k_b}{k_b + k_j} P_{ext} \qquad (8\text{-}7)$$

In terms of repeated stress, for use in Soderberg's equation in Sec. 6-6, we can write for the bolt

$$P_{av} = P_i + \frac{k_b}{2(k_b + k_j)} P_{ext} \qquad (8\text{-}8)$$

and

$$P_r = \frac{k_b}{2(k_b + k_j)} P_{ext} \qquad (8\text{-}9)$$

For the case in Fig. 8-17 where the spring rate of the joint is ten times that of the bolt, we find that $P_{av} = P_i + P_{ext}/22$ and $P_i = P_{ext}/22$ and the effect of the external load is almost negligible. This principle is applied in many situations, e.g., connecting rods of engines, cylinder heads of engines, hydraulic equipment, etc. In these cases, the required values of wrenching torque are usually specified by the manufacturer.

Several methods by which the flexibility of bolts may be increased without a decrease in static strength are shown in Fig. 8-18. The spring

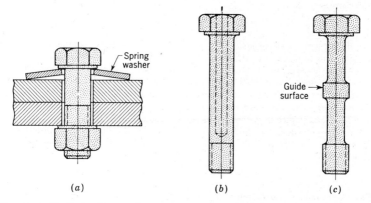

FIG. 8-18. Methods of increasing flexibility of a bolt relative to bolted materials.

washer a is particularly useful when a short bolt must be used. The reduced-diameter bolt c is preferred over the drilled bolt b because of the additional reduction in stress concentration under the head and at the end of the threads.

The methods illustrated in Fig. 8-18b and c are also important when bolts must carry impact loads. In this situation, the important factor is the ability of the bolt to absorb energy elastically. It can be shown that the energy absorbed during elastic deformation is a function of the square of the stress in the material and the volume of material under stress. Therefore, the maximum energy absorption will be reached when the areas of all cross sections are the same and the entire bolt is stressed to the same limiting value. The root section will most likely be the area upon which the design of the rest of the bolt will be based. Also, it is often practical to increase the volume of material under stress by making a design change to permit use of a longer bolt.

Example 8-1. In the design of a new-size reciprocating air compressor, it is proposed that eight UNC bolts be used to fasten the head to the cylinder. The cylinder-bore diameter will be 3.000 in., and the maximum working pressure will be 500 psi. The bolts are to be made from cold-drawn 1020 steel bar stock, and the threads will be machined.

We are asked to determine the bolt size required for a factor of safety of 3 under the following conditions:

(a) A compressible gasket seal will be used between the head and the cylinder. The gasket is made from a fiber sheet, and preliminary calculations have indicated that an initial compression load of about 14,000 lb is required on the gasket for a leakproof seal.[1]

(b) An O-ring seal is used with metal-to-metal contact between the head and the cylinder, and a torque wrench is used to screw up the nut to give the desired initial tension load.

Solution. The factor common to both conditions is the load per bolt due to the air pressure:

$$P = \frac{pA}{N} = \frac{500 \times \pi(3)^2/4}{8} = 442 \text{ lb}$$

(a) *With compressible gasket.* The bolt is relatively rigid in comparison with the joint, and the total load on the bolt will be considered to be the sum of the initial load and the external load due to the gas pressure. Since the unknown effects of surface finish and dimensional tolerances of the threads and the nut washer face preclude an exact analysis, either analytical or experimental, of the relationship of the initial load to the wrenching torque, it will be necessary to specify a value of torque great enough to ensure that the gasket compression is sufficient to seal properly. A conservative procedure will be to design the bolt on the basis of an initial tension 30 per cent greater than that actually required. Therefore, the initial tightening load per bolt is

$$P_i = \frac{1.3 \times 14,000}{8} = 2,275 \text{ lb}$$

The external load varies repeatedly from zero to 442 lb. Therefore, the design must be based on Eq. (6-23), which is

$$A = \frac{P_{av} + (s_y/s_e)K_f P_r}{s_y/\text{f.s.}} \quad \text{in.}^2$$

and A = stress area for bolt

$P_{av} = (P_{max} + P_{min})/2 = (2,717 + 2,275)/2 = 2,496 \text{ lb}$
$P_r = (P_{max} - P_{min})/2 = (2,717 - 2,275)/2 = 221 \text{ lb}$
$s_y = 66,000 \text{ psi, from Appendix D}$
$s_e = 31,000 \text{ psi, from Fig. 6-3 for } s_u = 78,000 \text{ and machined finish}$
$K_f = 2.5$, considering UNC and American National threads as similar in relation to stress concentration
f.s. = 3.0

Thus,

$$A = \frac{2,496 + (66,000/31,000) \times 2.5 \times 221}{66,000/3} = 0.1667 \text{ in.}^2$$

The smallest UNC bolt with an area equal to or greater than 0.1667 in.² is a 9/16-12

---

[1] See D. B. Rossheim and A. R. C. Markl, Gasket-loading Constants, *Mech. Eng.*, vol. 65, pp. 647–648, 1943.

UNC bolt with a stress area of 0.1821 in.^{2} (Table 8-1).   Therefore, we shall specify eight 9/16-12 UNC bolts and the torque necessary to give an initial tension load of 2,275 lb.   It should be noted that, since the initial tension may be expected to vary appreciably from the design value of 2,275 lb, the actual factor of safety will also vary from the design value of 3.   However, the design value was chosen large enough to compensate for such variations in conditions (see Sec. 6-18).

(b) *With O-ring seal, metal-to-metal contact, and initial tension.*   In this case the bolt is relatively flexible, and the total load on the bolt, until the external air-pressure load exceeds the initial tension load, will be considered to be the initial tension load. The bolt will be loaded statically, and Eq. (6-23) becomes

$$A = \frac{\text{f.s.}}{s_y} P \quad \text{in.}^2$$

However, we cannot simply substitute values from above, because the relationship of f.s. to P is not as clear as in the previous solution.   The initial tension load must, in effect, be three times the air-pressure load, on the basis of a rigid joint remaining closed, or the factor of safety of 3 will not exist.   Since the value of torque required to give this initial load cannot be calculated exactly, it is necessary to specify a value of torque great enough to ensure that the initial tension is at least three times the gas-pressure load and then to provide an additional margin of safety beyond this to ensure that the bolt does not yield during tightening.   In this case, a conservative design procedure will be to specify the torque required as 30 per cent greater than that calculated for the desired initial tension load and to select the screw size so that the 30 per cent greater stress will not exceed 75 per cent of the yield strength.   Therefore,

$$A = \frac{3 \times 1.3 \times 442}{0.75 \times 66,000} = 0.0347 \text{ in.}^2$$

and the minimum bolt size is 5/16-18 UNC, with a stress area of 0.0524 in.^{2}.   The initial tension should be $3 \times 1.3 \times 442 = 1,724$ lb.

At this point it appears that using a rigid joint instead of a flexible gasket will permit using $\frac{5}{16}$- rather than $\frac{9}{16}$-in. bolts.   However, this conclusion is not yet justified; there is still a major factor to consider.   The bolt will be stressed both in tension and in shear during the tightening operation.   This is a problem in combined stress, and Eq. (6-33) must be used to ensure that the bolt has a factor of safety greater than 1 when being tightened.   Actually, the factor of safety during tightening should be greater than 1, possibly 1.5 or so, to provide for removal after corrosion in service without automatically twisting the heads off.

The torque required for tightening the $\frac{9}{16}$-in. bolts is calculated in Example 10-1 and is found to be about 265 lb-in.   Carrying through the combined stress calculations indicates that for them the factor of safety will be about 2 during tightening.   Similar calculations for the $\frac{5}{16}$-in. bolt show that a torque of about 117 lb-in. will be required to give the initial tension of 1,724 lb.

The appropriate equation (6-33) is

$$\text{f.s.} = \frac{s_{sy}}{s_{s,\text{max}}} = \frac{s_{sy}}{\sqrt{(s/2)^2 + s_s^2}}$$

where $s_{sy} = s_y/2 = 33,000$ psi
$s = (P/A) = (1,724/0.0524) = 33,000$ psi
$s_s = (Tc/J) = (16T/\pi d_r^3) = (16 \times 108.2/\pi \times 0.2443^3) = 37,700$ psi

Thus, $\text{f.s.} = \dfrac{33,000}{\sqrt{(33,000/2)^2 + 37,700^2}} = \dfrac{33,000}{41,200} = 0.80$

and the bolt would fail by twisting before it could be tightened. Several of the possibilities now open to us are: (1) to specify heat-treated bolts with a yield strength of at least $1.5 \times 2 \times 41,200 = 123,600$ psi (allowing f.s. = 1.5 during tightening); (2) to decrease the tightening load, and with it the factor of safety of the bolt, to $(0.80/1.5)1,724 = 919$ lb; or (3) to specify larger bolts.

Neither 1 nor 2 is considered appropriate in this situation, and we shall "estimate" the size of bolt that should be specified. (The term estimate is desirable because the torque calculations will not be covered until Chap. 10.) However, from Eq. (10-18) we can see that, as a first approximation, the torque is a linear function of the pitch diameter of the thread. Then, a comparison of the relative magnitudes of the stress terms in the factor-of-safety calculation indicates that in this case the torque term is much more significant than the tension term. Furthermore, we can assume that, since the initial tension load will still be 1,724 lb, the tensile stress will decrease as $1/d_r{}^2$. For a given torque, the shear stress will decrease as $1/d_r{}^3$, and if, as a further approximation, we assume that the torque will increase linearly with d_r, we may say that both the tensile and torsional stresses—for this particular type of problem—will decrease approximately as $1/d_r{}^2$. Therefore, to decrease the maximum combined shear stress from 41,200 to $33,000/1.5 = 22,000$ psi, the bolt specified should have a root diameter of at least

$$\left(\frac{41,200}{22,000}\right)^{\frac{1}{2}} d_{r,\frac{5}{16}} = 1.368 \times 0.2443 = 0.334 \text{ in.}$$

From Table 8-1, we find that we can use 3/8-32 UNF or 7/16-14 UNC bolts, rather than the 5/16-18 UNC bolts selected on the basis of tension alone.

Closure. The final design values indicate that the reduction in size given by replacing a flexible joint with a rigid one is not so great as might be expected. Actually, this example has been concerned with relatively small bolts, for which the $1/d_r{}^3$ term in the torsional stress calculation becomes significant; and the benefit of using a flexible bolt with a rigid joint increases as the bolt size increases.

It should also be noted that the factor of safety in service will be considerably greater than the design value of 3 when either the $\frac{3}{8}$- or $\frac{7}{16}$-in. bolts are used with the rigid joint.

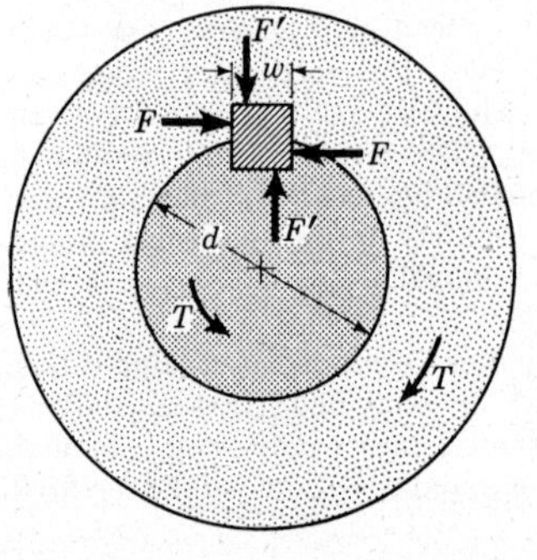

FIG. 8-19. Forces on key due to transmitted torque.

8-7. Keys and Splines. Keys and splines may be used in many forms, but in almost every case the major function is to prevent relative rotation between a shaft and the rotating members mounted on it. Axial motion may or may not be permitted.

Square keys are probably the most common type of key. The forces acting on a square key transmitting torque from the shaft to a hub are shown in Fig. 8-19. Since the points at which the resultant forces F act on the key cannot be exactly determined, the assumption is made that they act at the surface of the shaft. Thus,

$$T = \frac{Fd}{2} \qquad \text{approximately} \qquad (8\text{-}10)$$

The strength based on shearing of the key with length l is

$$T_s = \frac{Fd}{2} = \frac{wls_sd}{2} \tag{8-11}$$

and the strength based on crushing of the key is

$$T_c = \frac{Fd}{2} = \frac{wls_cd}{4} \tag{8-12}$$

It can be seen that, if the key is made out of steel which has the approximate relation of $s_s = \frac{1}{2}s_c$, the two strengths of a square key are the same. Standard square keys are specified as having sides equal to approximately one-fourth of the shaft diameter. Table 8-4 lists some of the

TABLE 8-4. STANDARD DIMENSIONS OF PLAIN PARALLEL KEYS

Shaft diameter, in. (inclusive)	Key dimensions, in.		
	Width	Thickness	
		Square key	Flat key
$\frac{1}{2} - \frac{9}{16}$	$\frac{1}{8}$	$\frac{1}{8}$	$\frac{3}{32}$
$\frac{5}{8} - \frac{7}{8}$	$\frac{3}{16}$	$\frac{3}{16}$	$\frac{1}{8}$
$\frac{15}{16}-1\frac{1}{4}$	$\frac{1}{4}$	$\frac{1}{4}$	$\frac{3}{16}$
$1\frac{5}{16}-1\frac{3}{8}$	$\frac{5}{16}$	$\frac{5}{16}$	$\frac{1}{4}$
$1\frac{7}{16}-1\frac{3}{4}$	$\frac{3}{8}$	$\frac{3}{8}$	$\frac{1}{4}$
$1\frac{13}{16}-2\frac{1}{4}$	$\frac{1}{2}$	$\frac{1}{2}$	$\frac{3}{8}$
$2\frac{5}{16}-2\frac{3}{4}$	$\frac{5}{8}$	$\frac{5}{8}$	$\frac{7}{16}$
$2\frac{7}{8}-3\frac{1}{4}$	$\frac{3}{4}$	$\frac{3}{4}$	$\frac{1}{2}$
$3\frac{3}{8}-3\frac{3}{4}$	$\frac{7}{8}$	$\frac{7}{8}$	$\frac{5}{8}$
$3\frac{7}{8}-4\frac{1}{2}$	1	1	$\frac{3}{4}$
$4\frac{3}{4}-5\frac{1}{2}$	$1\frac{1}{4}$	$1\frac{1}{4}$	$\frac{7}{8}$
$5\frac{3}{4}-6$	$1\frac{1}{2}$	$1\frac{1}{2}$	1

standard dimensions for square and flat plain parallel keys which may be cut to length from cold-finished stock and used without further finishing.

Flat keys are most often used on shafts which, because of a necessity for relatively great stiffness or rigidity, as in machine tools, happen to be much stronger than required on the basis of transmitted torque. The full depth of a square key may not be needed to prevent crushing of the key. Both square and flat keys should fit tightly in the keyways to prevent any tendency to rock or roll.

If, in the case of relatively small equipment, the hub is not press-fitted on the shaft and a set screw is used to keep the hub in its desired axial position, it is recommended that the set screw bear on the key and not on the shaft.   This will keep the key from working out of the end of the keyway and will prevent raising a burr on the shaft which would make it more difficult to remove the hub from the shaft at a future time.

Taper keys have a taper on only the surface projecting into the hub; the taper matches a corresponding taper in the hub keyway.   The wedging action as the key is driven into place forces the hub and shaft together on the side opposite the key and helps to keep the hub in its desired axial position.   The *gib-head taper* key in Fig. 8-20a has been used where it is

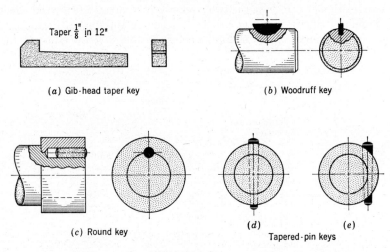

Taper $\frac{1}{8}$" in 12"

(a) Gib-head taper key

(b) Woodruff key

(c) Round key

(d)   (e)
Tapered-pin keys

Fig. 8-20. Keys.

impossible to drive out the key from the small end.   The protruding gib head is not desirable because of the hazard to personnel.

Woodruff keys have had their greatest use in the machine-tool and automotive industries.   The key (Fig. 8-20b), which fits in a keyway machined with a special side-milling cutter, cannot move axially but can adjust itself to the taper, if any, in the hub keyway.

Round keys and *pin* keys are usually tapered but may be cylindrical. The round key in Fig. 8-20c is pressed into a drilled and reamed hole. This type of key is usually considered to be most appropriate for low-power drives, but an adaptation has been used in multiple form to anchor the disk of a steam-turbine rotor to its shaft.   It is difficult to maintain interchangeability between different shafts and hubs because of the problem in controlling the tendency of a drill to drift into the softer material.

The pin keys in Fig. 8-20d and e are usually tapered and are held in place by the friction between the pin and the reamed, tapered hole.   Two commercially available pins made to fit tightly in cylindrical holes are the *grooved* pin in Fig. 8-21a and the *Sel-Lok spring* pin in Fig. 8-21b.   In each case, an elastic deformation of the pin provides the friction required to keep the pin tightly in place.

A common use of pins as keys is as *shear* pins, which are designed to protect equipment by failing when a specified maximum overload is applied.   Shear pins are discussed in more detail in Sec. 11-10.

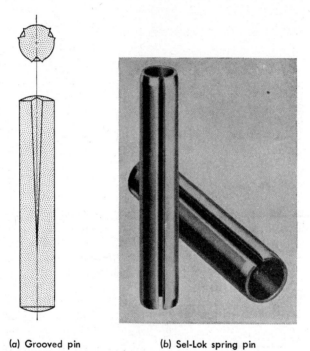

(a) Grooved pin   (b) Sel-Lok spring pin

Fig. 8-21. Special pins.   (b)   (*Courtesy Standard Pressed Steel Company.*)

Dowels, in the form of cylindrical or tapered pins, are widely used to ensure that apparently symmetrical parts, such as housings, cover plates, bearing caps, etc., which must be assembled in one particular orientation, can be assembled only as desired.   Figure 8-22 shows two applications of dowels.   Note the provision for withdrawing the dowel in Fig. 8-22b where a through hole could not be used.

Feather keys are used when there must be relative axial motion with no relative rotation between the hub and the shaft.   One of the numerous variations is shown in Fig. 8-23.   If a torque must be transmitted while there is relative axial motion, it is recommended that two diametrically opposite feather keys be used, because less axial force will be required to overcome friction.

Splines may be thought of as multiple feather keys with the keys machined integral with the shaft; they are particularly useful where radial

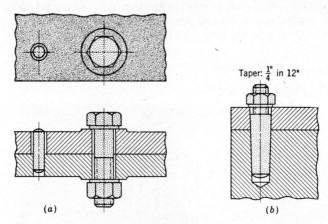

Fig. 8-22. Dowels.

space must be conserved. The Society of Automotive Engineers[1] has standardized the proportions of *parallel-side* splines with 4, 6, 10, and 16 splines. A six-spline fitting is shown in Fig. 8-24a. The splines are generally hobbed on the shaft and broached in the hub.

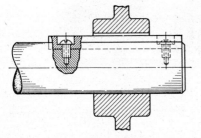

Fig. 8-23. Feather key.

A more recent development is the American Standard *involute* spline.[2] Involute splines (Fig. 8-24b) are closely related to external and internal involute gears and may be machined by the methods usually used in the production of gears. The nomenclature for involute splines corresponds closely with that for gears (Sec. 16-3). The pressure angle is 30°, and the splines are involute in form and based on diametral pitch, with the addendum equal to $0.5/P$ and the circular tooth thickness equal to $1.5708/P$.

[1] "1961 SAE Handbook," pp. 528–530, Society of Automotive Engineers, New York, 1961.

[2] Involute Splines, Serrations and Inspection, ASA B5.15–1960, American Society of Mechanical Engineers, New York, 1960.

The dedendum is 0.5/P or greater, depending on the type of fit specified. The tooth sizes are designated by two numbers such as 3/6, where the first number (3) is the diametral pitch P (number of teeth per inch of

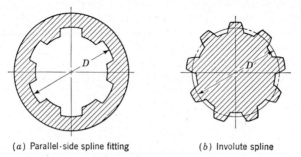

(a) Parallel-side spline fitting   (b) Involute spline

FIG. 8-24. Parallel-side and involute splines.

FIG. 8-25. Ball-bearing spline. (*Courtesy Saginaw Steering Gear Division, General Motors Corporation.*)

pitch diameter) used in determining the tooth dimensions. Standard pitches run from 1/2 to 48/96.

The *ball-bearing* spline in Fig. 8-25 is particularly useful where the operating forces must be as small as possible. Since the balls cannot roll indefinitely in the direction of motion, a return passage is provided so that the balls which have used up their rolling distance are free to return to their initial starting place where they again take up their share of the load. Additional applications of this principle of *recirculation* will be discussed in later sections in relation to ball-bearing translation screws and ball-bearing slides.

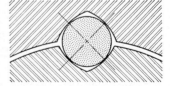

FIG. 8-26. Gothic-arch groove for ball-bearing splines. (*Courtesy Saginaw Steering Gear Division, General Motors Corporation.*)

The coefficient of friction for the ball-bearing spline is about 0.005 as compared with 0.2 for a regular spline with sliding contact.

Figure 8-26 shows the groove construction used by Saginaw. The

Gothic-arch (diverging arcs) design increases ball life[1] and eliminates ball hindrance due to contamination by providing clearance space into which the foreign matter may be pushed.

Involute serrations[2] are similar to involute splines except that the pressure angle is 45° and the serrations are specified for permanent fits, i.e., with no sliding. The serrations may be either parallel or tapered. The standard pitches run from 10/20 to 128/256.

8-8. Snap Rings. Figure 8-27 shows several types of external and internal *snap* or *retaining* rings and two applications in axially positioning

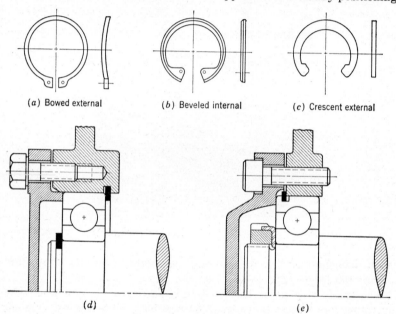

(*a*) Bowed external   (*b*) Beveled internal   (*c*) Crescent external

(*d*)   (*e*)

Fig. 8-27. Snap rings and applications to ball-bearing mountings.

a ball bearing on a shaft and in a housing. Snap rings are made of heat-treated spring steel and are inexpensive, and the required grooves are relatively easy to machine. One major consideration is that the grooves introduce rather severe stress concentration and should not be used in regions of high components of reversed stress. Also, since clearance between the groove and the snap ring is necessary for ease in assembly and the rings cannot be seated under axial load, they should not be used as in Fig. 8-27d unless some end play can be tolerated.

[1] See Chap. 18 for a discussion of the factors affecting the life of rolling-contact bearings.

[2] Involute Splines, Serrations and Inspection, ASA B5.15–1960, American Society of Mechanical Engineers, New York, 1960.

CHAPTER 9

SPRINGS

Machine members that are designed and constructed to give a relatively large elastic deflection under a given load are known as *springs*. The major uses of springs are as follows:

1. To apply force and to control motion. This group includes the majority of springs used in machinery, and typical uses are to provide the operating force in brakes and clutches, to provide the clamping force when a rigid joint with maximum connector flexibility is desired, and to keep cam followers in contact with the cam at high speeds, as in the internal-combustion-engine valve mechanism.

2. To alter the vibratory characteristics of a member, as in the coil-spring coupling (Fig. 11-14) and in the flexible mounting of compressors, motors, etc.

3. To reduce the magnitude of the transmitted force due to impact or shock loading, e.g., automobile springs, elevator buffer springs, in aircraft landing gears, and in artillery recoil systems.

4. To store energy, as in clock motors, circuit breakers, etc.

5. To measure force, as in scales.

Springs are often required to provide several of these functions at the same time. This is particularly true when vibration and shock characteristics are important.

Because of their superior strength at both normal and elevated temperatures and their relative freedom from creep or relaxation under load, metals are used in the manufacture of most springs. However, other resilient materials, e.g., rubber, cork, felt, various liquids, and air, are used where special properties such as high internal damping and low modulus of elasticity are desired. These materials are used mainly in vibration and shock control and will be discussed in more detail in Chap. 19.

This chapter will be devoted to some of the more common metallic springs. The most common type of spring, the helical spring, will be considered in some detail, but the reader is referred to the extensive

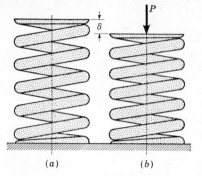

(a)   (b)

FIG. 9-1. Helical compression spring.

literature[1] for detailed information on the many varied types with special features and design problems.

9-1. Helical Compression Springs. A helical compression spring made from round wire is shown unloaded in Fig. 9-1a and carrying a load in Fig. 9-1b. The design of any spring must satisfy two requirements: (1) the spring must carry the service load without the stress exceeding a safe value; (2) the force-deflection characteristics, i.e., the *spring rate* P/δ, must be satisfactory for the given application.

A free-body diagram of a portion of the spring is shown in Fig. 9-2a. The shear-stress distribution across a section of the wire may be determined by superposition of the stress due to the torsional moment T and

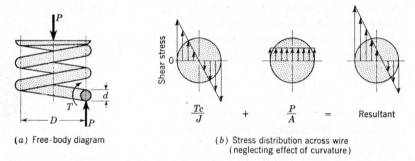

(a) Free-body diagram

(b) Stress distribution across wire
(neglecting effect of curvature)

$$\frac{Tc}{J} \quad + \quad \frac{P}{A} \quad = \quad \text{Resultant}$$

FIG. 9-2. Loading and stress distribution for helical compression springs

that due to the direct force P. If the effect of the curvature of the wire is neglected, the components and their sum will be as shown in Fig. 9-2b.

[1] "Handbook of Mechanical Spring Design," Associated Spring Corporation, Bristol, Conn., 1958.

N. P. Chironis (ed.), "Spring Design and Application," McGraw-Hill Book Company, Inc., New York, 1961.

A. M. Wahl, "Mechanical Springs," Penton Publishing Company, Cleveland, 1944.

C. Carmichael (ed.), "Kent's Mechanical Engineers' Handbook," "Design and Production" volume, 12th ed., pp. 11-02 to 11-40, John Wiley & Sons, Inc., New York, 1950.

L. S. Marks (ed.), revised by T. Baumeister, "Mechanical Engineers' Handbook," 6th ed., pp. 5-72 to 5-80, McGraw-Hill Book Company, Inc., New York, 1958.

The equation for the resultant shear stress is

$$s_s = \frac{Tc}{J} \pm \frac{P}{A} \tag{9-1}$$

After the proper substitutions in terms of mean coil diameter and wire diameter are made, the maximum stress is found to occur at the inner surface and is equal to

$$s_s = \frac{8PD}{\pi d^3} + \frac{4P}{\pi d^2} \quad \text{psi} \tag{9-2}$$

Equation (9-2) may also be written as

$$s_s = \frac{8PD}{\pi d^3} \left(1 + \frac{1}{2C} \right) \quad \text{psi} \tag{9-3}$$

where C is the *spring index* D/d.

Equation (9-3) is useful mainly in pointing out that the effect of the direct shear becomes more pronounced as the spring index becomes smaller.

The effect of the curvature of the coil on the stress distribution is shown in Fig. 9-3. The stress is in-creased at the inside surface in the same manner as previously discussed for curved bars in bending (Sec. 6-8), and the effect becomes more impor-tant as the curvature increases, i.e., for small values of the spring index.

Since both the direct shear stress and the increased torsional shear

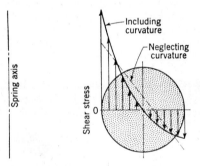

Fig. 9-3. Effect of curvature on stress distribution.

stress on the inner side of the wire may be related to the spring index, it is convenient to use a single factor K that will correct the nominal tor-sional stress for both effects. The situation now becomes analogous to that of stress concentration, and the stress equation becomes

$$s_s = K \frac{8PD}{\pi d^3} = K \frac{8PC}{\pi d^2} \quad \text{psi} \tag{9-4}$$

Wahl[1] first determined an approximate analytical solution for the stress factor K. His solution has been shown to be accurate within 2 per cent for springs with a spring index of 3 or greater.

[1] A. M. Wahl, Stresses in Heavy Closely Coiled Helical Springs, *Trans. ASME*, vol. 51, APM-51-17, 1929.

The Wahl stress factor K is given in Fig. 9-4 as a function of the spring index C. The curve should be used in all stress calculations for helical springs unless otherwise indicated.

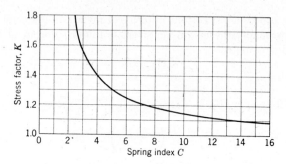

FIG. 9-4. Wahl stress factor for helical springs.

It can be shown that the deflection of a helical spring may be calculated with sufficient accuracy from

$$\delta = \frac{8PD^3N}{Gd^4} = \frac{8PC^3N}{Gd} \quad \text{in.} \qquad (9\text{-}5)$$

where N = number of active coils and G = shear modulus of elasticity, about 11,500,000 psi for spring steel.

Solving Eq. (9-5) for the spring rate gives

$$\frac{P}{\delta} = \frac{Gd^4}{8D^3N} = \frac{Gd}{8C^3N} \quad \text{lb/in.} \qquad (9\text{-}6)$$

In general, the active number of coils N does not equal the total number of coils N_t. This is largely because of the use of different types of end construction, shown in Fig. 9-5. The plain ends in Fig. 9-5a are the most

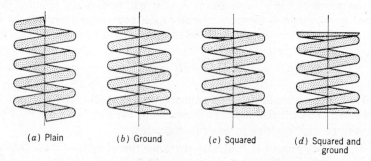

(a) Plain   (b) Ground   (c) Squared   (d) Squared and ground

FIG. 9-5. Types of ends for helical compression springs.

economical to produce.   But, since the spring bows sideways, the loading deviates considerably from the axial loading assumed in the derivation of the stress equations, and the stress on one side of the spring may be considerably higher than calculated.   The closest approach to axial loading is given by squared and ground ends (Fig. 9-5d).   Most highly stressed springs are made with this type of end.

The determination of the number of active turns is further complicated by the fact that the end coils slowly close as the load increases, thus further decreasing the number of active turns.

There is no way to determine an accurate value for the active number of coils except by calculation based upon an experimental determination of the spring rate.   However, for most design purposes, it is satisfactory to assume that the active number of coils is one-half coil less than the total number for plain ends, one coil less for plain ends ground, and one and three-quarters coils less for both types of springs with squared ends.

Because of the closing of the end coils and the errors in spacing of the individual coils, compression springs are seldom designed to close up under the maximum working load.   An allowance, known as the *clash allowance*, of 15 per cent of the maximum working deflection is usually adequate to prevent clashing of the coils during service.

Compression springs that have a free length greater than four times the mean diameter may be unstable, i.e., may buckle when compressed to the maximum working deflection.   Buckling can be prevented by installing the springs over a rod or in a tube, but this is generally undesirable because of the friction between the spring and the guide.

The curves in Fig. 9-6 should be used to check whether a spring will be

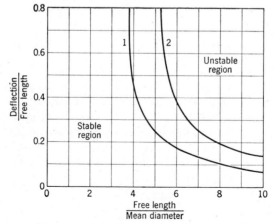

Fig. 9-6. Buckling conditions for helical compression springs.   (*From "Handbook of Mechanical Spring Design," Associated Spring Corporation, Bristol, Conn.*)

stable or will buckle. Buckling is likely if the operating point is above
and to the right of the appropriate curve. Curve 1 applies to a squared
and ground spring with one end on a flat surface and the other end on a
ball. Curve 2 applies to squared and ground springs with both ends
against parallel flat surfaces.

Time is required for a suddenly applied deflection of one end of the
spring to be equally divided among the several coils. The deflection

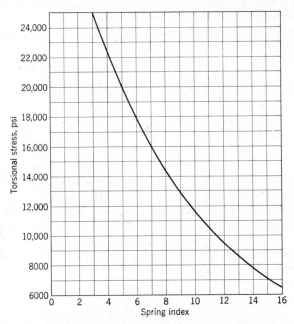

Fig. 9-7. Maximum torsional stress for initial tension in helical tension springs.
(*From "Handbook of Mechanical Spring Design," Associated Spring Corporation,
Bristol, Conn.*)

progresses from coil to coil as a wave progresses on the surface of a lake.
This wave, or surge, is ordinarily quickly damped out, and the spring
becomes motionless. If, however, the frequency of load applications
coincides with the natural frequency of surging of the spring, the motion
will rapidly build up instead of dying out, and the spring will cease to
perform as desired. An equation that may be used in calculating the
lowest critical or resonant frequency of a steel spring is

$$f_c = \frac{826,000d}{ND^2} \quad \text{cpm} \tag{9-7}$$

Since the spring will be in resonance also with frequencies which are
two, three, four, etc., times that given by Eq. (9-7), it may be necessary

to consider the higher harmonics of the disturbance.   This is particularly true for valve springs, where the harmonics of the cam displacement-time curve may become important.

9-2. Helical Tension Springs.   Helical tension or extension springs are similar to helical compression springs, except that the coils may be wound tightly together so that an initial force is required before extension

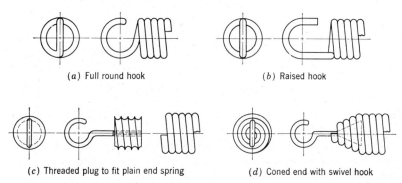

(*a*) Full round hook   (*b*) Raised hook

(*c*) Threaded plug to fit plain end spring   (*d*) Coned end with swivel hook

FIG. 9-8. Helical tension springs.

begins, and the stress concentration introduced by the shape of the ends must be considered in relation to its effect on the working stress.

Calculation of the maximum initial tension load that can be built into a steel spring may be based on the curve in Fig. 9-7, which relates the maximum initial stress to the spring index.   The initial stress does not include the Wahl stress factor, and the maximum possible initial tension load can be calculated from

$$P_{i,\max} = \frac{s_s d^3}{2.55D} \qquad \text{lb} \qquad (9\text{-}8)$$

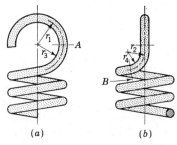

A few of the types of ends used on tension springs are shown in Fig. 9-8. Tension springs often fail at the hook because of stress concentration at the sharp bends.   Section A in Fig. 9-9 may be critical because of the bending moment $PD/2$ and the curvature of the loop.   The maximum bending stress is

(*a*)   (*b*)

FIG. 9-9

$$s_b = K_t \frac{16PD}{\pi d^3} + \frac{4P}{\pi d^2} \qquad \text{psi} \qquad (9\text{-}9)$$

K_t may be found in Fig. 6-20.   A simplified equation which gives stresses that are slightly higher (on the safe side) than Eq. (9-9) is

$$s_b = \frac{16PD}{\pi d^3} \frac{r_1}{r_3} \quad \text{psi} \qquad (9\text{-}10)$$

Section B in Fig. 9-9 may be critical because of the torsional moment $PD/2$. The maximum shear stress is

$$s_s = \frac{8PD}{\pi d^3} \frac{4C_1 - 1}{4C_1 - 4} \quad \text{psi} \qquad (9\text{-}11)$$

where $C_1 = 2r_2/d$.

A simplified equation, similar to that given above for the bending stress, is

$$s_s = \frac{8PD}{\pi d^3} \frac{r_2}{r_4} \quad \text{psi} \qquad (9\text{-}12)$$

It can be seen that, for most efficient use of the material, both r_3 and r_4 should be as nearly equal to $D/2$ as is possible.

9-3. Spring Materials and Design Stresses.   Unless special features such as better material properties at elevated temperatures, better corrosion resistance, and better electrical conductivity are required, springs are usually made from high-carbon (0.70 to 1.00 per cent) steels or from medium-carbon alloy steels such as SAE 6150 and 8660.[1]   These steels give predictable service at temperatures up to 350°F and are sometimes used at temperatures up to 400°F, but the creep or set under load will probably be excessive.   A stainless steel such as an 18-8 type (SAE 30302) or a type 316 (SAE 30316) can be used at temperatures up to 500°F.   For higher temperatures, the materials recommended include high-speed steels for up to 800°F, Inconel-X for up to 1000°F, and a cobalt superalloy NS-25 for up to 1400°F.

Type 18-8 stainless steel, spring brass, nickel silver, phosphor bronze, monel, Inconel, Permanickel, and other metal alloys are used for corrosion-resistant springs.

Beryllium copper is the best spring material with high electrical conductivity, although phosphor bronze and spring brass are extensively used where strength is not critical.

Small sizes of helical springs, i.e., with wire diameters under $\frac{3}{8}$ to $\frac{1}{2}$ in., are usually wound cold and are often given a low-temperature anneal or are heat-treated for the desired properties, depending upon the materials involved and quality of product desired.   The larger-diameter wires or rods cannot be wound cold but must be wound hot and heat-treated after winding.

Springs are normally designed with safety factors of 1.5 or less.   Using relatively low safety factors can be justified on the basis of criteria in

---

[1] For an excellent summary of spring-material properties, see H. C. R. Carlson, Selection and Application of Spring Materials, *Mech. Eng.*, vol. 78, pp. 331–334, 1956.

Sec. 6-18 for selecting a factor of safety. Of particular importance is the fact that most springs operate under well-defined deflections, and the corresponding loads and stresses can be accurately calculated. In most cases, an overload would simply close up a compression spring without a dangerous increase in deflection. The usual tension spring does not have this feature, and consideration should be given to the provision of overload stops or other devices, such as in Fig. 9-13, if damage to the machine or personnel can occur because of a failure of the spring. The material used in springs is carefully controlled in all stages of manufacture, and the thin and uniform wire cross section permits uniform heat-treatment and cold-working of the entire spring.

The fundamental approach for determining the values of stress for a given set of operating conditions should be based upon the discussion in Chap. 6. Since most springs operate between an initial load and a maximum load, the concept of a reversed stress superposed on a steady stress, as considered by the Soderberg equations, is of particular importance. However, the problem of spring design is complicated by the variation in material properties with size of the wire and the fact that many springs are required to operate for only a limited number of cycles.

Consequently, springs are often grouped into classes, according to service conditions, and tables of allowable stresses for different wire sizes and classes of service have been widely used for designing springs.

Light service consists of operation under essentially static conditions and where the maximum load is carried infrequently, i.e., a total of less than 1,000 times. Examples are safety valves, slip couplings, and bolted joints.

Average service includes springs subject to the maximum load 1,000 to 100,000 times during service life. Examples are engine governor springs, automobile suspension springs, and springs used in circuit-breaker mechanisms.

Severe service consists of a large number of cycles, i.e., greater than 1 million, of varying load where the minimum load is one-half or less of the maximum load. The common example of severe service is that of automotive valve springs.

These classifications are not explicit, and consequently any table of stresses based on them must be used with care. However, the design or allowable stresses in Table 9-1 have been used by Westinghouse Electric Corporation for general-purpose springs. The values are conservative and will give satisfactory results in most situations.

Some endurance data are available for springs that correspond closely to the use of the Soderberg equation for repeated loading. The terminology is a little different in that for springs it is more convenient to use the terms *minimum stress, maximum stress,* and *range of stress.* The range of

TABLE 9-1. HELICAL COMPRESSION SPRINGS: ALLOWABLE SHEAR STRESSES,* PSI†

Wire diameter, in.	Light service	Average service	Severe service
Up to 0.085	93,000	75,000	60,000
0.086–0.185	85,000	69,000	55,000
0.186–0.320	74,000	60,000	48,000
0.321–0.530	65,000	52,000	42,000
0.531–0.970	56,000	45,000	36,000
0.971–1.5	50,000	40,000	32,000

* For springs made of good quality steel. Can be used for springs made of music or oil-tempered wire in smaller sizes and for hot-wound springs, heat-treated after forming, in larger sizes. Use 50 per cent of these values for phosphor bronze and 75 per cent of these values for stainless steel.

† From A. M. Wahl, Helical Compression and Tension Springs, *Trans. ASME*, vol. 57, pp. A-35 to A-37, 1935.

stress is the difference between the maximum and minimum values. A typical plot of the allowable values for SAE 6150 steel is given in Fig. 9-10. Data as presented in Fig. 9-10 are very useful when the optimum design of a spring for severe service is desired. A comparison of the data in Fig. 9-10 with that in Table 9-1 points out that a considerably higher maximum stress is permitted when the maximum and minimum stresses are known and used.

The allowable maximum stress under severe service can be materially

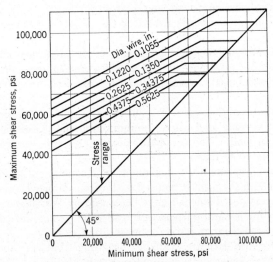

FIG. 9-10. Allowable torsional-stress range for chrome-vanadium steel (SAE 6150) wire. (*From W. M. Griffith, Standards for Spring Design*: II, *Product Eng., vol. 8, no. 4, pp. 140–143, 1937.*)

increased by shot-peening the surface of an otherwise finished spring. The impacts of the balls, or shot, intensively cold-work the surface and, in so doing, increase the mechanical properties of the material and introduce compressive residual stresses on the surface. The indications are that, with proper shot-peening, the short horizontal portions of the maximum stress curves in Fig. 9-10 are extended to the left to the intersection with the ordinate. The beneficial results of shot-peening will be decreased if the spring is subsequently heated above 500°F and will be completely lost at 825°F.

9-4. Helical-spring Design Procedure. The requirements of a spring are that it possess sufficient strength, have the desired load-deflection characteristics, not buckle under load, and not be in resonance with a cyclic disturbance.

After a material and its treatment are specified, the only variables are the mean coil diameter D, the wire diameter d, and the number of coils N. In general, no direct solution is possible, and there will be a number of satisfactory designs. If the space available for the spring is a consideration, there will be fewer possible answers; in fact, there may not be any.

Since the stress in the spring is independent of the number of coils, a logical procedure is to determine D and d on the basis of allowable stress. The first step should be the construction of a load-deflection curve, as illustrated in Fig. 9-11. Such a curve shows all the operating deflections and loads and helps to eliminate errors. The spring is usually designed

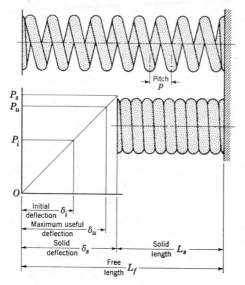

Fig. 9-11. Load-deflection diagram for helical compression springs.

to carry the specified maximum working load at the allowable stress for the appropriate class of service given in Table 9-1. It is usually not necessary to check whether the stress exceeds the elastic limit when the clash allowance is used up (spring closed) because of mishandling, etc.

The Wahl stress factor K should be considered in all stress calculations. However, it is not an independent variable but is dependent upon the spring index C. As shown in Fig. 9-4, the value of K is high for small values of C and rapidly becomes less as C increases. Experience has shown that values of C from 6 to 10 will result in a reasonably balanced design in relation to stability, size, and efficient use of material. If no other limitations need be considered, it is recommended that a spring index of 8 be used. After the spring index has been chosen, K can be taken from the curve in Fig. 9-4, and Eq. (9-4) can be solved for d. The problem here is that the allowable stress also depends on d; consequently, a trial-and-error solution must be used. Since d appears as a squared term if C is specified, the quickest solution will be reached by substituting values of s from Table 9-1 and solving for d. Then a check must be made to see if the value of d falls within the bracket corresponding to the allowable stress. If it does not, a new value of s corresponding to the first trial value of d should be used, and a second value of d should be calculated. With a little experience, the second trial is generally sufficient. Once a usable value of d is calculated, it is necessary to specify a commercially available wire size. Steel wire sizes are available in fractional dimensions and in accordance with Washburn and Moen (W & M) gauge numbers, as listed in Table 9-2. The nonferrous wire sizes are usually given by Brown and Sharpe gage numbers. To avoid confusion, *it is recommended that the wire size always be specified in decimals.*

After a standard wire size has been determined, the number of active coils is calculated by use of the deflection equation (9-5) or the spring-rate equation (9-6).

The total number of coils can then be calculated by adding to the active number the additional turns required by the type of end being used.

Several dimensions of the spring must be calculated so that the spring can be made and the designer can provide for adequate space. Some of the more useful dimensions for helical compression springs are

$$\text{Outside diameter OD} = D + d \qquad \text{in.} \qquad (9\text{-}13)$$

$$\text{Pitch } p = \frac{\delta_s}{N} + d \qquad \text{in.} \qquad (9\text{-}14)$$

$$\text{Solid length (approximate) } L_s = N_t d \qquad \text{in.} \qquad (9\text{-}15)$$

$$\text{Free length } L_f = L_s + \delta_s \qquad \text{in.} \qquad (9\text{-}16)$$

$$\text{Maximum working length } L_i = L_f - \delta_i \qquad \text{in.} \qquad (9\text{-}17)$$

$$\text{Minimum working length } L_u = L_f - \delta_u \qquad \text{in.} \qquad (9\text{-}18)$$

Table 9-2. Steel Wire Sizes

W & M* gage-number sizes	Fractional-inch sizes	Decimal equivalents	W & M* gage-number sizes	Fractional-inch sizes	Decimal equivalents
	1	1.0000	7	...	0.1770
	$\frac{31}{32}$	0.9688	8	...	0.1620
	$\frac{15}{16}$	0.9375		$\frac{5}{32}$	0.1563
	$\frac{29}{32}$	0.9063	9	...	0.1483
	$\frac{7}{8}$	0.8750	10	...	0.1350
	$\frac{27}{32}$	0.8438		$\frac{1}{8}$	0.1250
	$\frac{13}{16}$	0.8125	11	...	0.1205
	$\frac{25}{32}$	0.7813	12	...	0.1055
	$\frac{3}{4}$	0.7500		$\frac{3}{32}$	0.0938
	$\frac{23}{32}$	0.7188	13	...	0.0915
	$\frac{11}{16}$	0.6875	14	...	0.0800
	$\frac{21}{32}$	0.6563	15	...	0.0720
	$\frac{5}{8}$	0.6250	16	$\frac{1}{16}$	0.0625
	$\frac{19}{32}$	0.5938	17	...	0.0540
	$\frac{9}{16}$	0.5625	18	...	0.0475
	$\frac{17}{32}$	0.5313	19	...	0.0410
7-0	$\frac{1}{2}$	0.5000	20	...	0.0348
	....	0.4900	21	...	0.0317
6-0	$\frac{15}{32}$	0.4688		$\frac{1}{32}$	0.0313
	....	0.4615	22	...	0.0286
5-0	$\frac{7}{16}$	0.4375	23	...	0.0258
	....	0.4305	24	...	0.0230
4-0	$\frac{13}{32}$	0.4063	25	...	0.0204
	....	0.3938	26	...	0.0181
3-0	$\frac{3}{8}$	0.3750	27	...	0.0173
	....	0.3625	28	...	0.0162
2-0	$\frac{11}{32}$	0.3438		$\frac{1}{64}$	0.0156
	....	0.3310	29	...	0.0150
1-0	$\frac{5}{16}$	0.3125	30	...	0.0140
1	....	0.3065	31	...	0.0132
	....	0.2830	32	...	0.0128
	$\frac{9}{32}$	0.2813	33	...	0.0118
2	....	0.2625	34	...	0.0104
	$\frac{1}{4}$	0.2500	35	...	0.0095
3	....	0.2437	36	...	0.0090
4	....	0.2253	37	...	0.0085
	$\frac{7}{32}$	0.2188	38	...	0.0080
5	....	0.2070	39	...	0.0075
6	....	0.1920	40	...	0.0070
	$\frac{3}{16}$	0.1875	41	...	0.0066
			42	...	0.0062

* Washburn and Moen.

An additional factor that must often be considered is that the diameter of a helical spring increases under compression and decreases under tension. Adequate clearance should be provided to accommodate this motion without binding.

A similar analysis can be used to determine the dimensions of a tension spring.

Example 9-1. The spring in a toggle switch, similar to Fig. 2-18, must exert a force of 2 lb when at its maximum working length of $\frac{1}{2}$ in. The spring force must not exceed 4 lb when the spring is compressed to a length of $\frac{3}{8}$ in. during the switching operation.

We are asked to determine the dimensions of a steel spring with squared and ground ends for use in this application.

Solution. Service conditions will be considered as average (1,000 to 100,000 cycles of maximum load). Loads and deflections, related to Fig. 9-11, are

$$P_i = 2 \text{ lb}$$
$$P_u = 4 \text{ lb}$$
$$\delta_u - \delta_i = \frac{1}{8} \text{ in.}$$

Thus,
$$\frac{P}{\delta} = \frac{P_u - P_i}{\delta_u - \delta_i} = \frac{4 - 2}{\frac{1}{8}} = 16 \text{ lb/in.}$$

and
$$\delta_i = \frac{P_i}{P/\delta} = \frac{2}{16} = 0.125 \text{ in.}$$

$$\delta_u = \frac{P_u}{P/\delta} = \frac{4}{16} = 0.250 \text{ in.}$$

The free length will be

$$L_f = L_i + \delta_i = 0.5000 + 0.125 = 0.625 \text{ in.}$$

Wire size. The wire size will be determined so that the shear stress under the maximum operating load will not exceed the values in the average service column of Table 9-1. The appropriate equation (9-4) is

$$s_s = K \frac{8PC}{\pi d^2} \qquad \text{psi}$$

and s_s is assumed to be 75,000 psi for first trial because of the small forces involved
$K = 1.18$, from Fig. 9-4 for $C = 8$, as recommended for balanced design
$P = P_u = 4 \text{ lb}$
$C = 8$, as recommended for balanced design

Thus,
$$75,000 = 1.18 \frac{8 \times 4 \times 8}{\pi d^2}$$

from which $d = 0.0358$. The next-larger standard wire size is W & M No. 19, with a diameter of 0.0410 in.

Number of coils. The active number of coils will next be determined by use of Eq. (9-6):

$$\frac{P}{\delta} = \frac{Gd}{8C^3N} \qquad \text{lb/in.}$$

and $P/\delta = 16$ lb/in.

$\qquad G = 11,500,000$ psi

$\qquad d = 0.0410$ in.

$\qquad C = 8$

Thus,
$$16 = \frac{11,500,000 \times 0.0410}{8 \times (8)^3 \times N}$$

from which $N = 7.20$. The total number of coils for a compression spring with squared and ground ends is approximately

$$N_t = N + 1\tfrac{3}{4} = 7.20 + 1.75 = 8.95 \text{ coils}$$

or, for practical purposes, $N_t = 9$ coils and $N = 7\tfrac{1}{4}$ coils.

Checking for clashing of coils

$$\text{Approximate solid length} = N_t d = 9 \times 0.0410 = 0.369 \text{ in.}$$

and
$$\delta_s = L_f - L_s = 0.625 - 0.369 = 0.256 \text{ in.}$$

Hence
$$\frac{\delta_s}{\delta_u} = \frac{0.256}{0.250} = 1.024$$

Thus, on the basis of the approximate solid length of 0.369, clashing is a possibility. However, as shown in Fig. 9-5, the end coils of squared and ground springs are ground to a thickness of about $d/4$. Thus, a more realistic value for solid length would be

$$L_s = N_t d - \frac{d}{2} = 0.369 - 0.021 = 0.348 \text{ in.}$$

and
$$\delta_s = L_f - L_s = 0.625 - 0.348 = 0.277 \text{ in.}$$

Hence
$$\frac{\delta_s}{\delta_u} = \frac{0.277}{0.250} = 1.11$$

Thus, in reality, clashing will not be a problem. It should be noted that, if a clash allowance insufficient to allow for manufacturing tolerances were indicated, the designer could either use a larger spring index, e.g., 10, or smaller-diameter wire, e.g., 0.0348-in.-diameter W & M No. 20, at a stress level higher than the conservative values in Table 9-1, or ask that the specifications be modified so that they could be met.

Checking for buckling

$$\frac{\delta_u}{L_f} = \frac{0.250}{0.625} = 0.40$$

and
$$\frac{L_f}{D} = \frac{0.625}{Cd} = \frac{0.625}{8 \times 0.041} = 1.91$$

The point corresponding to these ratios is well within the stable region in Fig. 9-6, and buckling will not be a problem.

Manufacturing specifications

$\qquad$ Wire diameter = 0.041 in. (W & M No. 19)

$\qquad$ Outside diameter = $D + d = d(C + 1) = 0.041 \times 9 = 0.369$ in.

$\qquad$ Total coils = 9, with squared and ground ends

$\qquad$ Pitch = $\delta_s/N + d = 0.277/7.25 + 0.041 = 0.079$ in.

$\qquad$ Free length = 0.625 in. ($\tfrac{5}{8}$)

9-5. Variations in Metallic Springs. Helical springs are made from square and rectangular wire as well as from round wire   The advantages of square or rectangular wire are that a greater resilience can be obtained in a given space and the outside surface can be ground in order to achieve a desired spring rate with better accuracy.   Because relatively small volume of production and high cost limit the types and quality of materials that are readily available, the general use of springs made from other than round wire is not recommended.

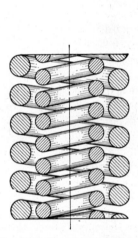

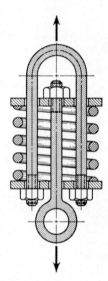

Fig. 9-12. Concentric springs.

Fig. 9-13. Compression spring used to carry a tension load.

A method of achieving a greater capacity when the OD is limited is to use *concentric* springs, illustrated in Fig. 9-12.   For maximum utilization of the material, both springs should be designed to work at the same relative stress level.   Adjacent springs should be wound with opposite-hand helices to prevent the locking of coils if one should move sideways. Some typical uses of concentric springs are the valve springs in aircraft and heavy-duty diesel engines (Fig. 2-9) and in railroad-car suspension systems.

One of the disadvantages of a helical tension spring is the possible disastrous result if the spring breaks.   Figure 9-13 illustrates a simple way to use a relatively safe compression spring to carry a tension load.

The torsion bar in Fig. 9-14 is a simple form of spring that has come into prominence because of its use in automobile suspension systems.

The springs shown in Fig. 9-15 are used in special applications where a telescoping spring or a spring with a spring rate that increases with load is desired.   The *conical* spring in Fig. 9-15a is wound with a uniform pitch, whereas the *volute* springs in Fig. 9-15b and c are wound in the form of a paraboloid with constant pitch and lead angles.   The springs may be made either partially or completely telescoping.   In either case, the number of active coils gradually decreases as the largest and most flexible coil

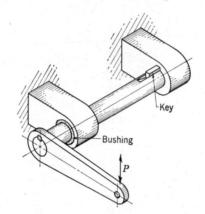

FIG. 9-14. Torsion bar

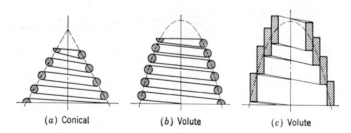

(a) Conical   (b) Volute   (c) Volute

FIG. 9-15. Conical and volute springs.

bottoms against another coil or the spring seat.   The decreasing number of coils results in an increasing spring rate.   This characteristic is sometimes utilized in vibration problems where springs are used to support a body that has a varying mass.

The *torsion* spring (Fig. 9-16) is widely used in hinges, brush holders, automobile starters, door locks, and door checks.   The name is somewhat misleading, because the wire is subjected to bending rather than torsional stress.   The clock or motor type of *power* spring is a special case of a torsion or spiral spring with a large number of turns of relatively thin flat tempered steel.

The *Neg'ator* is a type of spring with rather unusual characteristics.[1] Residual stresses are introduced in the forming process in such a manner that the slope of the force-deflection curve (spring rate) may be positive, zero, or even negative.

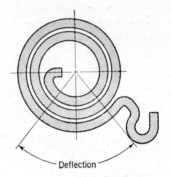

Fig. 9-16. Torsion spring.

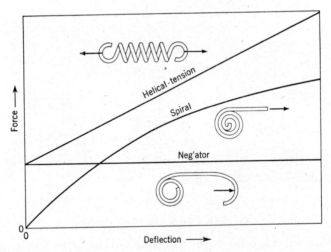

Fig. 9-17. Force-deflection curves for comparable helical tension, spiral, and Neg'ator springs.

Figure 9-17 illustrates the different force-deflection characteristics of comparable helical-tension (with initial tension), spiral, and Neg'ator springs. The possiblity of a constant force or torque, irrespective of the deflection, and the degree of control over its spring rate have resulted in Neg'ator springs and spring motors being used widely as constant-force

[1] F. A. Votta, Jr., The Theory and Design of Long-deflection Constant-force Spring Elements, *Trans. ASME*, vol. 74, pp. 439–450, 1952.

clamps (Fig. 9-18); in counterbalances for machine tools, vending machines, etc.; in power-feed mechanisms; in motor-brush holders to maintain a constant pressure between the brushes and the commutator; in reels (Fig. 9-19); and in spring motors for movie cameras.

Many springs are used in the form of cantilever or other simple beams.

Fig. 9-18. Neg'ator clamp.   (*Courtesy Hunter Spring Company.*)

Fig. 9-19. Constant-torque B motor as used in a reel.   (*Courtesy Hunter Spring Company.*)

The most common example is the automotive leaf spring. Figure 9-20a shows a typical semielliptical leaf spring, and Fig. 9-20b illustrates, approximately, the manner in which the lengths of the leaves are determined on the basis of a uniformly stressed flat spring.

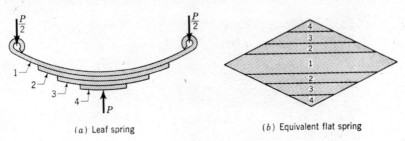

(a) Leaf spring

(b) Equivalent flat spring

FIG. 9-20

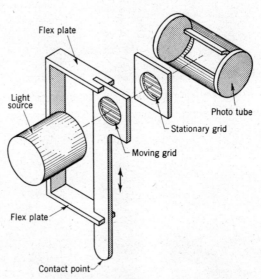

FIG. 9-21. Flexible hinges or flex plates.

Flat springs find wide application in small devices such as extensometers, etc., where they are used as pivots which are free from backlash and friction. Figure 9-21 shows the use of elastic hinges or "flex plates" in a photoelectric displacement-measuring device. The unit is used to measure motions in the order of ten-millionths of an inch, and consequently the support for the moving grid must not have any play or backlash.

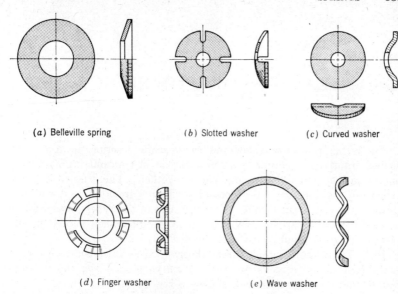

(a) Belleville spring   (b) Slotted washer   (c) Curved washer

(d) Finger washer   (e) Wave washer

FIG. 9-22 Miscellaneous springs.

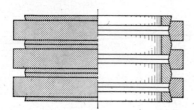

FIG. 9-23. Ring spring.

Spring washers, as shown in Fig. 9-22, are available in many different forms.   The *Belleville* spring *a* is used where space limitations require the use of high stresses and short range of motion.   The *wave* washer *e* is often used in disk clutches and brakes (Fig. 12-5) to give a positive separation of the unloaded disks to prevent dragging.

The *ring* spring in Fig. 9-23 is used where a particularly high spring rate is required.

CHAPTER 10

TRANSLATION OR POWER SCREWS

The terms *translation* screw and *power* screw describe the use of the helical translatory motion of a screw thread in transmitting motion or power rather than in clamping parts together. The Unified and the American National threads discussed in Chap. 8 are often used when the power involved is small, but, in general, one of the types of translation screws or a ball-bearing screw and nut is specified when maximum efficiency and a minimum wear rate are desired.

10-1. Translation Screw Thread Forms. The most commonly used translation screw threads are the *general-purpose Acme*, the *modified square*, and the *buttress* threads shown in Fig. 10-1.

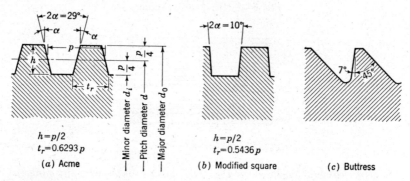

$h = p/2$
$t_r = 0.6293 p$

(a) Acme

$h = p/2$
$t_r = 0.5436 p$

(b) Modified square

(c) Buttress

FIG. 10-1. Thread forms for translation and power screws.

The Acme thread *a* is the most common form and has been widely used in machine tools. The 29° included angle of the thread is large enough to permit the use of an adjustable split nut for wear compensation while not so great as to decrease seriously the efficiency of the drive. Also, the thread may be economically produced by use of taps and dies and by thread milling.

Since each application is more or less special, it has not been desirable to set up complete standards in terms of numbers of threads and diameters, as for screw fasteners. The standards are concerned mainly with basic dimensions and the allowances and tolerances for the several types

186

of fits.[1]   However, certain diameter and pitch combinations are recommended for use under normal circumstances.   The standards and handbooks[2] should be consulted for more complete information, but Table 10-1 contains proportions for several of the recommended combinations for general-purpose Acme and modified square threads.

TABLE 10-1. NOMINAL BASIC DIMENSIONS OF TRANSLATION SCREW THREADS

Diameter, in.	General-purpose Acme			Modified square†		
	Threads per inch	Height of thread h, in.	Thickness at root t_r,* in.	Threads per inch	Height of thread h, in.	Thickness at root t_r,‡ in.
¼	16	0.0313	0.0393	10	0.0500	0.0544
⅜	12	0.0417	0.0524	8	0.0625	0.0680
½	10	0.0500	0.0629	6½	0.0769	0.0837
¾	6	0.0833	0.1049	5	0.1000	0.1087
1	5	0.1000	0.1259	4	0.1250	0.1359
1½	4	0.1250	0.1573	3	0.1667	0.1812
2	4	0.1250	0.1573	2¼	0.2222	0.2416
2½	3	0.1667	0.2098	2	0.2500	0.2718
3	2	0.2500	0.3147	1¾	0.2857	0.3106
4	2	0.2500	0.3147	1½	0.3333	0.3624
5	2	0.2500	0.3147			

* Neglecting the effect of major- and minor-diameter allowances, $t_r = 0.6293p$.

† Threads per inch correspond to the Sellers' square-thread system.

‡ Neglecting the effect of major- and minor-diameter allowances, $t_r = 0.5436p$.

True square threads with sides perpendicular to the thread axis have been used for many years.   The 0° included angle results in maximum efficiency and minimum radial or bursting pressure on the nut.   It is expensive to machine (usually turned with a single-point tool) and cannot easily be compensated for wear.   Consequently, it has been largely superseded by the modified square thread (Fig. 10-1b) which, because of its 10° included thread angle, is, for most practical purposes, the equivalent of a true square thread without its disadvantages.

The buttress thread[3] c is designed for use when large forces act along

[1] Acme Screw Threads, ASA B1.5-52, and Stub Acme Screw Threads, ASA B1.8-52, American Society of Mechanical Engineers, New York, 1952.

[2] For example, C. Carmichael (ed.), "Kent's Mechanical Engineers' Handbook," "Design and Production" volume, 12th ed., John Wiley & Sons, Inc., New York, 1950.

H. L. Horton (ed.), "Machinery's Handbook," 16th ed., The Industrial Press, New York, 1959.

[3] Buttress Screw Threads, ASA B1.9-1953, American Society of Mechanical Engineers, New York, 1953.

the screw axis in one direction only. The almost perpendicular thrust surface results in low bursting pressures, as for the modified square thread, and, in addition to its use in jacks, etc., it is ideally suited for connecting thin tubular members that must carry large forces, such as in attaching the barrel to the housing in antiaircraft guns, in the breech mechanisms of large guns, and in airplane propeller hubs.

10-2. Efficiency of Screws. Two considerations of special importance when a screw is being used to transmit motion and power are (1) the

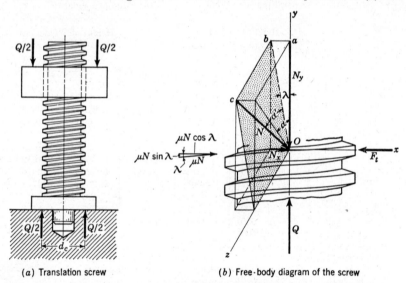

(a) Translation screw   (b) Free-body diagram of the screw

FIG. 10-2. Forces on a screw.

torque required to turn the screw or nut and (2) the mechanical efficiency of the drive.

In general, the torque is made up of two parts: that to raise the load,

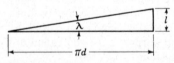

FIG. 10-3. Relationship of lead angle to lead and pitch diameter.

considering the screw as an inclined plane, and that to overcome friction. In Fig. 10-2, the screw rotates, and the nut moves axially against the load Q. The free-body diagram of the screw in Fig. 10-2a is illustrated in Fig. 10-2b. The angle λ is known as the *lead* angle and is defined as *the angle made by the helix of the thread with a plane perpendicular to the axis.* The relation of the lead angle to the pitch diameter d and lead l is shown in Fig. 10-3.[1] As shown in Fig. 10-1, α is the angle

[1] It should be recalled that the lead equals the pitch for a single-thread screw, the lead is twice the pitch for a double-thread screw, etc.

between the flank of the thread and a plane perpendicular to the screw axis. For symmetrical threads, α is one-half the included thread angle.

The angle α' is related to α and is the angle between the flank of the thread and a plane perpendicular to the helix of the thread.

The friction force is equal to the product of the normal force N and the coefficient of friction μ and always acts in the direction to oppose the motion. For convenience and clarity, the friction force and its x and y components have been shown to the left of the screw in Fig. 10-2b.

Summing up the forces in the x direction gives

$$\Sigma F_x = 0$$
$$N_x + \mu N \cos \lambda - F_t = 0 \tag{10-1}$$

However,
$$N_x = ba = ob \sin \lambda$$
$$= N \cos \alpha' \sin \lambda \tag{10-2}$$

Substituting Eq. (10-2) in (10-1) and solving for the tangential force F_t,

$$F_t = N(\cos \alpha' \sin \lambda + \mu \cos \lambda) \tag{10-3}$$

Similarly, the summation of forces in the y direction gives

$$\Sigma F_y = 0$$
$$Q - N_y + \mu N \sin \lambda = 0 \tag{10-4}$$

and following the procedure above to solve for Q, we find

$$Q = N(\cos \alpha' \cos \lambda - \mu \sin \lambda) \tag{10-5}$$

Dividing Eq. (10-5) by (10-3) and solving for F_t gives

$$F_t = Q \frac{\cos \alpha' \sin \lambda + \mu \cos \lambda}{\cos \alpha' \cos \lambda - \mu \sin \lambda} \tag{10-6}$$

which may be simplified by dividing both numerator and denominator by $\cos \lambda$ and then substituting $\tan \lambda$ for $\sin \lambda / \cos \lambda$ to

$$F_t = Q \frac{\cos \alpha' \tan \lambda + \mu}{\cos \alpha' - \mu \tan \lambda} \tag{10-7}$$

The tangential force acts at the pitch radius $d/2$ of the screw, and the *torque* T is

$$T = \frac{Qd}{2} \frac{\cos \alpha' \tan \lambda + \mu}{\cos \alpha' - \mu \tan \lambda} \tag{10-8}$$

The efficiency η for a mechanical process may be defined as the ratio of the work out to the work in. Considering one rotation of the screw, the efficiency becomes

$$\eta = \frac{W_{\text{out}}}{W_{\text{in}}} = \frac{Ql}{2\pi T} \tag{10-9}$$

After substituting from Eq. (10-8) for T in (10-9),

$$\eta = \tan \lambda \frac{\cos \alpha' - \mu \tan \lambda}{\cos \alpha' \tan \lambda + \mu} \tag{10-10}$$

Values of coefficient of friction μ may be estimated (Sec. 10-4), and from Fig. 10-3 $\tan \lambda = l/\pi d$. The main difficulty is in determining $\cos \alpha'$. However, for most practical purposes, α' may be assumed to equal α without introducing serious error.[1] In fact, very careful manipulation of a slide rule is required to detect any difference for an Acme screw with a lead angle of 30° or less. When a Unified thread screw (60° included angle) is used as a power screw, the difference is somewhat greater, but it is still less than the uncertainty as to the actual coefficient of friction. It should also be noted that α' is always less than α and that, consequently, the use of α will result in conservative answers, i.e., lower efficiencies than actually present.

Thus, for all practical purposes, Eqs. (10-8) and (10-10) become, respectively,

$$T = \frac{Qd}{2} \frac{\cos \alpha \tan \lambda + \mu}{\cos \alpha - \mu \tan \lambda} \quad \text{lb-in.} \tag{10-15}$$

and

$$\eta = \tan \lambda \frac{\cos \alpha - \mu \tan \lambda}{\cos \alpha \tan \lambda + \mu} \tag{10-16}$$

---

[1] From Fig. 10-2b,

$$\tan \alpha' = \frac{cb}{ob} = \frac{oa \tan \alpha}{oa/\cos \lambda} = \cos \lambda \tan \alpha \tag{10-11}$$

The relationship in Eq. (10-11) may be represented by the triangle in Fig. 10-4.

$$\cos \alpha' = \frac{1}{\sqrt{1 + \cos^2 \lambda \tan^2 \alpha}} \tag{10-12}$$

However, from Fig. 10-3,

$$\cos^2 \lambda = \frac{(\pi d)^2}{l^2 + (\pi d)^2} \tag{10-13}$$

Noting that l^2 is usually negligible in comparison with $(\pi d)^2$, we may let $\cos^2 \lambda = 1$.

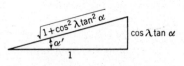

FIG. 10-4

Equation (10-12) becomes

$$\cos \alpha' = \frac{1}{\sqrt{1 + \tan^2 \alpha}} = \frac{1}{\sec \alpha} = \cos \alpha \tag{10-14}$$

and

$$\alpha' = \alpha$$

Figure 10-5 shows the variation of efficiency with lead angle for an Acme thread when the coefficient of friction is 0.10 and 0.15. It should be noted that the efficiency is low when the lead angle is either very small or very large. For example, a single-thread 1-in. Acme thread (Table 10-1) has a lead angle of 4°3′, and, with $\mu = 0.1$, the efficiency is about 40 per cent. If a quadruple thread is used instead of the single thread, the lead angle is 15°48′, and the efficiency is increased to about 71 per cent.

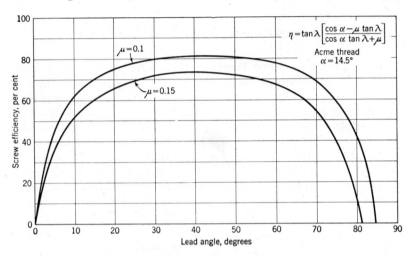

$$\eta = \tan \lambda \left[\frac{\cos \alpha - \mu \tan \lambda}{\cos \alpha \tan \lambda + \mu} \right]$$

Acme thread
$\alpha = 14.5°$

FIG. 10-5. Efficiency of Acme screw threads.

At the same time, it should be appreciated that the torque requirement is usually the major consideration and that efficiency is secondary. If both minimum torque and maximum efficiency are of utmost importance, the solution is probably a ball-bearing screw, as will be discussed in Sec. 10-5.

The preceding discussion has considered the effect of screw-thread friction only, and no mention has been made of the friction in the thrust bearing which supports the rotating member under the load Q. If a ball or roller thrust bearing (Secs. 18-17 and 18-18) is being used, the friction will be negligible in relation to that of the threads and need not be considered. However, if a simple sliding bearing, or thrust collar, is used, the thrust-bearing friction may exceed that of the screw itself.

If the collar friction is considered to act at the mean radius of the collar,[1] the collar-friction torque becomes

$$T_c = \frac{\mu Q d_c}{2} \qquad \text{lb-in.} \qquad (10\text{-}17)$$

[1] See Sec. 12-3 for a discussion of the concept and application of the mean, or friction, radius.

where d_c = mean diameter of collar, in. (Fig. 10-2a).   The total torque for the screw and thrust bearing is

$$T = Q \frac{d}{2} \left(\frac{\cos \alpha \tan \lambda + \mu}{\cos \alpha - \mu \tan \lambda} + \frac{d_c}{d} \mu_c \right) \quad \text{lb-in.} \qquad (10\text{-}18)$$

and the efficiency becomes

$$\eta = \frac{\tan \lambda}{[(\cos \alpha \tan \lambda + \mu)/(\cos \alpha - \mu \tan \lambda)] + (d_c/d)\mu_c} \qquad (10\text{-}19)$$

10-3. Self-locking Screw Threads.   Since a friction force always acts in the direction to oppose the sliding motion, the direction of the load relative to the axial motion of the nut or screw becomes an important consideration in situations where the load acts always in one direction. For example, if the screw is part of a jack and the load is a dead weight, the directions of the forces in Fig. 10-2b apply when the load is being raised.   If the direction of rotation of the screw is reversed to lower the load, the friction force $F_f = \mu N$ and the applied force F_t would reverse their directions while N and Q remain as before.   The equations for the summation of forces become

$$\Sigma F_x = 0$$
$$N_x - \mu N \cos \lambda + F_t = 0 \qquad (10\text{-}20)$$
and
$$\Sigma F_y = 0$$
$$Q - N_y - \mu N \sin \lambda = 0 \qquad (10\text{-}21)$$

Similarly, the torque equation, neglecting thrust-bearing friction, becomes

$$T = Q \frac{d}{2} \frac{- \cos \alpha \tan \lambda + \mu}{\cos \alpha + \mu \tan \lambda} \quad \text{lb-in.} \qquad (10\text{-}22)$$

As can be seen in Eq. (10-22), if $\mu < \cos \alpha \tan \lambda$, the torque to lower the load will be negative; i.e., the load will descend unless a restraining torque is applied.

If, however, $\mu > \cos \alpha \tan \lambda$, a positive torque is required to lower the load, and the screw is considered to be *self-locking*.

As a matter of convenience, an easily remembered rule is the following: *A screw will be self-locking if the coefficient of friction is equal to or greater than the tangent of the lead angle.*

Figure 10-6 shows a 35-ton-capacity screw jack that uses a double-thread, non-self-locking screw so that the load will be self-lowering.   The lowering speed is limited to a maximum of $\frac{1}{2}$ in./sec by the action of the centrifugal brake or governor.   A thumb-screw-operated shoe brake (Sec. 13-1) is provided for controlling the rate of descent from zero to the maximum value permitted by the centrifugal brake.   The governor-brake mechanism is located at the high-speed end of a gear train so that the effect of the friction torque due to the centrifugal and shoe brakes will

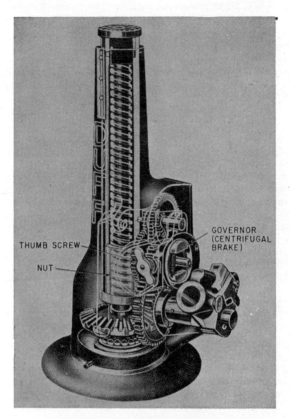

THUMB SCREW

NUT

GOVERNOR
(CENTRIFUGAL
BRAKE)

FIG. 10-6. Governor-controlled, self-lowering screw jack. (*Courtesy Duff-Norton Company.*)

be multiplied at the screw.   A ratchet is provided to permit overrunning the brake when the jack is being raised.[1]   The threads are modified square threads, the screw is made of a heat-treated alloy steel (approximately 8647), the nut is made of phosphor bronze, and a ball thrust bearing supports the load.

[1] See Secs. 12-4 and 12-6 for a discussion of ratchets and overruning or one-way clutches.

10-4. Coefficient of Friction.   Table 10-2 contains coefficients of friction for screw threads and thrust collars obtained from tests made over a wide range of operating conditions.   Ham and Ryan found the coefficients to be practically independent of load and to vary only slightly with different combinations of materials and rubbing speed.

TABLE 10-2. COEFFICIENTS OF FRICTION FOR SCREW THREADS AND
THRUST COLLARS*

Steel screw and bronze or cast-iron nut			Thrust-collar friction		
Conditions	Average coefficient of friction, μ		Materials	Average coefficient of friction, μ_c	
	Starting	Running		Starting	Running
High-grade materials and workmanship and best running conditions...	0.14	0.10	Soft steel on cast iron..............	0.17	0.12
Average quality of materials and workmanship and average running conditions...........	0.18	0.13	Hardened steel on cast iron......... Soft steel on bronze Hardened steel on bronze...........	0.15 0.10 0.08	0.09 0.08 0.06
Poor workmanship or very slow and infrequent motion with indifferent lubrication or newly machined surfaces................	0.21	0.15			

* After C. W. Ham and D. G. Ryan, An Experimental Investigation of the Friction of Screw Threads, *Univ. Illinois Eng. Expt. Sta. Bull.* 247, 1932.

Example 10-1.   Example 8-1 was concerned with the selection of bolts to be used in fastening the cylinder head to the cylinder of an air compressor.   It was found that, when a flexible gasket was used to seal the joint, eight 9/16-12 UNC bolts were required.   The problem here is to determine the approximate value of torque necessary to tighten the bolt to an initial tension load of 2,275 lb.

Solution.   The applicable equation is Eq. (10-18):

$$T = \frac{Qd}{2}\left(\frac{\cos\alpha\tan\lambda + \mu}{\cos\alpha - \mu\tan\lambda} + \frac{d_c}{d}\mu_c\right)$$

and $Q = 2,275$ lb

$d = \frac{9}{16} - 2 \times \frac{3}{8}h$ (from Fig. 8-3) $= 0.5625 - 2 \times \frac{3}{8} \times 0.86603 \times \frac{1}{12}$
$= 0.5084$ in.

$\cos\alpha = \cos 30° = 0.866$ $\alpha = 30°$, from Fig. 8-3

$\tan\lambda = l/\pi d = (\frac{1}{12})/(\pi \times 0.5084) = 0.0522$

$\mu = 0.15$, dry steel on steel[1]

$d_c = \frac{1}{2} \times (0.5625 + 0.8125) = 0.6875$ in., considering d_c as mean of bolt diameter and width across flats of the bolt head, from Table 8-3

$\mu_c = 0.17$, soft steel on cast iron, from Table 10-2

[1] Carmichael, *op. cit.*, pp. 7–28.

Thus,

$$T = \frac{2{,}275 \times 0.5084}{2}\left(\frac{0.866 \times 0.0522 + 0.15}{0.866 - 0.15 \times 0.0522} + \frac{0.6875}{0.5084} \times 0.17\right) = 265 \text{ lb-in. or}$$
$$22 \text{ lb-ft}$$

Therefore, the installation instructions would specify tightening the head bolts with a torque wrench to a torque of 22 lb-ft.

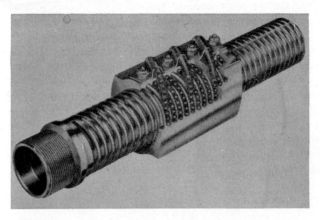

Fig. 10-7. Ball-bearing screw. (*Courtesy Saginaw Steering Gear Division, General Motors Corporation.*)

10-5. Ball-bearing Screws. The ball-bearing screw in Fig. 10-7 uses the recirculating-ball principle, discussed in Sec. 8-7 in relation to the ball-bearing spline, in substituting rolling friction for the sliding friction of the ordinary screw. The ball-bearing screw was originally developed for the steering mechanisms of passenger cars, trucks, and buses but is now used also in a wide variety of applications where maximum efficiency is required, e.g., the actuation of landing gears and control surfaces on aircraft and guided missiles, machine tools, automobile bumper jacks, and power actuators.

Figure 10-8 compares the efficiencies of ball-bearing and Acme screws when used to convert rotary motion into axial motion. The difference is particularly marked when the lead angle

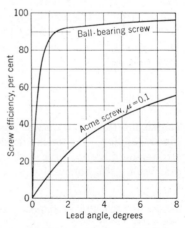

Fig. 10-8. Efficiencies of ball-bearing and Acme screws. (*Courtesy Saginaw Steering Gear Division, General Motors Corporation.*)

is small, as is the usual case.[1]   The efficiency of the ball-bearing screw in converting axial motion to rotary motion may be over 80 per cent, whereas the comparable Acme screw will probably be self-locking and consequently cannot be used to convert axial motion into rotary motion.

10-6. Screw and Nut Design.   The screw must have adequate strength to withstand the applied torque and to carry the load.   The length of engagement for the screw and nut threads must be sufficient to prevent shearing of the threads and a too-rapid rate of wear of the sliding surfaces.

Since practically all translation and power screws are made of steel, the following discussion will consider only steel screws.

The minor-diameter cross section of the portion of the screw between the nut and the thrust bearing is subjected to a *biaxial stress* consisting of a tensile or compressive stress due to the axial force and a shear stress due to the torque.   When the unsupported length of the screw is quite short or is loaded in tension, failure will take place when the maximum shear stress is equal to the shear yield strength of the material.   In this case, the maximum shear stress on the minor-diameter section[2] of the screw and the factor of safety for the screw are, respectively,

$$s_{s,\text{max}} = \sqrt{\left(\frac{s}{2}\right)^2 + (s_s)^2} \qquad \text{psi} \qquad (10\text{-}23)$$

and
$$\text{f.s.} = \frac{s_{sy}}{s_{s,\text{max}}} = \frac{s_{sy}}{\sqrt{(s/2)^2 + (s_s)^2}} \qquad (10\text{-}24)$$

where $s = Q/A$, the normal stress on the minor-diameter section, psi, and $s_s = Tc/J$, the maximum torsional shear stress on the minor-diameter section, psi.   Equation (10-24) may be rewritten in terms of the yield strength in tension or compression, by substituting $s_y/2$ for s_{sy} (Sec. 6-5), as

$$\text{f.s.} = \frac{s_y/2}{\sqrt{(s/2)^2 + (s_s)^2}} \qquad (10\text{-}25)$$

In Sec. 6-6 this equation was applied to the design of shafts under combined bending and torsion loads.   Since both the normal and shear stresses on a section of a shaft are functions of the same dimension (or dimensions, if a hollow shaft), it was possible to write the equation in a

---

[1] The lead angles for single-thread Acme screws with the diameter-pitch combinations in Table 10-1 range from 5°12' for the ¼-in. screw down to 1°55' for the 5-in. screw.

[2] The indeterminate additional strength and rigidity due to the threads are neglected, and only the minor-diameter cross section is considered in screw calculations.

form such that a direct solution could be made for the shaft diameter. Unfortunately, this is not the case for translation screws.   The difficulty is that the section area is a function of the minor diameter d_i; the torque is a function of the pitch diameter d, the lead angle λ, and the thrust-bearing friction radius $d_c/2$; and the lead angle cannot be calculated until the pitch diameter and the lead are known.   Consequently, Eqs. (10-24) and (10-25) are useful only in checking to see whether or not a given screw will be satisfactory, and the screw diameter, in general, must be selected by a trial-and-error solution.

The one case that does not require a trial-and-error solution is that where the only torque on the part of the screw between the nut and the thrust bearing is the negligible friction torque of a rolling-contact thrust bearing.

In a practical problem, a trial solution for the diameter of the minor-diameter cross section of the screw is made by considering only the normal stress, i.e.,

$$A_i = \frac{Q}{s_y/\text{f.s.}} \qquad \text{in.}^2 \tag{10-26}$$

Then a nominal thread is selected that will have a minor-diameter section area equal to or greater than that calculated by Eq. (10-26).   After the dimensions and materials for the screw, nut, and thrust bearing have been determined, the torques may be calculated.

When the torques, the load, and the dimensions are known, it is a simple matter to calculate s and s_s and the corresponding factor of safety [Eq. (10-24) or (10-25)].   The new value of factor of safety is next compared with the originally selected design value, and a decision must be made as to whether the agreement is adequate or whether another solution must be made.

When a nominal-size screw is selected, the "rounding off" to the next larger size will, in most cases, result in a sufficient increase in area so that the first trial solution will be adequate if a reasonably high factor of safety was used.

When the screw is axially loaded in compression and the unsupported length is too great for the screw to be considered a simple compression member, the design must be based on the column theory in Secs. 6-10 to 6-12.   In this case, the minor-diameter section area will probably be calculated by use of the J. B. Johnson formula [Eq. (6-59)], but a check should be made to ensure that it and not the Euler formula applies to the situation.   The value of s to be used in Eqs. (10-24) and (10-25) is not the average stress Q/A but must be an equivalent column stress based on the appropriate column formula.   The equivalent stresses for the J. B. Johnson and Euler formulas become, respectively,

$$s_{\text{equiv}} = \frac{Q}{A_i} \frac{1}{1 - [s_y(L/k)^2/4C\pi^2 E]} \quad \text{psi} \tag{10-27}$$

and

$$s_{\text{equiv}} = \frac{Q}{A_i} \frac{s_y(L/k)^2}{C\pi^2 E} \quad \text{psi} \tag{10-28}$$

After the thread dimensions have been selected as outlined above, the length of thread engagement of the screw and nut must be determined on the bases of shear strength of the threads and wear of the screw and nut.

If it is assumed that the load is distributed uniformly over the threads in contact, the required length of engagement L_e for adequate shear strength becomes, in general form,

$$L_e = \text{f.s.} \frac{pQ}{A_s s_s} \quad \text{in.} \tag{10-29}$$

where p = pitch of thread (Sec. 8-1), in.

A_s = shear area of thread, in.²

s_s = applicable shear strength, for example, s_{sy} for a steel screw under static loading, psi

Since the shear area will be less for the thread on the screw than for the thread in the nut and since the screw and nut are usually made of different materials, the nut material having the lower mechanical properties, it becomes necessary to calculate separately the length of engagement required for the screw and for the nut.

Neglecting the radial clearance between threads, or allowance at the major and minor diameters, and considering the threads as a series of collars, the equations for thread engagement become

$$L_{e,\text{screw}} = \text{f.s.} \frac{pQ}{\pi d_i t_i s_{s,\text{screw}}} \quad \text{in.} \tag{10-30}$$

$$L_{e,\text{nut}} = \text{f.s.} \frac{pQ}{\pi d_o t_o s_{s,\text{nut}}} \quad \text{in.} \tag{10-31}$$

where d_i = minor diameter, in.

t_i = thread thickness at minor diameter, in.   (t_r in Fig. 10-1)

d_o = major diameter, in.

t_o = thread thickness at major diameter, in.

Except for the Unified thread system, the thread thickness is nominally the same at the major and minor diameters, that is, $t_i = t_o = t_r$. Therefore, it is usually necessary to solve for the thread engagement for only the screw or nut, whichever has the smaller product ds_s.

The rate at which rubbing, or sliding, surfaces wear is found to be proportional to the *work of friction*. In turn, for a given combination of contacting materials and degree of lubrication, the work of friction is proportional to the product of the contact, or bearing, pressure and the sliding velocity.   This principle has already been mentioned (Sec. 3-11) relative to the transmission of motion and force through direct contact, as with a cam and follower, and will be encountered in later chapters on clutches, brakes, and bearings and lubrication.

The small amount of available design information is based almost entirely on experience, and the design criterion is generally given as an allowable value of pV, where p is the contact pressure in pounds per square inch and V is the sliding velocity in feet per minute.   However, since p is normally used to designate the pitch of the thread, to avoid confusion we shall use s_b rather than p for the bearing or contact pressure between screw threads.

The sliding velocity for translation screws is generally low, and the common practice is to calculate the length of thread engagement necessary to limit the bearing pressure between the contacting surfaces to a value low enough to give a satisfactory service life.   The required length of engagement is

$$L_e = \frac{4pQ}{\pi(d_o{}^2 - d_i{}^2)s_b} \quad \text{in.} \tag{10-32}$$

where s_b = allowable bearing pressure, psi.

Suggested values of allowable bearing pressures for several combinations of materials and speeds are given in Table 10-3.

TABLE 10-3. ALLOWABLE BEARING PRESSURES FOR SCREWS

Application	Material		Allowable bearing pressure, psi	Sliding speed at thread pitch diameter
	Screw	Nut		
Hand press	Steel	Bronze	2,500–3,500	Low speed, well lubricated
Jack screw	Steel	Cast iron	1,800–2,500	Low speed, 8 fpm
Jack screw	Steel	Bronze	1,600–2,500	Low speed, 10 fpm
Hoisting screw	Steel	Cast iron	600–1,000	Medium speed, 20–40 fpm
Hoisting screw	Steel	Bronze	800–1,400	Medium speed, 20–40 fpm
Lead screw	Steel	Bronze	150–240	High speed, 50 fpm

The length of thread engagement to be specified will be the greatest value given by Eq. (10-30), (10-31), or (10-32).

The service or wear life of a ball-bearing screw will be defined in terms of a finite number of cycles of operation in the same manner as for other rolling-contact bearings (Secs. 18-21 and 18-22).   The designer should consult either the manufacturers or their catalogues for specific information.

Example 10-2   A screw-operated positioning device is being designed to raise a load of 10,000 lb a distance of 15 in.   The proposed design is shown schematically in Fig. 10-9.   It is proposed to use modified square threads.   The nut material will be bronze, and the screw material will be AISI 3140 steel oil-quenched and tempered at 1000°F.   All thrust and guide bearings will be rolling-contact bearings, and their friction may be neglected.

We are asked to determine (a) the dimensions of the screw and nut for a factor of safety of 2, (b) the time required to raise the load, and (c) the horsepower of the electric motor.

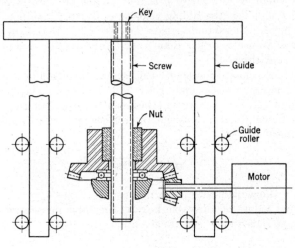

FIG. 10-9

Solution.   The first step will be to determine the required screw diameter by considering the screw as a column, as in Secs. 6-10 to 6-12.   The J. B. Johnson formula [Eq. (6-59)] applies.   It is found that a 1-in. four-threads-per-inch screw is the smallest modified square thread in Table 10-1 that will be satisfactory.   However, the unsupported length must carry also the screw torque, and the screw strength must be checked under the combined stress of the column and torque loads.

Screw torque.   The screw torque will be determined by the use of Eq. (10-15):

$$T = \frac{Qd}{2} \frac{\cos \alpha \tan \lambda + \mu}{\cos \alpha - \mu \tan \lambda} \quad \text{lb-in.}$$

and $Q = 10,000$ lb

$d = 1.000 - h = 1 - 0.125 = 0.875$ in.

$\cos \alpha = \cos 5° = 0.9962$, $\alpha = 5°$ from Fig. 10-1b

$\tan \lambda = l/\pi d = \frac{1}{4}/(\pi \times 0.875) = 0.0909$ (single-thread screw)

$\mu = 0.14$, from Table 10-2, starting friction for high-grade materials and workmanship and best running conditions

Thus, $T = \dfrac{10,000 \times 0.875}{2} \dfrac{0.9962 \times 0.0909 + 0.14}{0.9962 - 0.14 \times 0.0909} = 1,163$ lb-in.

Checking screw strength under combined stress. The check will be made by comparing the factor of safety under combined stress from Eq. (10-25) with the specified value of 2.

$$\text{f.s.} = \frac{s_y/2}{\sqrt{(s/2)^2 + (s_s)^2}}$$

Equation (10-27) will be used to calculate an equivalent normal stress for the column load:

$$s_{\text{equiv}} = \frac{Q}{A_i} \frac{1}{1 - [s_y(L/k)^2/4C\pi^2 E]}$$

and $Q = 10,000$ lb

$A_i = \pi d_i^2/4 = \pi(0.750)^2/4 = 0.442$ in.^{2}

$s_y = 132,000$ psi, from Appendix E

$L = 15$ in.

$k = d_i/4 = 0.750/4 = 0.1875$ in.

$C = 3$, from Fig. 6-26, on the basis that the screw is essentially a column with fixed ends but that the normal clearance between the threads at the nut will give a slight decrease in the rigidity at that end

$E = 30,000,000$ psi

Thus,

$$s_{\text{equiv}} = \frac{10,000}{0.442} \frac{1}{1 - \{[132,000 \times (15/0.1875)^2]/(4 \times 3 \times \pi^2 \times 30,000,000)\}}$$
$$= 29,700 \text{ psi}$$

Equation (6-27) will be used to calculate the shear stress due to the torque:

$$s_s = \frac{Tc}{J} = \frac{T}{J/c}$$

and $T = 1,163$ lb-in.

$J/c = \pi d_i^3/16 = [\pi \times (0.750)^3]/16 = 0.0828$

Thus,

$$s_s = \frac{1,163}{0.0828} = 14,000 \text{ psi}$$

Solving for the factor of safety, we find

$$\text{f.s.} = \frac{132,000/2}{\sqrt{(29,700/2)^2 + (14,000)^2}} = 3.23$$

Therefore, since 3.23 is greater than 2, the screw has more than adequate strength as a column.

Length of thread engagement. The specified length of thread engagement will be the greatest of the lengths required for adequate shear strength of the screw threads, adequate shear strength of the nut threads, and satisfactory wear life. In this case, we are interested in raising the load as rapidly as possible, and the rate of wear will probably be the most critical factor. The procedure will be to calculate the lengths of engagement required for adequate strength, then to select a length at least equal to the larger of the two, and to determine the maximum operating speed of the nut for a reasonable wear rate.

Shear of screw threads [Eq. (10-30)]

$$L_{e,\text{screw}} = \text{f.s.} \frac{pQ}{\pi d_i l_i s_{s,\text{screw}}}$$

and f.s. $= 2$

$\qquad p = \frac{1}{4} = 0.250$ in.

$\qquad Q = 10,000$ lb

$\qquad d_i = 0.750$ in.

$\qquad t_i = t_r = 0.1359$ in., from Table 10-1

$\qquad s_s = s_{sy} = s_y/2 = 132,000/2 = 66,000$ psi

Thus,
$$L_{e,\text{screw}} = 2\,\frac{0.250 \times 10,000}{\pi \times 0.750 \times 0.1359 \times 66,000} = 0.237 \text{ in.}$$

Shear of nut threads [Eq. (10-31)]

$$L_{e,\text{nut}} = \text{f.s.}\,\frac{pQ}{\pi d_o t_o s_{s,\text{nut}}}$$

and all values are the same as for the screw except that

f.s. $= 4$, increased because design will be based upon ultimate shearing strength of bronze nut

$\qquad d_o = 1.000$ in.

$\qquad s_s = s_{su} = 35,000$ psi, from Appendix D

Thus,
$$L_{e,\text{nut}} = 4\,\frac{0.250 \times 10,000}{\pi \times 1.000 \times 0.1359 \times 35,000} = 0.669 \text{ in.}$$

Therefore, any length of thread engagement equal to or greater than 0.669 in. will result in adequate shear strength of the threads. Since wear is a function of the product of the bearing pressure and sliding velocity, it is evident that the maximum speed of operation will be possible when the bearing pressure is a minimum. Thus, the length of engagement should be as large as practical. The load will not be uniformly distributed over the several threads, being carried mostly by the first thread or two, and consequently there is little to be gained by using a length of engagement greater than one or one and one-half times the screw diameter. Here, we shall use $L_e = 1.5 \times 1 = 1.5$ in., calculate the bearing pressure, and see what sliding speed will be permissible on the basis of the values in Table 10-3. Equation (10-32) is

$$L_e = \frac{4pQ}{\pi(d_o{}^2 - d_i{}^2)s_b} \quad \text{in.}$$

Thus,
$$1.5 = \frac{4 \times \frac{1}{4} \times 10,000}{\pi \times (1.000^2 - 0.750^2)s_b}$$

Solving for s_b, we find

$$s_b = 4,850 \text{ psi}$$

which is higher than recommended for even low-speed, well-lubricated operation. Therefore, a new approach to the design is required. Nevertheless, our efforts have not been wasted entirely because we now know that any screw selected for this application on the basis of wear will be more than adequately strong. The difference is so great that, except for the increased resistance of the hardened surface to wear, there is little to be gained in using the heat-treated alloy steel, and, if conditions warrant, the material may be changed to cold-drawn low-carbon bar stock, 1020 or equivalent.

Redesign. The design will now be based entirely on wear considerations. A reasonable combination of allowable bearing pressure and sliding speed is 800 psi at 40 fpm, from Table 10-3. The size of screw that will give a bearing pressure of

800 psi or less, with a length of engagement of 1.5 times the nominal diameter d_o, must be determined by trial and error, as in the table below.

d_o, in.	d_i,* in.	p,* in.	L_e, in.	s_b, psi
2.000	1.5556	1/2.25	3.00	1,194
2.500	2.000	½	3.75	755

* Values from Table 10-1.

Therefore, a 2½-in. two-threads-per-inch modified square thread will be satisfactory.
Time required to raise the load.   The pitch-line sliding velocity is

$$V = \frac{\pi d n}{12} \quad \text{fpm}$$

and $V = 40$ fpm
  $d = 2.500 - h = 2.500 - 0.250 = 2.250$ in.
  $n =$ rpm of nut
Solving for n, we find it equal to 67.9 rpm. Each revolution will raise the load a distance equal to the lead. If a single-thread screw is used, the lead will be 0.500 in., and the lifting speed will be

$$V_{\text{load}} = nl = 67.9 \times 0.500 = 33.8 \text{ in./min}$$

and the time to raise the load 15 in. will be

$$t = \frac{S}{V} = \frac{15}{33.8} = 0.444 \text{ min} \quad \text{or} \quad 26.6 \text{ sec}$$

If 26.6 sec is too long, a double-thread screw will require only 13.3 sec, a triple-thread screw will require only 8.9 sec, etc.
  Motor horsepower requirements. If a lift time of 13.3 sec for the double-thread screw is considered reasonable, the drive horsepower will now be determined. In terms of torque and rotative speed, horsepower may be calculated from

$$\text{hp} = \frac{Tn}{63,000}$$

where $T =$ torque, lb-in., and $n =$ rpm.
  The torque calculated in the first part of the solution is no longer valid, because both the lead and the diameter have been changed, and another calculation must be made. Since the motor will probably have a starting torque in the order of 200 per cent of rated torque, there is considerable justification for determining the power requirements under running conditions. However, there is an appreciable difference between the starting and running coefficients of friction, 0.14 and 0.10, respectively, from Table 10-2, and a slight change in lubrication conditions could increase the coefficient of friction to the point where overheating of the motor might become a problem. Thus, the power requirement will be based on the starting coefficient of friction. The excess capacity of the motor will result in a more rapid acceleration of the load.

$$T = \frac{Qd}{2} \frac{\cos \alpha \tan \lambda + \mu}{\cos \alpha - \mu \tan \lambda} \quad \text{lb-in.}$$

and $Q = 10,000$ lb

$\quad d = 2.250$ in.

$\cos \alpha = 0.9962$

$\tan \lambda = l/\pi d = (2 \times \frac{1}{2})/(\pi \times 2.250) = 0.1415$

$\quad \mu = 0.14$

Thus,

$$T = \frac{10,000 \times 2.250}{2} \frac{0.9962 \times 0.1415 + 0.14}{0.9962 - 0.14 \times 0.1415} = 3{,}960 \text{ lb-in.}$$

Therefore, if the relatively small power losses in the thrust bearings, the guide bearing, and the gear train between the motor and the nut are neglected, the power required is

$$\text{hp} = \frac{3{,}960 \times 67.9}{63{,}000} = 4.27$$

Thus, a 5-hp motor will be specified. It should be noted that, for the double-thread screw, $\mu = 0.14$ and $\cos \alpha \tan \lambda = 0.1409$, and, therefore, the screw is not self-locking. The guide bearings, gears, etc., might possibly provide enough additional friction to hold the load at any point under starting or static friction. However, if for any reason, such as a pronounced vibration, the friction coefficient decreases to the running value, the load would rapidly lower itself. The solution is to use either the self-locking single-thread screw with its slower motion (and smaller power requirement) or to use a brake of some type, such as that in Fig. 10-6. In this design it will be assumed that the saving in lift time will justify the additional complication of adding a brake.

Specifications

Screw and nut: double-thread, modified-square-thread, 2½-in.-diameter, four-threads-per-inch, low-carbon-steel screw and bronze nut with length of engagement (nut height) of 3¾ in.

Operation time: 13.3 sec (neglecting acceleration and deceleration).

Motor: 5 hp.

COUPLINGS

Couplings are used to join one shaft to another. The joining of shafts may be necessary because of special requirements in manufacturing, shipping, installation, service, or operation of a machine. The most common application is the joining of two or more separately built or purchased units so that a new machine is formed, for instance, joining an electric-motor shaft to the input shaft of a hydraulic pump. The coupling may be required to join shafts at angles, to compensate for misalignment between shafts, to prevent the transmission of overload power, and to alter the vibration and shock characteristics of the drive. With few exceptions, the term *coupling* describes a device used to make a permanent or semipermanent connection, and the term *clutch* describes those devices which permit rapid connection or disconnection. Clutches will be discussed in Chap. 12.

11-1. Misalignment of Shafts. It is almost impossible to line up connecting shafts so that their axes are exactly collinear, i.e., with none of the types of misalignment shown in Fig. 11-1. Even where the bed plates are mounted on rigid structures and great care is taken in shimming the

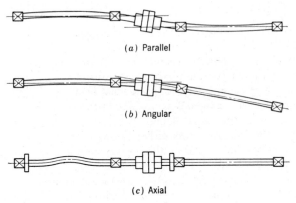

(*a*) Parallel

(*b*) Angular

(*c*) Axial

Fig. 11-1. Types of shaft misalignments.

equipment, there must be some provision for accommodating axial displacements due to thermal expansion.   This axial displacement, or misalignment, may be eliminated in the case of a complete design, such as a steam turbine driving a generator and exciter, by permitting only one of the several bearings to resist axial motion or thrust forces.

However, standard commercial units such as electric motors and gear reduction units are necessarily designed so that each piece of equipment contains at least one bearing which will resist axial motion of the shaft. Therefore, a coupling must usually accommodate some axial misalignment.

In the remainder of this chapter the several types of couplings will be classified according to the type and degree of misalignment permitted and according to their other useful characteristics.   The reader is referred to manufacturers' catalogues and the handbooks for specific data as to size, capacity, etc.

11-2. Rigid Couplings.   Rigid couplings do not accommodate misalignment and consequently should not be used indiscriminately.   The *flanged* shaft coupling in Fig. 11-2 is a widely used rigid coupling which

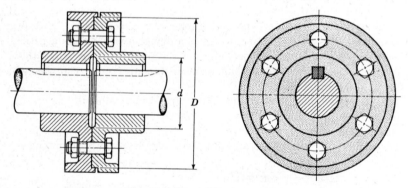

Fig. 11-2. Flanged shaft coupling.

is capable of transmitting large torques.   If the bolts are fitted in line-reamed holes, the torque is assumed to be transmitted by each bolt carrying its share of the load.   The coupling must be designed to avoid failure of the keys, shearing of the bolts, crushing of the bolts or flange, and failure by shearing of the radial flange at its juncture with the hub.   The use of line-reamed holes requires either a dowel or an unsymmetrical location of one of the holes so that the corresponding holes always line up correctly during assembly of the coupling.

In order to eliminate the necessity for using ground bolts and line-reamed holes, the coupling may be designed on the basis of transmitting the torque by friction between the contacting faces of the coupling.

Interchangeability is then no problem, because the bolts can be used in clearance holes and are loaded only in tension. The friction-torque capacity is based upon the concept of the friction force acting at the mean radius (friction radius) of the friction surfaces and is given by

$$T = N_b \mu P \frac{D + d}{4} \quad \text{lb-in.} \tag{11-1}$$

where N_b = number of bolts
  μ = coefficient of friction
  P = tension load in each bolt, lb
  D = outside diameter of friction face, in.
  d = inside diameter of friction face, in.

A conservative design value for the coefficient of friction μ is 0.1.

Alignment of the two halves of the coupling is ensured by providing some form of circumferential pilot surfaces, as at diameter D in Fig. 11-2, or, if the connecting shafts are the same diameter, by simply extending one shaft beyond its hub a short distance into the other hub.

The circumferential rims on the flanges are provided in the interest of safety to cover the protruding bolt heads and nuts.

11-3. Flexible Couplings. The term *flexible coupling* includes couplings which will operate satisfactorily under one or more of the types of misalignment in Fig. 11-1 and couplings which may or may not accommodate misalignment but, because of their torsional flexibility or resilience, assist in lessening the effects of shock loads or vibration which may be present.

The somewhat special cases of the spring coupling, the universal joint, the fluid coupling, and the electrical couplings will be considered in subsequent sections.

11-4. Flexible Couplings with No Torsional Flexibility. *Floating-center* couplings, typified by the *Oldham* in Fig. 11-3 and the *American*

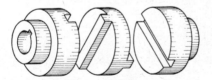

Fig. 11-3. Oldham coupling.

in Fig. 11-4, will accommodate considerable parallel and axial and some angular misalignment without play or backlash except that due to manufacturing tolerances and wear. All floating-center couplings require lubrication of the sliding surfaces. The American flexible coupling has replaceable wearing surfaces and is lubricated by grease flowing through

drilled passages from the center cavity to the points of wear. Standard
units are available with rated capacities up to 4,600 hp.

Gear couplings, as in Fig. 11-5, are best adapted for accommodating
axial misalignment, but clearance is usually built into the coupling so

Fig. 11-4. American flexible coupling. (*Courtesy Mechanical Power Transmission
Division, Zurn Industries, Inc.*)

Fig. 11-5. Amerigear flexible coupling. (*Courtesy Mechanical Power Transmission
Division, Zurn Industries, Inc.*)

that some angular and offset misalignments can be tolerated. The torque
is transmitted through the teeth, or splines, from one member (*A*) with
external teeth to the members (*B*) with internal teeth and then to the
remaining member (*C*) with external teeth.

The hub teeth in the *Amerigear* coupling in Fig. 11-5 are made with curved tips and faces, as shown in Fig. 11-6.   This design permits use under angular misalignments up to 11° with a minimum backlash.   The maximum rated capacity of a standard Amerigear coupling exceeds 65,000 hp at 100 rpm.   Lubrication is provided by oil under centrifugal force.

The *chain* coupling (Fig. 11-7) consists of two sprockets joined by an end-less double-width roller chain or inverted-tooth chain (Chap. 15).   These

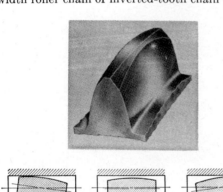

FIG. 11-6. Amerigear flexible-coupling hub teeth.   (*Courtesy Mechanical Power Transmission Division, Zurn Industries, Inc.*)

FIG. 11-7. Roller-chain coupling.   (*Courtesy Morse Chain Company, Division of the Borg-Warner Corporation.*)

couplings have some end play for axial misalignment, but they are particularly useful because the two sprockets may be readily connected or disconnected by inserting or removing one of the chain pins. The rollers are sometimes barrel-shaped to permit greater angular misalignment.

However, in general, the flexibility for offset and angular misalignments is limited to that provided by clearances. The coupling is usually enclosed in a grease-tight case.

The *flexible-ring* coupling in Fig. 11-8 requires no lubrication and accommodates all three types of misalignment. Each laminated flexible ring is attached to a hub and to the center ring at alternate 90° points, and there is no rotational backlash. The permissible misalignment varies with the size and capacity of the coupling; the ranges for Thomas couplings are

FIG. 11-8. Flexible-ring coupling. (*Courtesy Thomas Flexible Coupling Company.*)

parallel, $\frac{1}{64}$ to $\frac{5}{64}$ in.; angular, 1 to 2°; and axial, $\frac{1}{32}$ to $\frac{1}{4}$ in. Two couplings joined by an intermediate shaft may be used with greater parallel misalignment. These couplings have been used in drives with capacities from $\frac{1}{2}$ to 40,000 hp. Small sizes of this type of coupling have been used in potentiometer drives in precision electronic apparatus where torsional flexibility cannot be tolerated and the potentiometer cannot be located directly on the case.

11-5. Flexible Couplings with Small Torsional Flexibility. There are many variations of the *cushion* type of flexible coupling illustrated in Fig. 11-9. These couplings are relatively inexpensive and, though used

FIG. 11-9. Cushion-type flexible coupling. (*Courtesy Boston Gear Works.*)

mainly in fractional-horsepower drives, are available as stock items with capacities greater than 20 hp at 1,800 rpm.   The cushion is usually made of an oil-resistant synthetic rubber, although natural rubber and leather are also used.

Another group of couplings is made with steel pins in resilient bushings. If the steel pins are mounted in an axial direction, as for the *Ajax* coupling

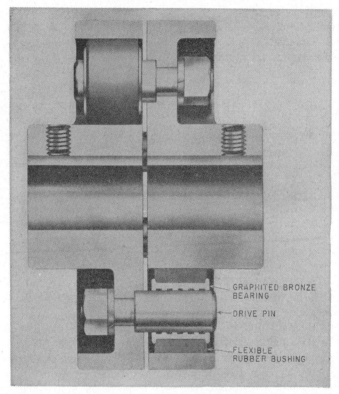

GRAPHITED BRONZE BEARING

DRIVE PIN

FLEXIBLE RUBBER BUSHING

FIG. 11-10. Ajax flexible coupling.   (*Courtesy Ajax Flexible Coupling Company, Inc.*)

in Fig. 11-10, the coupling can be used with relatively large axial misalignments.   In this case, the pins ride in self-lubricating bronze bushings which are bonded to the rubber bushing.   Standard Ajax couplings are made with capacities up to 15,000 hp at 2,500 rpm.   If the pins are mounted in a radial direction, all the flexibility must be given by deflection of the rubber bushings.

11-6. Flexible Couplings with High Torsional Flexibility.   The *bonded-neoprene Lord flexible* couplings in Fig. 11-11 may operate with $\frac{1}{32}$-in.

parallel and 2° angular misalignments and have a 15° twist angle under rated torque. They are particularly useful in minimizing the transmission of torsional vibrations from one shaft to the other (Sec. 19-4). The one-piece coupling *a* is manufactured for shafts from ³⁄₁₆- to 1-in. diameter and with capacities up to 1 hp at 1,750 rpm. In the larger sizes *b*, 2 to 15 hp at 1,750 rpm, the coupling is made in three pieces. This construction permits the installation or removal of the flexing element without moving the equipment. This is a matter both of convenience with heavy equipment and of economy with large couplings. The largest sizes, 20 to 100 hp at 1,750 rpm, are stiffer; the twist angle is only 2°, and the maximum angular misalignment is 1°.

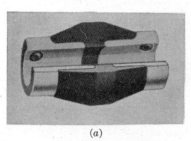

(a)   (b)

Fig. 11-11. Lord bonded-neoprene flexible couplings.   (*Courtesy Lord Manufacturing Company.*)

The *Morflex* coupling (Fig. 11-12) can be used with angular misalignment up to 5° and has a torsional deflection up to approximately 3.5°, depending upon its size. When a double coupling is used, the corresponding deflections are doubled. The preloaded rubber biscuits are designed to make the torsional deflection a linear function of the torque. The couplings are available with bore diameters of ¼ to 3 in. and in capacities up to about 250 hp at 1,750 rpm.

When the power transmitted is very small, a piece of rubber tubing clamped to the ends of the two shafts makes a simple and satisfactory connection.

The *Falk Steelflex* coupling (Fig. 11-13) has a chrome-alloy-steel grid spring for its resilient member. The tapered grooves automatically shorten the unsupported length of the spring, thus increasing the torsional rigidity of the coupling as the deflection increases. This feature protects the coupling under severe overload because the spring length approaches zero and the force is transmitted in almost pure shear rather than in bending. The capacity is related to conditions of use, particularly to the degree of torque pulsations. Standard couplings are available in a

Fɪɢ. 11-12. Morflex coupling.   (*Courtesy Morse Chain Company, Division of the Borg-Warner Corporation.*)

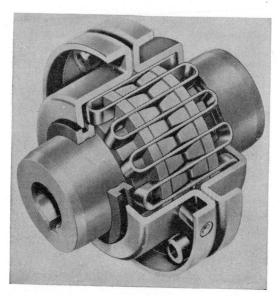

Fɪɢ. 11-13. Falk Steelflex coupling.   (*Courtesy the Falk Corporation.*)

wide range of sizes from a basic rating capacity of 0.3 hp to 130,000 hp at 100 rpm. The housing is filled with grease to minimize wear resulting from the spring sliding in the grooves while under load.

11-7. Couplings with Torsional Flexibility Only. The *coil-spring* coupling in Fig. 11-14 is used only to minimize the transmission of shock

FIG. 11-14. Coil-spring coupling. (*Courtesy Westinghouse Electric Corporation.*)

loads or to alter the torsional-vibration characteristics of the drive. The coil springs are loaded in compression whenever one hub is rotated relative to the other. Important applications of the spring coupling are between the gears and propeller shaft of diesel-powered boats, in automotive clutches, in aircraft-engine auxiliary drives, and between the gears and wheels of electric locomotives. In most cases, a compact unit is achieved by making the coupling an integral part of the gear or clutch.

Hydraulic or *fluid* couplings (Fig. 11-15) may be considered to be flexible couplings in that a major function is to reduce the transmission of shock and vibration. However, none of the couplings previously discussed alters the torque-speed characteristics or offers any protection against overloads, while the fluid coupling has a marked effect on both.

Figure 11-15b illustrates the manner in which the fluid picks up energy as its velocity increases because of its outward radial flow in the primary half of the coupling. The fluid then flows axially to the more slowly rotating secondary and gives up its energy as it flows inward to the point where its flow is directed axially again to the primary impeller, and the cycle starts again. There can be no transfer of energy from the input to the output shaft unless the secondary is rotating at a lower speed than the primary. The difference between the primary and secondary speeds is

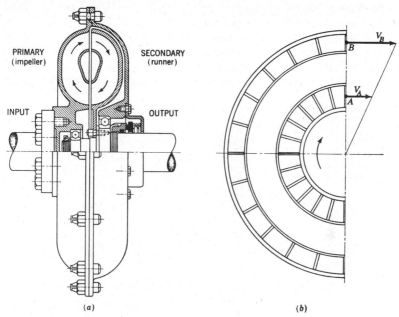

Fig. 11-15. Hydraulic coupling.

expressed as *per cent slip*, or

$$s = \frac{n_p - n_s}{n_p} \times 100 \qquad \text{per cent} \tag{11-2}$$

where n_p = primary speed, rpm, and n_s = secondary speed, rpm.

The ordinary fluid coupling, consisting of only a primary and secondary set of blades, can have no effect on the torque being transmitted; i.e., *the output torque always equals the input torque*.   Therefore, since horsepower is

$$\text{hp} = \frac{Tn}{63,000} \tag{11-3}$$

where T = torque, lb-in., and n = speed, rpm, and efficiency is defined as

$$\eta = \frac{\text{power output}}{\text{power input}} \times 100 \qquad \text{per cent} \tag{11-4}$$

the efficiency is directly related to slip and becomes

$$\eta = 100 - s \qquad \text{per cent} \tag{11-5}$$

Fluid couplings are sometimes operated with as little as 1 per cent slip, but usually the slip is 3 to 6 per cent.

Typical curves showing the torque-speed characteristics for an electric motor and a hydraulic coupling are given in Fig. 11-16. For this particular set of curves, the design operating point at 100 per cent load will require a motor speed of 1,750 rpm and a slip of about 3.6 per cent. Thus

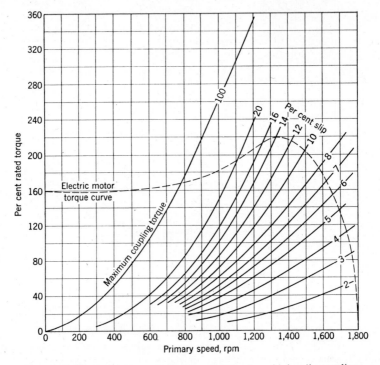

FIG. 11-16. Typical torque-speed characteristics for a hydraulic coupling.

the output speed will be $1,750 \times (100 - 3.6)/100 = 1,690$ rpm, and the coupling efficiency is 96.4 per cent. If the load torque increases to 200 per cent of rated motor torque, the motor speed drops to 1,510 rpm, the output speed to 1,370 rpm, and the coupling efficiency to 90.4 per cent. If the load torque exceeds 218 per cent of rated motor torque, the coupling stalls, and the motor speed drops to 780 rpm and continues to run at that speed until either the overload is removed or a thermal protective device cuts off the current to the motor.

With slight modification, fluid couplings are used as variable-speed drives. This application is discussed in more detail in Sec. 17-6.

A *torque converter* is similar to a hydraulic coupling except that additional rows of stationary blades are used to change the direction of flow and thus get a reaction from the frame so that there can be a multiplication of torque.

Electromagnetic couplings of the type illustrated in Fig. 11-17 have been used to reduce peak stresses in diesel-engine marine drives and as clutches to permit engagement or disengagement of one or more engines.[1] The rotating primary winding has radial poles and is supplied with direct current through the slip rings. The secondary winding is similar to the rotor of a squirrel-cage induction motor and, because of its rugged construction, is generally connected to the engine, or disturbance end of the drive.

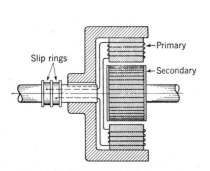

FIG. 11-17. Electromagnetic coupling.   FIG. 11-18. Torque-slip curves for hydraulic and electromagnetic couplings.

The major difference between hydraulic and electromagnetic couplings is in their torque-slip characteristics, as shown in Fig. 11-18 for couplings with the same maximum torque capacity. The electric coupling has much the same characteristics as an induction motor, in that the speed of the secondary must be within about 10 or 15 per cent of the synchronous speed or pull-out may occur. It is thus necessary for the maximum coupling torque to exceed maximum engine torque by some value, e.g., about 75 per cent, to ensure stable operation. The electromagnetic coupling transmits a much larger torque at low values of slip and consequently is more efficient than the hydraulic coupling. Because of the power supplied to the primary winding, the actual efficiency of the electromagnetic coupling is somewhat less than calculated when slip only is considered. This loss is small, and efficiencies of 99 per cent have been claimed for conservatively rated couplings.

[1] Additional types of electromagnetic couplings better suited for use as clutches and variable-speed drives will be discussed in Secs. 12-5 and 17-6.

11-8. Universal Joints. The term *universal joint* describes a spherical linkage which permits relatively large angular motions in every direction. The joint itself has no torsional flexibility nor can it accommodate any parallel misalignment; i.e., shafts must intersect. Two universal joints with an intermediate shaft can allow much greater parallel misalignment than any flexible coupling. If the offset changes during operation, as for the automobile propeller-shaft drive, it is necessary to provide some freedom for axial motion in the form of a splined or some similar connection.

The *Hooke's coupling* or *Cardan joint* in Fig. 11-19a is the most common form of universal joint and consists of two forks and an intermediate

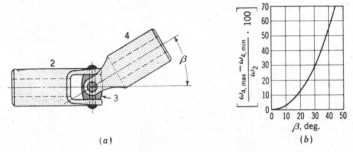

(a)   (b)

FIG. 11-19. Hooke's coupling.

block containing pins in the form of a cross. Figure 11-20 is an exploded view of a universal joint which uses needle bearings (Sec. 18-18) to substitute rolling friction for the sliding friction of the joint in Fig. 11-19a. Standard needle-bearing universal joints are available with torque capacities up to 70,000 lb-in., and standard plain journal-bearing joints are available with capacities up to about 90,000 lb-in. for continuous duty and 500,000 lb-in. for momentary loads. It is shown in kinematics textbooks[1] that, while both yokes of the joint must make complete revolutions in the same time, the ratio of the angular velocity of the driven yoke to that of the driving yoke will not be constant during a revolution. The variation from a uniform-velocity ratio is a function of the angle of intersection of the two shafts and is so presented in Fig. 11-19b. Since this cyclic variation results in the introduction of a torsional vibration, it is recommended that the working angle be kept below 15°. Angles up to 45° may be used for hand or very slow-speed operation. The torsional acceleration due to the varying angular-velocity ratio can be calculated, and single universal joints are sometimes used in calibrating angular accelerometers and in fatigue tests involving combined bending and torsion loads.

[1] C. W. Ham, E. J. Crane, and W. L. Rogers, "Mechanics of Machinery," 4th ed., pp. 40–42, McGraw-Hill Book Company, Inc., New York, 1958.

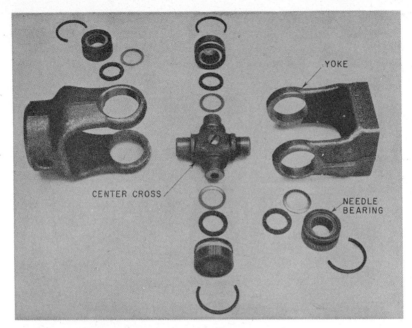

Fig. 11-20. Needle-bearing universal joint. (*Courtesy Blood Brothers Machine Division, Rockwell Spring and Axle Company.*)

In order to transmit a constant angular-velocity ratio, it is necessary to use two Hooke's couplings and an intermediate shaft so arranged that the variation between the input and the intermediate shafts is exactly compensated for by an equal but opposite variation between the intermediate and output shafts.   The necessary conditions are as follows:

1. The angle of intersection of the input and intermediate shafts must equal the angle of intersection of the intermediate and output shafts.

2. The forks at each end of the intermediate shaft must simultaneously lie in the planes of their respective intersecting shafts.

In practically all applications, the input and output shafts are parallel, and condition 2 simplifies to requiring that the forks at the ends of the intermediate shaft be parallel, as illustrated in Fig. 11-21.

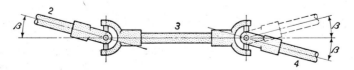

Fig. 11-21

The fact that it is not practical to use two Hooke's couplings and an intermediate shaft in driving the front wheels of military and construction vehicles and sport cars led to the development of the *constant-velocity* universal joints.   The principle of operation is shown in Fig. 11-22.   The

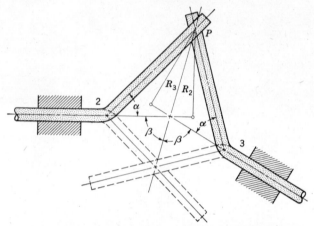

FIG. 11-22. Simple constant-velocity universal joint.

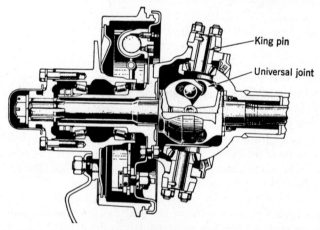

FIG. 11-23. Vehicle front-wheel drive.   (*Courtesy Bendix Products Division, Bendix Aviation Corporation.*)

point of contact between the two members must always lie in the plane bisecting the angle between the shafts.   The radius to the point of contact will be the same for both members, and, therefore, the angular velocities of the two shafts will be equal at all times.   The plane of motion transmission is called the *homokinetic* plane.

Figure 11-23 shows the *Bendix-Weiss constant-velocity joint* as used in a front-wheel drive. Both motion and force are transmitted through four steel balls which are constrained by the intersection of curved grooves, or races, in the yokes to lie in the homokinetic plane, as shown in Fig. 11-24.

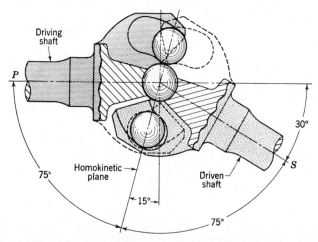

Fig. 11-24. Bendix-Weiss constant-velocity universal joint. (*Courtesy Bendix Products Division, Bendix Aviation Corporation.*)

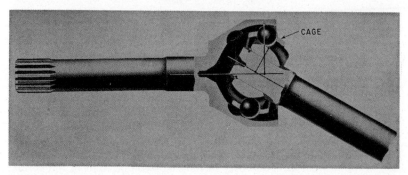

Fig. 11-25. Rzeppa constant-velocity universal joint. (*Courtesy Con-Vel Division, Dana Corporation.*)

A major advantage of the Bendix-Weiss joint is that the balls may roll and accommodate endways motion without a sliding-spline connection.

The *Rzeppa* (pronounced sheppa) *constant-velocity universal joint* (Fig. 11-25) utilizes a cage to keep the six balls in the homokinetic plane at all times. As shown in the figure, the cage is piloted on spherical surfaces on the outer and inner members, the center being at the joint center, i.e., the point of intersection of the shaft axes. The centers of curvature for the grooved races are offset, in opposite directions from the joint center along the axes of the shafts.   The resultant wedging action automatically forces the balls and cage into their correct positions.

The load-carrying capacity of both types of constant-velocity joint is dependent upon the same factors as for rolling-contact bearings (Secs. 18-21 and 18-22), and, in general, the joint capacity is based on a limited number of cycles of operation.   Since the rolling motion of the balls is a function of both the speed and the angle between the shafts, both factors become important considerations in determining the size of joint required. The capacity for continuous service with a working angle of 35° is only 30 per cent of that for a working angle of 3° or less.   Standard Rzeppa joints are available with basic rated torque capacities up to 56,700 lb-in. for continuous duty.

11-9. Flexible Shafts.   Flexible shafting is extensively used for low-power drives, such as speedometer drives, positioning devices, and portable grinders, where the utmost flexibility is desired to accommodate large angular and offset misalignments or simply to decrease the complexity of the drive.   A typical shaft (Fig. 11-26) consists of an inner core of wire

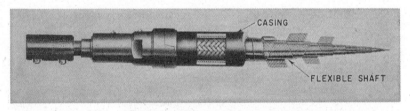

FIG. 11-26. Flexible shaft.   (*Courtesy Stow Manufacturing Company.*)

coils surrounded by a casing which protects the shaft, retains the lubricant, and prevents the shaft from whipping or winding up on itself.

The torque capacity of a given flexible shaft is dependent mainly upon the direction of rotation and the minimum radius of curvature of the shaft.   Catalogue ratings are given for rotation in the wind-up direction, i.e., tending to tighten the outer layer of wires, and the capacity for rotation in the unwind direction may be only 50 to 80 per cent of the normal rating.   A shaft rated for 10 hp at 1,800 rpm, when the radius of curvature is over 50 in., will carry only 3 hp when the radius of curvature is 5 in.

11-10. Overload Release Couplings.   Overload protection is recommended in applications where the motor armature inertia is a major portion of the total machine inertia; where light accessory drives are taking power from a heavy-duty main drive; where jamming may easily occur, as in tapping machines and packaging machines; and where limit switches or releases are used to stop or reverse motion.   In general, the release device should be located so that the effects of inertia will be minimized.

The *shear pin*, shown in Fig. 11-27 as applied to an outboard-motor propeller shaft, is the simplest and cheapest overload protection.   It is not recommended for general use because it is not very sensitive and must be replaced after failure.   It is important to educate the operators not to replace the necked-down soft pin with a full-diameter hardened pin.

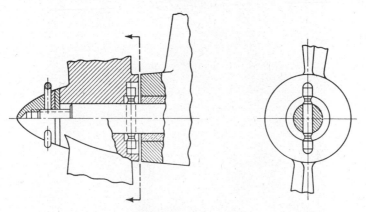

Fig. 11-27. Shear pin.

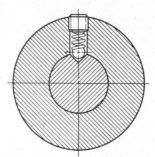

Fig. 11-28. Ball detent.

The *spring-loaded ball detent* in Fig. 11-28 is another simple device that is suitable for light loads.   It may be designed as a trip clutch to release the load completely until reset or, as illustrated, with a simple spring-loaded ball which will chatter until the load drops below the design limit

Overload release couplings that transmit torque by means of the friction between spring-loaded surfaces are commonly known as *slip couplings* or *slip clutches*.   While both devices are actual couplings, as defined in the introduction to this chapter, the manufacturers have found it convenient to use the term *coupling* when two shafts are being connected and the term *clutch* (Fig. 11-29) when a sheave, sprocket, etc., is being connected to the shaft.   In Fig. 11-29, the driving member would usually be mounted on the sleeve.   The equation for the capacity of a slip coupling

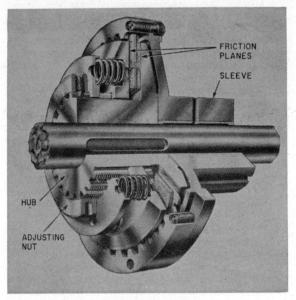

FIG. 11-29. Single-plate slip clutch.   (*Courtesy the Hilliard Corporation.*)

is basically the same as Eq. (11-1), which was derived for the rigid flanged-shaft coupling transmitting torque by means of friction:

$$T = N \mu P \, \frac{D + d}{4} \quad \text{lb-in.} \tag{11-6}$$

where N = number of friction planes
μ = coefficient of friction
P = axial force, lb
D = outside diameter of friction surfaces, in.
d = inside diameter of friction surfaces, in.

Since the material requirements and design details for slip couplings are so similar to those for clutches, the reader is referred to the more extensive discussion in Chap. 12 for additional information.   One consideration of major importance is that the difference between input and output horsepower must be dissipated as heat, the worst case being when the output shaft stops completely.   The result will be a very rapid rise in temperature and possible destruction of the friction surfaces unless a warning device or thermal cutoff is provided.

CHAPTER 12

CLUTCHES AND OTHER INTERMITTENT-MOTION MECHANISMS

Clutches are used to connect or disconnect shafts at will. The principal types are positive, friction, hydraulic, and electromagnetic clutches. In addition to these, the intermittent-motion and unidirectional mechanisms may be considered as special types of clutches. The particular operational characteristics of hydraulic and eddy-current clutches are advantageous in many applications where either automatic or manual speed control is required. They may be used as simple connect-disconnect clutches, but since they are normally used in variable-speed operation, the discussion of these types of clutches will be left for Chap. 17.

12-1. Positive Clutches. Two common forms of positive clutches are the *spiral-jaw* clutch in Fig. 12-1a and the *square-jaw* clutch in Fig. 12-1b.

Fig. 12-1. Jaw clutches. (a) Spiral-jaw; (b) square-jaw. (*Courtesy Link-Belt Company.*)

One of the members must always slide axially on feather keys or splines to engage and disengage the clutch. The square-jaw clutch is the simplest form and can theoretically transmit torque in either direction without introducing an axial component of force. The spiral-jaw clutch can be engaged at somewhat higher speeds without serious clashing, but it

225

can transmit torque in only one direction without requiring an external axial force to maintain the engagement.   Frequent engagement may cause sufficient wear of the jaws so that either type of jaw clutch will develop an axial-force component.

Straight splines (Sec. 8-7) are also widely used to give positive engagement with no tendency to develop an axial force.   The most common application of the spline clutch is in automotive transmissions, as in Fig. 12-8.

12-2. Friction Clutches.   Friction clutches are composed of two or more rotating concentric surfaces which are forced together so that the tangential friction forces transmit torque from the input to the output shafts.   The principal advantage of friction clutches is that the engagement may be smoothly made by a gradual increase in the normal force. Since slipping normally occurs only during engagement and, to a lesser extent, disengagement, the power loss and consequent heating become problems only when the operation requires frequent engagement.   Friction clutches are almost a necessity when low-torque prime movers, such as the internal-combustion engine and certain electric motors, must start a load moving from rest.   They are also frequently used when it is desirable to have one source of power drive two or more shafts which have different cycles of starting and stopping.

12-3. Plate Clutches.   A single-plate truck, or automotive-type, clutch is illustrated in Fig. 12-2.   The single driven disk is surfaced on both sides with friction material and is free to move axially on the splines through which it transmits the torque to the output shaft.   Both the flywheel and pressure plate rotate with the engine crankshaft.   The pressure plate is spring-loaded and forces the driven plate into contact with the flywheel. Thus, the clutch is engaged at all times except when the clutch fork engages the release bearing and moves it to the left.   The linkage pulls back the pressure plate against the spring force and releases the clutch.

It is important to note the following:

1. That an odd number (three) of disks is used and that the axial force during engagement is entirely internal.   This simplifies the design in that the bearings are not required to transmit a thrust force between the frame and the rotating members except during the relatively short periods in which the clutch is disengaged.

2. That each surface of the driven disk is subject to the same normal and friction forces.   Therefore, *the number of friction planes* or pairs of friction surfaces, *not the number of disks, determines the capacity of the clutch.*

The design of plate clutches is based upon one of three premises:

1. Uniform pressure
2. Uniform axial wear
3. Average friction radius

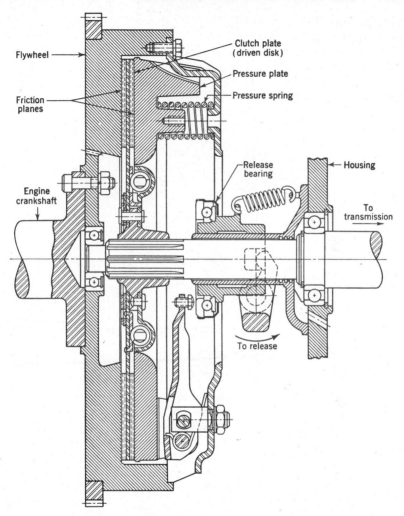

Flywheel

Clutch plate
(driven disk)

Pressure plate

Pressure spring

Friction
planes

Release
bearing

Housing

Engine
crankshaft

To
transmission

To release

FIG. 12-2. Automotive-type single-plate clutch.   (*Courtesy Rockford Clutch Division, Borg-Warner Corporation.*)

The friction torque given by the single friction surface in Fig. 12-3 is found by writing equations for the tangential friction force F and the friction torque T for the elemental area and then integrating.   In general,

$$dF = \mu \, dP \qquad F = 2\pi\mu \int_{d/2}^{D/2} pr \, dr \tag{12-1}$$

and

$$dT = r \, dF \qquad T = 2\pi\mu \int_{d/2}^{D/2} pr^2 \, dr \tag{12-2}$$

where μ = coefficient of friction

P = total operating force, lb

p = pressure on elemental area, psi

Uniform Pressure.   For rigidly mounted, unworn friction surfaces the usual assumption is that the force P is uniformly distributed over the

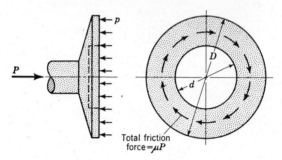

FIG. 12-3

entire area, or

$$p = \frac{P}{A} = \frac{4P}{\pi(D^2 - d^2)} \quad \text{psi} \tag{12-3}$$

Substituting from (12-3) for p in (12-2) and integrating gives, for a single friction plane,

$$T = \frac{\mu P}{3} \frac{D^3 - d^3}{D^2 - d^2} \quad \text{lb-in.} \tag{12-4}$$

Uniform Axial Wear.   As discussed in Secs. 3-11 and 10-6, the basic principle for the design of machine parts that are subject to wear due to sliding friction is that the normal wear is proportional to the work of friction.   The work of friction is proportional to the product of the normal pressure and the sliding velocity.   Therefore,

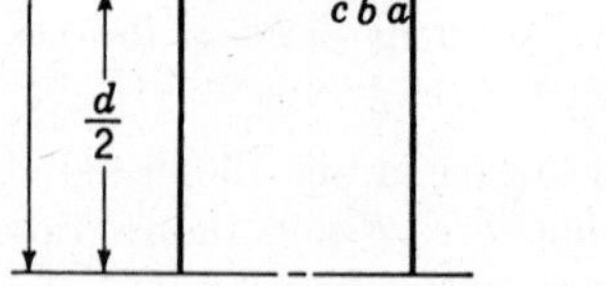

$$\text{Wear} \propto pV$$

and a friction surface which starts with a uniform pressure distribution will wear most rapidly where the sliding velocity is the greatest and in so doing will reduce the pressure between the friction surfaces.   This wearing-in process continues until the product pV is a constant over the entire surface, after which time the wear proceeds uniformly, as illustrated in Fig. 12-4.

FIG. 12-4

After the initial wear has ended,

$$p = \frac{C}{V} = \frac{C'}{r} \tag{12-5}$$

where C and C' are constants. The total operating force is

$$P = \int p \, dA \tag{12-6}$$

Substituting from (12-5) for p in (12-6) and integrating gives

$$P = \pi C'(D - d) \tag{12-7}$$

and

$$C' = \frac{P}{\pi(D - d)} \tag{12-8}$$

Then substituting from (12-5) for p in (12-2) and integrating,

$$T = \pi \mu C' \frac{D^2 - d^2}{4} \tag{12-9}$$

Substitution from (12-8) for C' into (12-9) results in, for a single friction plane,

$$T = \mu P \frac{D + d}{4} \qquad \text{lb-in.} \tag{12-10}$$

Average Friction Radius. The concept of the resultant tangential friction force being concentrated at the average friction radius has been used in the sections on rigid couplings (Sec. 11-2) and slip couplings (Sec. 11-10). The equation for the torque transmitted by a single friction plane is

$$T = \mu P \frac{D + d}{4} \qquad \text{lb-in.} \tag{12-11}$$

For almost all practical purposes, it is both satisfactory and desirable to design a clutch on the basis of uniform axial wear or its equivalent, the average friction radius. The torque capacity given by Eq. (12-11) will be conservative for a new, unworn clutch and will be correct after the initial wear has taken place. Since that part of the friction material located at a small radius does not contribute very much to the torque capacity of the clutch, d is seldom less than $0.5D$ and is usually between 0.6 and $0.7D$. Consequently, the difference between the capacity given by the several equations is less than the variation in accepted design values of coefficient of friction μ. Therefore, the capacity of a plate clutch with N friction planes should be based upon

$$T = N \mu P \frac{D + d}{4} \tag{12-12}$$

The limiting factor in clutch design is wear of the friction surfaces. For a given capacity, the rate of wear will be dependent upon the properties of the materials, the area of the friction surfaces, and the frequency of operation. These factors are included in manufacturers' recommendations which specify allowable pressures calculated as uniformly distributed pressure; that is, $p = P/A$.

The majority of dry plate clutches depend upon either natural or forced convection for cooling and generally use an asbestos composition in contact with cast iron.   However, bronze plates in contact with steel plates and metal-ceramic, leather, cork, and wood friction surfaces are also used successfully.

In many situations, it is difficult to seal off the clutch from entry of oil or grease from bearings or other parts of the machine, e.g., when a number of clutches are used in one machine, such as in machine-tool and automobile automatic transmissions.   In this case, it is usually advisable to eliminate the problem, and the need for some of the seals, by using a wet clutch.

A copious supply of oil is sometimes used to assist in transferring heat from the clutch, but, except in unusual circumstances, the term *wet clutch* does not mean that it operates immersed in oil.   In general, *a wet clutch is one designed for use in an oil atmosphere*.   The gains in simplicity of design and a more rapid heat-transfer rate are offset by the decrease in coefficient of friction—from 0.3 or better for dry operation to less than 0.1 for wet operation.

The *Maxitorq* clutch in Fig. 12-5 is a typical *multiple-disk* clutch designed

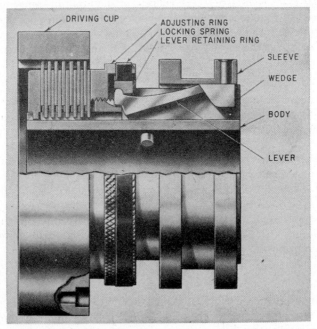

FIG. 12-5. Maxitorq multiple-disk clutch.   (*Courtesy the Carlyle Johnson Machine Company.*)

for either wet or dry operation.   For wet operation the plates are either all hardened steel or alternately steel and steel-backed bronze.   For dry operation the plates are alternately steel and steel-backed self-lubricating bronze.   Cork and cork-rubber facings are also commonly used in wet clutches.

The clutch in Fig. 12-5 has 15 disks, seven keyed to the driving cup and eight keyed to the body, and there are 14 friction planes.   When the sleeve is moved to the left, the wedges force the levers down and the resulting lever rotation acts to press the plates together.   Since the lever motion is dependent only upon the dimensions of the parts, the axial pressure—and therefore the torque capacity—on the plates during engagement may be controlled by adjustment of the clearance between plates when the clutch is disengaged.   Undesired rotation of the adjusting ring is prevented by the locking spring which passes through a slot in the adjusting ring and engages one of the numerous slots in the lever retaining ring. Wave washers (Fig. 9-22e) are used between the disks keyed to the body as separator springs to keep the plates from rubbing when the clutch is disengaged.

Meager design data are available in relation to values of coefficient of friction and allowable pressure for the different combinations of materials. Automotive clutches, which use an asbestos composition in contact with cast iron, are designed on the basis of a coefficient of friction between 0.30 and 0.40.   The allowable pressure varies from 15 psi for large heavy-duty double-plate clutches to 35 psi for the average passenger-car clutch. Commercial units are rated by the manufacturer for operation under specified conditions.   In order to account for the usual variations from standard operating conditions, service factors are applied to the nominal drive torque to determine the required rated capacity of a standard unit. These service factors are based upon type of prime mover, type of load, hours of daily use, and frequency of engagement.   The allowable pressure for clutches with metal plates (Fig. 12-5) is in the order of 100 to 150 psi.

The operating force that presses the friction surfaces together may come from springs, mechanical toggle linkages, electromagnets, or hydraulic or pneumatic pressure.

12-4. Miscellaneous Friction Clutches.   The *Fawick Airflex* clutch in Fig. 12-6 uses compressed air to expand the actuating tube and force the friction shoes against the drum on the driven member.[1]   The torque is transmitted through the actuating tube to which the shoes are attached, and the flexibility of the actuating tube helps to minimize the transmission of shock or vibration and to accommodate unavoidable misalignment. Since the torque capacity is a function of the air pressure, the clutch is also well suited for overload protection service.   Standard units are avail-

---

[1] See Chap. 13 for discussion of the use of shoes as friction members.

able with torque capacities up to 1,130,000 lb-in. at 75 psi air pressure and 100 rpm.   Because of the effect of centrifugal force on the actuating tube and shoes, the torque capacity of the clutch in Fig. 12-6 decreases with increasing speed.   The largest standard unit has a maximum rated capacity of 4,407 hp with 75 psi air pressure at 400 rpm.   Increasing the air pressure to 100 psi raises the capacity to 6,969 hp.

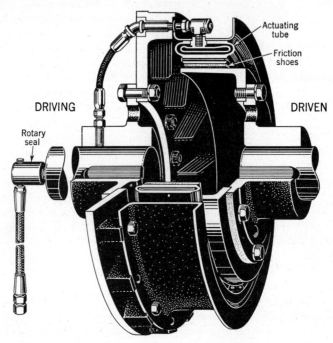

FIG. 12-6. Fawick Airflex clutch.   (*Courtesy Fawick Airflex Division, Fawick Corporation.*)

Figure 12-7 shows a combination of a flywheel, a clutch, a spring-operated air-released brake, a rotary timing switch, and electrically operated air valves that is supplied as a standard unit for applications where a large force is required during a short period of time, as for punch presses. In this case, it is uneconomical to use a motor with a rated torque sufficient to supply the peak load at rated speed.   The best solution is to use a flywheel to supply the energy for the short-time peak loads.   The motor then has a relatively long period of time before the next peak load in which to bring the flywheel back up to its desired speed, thus restoring the energy given up by the flywheel during the working period.   While the selection of the proper electric motor cannot be considered in detail here, it is worthwhile to point out that in this case NEMA Design D, or high-slip, motors are normally used in preference to the more common class A or B motors.[1] The speed of a high-slip motor drops rapidly with an increase in torque and a relatively small increase in motor current.   Thus, the torque impulse

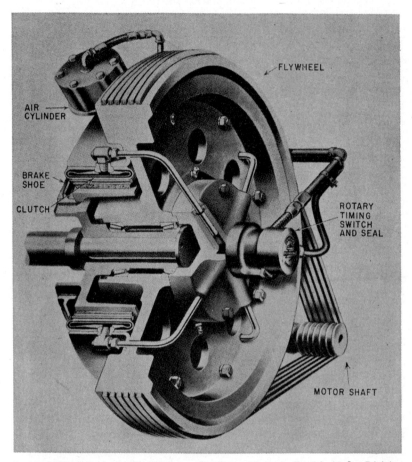

FLYWHEEL

AIR CYLINDER

BRAKE SHOE

CLUTCH

ROTARY TIMING SWITCH AND SEAL

MOTOR SHAFT

FIG. 12-7. Fawick standardized press application.   (*Courtesy Fawick Airflex Division, Fawick Corporation.*)

will be supplied mainly by the flywheel, and the motor torque need be adequate only to bring the flywheel back up to speed in the time available before the next stroke.

[1] C. C. Libby, "Motor Selection and Application," chap. 5, McGraw-Hill Book Company, Inc., New York, 1960.

In Fig. 12-7 the timing switch controls the sequential operation of air valves which in turn disengage the brake, engage the clutch, disengage the clutch, and engage the brake.   The brake is necessary to ensure that the operating mechanism quickly stops at the correct position for unloading and loading the machine in preparation for the next cycle of operation.

Conical clutches were extensively used as the main clutches in automobiles and other machines.   In recent years, the plate clutch has almost completely displaced the conical clutch except for some minor uses, such as in the synchronizing device in Fig. 12-8.

Figure 12-8 shows the part of a standard synchromesh automobile transmission used in shifting into second and third gears.   The other gears are

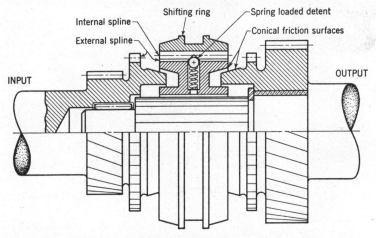

FIG. 12-8. Synchromesh spline clutch.

not shown (see Fig. 17-12), but the spline clutch is shifted to the right to engage the gear to the output shaft for second gear and to the left for direct engagement of the input shaft to the output shaft for third gear.   Without the conical synchronizing clutches, the relative motion of the internal and external splines would result in clashing of the teeth, which is both noisy and destructive.   In operation, the shifting-ring motion is transmitted by the spring-loaded ball detent to bring one of the internal conical friction surfaces into contact with the desired external cone.   The friction force is sufficient to accelerate the idling parts, i.e., the clutch is disengaged, until the speeds of the splines are the same, or synchronized.   Continued motion of the shifting ring overcomes the ball-detent resistance, and the internal spline of the shifting ring engages the external spline without clashing.   Since the teeth may not line up exactly when the speeds are synchronized, the spline teeth are rounded or pointed so that the positions may adjust slightly during the engagement.   It should be noted that power is transmitted only through direct engagement of the splines.

If the resultant tangential friction force is considered to act at the average friction radius, the equation for the capacity of the conical clutch in Fig. 12-9 is

$$T = \frac{\mu P_a}{4 \sin \alpha} (D + d) \qquad \text{lb-in.} \qquad (12\text{-}13)$$

where P_a = axial (operating) force, lb, and α = cone-face angle.

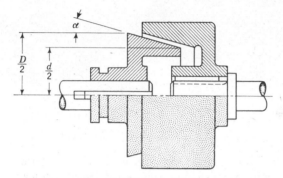

Fig. 12-9. Conical clutch.

Analysis of Eq. (12-13) shows that the capacity of a conical clutch can be increased by decreasing the cone-face angle.   Some practical considerations which limit the use of this principle are the difficulty in compensating for the effect of wear on the axial location of the parts, the necessity for supplying the axial force to maintain engagement, and the additional axial force required to overcome friction while engaging and disengaging the clutch.   A practical value for α is 12.5°.   It should be noted that, if $\alpha = 90°$, the capacity becomes the same as for a single-friction-plane plate clutch.

Centrifugal clutches are designed to give automatic and smooth engagement of the load to the driving member whenever the driving member is running at or above some minimum speed.   Since the operating centrifugal force is a function of the square of the angular velocity, the friction torque available for accelerating a load is also a function of the square of the speed of the driving member.   This characteristic makes the centrifugal clutch ideal for applications requiring smooth load pickup, as for textile and paper machinery, or where the motor starting torque is relatively low, as for a synchronous motor driving an inertia load.

Centrifugal clutches are particularly useful with internal-combustion engines, which cannot be started under load.   The need for a manually operated clutch is eliminated, and the engine will not stall if it is subjected

FIG. 12-10. Centrifugal clutch with two sets of shoes.   (*Courtesy Formsprag Company.*)

to an overload, because the clutch will slip after the speed drops below that required to transmit the torque demanded by the load.   These features have been responsible for the widespread application of the centrifugal clutch to gasoline motor bicycles, gasoline-engine-powered chain saws, and helicopter rotor drives.

The simplest form consists of a single ring of rotating shoes which are pressed against the cylindrical drum by centrifugal force.   A variation, shown in Fig. 12-10, has two sets of shoes.   The inner shoes on the driven half engage the driving drum after the driven half is brought up to speed by the outer shoes on the driving half.   Figure 12-11 shows typical torque-

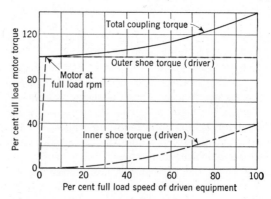

FIG. 12-11. Torque-speed curves for a typical centrifugal clutch with two sets of shoes.

speed curves for this type of clutch. Note that the motor comes up to full speed almost immediately, and the full outer-shoe (driver) torque is quickly available for accelerating the load. The total torque capacity then increases as the driven member comes up to speed. The net result is a well-balanced combination of smooth pickup and overload capacity at rated speed.

Standard couplings are available in a number of sizes with capacities up to 550 hp at 690 rpm and 160 hp at 1,750 rpm. Also available are couplings in which springs keep the shoes from contacting the drum until a predetermined speed is reached. In this case, only the outer shoes are used. This is advantageous when an engine is the prime mover, because it permits warming up and idling of the engine without friction between the shoes and the drum.

The clutch in Fig. 12-12 is operated by mercury under centrifugal pressure. As the driving member comes up to speed, the mercury moves out-

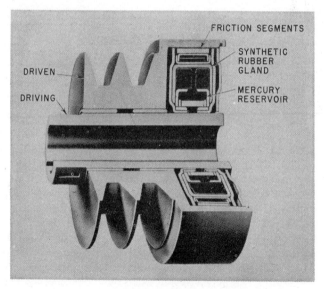

FIG. 12-12. Mercury centrifugal clutch. (*Courtesy Automatic Steel Products, Inc.*)

ward under centrifugal pressure and forces the friction segments, or shoes, against the drum to which the V-belt sheave is attached. Mercury clutches are available in standard models with capacities up to 40 hp at 1,800 rpm, and special units have been made with capacities up to 1,000 hp. Retracting springs are used to keep the clutch disengaged below a predetermined (idling) speed.

Overrunning or *freewheeling* clutches are used where it is desirable to transmit torque in only one direction.   Overrunning clutches have been used in bicycle coaster brakes, in automobile freewheeling, in torque converters to permit the "stationary"-blade elements to freewheel when necessary, in helicopter drives to permit rotor freewheeling, and as high-speed or fine-adjustment ratchets in intermittent-motion applications such as feed mechanisms.

Figure 12-13 shows a light-duty (fractional-horsepower) overrunning clutch.   The small angle between the inner cam flat and the tangent at

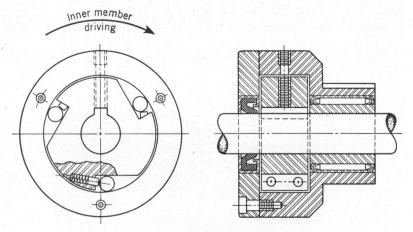

Fig. 12-13. Hilliard overrunning clutch.   (*Courtesy the Hilliard Corporation.*)

the point of contact of the rollers on the outer race results in a wedging action when the rotation of the outer race is counterclockwise relative to the inner race.   When the outer race is rotating clockwise relative to the inner race, the rollers roll out of the wedge, and the clutch freewheels. Either member may be the driver, but it is recommended that the outer member rotate during freewheeling periods; for example, a typical clutch may be used with an overrunning speed of 2,400 rpm when the outer member is rotating and only 410 rpm when the inner member is rotating. Standard Hilliard clutches are available with torque capacities up to 248,000 lb-in.

The *Formsprag* overrunning clutch in Fig. 12-14 uses specially shaped rockers, or *sprags*, located between concentric circular races.   The principle of operation is shown in Fig. 12-15.   The energizing garter spring pushes outward and maintains the sprags in contact with both races at all times.   Standard Formsprag clutches are available with torque capacities up to 139,000 lb-in.

Since no circumferential space is taken up by cam flats and springs, more sprags can be used, and a given-diameter sprag or cam clutch will have a greater torque capacity than the same-diameter roller clutch.

FIG. 12-14. Formsprag overrunning clutch.   (*Courtesy Formsprag Company.*)

Other differences in operation are that the rollers have a combination rolling and sliding action when freewheeling, whereas the sprags have pure sliding; the roller wear is distributed evenly around the roller, whereas sprag wear is localized; and the wear of the cam flats in roller clutches is localized, whereas the wear of the races in sprag clutches is evenly distributed. Both types require careful attention to lubrication.

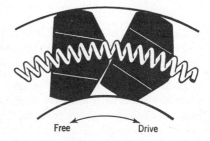

FIG. 12-15. Sprag operation.   (*Courtesy Formsprag Company.*)

Example 12-1. A lathe is being designed for automatic operation. There is to be no shifting of gears in the headstock, and all speed changes will be made by using hydraulically operated clutches to connect the proper gear train to the output shaft.

We are asked to determine the number of plates and the operating force required for a clutch that is to transmit a torque of 300 lb-in. under normal operating conditions. The clutch is to be designed to slip under 300 per cent of rated torque to protect the gears and other parts of the drive. Space limitations dictate an upper limit of 4 in.

and a lower limit of $2\frac{1}{2}$ in. for the diameters of the friction surfaces. The clutch will operate in an oil atmosphere.

Solution. We shall use a wet clutch with hardened steel plates. The design equation (12-12) is

$$T = N\mu P \frac{D + d}{4}$$

and $T = 3 \times 300 = 900$ lb-in.   (300% rated torque)
  N = number of friction planes
  $\mu = 0.1$
  $P = pA = 100\pi[(D^2 - d^2)/4] = 100\pi[(4^2 - 2.5^2)/4] = 766$ lb, based on $p = 100$ psi for long wear life with frequent use
  $D = 4$ in., maximum for space limitation
  $d = 2.5$ in., minimum for space limitation

Thus, $$900 = N \times 0.1 \times 766 \times \frac{4 + 2.5}{4}.$$

Solving for N, we find that it equals 7.23. The next larger even whole number of friction planes is 8. Therefore, eight friction planes and nine plates will be specified.

We now have the choice of either reducing the operating force, and thus the pressure on the plates, by the ratio 7.23/8 or keeping the pressure between the plates at 100 psi and reducing the outer diameter of the plates. In this case, space is an important consideration. Therefore, we shall determine the outer diameter required when $p = 100$ psi and $N = 8$. Substituting $pA = p\pi[(D^2 - d^2)/4]$ for P in Eq. (12-12) gives

$$T = N\mu p\pi \frac{D^2 - d^2}{4} \frac{D + d}{4}$$

Thus $$900 = 8 \times 0.1 \times 100 \frac{D^2 - 2.5^2}{4} \frac{D + 2.5}{4}$$

Solving the resulting cubic equation for D, we find it to be 3.90 in. Solving for P, we find it to be 704 lb.

Specifications

Plates: 9 (eight friction planes), hardened steel, outer diameter of friction surface = 3.90 in., and inner diameter of friction surface = 2.50 in.
Operating force: 704 lb.

12-5. Magnetic-particle Clutches. Magnetic-particle clutches depend upon the effect of a magnetic field on the consistency, or viscosity, of a medium of ferromagnetic particles. The particles may be in an oil slurry or in a dry mixture. Figure 12-16 shows a Vickers flange-mounted Magneclutch for use on a standard C face motor. The inner rotating member mounts on the motor shaft. The shearing resistance of the magnetic medium, and thus the torque capacity of the clutch, is a function of the current in the field coil. The major differences between the magnetic-particle and other electromagnetic clutches (Figs. 11-17, 17-31, and 17-32) are that the magnetic-particle clutch transmits rated torque with zero slip and that the magnetic-particle medium results in a drag torque without excitation of about 1 to 2 per cent of rated torque.

The magnetic-particle clutch is also suitable for use as a variable-speed device, but the compactness of the unit makes it difficult to dissipate the heat generated when the clutch slips. For example, an air-cooled unit with a rated capacity of about 70 hp at 3,600 rpm will transmit 35 hp but can dissipate continuously only slightly over $\frac{1}{2}$ hp at 1,800 rpm.

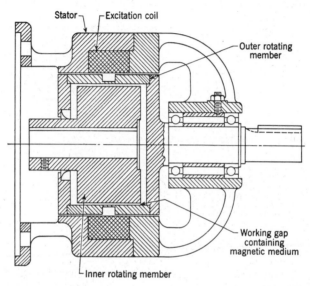

FIG. 12-16. Magnetic-particle clutch. (*Courtesy Vickers Electric Division, Vickers Incorporated, Sperry Rand Corporation.*)

12-6. Intermittent-motion Mechanisms.   The term *intermittent-motion mechanisms* properly includes clutches, indexing devices, and reversing mechanisms.   However, in this discussion the term will be used to describe those devices which automatically engage and disengage one shaft to another with some specified cycle of operation and without the use of an external force or timing signal.

There are many types and variations of mechanisms useful as intermittent-motion devices.   This discussion will be limited to several of the basic types, and the reader is referred to other, more detailed books[1] for additional information.

---

[1] "Ingenious Mechanisms for Designers and Inventors," vols. I, II, and III, The Industrial Press, New York, 1930, 1936, 1951.

C. D. Albert and F. S. Rogers, "Kinematics of Machinery," John Wiley & Sons, Inc., New York, 1931.

The *ratchet* mechanism in Fig. 12-17 is used to convert oscillation of the arm 3 into intermittent clockwise rotation of the toothed wheel 2. The pawl 4 transmits a force to the wheel when it moves to the right and slides over the teeth when returning to the left. The pawl 5 prevents the ratchet wheel 2 from rotating in a counterclockwise direction. The normal to the contact surfaces between the pawl 4 and the ratchet teeth should pass between the pawl and ratchet-wheel centers to ensure that the pawl remains engaged under load. It is usually desirable at low speeds, and

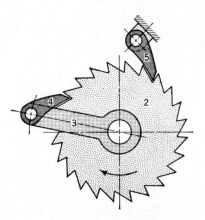

Fig. 12-17. Ratchet mechanism.

necessary at high speeds, to spring-load the pawls to ensure correct engagement without skipping.

The usefulness of a ratchet mechanism may depend upon the number and magnitude of the possible increments of rotation. Since the strength requirements may dictate the minimum tooth size, and thus tooth spacing, it is sometimes necessary to use several different-length pawls operating side by side.

As discussed in Sec. 12-4, overrunning clutches are particularly well adapted for use as high-speed, infinitely variable, and quietly operating ratchet mechanisms.

Escapements are self-actuating ratchets that alternately release and stop the driven member. The most common application has been in clocks and watches where an escapement controls the driving energy from a wound spring or a falling weight so that the parts move at regular intervals of time and in definite steps.

In Fig. 12-18 the escape wheel is driven in a clockwise direction. The double pawl 3 oscillates at the natural frequency of the pendulum system. The escape wheel is alternately released and stopped by the pawl ends, or pallets, A and B as they break and make contact with the pointed teeth. In the phase shown, the pendulum is moving to the right (ccw), and the escape-wheel tooth C gives an impulse to the pendulum as the tip of the tooth slides across the inclined face of pallet B. If the pendulum weight is correctly related to the rotative effort of the escape wheel, the right amount of energy will be supplied by the escape wheel to overcome the frictional losses and maintain the amplitude of vibration or oscillation of

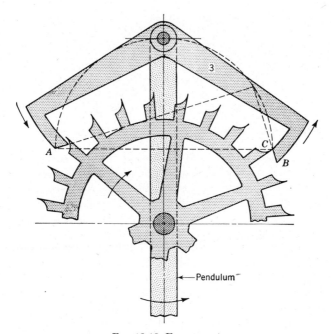

FIG. 12-18. Escapement.

the pendulum.   More compact escapements use a balance wheel or torsional pendulum instead of a common pendulum to regulate the motion of the escape wheel.

Intermittent gearing is used to convert continuous rotation of the driving member into intermittent rotation of the driven member.   Applications are counting mechanisms, motion-picture cameras and projectors, feed mechanisms, indexing devices, etc.   The form of intermittent gearing in Fig. 12-19 uses ordinary gear teeth and may be easily adapted to fit unusual conditions.   In Fig. 12-19, member 2 must always drive and will make 10 revolutions while the driven member 3 makes 1 revolution.   It should be noted that the driven member is locked against rotation except when being positively driven. This particular example could be used in a mechanical counter. The meshing teeth may be arranged in other ways to give different relative motions of the driver and follower. The main disadvantage of this toothed form of intermittent gearing is that the sudden engagement of the teeth and the resulting impact limit its use to relatively low speeds.

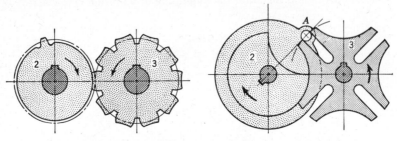

FIG. 12-19. Intermittent gearing.   FIG. 12-20. Geneva wheel.

The *Geneva wheel* in Fig. 12-20 is a form of intermittent gearing that is satisfactory for relatively high-speed operation and has been extensively used in the intermittent transporting of film in motion-picture equipment. Member 2 must be the driver, and it makes 4 revolutions for each revolution of the driven wheel 3.   When properly designed, as in Fig. 12-20, the pin A is moving parallel to the slots when contact is being made or broken. This is advantageous in that the motion is relatively smooth with reasonable accelerations and little trouble from impact.

The *Geneva stop* is a useful variation of the Geneva wheel in which one of the slots in the driven member is left closed, thus permitting only a certain number of revolutions (for Fig. 12-20, about $3\frac{1}{2}$) before the pin hits the closed slot and stops the motion.   Geneva stops have been used to prevent overwinding of clocks, watches, spring motors, etc.

CHAPTER 13

BRAKES

A brake is a machine element whose primary purpose is to absorb energy. Although most brakes are used to absorb kinetic energy by stopping or slowing down some moving part, many brakes are used to absorb the potential energy given up by objects being lowered by hoists, elevators, etc. Additional uses of brakes are to hold parts in position at rest and to prevent an unwanted reversal of the direction of rotation. The major functional difference between a clutch and a brake is that a clutch connects one moving member to another, whereas a brake connects a moving member to a stationary frame. The same basic principles apply to both, but the dissipation of the energy absorbed as heat is a greater problem for brakes than for clutches.

13-1. Shoe Brakes. The most common forms of *shoe* or *block* brakes, shown in Fig. 13-1, consist of relatively rigid shoes pressed against the

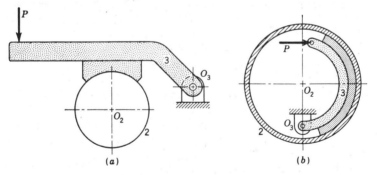

Fig. 13-1. Lever-mounted shoe brakes. (a) External; (b) internal.

inside or outside surfaces of a rotating cylinder or *brake drum*. The tangential friction between the shoes and the drum gives the braking force and thus the braking torque.

As for clutches, the major consideration in brake problems is wear, in particular its effect on the pressure distribution and service life. Again, the assumption is that wear is proportional to the product pV. However, since the brake drum is a cylinder, the velocity is uniform and the pressure distribution can be determined only by consideration of the motion of the shoe relative to the drum.

For the external, lever-mounted shoe brake in Fig. 13-1a, the force P acts to rotate the lever in a ccw direction about its center of rotation O_3. If the drum were not present, the motion of the shoe for rotation of the lever through an angle α would be from the solid to the dotted position in Fig. 13-2. However, the drum is there and something must give. If we

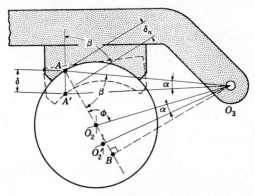

FIG. 13-2

assume that everything *except the brake lining* has infinite rigidity, the entire deformation to accommodate the motion must take place in the lining. This is illustrated for one point where A (on the rigid shoe) moves to A' and the compression of the lining is δ.

Brake linings will not, in general, have linear stress-strain curves, but the error introduced by assuming a linear relationship will probably be small in comparison with other uncertainties, particularly the value of coefficient of friction, and is far outweighed by the convenience of being able to obtain an analytical expression in closed form. Therefore, we shall assume a linear relationship between pressure and deflection. Thus,

$$p = C\delta \tag{13-1}$$

where C = a constant of proportionality.

Friction forces result from normal forces, or, in this case, from the normal pressure p, which, in turn, results from the component of the deflection in the direction normal to the drum. As shown in Fig. 13-2, for small angles of rotation of the lever, the normal component of the deflection for any point, such as A, can be expressed as

$$\delta_n = \delta \sin \beta \tag{13-2}$$

Also
$$\delta = O_3A\,\alpha \tag{13-3}$$

Substituting Eq. (13-3) in Eq. (13-2) gives

$$\delta_n = O_3A \; \alpha \sin \beta \qquad (13\text{-}4)$$

In many design problems the diameter of the drum, the position of the drum center of rotation O_2, the position of the lever pivot O_3, and the desired shape of the shoe and its position are known. Thus, it will be desirable to derive our equations in terms of these known quantities. In Fig. 13-2, A is on the radius that makes the angle ϕ with the line of centers O_2O_3.

From the figure we can write

$$\sin \beta = \frac{O_3B}{O_3A} \qquad (13\text{-}5)$$

where B is at the intersection of AO_2 and a line dropped from O_3 perpendicular to AO_2.

Also

$$O_3B = O_2O_3 \sin (180° - \phi) = O_2O_3 \sin \phi \qquad (13\text{-}6)$$

Substituting from Eq. (13-6) for O_3B in Eq. (13-5) and then substituting the new Eq. (13-5) for $\sin \beta$ in Eq. (13-4) results in

$$\delta_n = O_2O_3 \; \alpha \sin \phi \qquad (13\text{-}7)$$

and, in terms of pressure between the lining and the drum, from Eq. (13-1),

$$p = C\delta_n = C \, O_2O_3 \, \alpha \sin \phi \qquad (13\text{-}8)$$

The value of the constant C is almost never known, but some information related to allowable pressures can be found in catalogues and in Sec. 13-5. Equation (13-8) shows that the maximum pressure p_{max} occurs where $\sin \phi$ is a maximum; i.e.,

$$p_{max} = C \, O_2O_3 \, \alpha \, (\sin \phi)_{max} \qquad (13\text{-}9)$$

Dividing Eq. (13-8) by Eq. (13-9) and rearranging gives

$$p = p_{max} \frac{\sin \phi}{(\sin \phi)_{max}} \qquad (13\text{-}10)$$

In most designs ϕ includes the angle 90° and Eq. (13-10) becomes

$$p = p_{max} \sin \phi \qquad (13\text{-}11)$$

Figure 13-3 is a partial free-body diagram of the lever and shoe showing the forces acting on an elemental area of the shoe when the drum is rotating ccw. Considering the lever as a free body, we see that the summation of moments about O_3 includes (1) a term M_P due to the actuating force P, (2) a term M_n due to the normal pressure, and (3) a term M_μ due to the

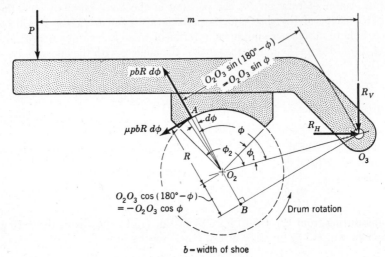

b = width of shoe

FIG. 13-3

friction forces. Thus,

$$\Sigma M_{O_3} = 0$$
$$M_P + M_n + M_\mu = 0 \qquad (13\text{-}12)$$

Designating ccw moments as positive and considering each term separately, we find that

1. Owing to the actuating force

$$M_P = Pm \qquad (13\text{-}13)$$

2. Owing to the normal pressure

$$M_n = - \int_{\phi_1}^{\phi_2} pbR \, O_2O_3 \sin \phi \, d\phi \qquad (13\text{-}14)$$

Substituting from Eq. (13-10) for p and rearranging Eq. (13-14) gives

$$M_n = - \frac{p_{\max}bR \, O_2O_3}{(\sin \phi)_{\max}} \int_{\phi_1}^{\phi_2} \sin^2 \phi \, d\phi \qquad (13\text{-}15)$$

Upon integration and substitution of limits,

$$M_n = - \frac{p_{\max}bR \, O_2O_3}{4(\sin \phi)_{\max}} [2(\phi_2 - \phi_1) - \sin 2\phi_2 + \sin 2\phi_1] \qquad (13\text{-}16)$$

3. Owing to the friction forces

$$M_\mu = \int_{\phi_1}^{\phi_2} \mu p b R (R - O_2 O_3 \cos \phi) \, d\phi \qquad (13\text{-}17)$$

Substituting from Eq. (13-10) for p and rearranging Eq. (13-17) gives

$$M_\mu = \frac{\mu p_{max} b R}{2(\sin \phi)_{max}} \int_{\phi_1}^{\phi_2} (2R \sin \phi - O_2 O_3 \sin 2\phi) \, d\phi \qquad (13\text{-}18)$$

Upon integration and substitution of limits,

$$M_\mu = \frac{\mu p_{max} b R}{4(\sin \phi)_{max}} [4R(\cos \phi_1 - \cos \phi_2)$$
$$+ O_2 O_3 (\cos 2\phi_2 - \cos 2\phi_1)] \quad (13\text{-}19)$$

It should be noted that the derivation of the moment equations has not been completely general. The M_P and M_n terms offer no difficulty, because they must always be opposed and the correct signs can be determined by inspection. However, the sign of the friction moment depends not only upon the geometry of the brake but also upon the direction of rotation. For example, Eq. (13-17) is based on the elemental friction force giving a positive moment about O_3 with ccw drum rotation. Therefore, a positive value of M_μ from the solution of Eq. (13-19) means that M_μ acts in a ccw direction and a negative value means that M_μ acts in a cw direction. If the direction of rotation of the drum is reversed, the right-hand side of Eq. (13-19) should be multiplied by -1 to maintain a consistent sign convention.

The preceding discussion has related the actuating force to the normal pressure and friction forces. However, normally the most important specification is the torque the brake must exert on the drum. For the brake in Fig. 13-3, the torque on the drum is

$$T = - \int_{\phi_1}^{\phi_2} \mu p b R^2 \, d\phi \qquad (13\text{-}20)$$

Substituting for p from Eq. (13-10) in Eq. (13-20) and rearranging gives

$$T = - \frac{\mu p_{max} b R^2}{(\sin \phi)_{max}} \int_{\phi_1}^{\phi_2} \sin \phi \, d\phi \qquad (13\text{-}21)$$

Upon integration and substitution of the limits,

$$T = \frac{\mu p_{max} b R^2}{(\sin \phi)_{max}} (\cos \phi_2 - \cos \phi_1) \qquad (13\text{-}22)$$

The application of Eqs. (13-22), (13-13), (13-16), and (13-19) to a design situation will be illustrated in Example 13-1.

When the friction assists in applying the brake, i.e., when M_μ acts in the same direction as M_P, the brake is considered to be *self-energizing*. This is the same as saying that the brake is self-energizing when M_μ and M_n have opposite signs.   It is interesting to note the behavior of a self-energizing brake as the magnitude of the friction moment increases from zero.   Referring to Eq. (13-12) and the preceding discussion, it can be seen that, when $M_\mu = 0$, i.e., there is zero self-energization, the direction of rotation has no effect on the relationship of p, and thus μp, to P.   Then, assuming that the braking torque, and thus p, is to be kept constant, it can be seen that, as M_μ increases in magnitude and the degree of self-energization increases, the required magnitude of P will decrease.   When the magnitude of M_μ equals that of M_n, P becomes zero.   This means that the self-energization is so great that irrespective of the magnitude of the torque, no external force is required.   In practice the brake would grab or lock almost instantaneously when the shoe touches the drum. This degree of self-energization is known as *self-locking* and occurs when (1) the brake is self-energizing *and* (2) $M_\mu \geqq M_n$.

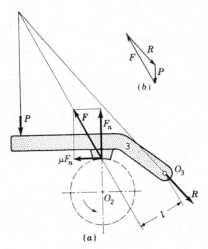

(b)

(a)

Fig. 13-4. Free-body diagram of an external, lever-mounted shoe brake with a short shoe.

Self-locking is a rather remote possibility for external brakes but may be achieved readily with internal shoe brakes, such as in Figs. 13-1b and 13-10, and with band brakes.   The few instances in which self-locking is desirable will be discussed in Sec. 13-2 on band brakes.

In many cases the shoe is relatively short, for example, $\phi_2 - \phi_1 \leqq 45°$, and the pressure distribution may be considered to be uniform without introducing serious error.   Then, a further approximation permits assuming that the resultant normal and friction forces act at the center of the shoe.   Figure 13-4a shows the forces acting on the lever of a short-shoe brake.   The magnitudes can be found by using the ordinary conditions of static equilibrium, that is, $\Sigma M_{O_3} = 0$, $\Sigma F_V = 0$, and $\Sigma F_H = 0$, or by using a graphical vector solution for $\Sigma F = 0$, as illustrated in Fig. 13-4b.

As shown, the brake is self-energizing but not self-locking.   The reader should prove to himself that the condition for self-locking for this general configuration would be when the line of action of the resultant F of normal and friction forces passes through or to the right of O_3.

The pivoted-shoe brake in Fig. 13-5 is of more academic than practical value. It can be shown that if

$$h = R \frac{4 \sin \theta/2}{\theta + \sin \theta} \tag{13-23}$$

the resultant normal and friction forces pass through the pivot. Consequently, there will be no moment acting on the shoe, and the pressure

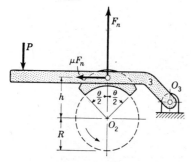

FIG. 13-5. Pivoted-shoe brake.

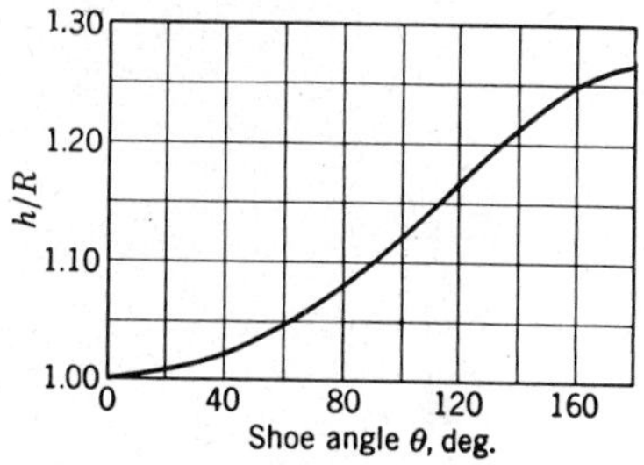

FIG. 13-6

distribution and, therefore, the wear of the lining will be as uniform as possible.

The braking torque is

$$T = \mu F_n h \tag{13-24}$$

For design purposes, the relationship between h and R, Eq. (13-23), can be taken from the curve in Fig. 13-6.

The reasons pivoted-shoe brakes are seldom used, although many brakes appear at first glance to have pivoted shoes, are the physical problem inherent in locating the pivot so close to the drum surface and the impracticality of maintaining the correct position as the lining wears. Rapid wear at the leading edge (toe) or trailing edge (heel) will result from the moment introduced by unavoidable errors in the pivot location.

Whenever possible, it is desirable to minimize the additional shaft and bearing loads due to the brake operating forces by using two opposed shoes, as in Fig. 13-7. The arrangement of the pivots in Fig. 13-7 is such that, with counterclockwise rotation of the drum, the right-hand brake

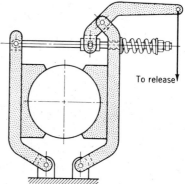

FIG. 13-7. Double-shoe brake.

contributes more than half of the braking effort, and the reverse is true with clockwise rotation.   If the brake is to be used as much in one direction as in the other, the two shoes will tend to have the same rate of wear.

(a)

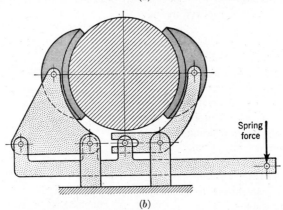

Spring force

(b)

Fig. 13-8. (a) Double-shoe brake; (b) schematic diagram of operating linkage. (*Courtesy Westinghouse Electric Corporation.*)

However, if the brake is used mainly in one direction, the shoe contributing most to the braking will wear more rapidly.   The service life of the shoes may be balanced by (1) designing the linkage so that neither shoe is self-energizing, (2) designing the linkage to apply normal forces that result in equal braking torques, (3) using a poorer-quality material with a higher coefficient of friction on the lightly loaded shoe, or (4) using a shorter or narrower lightly loaded shoe.

The d-c magnetic brake in Fig. 13-8 is designed for floor mounting with the brake wheel mounted on the motor-shaft extension. The application force is supplied by a compressed spring, and the brake is released when an electromagnet is energized. The linkage is underneath the brake wheel,

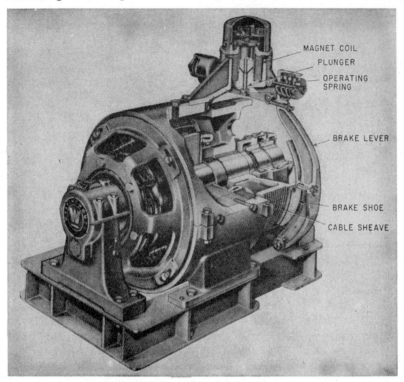

MAGNET COIL

PLUNGER

OPERATING SPRING

BRAKE LEVER

BRAKE SHOE

CABLE SHEAVE

Fig. 13-9. Variable-voltage gearless elevator machine. (*Courtesy Westinghouse Electric Corporation.*)

and the motor may be removed without dismantling any part of the brake. It is interesting to note that, in this case as in most other cases, the brake is actually a lever-mounted shoe brake and not a pivoted-shoe brake. The brake-shoe pivot bolts are provided to permit the initial alignment of the shoes on the wheel. The bolts are then tightened to hold the shoes in position and prevent the tips from dragging.

Figure 13-9 is a cutaway view of a variable-voltage gearless elevator machine that is, essentially, an electric motor combined with a brake and a cable sheave mounted directly on the armature shaft. The double-shoe

brake is spring-set and electrically released.   The spring force on each
lever may be adjusted to stop and hold the load while the other brake shoe
is removed for inspection or repair.

The typical automotive brakes shown in Fig. 13-10 are hydraulically
operated, internal, lever-mounted shoe brakes.   In the normal operation
of an automobile, the braking of forward motion is more frequent and
requires greater braking capacity.   An additional factor that must be considered is that the inertia of the car during braking unbalances the wheel
loads so that the front wheels carry about 55 to 60 per cent of the weight

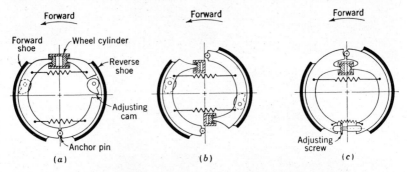

FIG. 13-10. Automotive internal shoe brakes.

of the automobile and thus can do about 55 to 60 per cent of the braking.
In view of these considerations, it is evident that the single-cylinder brake
a, with one shoe self-energized during forward motion and the other self-
energized during reverse motion, is more suited for use on the rear wheels
than on the front whereas the double-cylinder brake *b*, with both shoes
self-energized during forward motion, is ideal for use on the front wheels
but will contribute very little braking during reverse motion.   The brake
c utilizes two shoes, connected by the adjusting-screw link at the bottom,
and a single anchor pin.   The shoes are free to move circumferentially
until contact is made with the anchor pin.   Thus, both shoes are always
self-energized; therefore, the brake can be designed to be equally effective
in either direction of rotation.

The actuating force on the secondary shoe is the reaction from the self-
energizing primary shoe, and since this reaction is greater than the actuating force on the primary shoe, there is a multiplying effect.   Thus, for
a given actuating force from the hydraulic cylinders, this brake will have
considerably greater torque capacity than brake *b*.   However, the advantages of increased capacity and the possibility of equal effectiveness in both
directions must be weighed against the problems resulting from nonuniform wear rates of the shoes and an increased sensitivity to variations in
the coefficient of friction.   It should be noted that the theory developed

earlier in this section does not apply directly to the brakes shown in Fig.
13-10, because the levers do not have fixed centers of rotation and the compression of the lining cannot be related simply to the angle of rotation of

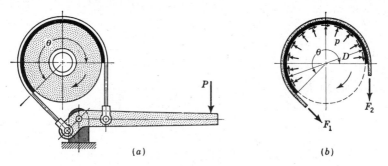

(a)   (b)

Fig. 13-11. Band brake.

the levers.   Positive release is provided by tension springs, and the brakes
can be adjusted, after moderate wear has taken place, by use of cams a
and b or adjusting screws c which con-
trol the released position of the shoes.

13-2. Band Brakes.   The *band* brake
in Fig. 13-11a consists of a steel band or
strap lined on one side with leather,
wooden blocks, or, more probably, with
an asbestos composition material, as
shown in the figure.   The operating force
must be applied in the direction such
that the lever motion tends to stretch
or tighten the band.

Figure 13-11b is a free-body diagram
for the band in Fig. 13-11a.   The band
must be in tension, and the difference in
the tension forces, $F_1 - F_2$, will be the
total friction force on the band.   The

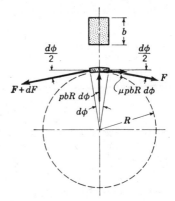

Fig. 13-12. Forces on an element of
the band.

friction, or braking, force on the drum will be in the direction of F_1 and
will be $F_1 - F_2$; therefore, the braking torque will be

$$T = (F_1 - F_2)\frac{D}{2}$$

or this may be rewritten as

$$F_1 - F_2 = \frac{2T}{D} \tag{13-25}$$

In order to get another relationship between F_1 and F_2, the forces acting
on the elemental length of band $R\, d\phi$ in Fig. 13-12 will be considered.

Summation of the radial forces gives

$$\Sigma F_r = 0$$

or
$$-(F + dF) \sin \frac{d\phi}{2} + pbR \, d\phi - F \sin \frac{d\phi}{2} = 0 \qquad (13\text{-}26)$$

Equation (13-26) can be simplified to

$$pbR \, d\phi - 2F \sin \frac{d\phi}{2} - dF \sin \frac{d\phi}{2} = 0 \qquad (13\text{-}27)$$

which, in the limit, as $d\phi \to 0$, becomes

$$pbR - F = 0 \qquad (13\text{-}28)$$

or
$$p = \frac{F}{bR} \qquad (13\text{-}29)$$

Summation of the tangential forces gives

$$\Sigma F_t = 0$$

or
$$-(F + dF) \cos \frac{d\phi}{2} + \mu pbR \, d\phi + F \cos \frac{d\phi}{2} = 0 \qquad (13\text{-}30)$$

which, in the limit, becomes

$$-dF + \mu pbR \, d\phi = 0$$
or
$$dF = \mu pbR \, d\phi \qquad (13\text{-}31)$$

Substituting from Eq. (13-29) for p in (13-31), simplifying, and rearranging results in

$$\frac{dF}{F} = \mu \, d\phi \qquad (13\text{-}32)$$

The value of F increases from F_2 to F_1 while the angle increases from 0 to θ. Thus,

$$\int_{F_2}^{F_1} \frac{dF}{F} = \mu \int_0^\theta d\phi \qquad (13\text{-}33)$$

which, after integration and substitution of limits, becomes

$$\ln \frac{F_1}{F_2} = \mu\theta$$

or
$$\frac{F_1}{F_2} = e^{\mu\theta} \qquad (13\text{-}34)$$

Thus, for given values of braking torque T, wrap angle θ, and coefficient of friction μ, the required values of F_1 and F_2 may be found by simultaneous solution of Eqs. (13-25) and (13-34). The value of θ must be in radians.

The free-body diagram of the lever in Fig. 13-11a is shown in Fig. 13-13.

$$\Sigma M_o = 0$$
$$Pm + F_1 q - F_2 n = 0 \qquad (13\text{-}35)$$

As discussed in Sec. 13-1, a brake is self-energizing when the friction force acts to apply the brake.   Recalling that the friction force is $F_1 - F_2$ and observing that it always acts in the direction of F_1, it becomes apparent that *a band brake will be self-energized when F_1 acts to apply the brake.*   As study of the geometry in Fig. 13-13 will show, F_1 will act to apply the brake when its moment arm is smaller than that for F_2.   Thus, the brake shown in Figs. 13-11 and 13-13 will be self-energized when the drum is rotating in the cw direction.

If, when F_1 is acting to apply the brake, the pivot is so located that $F_1q > F_2n$, the self-energization will be so great that, once the band touches the drum, the brake will be applied without the external operating force P. Under these conditions, the brake is self-locking.   Except for those few cases in which a *backstop* is required to prevent rotation in the wrong direction, it is usually desirable to keep far enough away from the self-locking

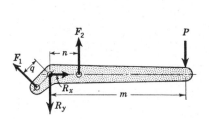

Fig. 13-13. Free-body diagram for band-brake lever.

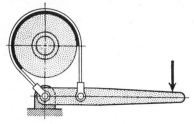

Fig. 13-14. Band brake with zero self-energization.

condition so that small changes in the coefficient of friction due to a change in climatic conditions, lining temperature, dust, wear, or splashing oil or water will not take control of the braking away from the operator.

If the direction of drum rotation is reversed, F_1 and F_2 change places, because F_1 must always be in the direction to resist motion of the drum. This is usually not desirable, because a greater operating force P would be required than in the original case.

If the brake is to be used with drum rotation in both directions or if it is important to preclude the possibility of self-locking, one side of the band is attached at the lever pivot, as in Fig. 13-14.

13-3. Disk Brakes.   The equations derived in Chap. 12 for plate clutches apply to disk brakes.   The common bicycle coaster brake is an example of a multidisk brake.   Disk brakes are also used on airplanes, trains, trucks, and automobiles.   Advantages claimed are increased braking surface and better heat dissipation.   It has become common practice to have a brake built into the electric motor of a machine drive so that the brake engages and quickly stops the machine when the electric power to the motor is cut off.   The brake is usually spring-loaded, and release is accomplished by direct pull of electromagnets or by use of a solenoid acting through a linkage.   Figure 13-15 shows a brake motor with a spring-set brake that uses the direct pull of electromagnets to release the brake when line voltage is applied to the motor.   Note that there are three sta-

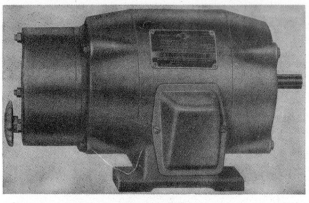

(a)

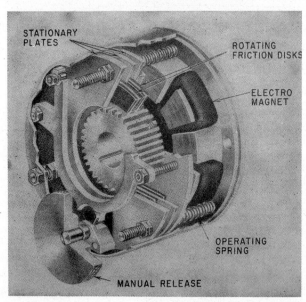

STATIONARY
PLATES

ROTATING
FRICTION DISKS

ELECTRO
MAGNET

OPERATING
SPRING

MANUAL RELEASE

(b)

FIG. 13-15. (a) Brake motor; (b) cutaway view of brake.   (*Courtesy Safety Electrical Equipment Corporation.*)

tionary plates and that two friction disks are spline-mounted on the motor shaft.

13-4. Hydrodynamic Brakes.   The hydrodynamic brake is essentially a fluid coupling with the output rotor stationary so that the coupling operates with 100 per cent slip at all times.   Water is generally used as the fluid.   The braking effort is controlled by regulating the amount of water in the brake, and the energy absorbed in the form of heat is removed by circulating the water through a heat exchanger or radiator.   Hydrodynamic brakes have been used in laboratories as dynamometers; in oil fields to absorb the potential energy given up as the tools, drill pipe, etc., are lowered into the well; and on trucks to limit the speed on long down grades. The Hydrotarder used on trucks circulates the water through the engine radiator.   The most important consideration in the use of a hydrodynamic brake is that it cannot stop motion entirely and a shoe or band brake is required to stop the motion and to hold the member in position.

13-5. Basic Design Data.   The capacity of a brake is ultimately based upon its ability to absorb the energy and then dissipate it without permitting its temperature to rise so high that the brake lining or, in extreme cases, the drum surface will be damaged.   Most linings show a decrease in coefficient of friction with increasing temperature, and it is important to keep the surface temperature below the point at which the brakes start to "fade."   Fading is particularly critical for automobiles and trucks operating in hilly or mountainous country.

Even with a good knowledge of the principles of heat transfer, it is difficult to calculate the friction-surface temperature during braking.   At the same time, there is very little reliable information available as to limiting temperatures.   Some typical values that may be used are the following:

$$\begin{array}{ll} \text{Leather, fiber, and wood facing} \dots \dots & 150\text{--}160°\text{F} \\ \text{Asbestos} \dots \dots \dots \dots \dots \dots & 200\text{--}220°\text{F} \\ \text{Automotive asbestos block lining} \dots \dots & 400\text{--}500°\text{F} \end{array}$$

As a logical approach, the designer can rely on the principle that wear is proportional to the work of friction.   Experience has shown that values can be set for the product pV, where p is the pressure in pounds per square inch and V is the sliding velocity in feet per minute, that will give reasonable wear life for the brake.   Typical recommendations are as follows:

$pV \leqq 30,000$ for continuous application of load and poor dissipation of heat

$pV \leqq 60,000$ for intermittent application of load, comparatively long periods of rest, and poor dissipation of heat

$pV \leqq 84,000$ for continuous application of load and good dissipation of heat, as in an oil bath

Table 13-1 contains values for coefficient of friction and allowable pressures for some of the common combinations of materials used in clutches and brakes.

TABLE 13-1. COEFFICIENTS OF FRICTION AND ALLOWABLE PRESSURES

Material	μ	p, psi
Asbestos in rubber compound, on metal....	0.3–0.4	75–100
Asbestos in resin binder, on metal:		
Dry.................................	0.3–0.4	75–100
In oil..............................	0.10	600
Sintered metal on cast iron:		
Dry.................................	0.20–0.40	400
In oil..............................	0.05–0.08	

Example 13-1. A small hoist is being designed for use primarily in the construction industry. Specifically, it is to be used to lift a cubic yard of concrete at the rate of 200 fpm. A yard of concrete weighs approximately 4,000 lb, and the bucket weighs 1,250 lb. Since the contractor may wish to use the hoist for other purposes, it will be designed with a rated capacity of 3 tons.

There will be no counterweights, and the cable drum will be connected to an electric motor through a reduction gear train. Lowering is to be controlled by manual operation of a brake. The brake must automatically hold the load at any position when the motor is not driving.

One of the methods proposed is to use a spring-loaded shoe brake that will be manually released by the operator during the lowering operation. An overrunning clutch is to be provided to disengage the brake automatically when the torque from the motor through the gear train is sufficient to raise the load.

We are asked to design the brake using asbestos-in-resin-binder brake linings and a cast-iron brake drum. We shall assume the dimensions and relative proportions shown in Fig. 13-16 for a trial design.

At this point, the major problems are to determine values for D and the spring force required to apply the brake.

Solution. From the figure, the required torque capacity is

$$T = F\frac{D}{2} = 6,000 \times \frac{24}{2} = 72,000 \text{ lb-in.}$$

and will be the sum of the torques from the right and left shoes; i.e.,

$$T = T_R + T_L$$

The appropriate equation (13-22) is

$$T = \frac{\mu p_{max} b R^2}{(\sin \phi)_{max}} (\cos \phi_2 - \cos \phi_1)$$

T_R and T_L are related to each other because the actuating force from the spring is the same on both levers. The limiting factor for the design will be the value of p_{max}. Thus, since one shoe may be expected to be self-energized and the other not, with the one that is self-energized having the greater value of p_{max} for a given spring force, it will be desirable to assume short shoes and make simple free-body diagrams to help us decide which shoe will be the critical one.   In Fig. 13-17, μF_R acts to apply the brake and μF_L acts to release the brake.   Therefore, we may expect the right shoe to be self-energized and, consequently, to have the greater value of p_{max}.

Considering the right shoe first, from the geometry of the brake we find that $\phi_1 = 25°$ or 0.436 radian, $\phi_2 = 115°$ or 2.01 radians, and $(\sin \phi)_{max} = \sin 90° = 1.00$.

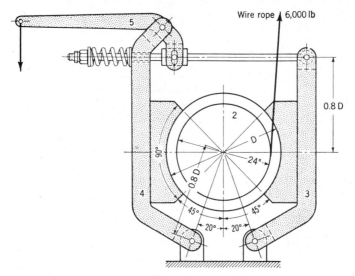

FIG. 13-16

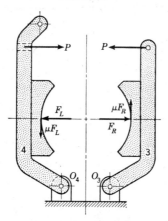

FIG. 13-17

From Table 13-1, assuming the lower limits for conservative design, we find that $\mu = 0.3$ and $p = 75$ psi.   Since the brake will be used in many different situations— it may even be used almost continuously as a general-purpose hoist—we must consider the problems of wear and heat dissipation.   We shall assume that $pV = 30,000$

is appropriate here. For $p = 75$ psi

$$V = \frac{pV}{p} = \frac{30,000}{75} = 400 \text{ fpm}$$

Although this value applies to the sliding between the lining and the drum, it is probably in the same order of magnitude as the cable speed, and in comparison with the lifting speed of 200 fpm, it appears to be reasonable.

The width b of the shoe is usually between one-fourth and one-half of the drum diameter. Narrower shoes will result in an unnecessarily large brake, and wider shoes increase the difficulty of maintaining a uniform pressure distribution across the face of the lining. Since size is no particular problem here and since rather rough treatment is to be expected, we shall compromise in the direction of minimizing the alignment problem and use $b = D/3$.

Substituting these values in the equation, we find for the right shoe that

$$T_R = 0.3 \times 75 \times \frac{D}{3} \times \left(\frac{D}{2}\right)^2 (\cos 115° - \cos 25°)$$
$$= 1.875D^3(-0.423 - 0.906) = -2.49D^3$$

The minus sign tells us, as can be seen in Fig. 13-17, that the torque on the drum is in the clockwise direction. For our purposes, the magnitude only will be sufficient, and thus

$$T_R = 2.49D^3 \tag{a}$$

Now, considering the right lever, we can write

$$\Sigma M_{o_3} = 0$$
$$M_P + M_n + M_\mu = 0 \tag{b}$$

where, from Eqs. (13-13), (13-16), and (13-19), respectively,

$$M_P = P(0.8D + 0.8D \cos 20°) = P \times 1.552D$$
$$M_n = -\frac{p_{max}bR \ O_2O_3}{4} [2(\phi_2 - \phi_1) - \sin 2\phi_2 + \sin 2\phi_1]$$
$$= -\frac{75 \times D/3 \times D/2 \times 0.8D}{4} [2(2.01 - 0.436) - \sin 230° + \sin 50°]$$
$$= -2.5D^3(3.148 + 0.766 + 0.766) = -11.71D^3$$

and

$$M_\mu = \frac{\mu p_{max}bR}{4} [4R(\cos \phi_1 - \cos \phi_2) + O_2O_3(\cos 2\phi_2 - \cos 2\phi_1)]$$
$$= \frac{0.3 \times 75 \times D/3 \times D/2}{4} \left[4\frac{D}{2}(\cos 25° - \cos 115°) \right.$$
$$\left. + 0.8D(\cos 230° - \cos 50°)\right]$$
$$= 0.937D^2[2D(0.906 + 0.423) + 0.8D(-0.643 - 0.643)] = 1.528D^3$$

Substituting for these values in (b) gives

$$P \times 1.553D - 11.71D^3 + 1.528D^3 = 0$$

or

$$P = \frac{10.18D^3}{1.553D} = 6.55D^2 \tag{c}$$

There are several possible paths to follow from this point in the solution. Of these, the most straightforward approach is to assume a value of p_{max} for the left shoe and then work back up to the spring force in the same manner as for the right shoe. If we call this value $p_{L,max}$ and switch sign conventions, i.e., designate cw angles and moments as positive, the procedure just outlined for the right shoe may be applied directly to the left-hand shoe. In fact, because of symmetry, the only difference is that M_μ will now oppose M_P. Therefore,

$$T_L = \frac{p_{L,max}}{75} 2.49D^3 \qquad (d)$$

$$M_{P,L} = P \times 1.553D$$

$$M_{n,L} = -\frac{p_{L,max}}{75} 11.71D^3$$

$$M_{\mu,L} = -\frac{p_{L,max}}{75} 1.528D^3$$

and

$$\Sigma M_{o_4} = 0$$

$$P \times 1.553D - \frac{p_{L,max}}{75} 11.71D^3 - \frac{p_{L,max}}{75} 1.528D^3 = 0$$

or

$$P = \frac{0.1766 p_{L,max}D^3}{1.553D} = 0.1137 p_{L,max}D^2 \qquad (e)$$

Substituting for P from (c), (e) becomes

$$6.55D^2 = 0.1137 p_{L,max}D^2$$

and

$$p_{L,max} = \frac{6.55}{0.1137} = 57.7 \quad \text{psi} \qquad (f)$$

Substituting 57.7 psi for $p_{L,max}$ in (d) gives

$$T_L = \frac{57.7}{75} 2.49D^3 = 1.917D^3 \qquad (g)$$

Since $T_R + T_L = 72,000$ lb-in., from (a) and (g) we can write

$$2.49D^3 + 1.917D^3 = 72,000$$

from which

$$D = 25.4 \text{ or } 26 \text{ in.}$$

Also,

$$b = \frac{D}{3} = \frac{25.4}{3} = 8.45 \text{ or } 8\tfrac{1}{2} \text{ in.}$$

and from (c)

$$P = 6.55D^2 = 6.55(25.4)^2 = 4,230 \text{ lb}$$

The maximum lowering speed will be

$$400 \frac{D_{\text{rope drum}}}{D} = 400 \frac{24}{26} = 369 \text{ fpm}$$

Thus, our recommendations will be:

$$\text{Drum diameter} = 26 \text{ in.}$$
$$\text{Width of the shoes} = 8\tfrac{1}{2} \text{ in.}$$
$$\text{Min spring force} = 4,230 \text{ lb}$$

Closure. The discerning reader has probably wondered why there has been no mention of the necessity for providing some excess braking torque so that the moving weight can be stopped as well as held in position after it is stopped. To stop the weight will require additional torque, and the spring force should be increased. The question then becomes, "How much should it be increased?" The primary factor will be the rate of acceleration that can be permitted or is desired. When lowering at a constant speed, the brake will be operating with the spring force on the levers necessary to give a torque equal to that required to hold the weight at rest. Then, if the operator suddenly releases the lever, the total spring force will act and the excess, over that for the at-rest case, will decelerate the moving mass. For example, the acceleration will be

$$A = - \left(\frac{1.42P_i}{W} - 1\right) 32.2 \quad \text{ft/sec}^2$$

where W = weight on hoist, lb, and P_i = initial compression of spring, lb. If P_i = 5,000 lb and the hoist is fully loaded, the acceleration will be -5.86 ft/sec^2 or -0.182 g. Combining Eqs. (5-2) and (5-3) for uniformly accelerated motion leads to

$$S = \frac{V^2}{2A}$$

For a hoist speed of 369 fpm (6.15 fps) and for the acceleration of -5.86 ft/sec^2, the hoist would come to rest in 3.23 ft. For loads of less than rated capacity the stopping distance would be even less.

The reader should note that, although p_{max} will be greater than 75 psi when at rest or stopping the load, it will not be greater when lowering the load at a constant speed, which will be the conditions under which most of the wear takes place. Also, it will be difficult to overcome a spring force in the order of 5,000 lb by use of the simple lever indicated in Fig. 13-16, and further thought will have to be given to this problem before the over-all design can be considered complete.

CHAPTER 14

BELT AND ROPE DRIVES

Mechanical power is generally transmitted between two shafts by means of belts, chains, or gears. Each of these types of mechanical drives has specific features that often dictate its selection in a particular situation. In many cases, however, the overlapping of characteristics makes two or more of the drives competitive, and the final choice will be based upon considerations such as first cost, maintenance, reliability, availability of replacement parts, and appearance.

Operational characteristics and the design of the several drives will be discussed in this and the next two chapters.

14-1. Belts and Ropes. A *belt* or *rope* drive has several useful characteristics. It can be used to connect widely or closely spaced shafts; it does not require precise alignment of the shafts and pulleys; it normally requires only periodic adjustment to compensate for wear and stretching; it is relatively quiet in operation; and it can be designed to slip under excessive load in the same manner as a slip coupling (Sec. 11-10). One important limitation to its use is that, since it is not a positive drive, slip and creep will occur, and it cannot be used where either positive timing or an exact velocity ratio is necessary.

Before the development and widespread use of electric power, most machines were driven by flat belts operating from a line shaft which, in turn, was driven by a flat belt or rope drive from a steam engine or water wheel. At the present time, the use of rope drives is practically limited to those applications involving hoisting or otherwise moving materials. Manila, cotton, and wire ropes have been successfully used, but a brief discussion of only wire rope as used in hoisting applications is considered pertinent enough to be included in this book. For additional details, the reader should refer to the handbooks.[1]

Even though belts are now seldom used to transmit power between

[1] L. S. Marks (ed.), revised by T. Baumeister, "Mechanical Engineers' Handbook," 6th ed., pp. 8-65 to 8-82, McGraw-Hill Book Company, Inc., New York, 1958.

C. Carmichael (ed.), "Kent's Mechanical Engineers' Handbook," "Design and Production" volume, 12th ed., pp. 15-44 to 15-67, John Wiley & Sons, Inc., New York, 1950.

widely separated shafts and most new machines have individual motor drives, belts still have an important—in some instances, increasingly so—place in machine design.   Flat belts are still widely used, but V belts are used in the majority of new applications.

Flat belts used for transmission of appreciable power are made of leather, canvas (often rubberized or treated with special materials), or steel.   The best leather belts are made from 4- to 5-ft-long strips cut from either side of the backbone of top-grade steer hides.   The leather may be either oak-tanned or mineral-salts-tanned, e.g., chrome-tanned. In order to have a practical belt, it is necessary to cement strips together so that the belt size is not limited to that which could be made from a single piece of leather.   Belts are specified according to the number of layers, e.g., single-, double-, or triple-ply, and according to the thickness of the hides used, e.g., light, medium, or heavy.   The strips are overlapped and cemented to make wider belts.

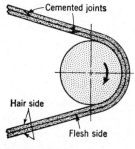

FIG. 14-1. Double-ply leather belt.

Figure 14-1 shows a double-ply belt with cemented joints and illustrates the desirable relationship between the scarfing or beveling and the direction of motion, so that a loose edge would be flattened down rather than torn loose by striking an object such as a guard. The cemented joint is as strong as the rest of the belt.   However, as a matter of convenience, rawhide lacing, wire loops, or other commercial fasteners are often used to join the ends of the belt.   Since belts are usually selected for a satisfactory wear life, the lower strength of the joints is neglected in the design calculations.

The hair side of the belt should be placed next to the pulley to protect the stronger flesh side from wear.   Leather belts must be periodically cleaned and dressed or treated with a compound or dressing containing neatsfoot or other suitable oils so that the belt will remain soft and flexible.

Most fabric belts are made with textile or wire cords and cotton-duck canvas impregnated with and separated by rubber.   The fibers or cords carry most of the tension load; the rubber protects the fibers or cords from moisture, abrasion, etc., and improves the frictional contact with the pulley.   Rubber belts may be made endless by splicing and vulcanizing, or the ends may be joined by metal fasteners.   A comparison of the characteristics of leather belts with rubberized canvas belts shows that the leather belt has a higher coefficient of friction in contact with a pulley, while the rubberized canvas belt is more resistant to moisture and can operate at higher temperatures.

The V belt (Fig. 14-2) is a widely used form of rubber-covered and -impregnated belt. The shape gives a wedging action in the groove of the pulley, or *sheave*, as it is usually known, which results in an increased effective coefficient of friction (Sec. 14-5). A V-belt sheave with three

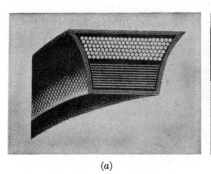
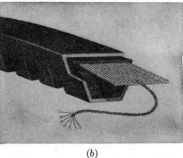

(a)   *(b)*

FIG. 14-2. V belts. *(a)* Textile-cord load-carrying members. (*Courtesy Gates Rubber Company.*) *(b)* Steel-cord load-carrying members. (*Courtesy Browning Manufacturing Company.*)

grooves is shown in Fig. 14-3. Split taper bushings are available with a wide range of bore diameters, but all have the same tapered outside surface that will fit into a standard hub bore. Thus, the manufacturer or distributor can supply a sheave and bushing combination for most shaft diameters from a small stock of expensive sheaves and a large stock of relatively inexpensive bushings. The use of a split taper bushing also simplifies the installation and removal of a sheave from a shaft. V belts are generally molded as endless rubber and fabric belts with a core of textile- (Fig. 14-2a) or steel-cord (Fig. 14-2b) load-carrying members located near the neutral axis of the belt. These belts may be used singly or in multiple and are available from stock in standard lengths and cross sections.

In a multiple V-belt drive it is important that all belts have the same length so that the load is evenly divided. It is desirable, but not always necessary, selectively to match new belts into sets of the same length. However, if one of a set of

FIG. 14-3. V-belt sheave with malleable-iron split taper bushing. (*Courtesy Browning Manufacturing Company.*)

belts breaks, all should be replaced at the same time. If only one belt is replaced, the new unworn and unstretched belt will be more tightly stretched and will move with a different velocity, because it will ride higher in the groove and thus at a greater radius. This "fighting" with old worn and stretched belts means that the new belt will carry more than its share of the load and will deteriorate rapidly. It is not always necessary to replace a set of belts the instant one breaks or becomes too badly worn for use. Usually, the remaining belts will transmit the required load at the expense of an increased rate of wear. Often it is practical to save the old belts for use as replacements in a set of belts with the same amount of wear and permanent stretch.

When a V belt is bent in passing around the sheave, the Poisson's-ratio effect results in a decrease of width at the top and an increase of width at the bottom of the belt, this effect being more pronounced with smaller-diameter sheaves. The angle included between the belt sides is generally 40°; the angle between the sheave-groove sides is varied between 30 and 38°, depending upon the diameter of sheave and type of belt, to ensure more uniform contact by compensating for the changes in width. Figure 14-2 shows two methods used by manufacturers to achieve more uniform contact between the belt and the sheave when small-diameter sheaves are required.

Several special types of V belts are shown in Fig. 14-4. The double-V belt *a* is used only when the drive cannot be arranged so that an ordinary V belt will work; the wide V belt *b* is used in variable-speed drives (Sec. 17-6); and the wide multiple-V belt *c* is a new development, designed to replace the several belts in a conventional multiple-V-belt drive with a single, more compact belt containing the same number of wedging friction surfaces.

Thin woven fabric belts are used for high-velocity applications, such as grinding-wheel drives, and in applications where the utmost flexibility is required, such as driving the part to be balanced in a balancing machine.

Mylar belts are also used where extreme flexibility and relatively high strength are required, such as in the drive of tape recorders to be used in satellites.

Steel belts have been used in Europe for large-capacity high-velocity drives. The necessity for precise alignment, the low coefficient of friction, and the difficulty in maintaining the correct tension have limited their use in this country to minor applications such as positioning devices in drafting machines (Fig. 2-33), involute-gear generators, etc. Small round leather belts have been used in fractional-horsepower drives such as sewing machines. Long continuous helical springs are used as belts in small mechanisms, such as home movie projectors, where the elasticity of the belt eliminates the need to provide for adjustment.

14-2. Belt-drive Kinematics. Belts are generally used to connect parallel shafts so that the pulleys rotate in the same direction, as in the *open*-belt drive in Fig. 14-5*a*, or in opposite directions, as in the *crossed*-belt drive *b*.

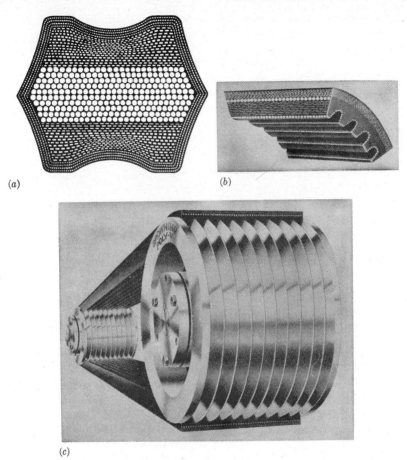

FIG. 14-4. (a) Double-V belt. (*Courtesy Gates Rubber Company.*)   (b) Wide V belt. (*Courtesy Lovejoy Flexible Coupling Company.*)   (c) Poly-V belt.   (*Courtesy Browning Manufacturing Company.*)

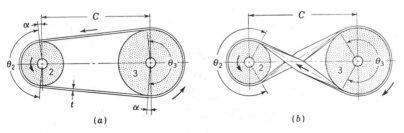

FIG. 14-5. Open- and crossed-belt drives.

The *velocity ratio* of a belt drive is defined as the ratio of the angular velocity of the driving pulley to the angular velocity of the driven pulley. Since most belt drives are used to reduce the speed of a relatively high-speed prime mover, such as an electric motor, to a lower speed required by the driven machine, the velocity ratio is usually, but not necessarily, greater than 1. Assuming that there is no slipping of the belt on the pulleys and that the belt does not stretch under load, the middle of the flat belt section will have a constant linear speed regardless of whether it is passing around a pulley or is between the pulleys. The radius to the center of the belt is

$$R = \frac{D}{2} + \frac{t}{2} \tag{14-1}$$

The velocity of the belt is

$$V = 2\pi n_2 \left(\frac{D_2}{2} + \frac{t}{2}\right) = 2\pi n_3 \left(\frac{D_3}{2} + \frac{t}{2}\right) \tag{14-2}$$

and

$$n_2(D_2 + t) = n_3(D_3 + t) \tag{14-3}$$

Therefore,

$$\frac{n_2}{n_3} = \frac{D_3 + t}{D_2 + t} \tag{14-4}$$

However, in a practical case, the thickness of the belt is small compared with the diameter of the pulley, and, considering that the belt both slips and creeps (Sec. 14-3), it is sufficiently accurate to say that

$$\frac{n_2}{n_3} = \frac{D_3}{D_2} \tag{14-5}$$

When this is applied to a V-belt drive where the thickness of the belt may not be small as compared with the sheave diameter, it is necessary to use the pitch diameters of the sheaves in Eq. (14-5). The pitch diameter will not be to the center of the belt but will be approximately to the neutral axis of the belt when it is bent around the sheave.

The *angle of wrap* is defined as the angle subtended by the arc over which the belt contacts the pulley. For the open-belt drive, the angles of wrap are

$$\theta_2 = \pi - 2\alpha = \pi - 2 \sin^{-1} \frac{D_3 - D_2}{2C} \qquad \text{radians} \tag{14-6}$$

and

$$\theta_3 = \pi + 2\alpha = \pi + 2 \sin^{-1} \frac{D_3 - D_2}{2C} \qquad \text{radians} \tag{14-7}$$

For the crossed-belt drive, the angles of wrap are the same for both pulleys and are

$$\theta_2 = \theta_3 = \pi + 2 \sin^{-1} \frac{D_3 + D_2}{2C} \qquad \text{radians} \tag{14-8}$$

The required length of belt for any drive other than a simple open- or crossed-belt drive may most readily be determined from a scale layout of the drive.   The length of belt for an open-belt drive is approximately

$$L = 2C + \frac{\pi}{2}(D_3 + D_2) + \frac{(D_3 - D_2)^2}{4C} \qquad (14\text{-}9)$$

and for the crossed-belt drive is

$$L = 2C + \frac{\pi}{2}(D_3 + D_2) + \frac{(D_3 + D_2)^2}{4C} \qquad (14\text{-}10)$$

The pulleys for flat-belt drives are usually crowned, as shown in Fig. 14-6a or b, to prevent the belt from running off the pulley.   In Fig. 14-6c

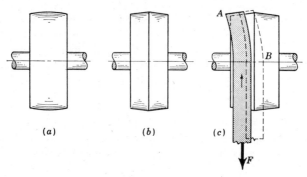

(a)   (b)   (c)

FIG. 14-6. Crowned pulleys.

the belt under tension on the double conical pulley tries to lie flat on the pulley at A.   At the same time, it resists the necessary lateral bending, and the oncoming belt is thrown toward the right at B.   Thus, even with a considerable misalignment, the belt will reach an equilibrium running position near the mid-plane of the pulley.

Flat belts have often been used for drives between shafts which are not parallel.   To do so, the pulleys must be located so that the centerline of the *approaching* belt will lie in the central plane of the pulley.   The example in Fig. 14-7 is known as a *quarter-turn* drive.   It is apparent that, if the direction of belt motion were reversed, the belts would im-

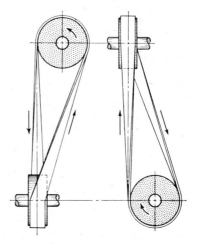

FIG. 14-7. Quarter-turn belt drive.

mediately run off the pulleys. Even with a drive that normally runs in only one direction, the slightest backward motion upon stopping would cause trouble. At least one idler or guide pulley is required to make a reversible flat-belt drive between nonparallel shafts.

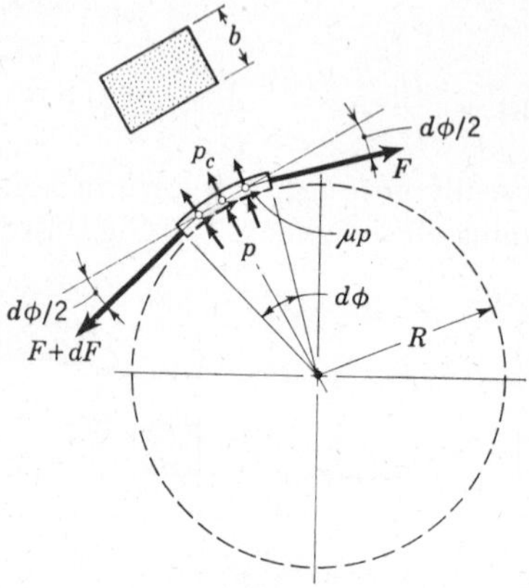

Because of the guidance of the grooves in the sheaves, a V-belt drive may be used as a quarter-turn drive without the belt slipping off.

14-3. Belt Forces. The analysis of a belt drive is similar to that of a band brake (Sec. 13-2) except that the velocity of the belt around the pulley introduces an additional term, because of centrifugal force. The free-body diagram of an element of a belt is shown in Fig. 14-8. The unit centrifugal force p_c equals mV^2/R, where m is the mass beneath a unit area of belt surface, or

FIG. 14-8. Forces on an element of a flat belt.

$m = \gamma t/g$, where γ is the weight density of the belt material and t is the thickness. If all quantities, except V, are in pound-inch-second units and V is in feet per minute,

$$p_c = 0.000208 \frac{\gamma t V^2}{D} \qquad \text{psi} \qquad (14\text{-}11)$$

Then, summing forces radially,

$$\Sigma F_r = 0$$

$$(p_c + p)bR \, d\phi - F \sin \frac{d\phi}{2} - (F + dF) \sin \frac{d\phi}{2} = 0 \qquad (14\text{-}12)$$

which, in the limit, becomes

$$(p_c + p)bR - F = 0 \qquad (14\text{-}13)$$

If only centrifugal force acts on the belt, $p = 0$ and $F = F_c$; thus,

$$F_c = p_c bR = 0.000104 \gamma b t V^2 \qquad (14\text{-}14)$$

Then, substituting for p_c from (14-14) into (14-13) and rearranging,

$$p = \frac{F - F_c}{bR} \qquad (14\text{-}15)$$

Summation of the tangential forces gives

$$\Sigma F_t = 0$$

$$(F + dF) \cos \frac{d\phi}{2} - F \cos \frac{d\phi}{2} - \mu p bR \, d\phi = 0 \qquad (14\text{-}16)$$

which becomes, in the limit,

$$dF - \mu pbR \, d\phi = 0 \qquad (14\text{-}17)$$

Substituting for p from (14-15) into (14-17) and rearranging gives

$$dF = \mu(F - F_c) \, d\phi \qquad \text{or} \qquad \frac{dF}{F - F_c} = \mu \, d\phi \qquad (14\text{-}18)$$

Then, if F changes from F_2 to F_1 as ϕ changes from 0 to θ, integration between these limits gives

$$\int_{F_2}^{F_1} \frac{dF}{F - F_c} = \mu \int_0^\theta d\phi \qquad (14\text{-}19)$$

and
$$\ln \frac{F_1 - F_c}{F_2 - F_c} = \mu\theta$$

or
$$\frac{F_1 - F_c}{F_2 - F_c} = e^{\mu\theta} \qquad (14\text{-}20)$$

Figure 14-9 shows the belt forces acting when the small pulley is driving in the direction indicated. The torque acting on the small pulley

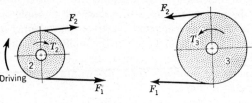

Fig. 14-9

is
$$T_2 = (F_1 - F_2) \frac{D_2}{2} \qquad (14\text{-}21)$$

and on the large pulley,
$$T_3 = (F_1 - F_2) \frac{D_3}{2} \qquad (14\text{-}22)$$

The term $F_1 - F_2$ is known as the *net belt pull* and may be related to the horsepower by

$$\text{hp} = \frac{(F_1 - F_2)V}{33,000} \qquad (14\text{-}23)$$

or
$$F_1 - F_2 = \frac{33,000 \times \text{hp}}{V} \quad \text{lb} \qquad (14\text{-}24)$$

where V is in feet per minute.

Equations (14-20) and (14-24) are fundamental in the solution of a belt-drive problem. However, several additional considerations are worth noting.

When the belt moves from the loose side (F_2) to the tight side (F_1), the belt must stretch as it moves around the pulley. Conversely, when moving from the tight to the loose side, the belt contracts as it moves around the pulley. This phenomenon of elongation and contraction of the belt is known as *creep* and results in a decrease in the angular velocity of the driven pulley from that expected by considering only the diameters of the pulleys. Both creep and slip result in a lower-than-expected angular velocity of the driven member and have the same effect on efficiency as slip does in a fluid coupling. The loss due to creep is in the order of 1 to 2 per cent, and slippage of a heavily loaded drive may result in an additional 1 to 2 per cent loss. These losses are usually neglected in calculations of the belt forces.

Since the angle of wrap θ affects the capacity of the drive, it is desirable to have the loose side on top, as in Fig. 14-9, so that sagging of the belt acts to increase the angle of wrap. It should be noted that, for pulleys of the same material, the capacity of the drive is limited by the angle of wrap on the small pulley and that the effect is more pronounced as the center distance decreases. The decrease in angle of wrap on the small pulley is often compensated for by using a paper pulley with a greater coefficient of friction between the pulley and the belt. Short-center drives are those with a center distance of less than twice the diameter of the large pulley. These drives are popular because of the saving in belt cost and space.

In a fixed-center drive, the belt must be installed under an initial-tension load such that F_1 and F_2 may vary according to Eqs. (14-20) and (14-24) without F_2 becoming zero. The relationship between the *initial tension* F_0 and the load forces F_1 and F_2 depends upon the stress-strain characteristics of the belt material, but for most purposes, the simple relationship

$$2F_0 = F_1 + F_2 \tag{14-25}$$

will be sufficiently accurate.

If F_1 and F_2 are calculated for the maximum capacity of the drive and F_0 is determined by use of Eq. (14-25), the inevitable stretching of the belt will result in a decreased F_0 and thus a decreased capacity of the drive. To eliminate the necessity for the frequent adjustments of fixed-center drives, it is desirable to make the initial tension greater than required by Eq. (14-25).

The problems of maintaining sufficient belt tension as the belt stretches and of increasing the angle of wrap on the small pulley of a short-center drive have resulted in the use of pivoted motor bases and idler pulleys, as shown in Fig. 14-10. Vertical belt drives should not be used because of the increased difficulty in maintaining the proper belt tension.

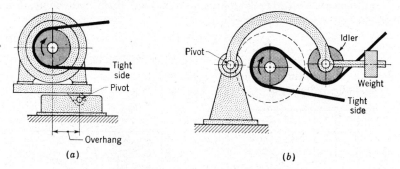

FIG. 14-10. Methods of maintaining belt tension and increasing the angle of wrap. (a) Pivoted motor base or Rockwood-type drive; (b) weighted idler pulley.

14-4. Selection of Flat Belts. The general solution of the problem of determining a size of belt for a given application is based upon calculating the cross-sectional area required to limit the stress due to the maximum belt force F_1 to an allowable value. Equations (14-20) and (14-24) can be combined to give

$$F_1 = \frac{33,000 \times hp}{V} \frac{e^{\mu\theta}}{e^{\mu\theta} - 1} + F_c \quad \text{lb} \quad (14\text{-}26)$$

Substituting $s_1 A$ for F_1 and $s_c A$ for F_c into Eq. (14-26) and solving for the cross-sectional area A of the belt,

$$A = \frac{33,000 \times hp}{V(s_1 - s_c)} \frac{e^{\mu\theta}}{e^{\mu\theta} - 1} \quad \text{in.}^2 \quad (14\text{-}27)$$

Rearranging (14-27) to solve for horsepower per unit area gives

$$\frac{hp}{A} = \frac{V(s_1 - s_c)}{33,000} \frac{e^{\mu\theta} - 1}{e^{\mu\theta}} \quad (14\text{-}28)$$

where, from Eq. (14-14) and considering that $A = bt$,

$$s_c = \frac{F_c}{A} = 0.000104 \gamma V^2 \quad \text{psi} \quad (14\text{-}29)$$

The following procedure for selecting a flat leather belt can be used as a guide in the design of drives with any type of flat belt. However, the manufacturers' catalogues should be consulted for the appropriate data as to allowable stress, coefficient of friction, weight density, and standard dimensions.

The breaking strength of belt leather varies from 3,000 to 5,000 psi. However, the wear life of a belt is more important than actual strength, and experience has shown that under average conditions an allowable stress of 400 psi or less will give a reasonable belt life. An allowable stress of 250 psi may be expected to give a belt life of about 15 years.

The density of leather is approximately 0.035 lb/in.[3]; therefore, Eq. (14-29) becomes, for a leather belt,

$$s_c = 3.64 \times 10^{-6} V^2 \qquad \text{psi} \qquad (14\text{-}30)$$

Applying the usual methods for determining maxima to Eq. (14-28) will show that the optimum velocity varies only with the allowable value of s_1, as indicated in Fig. 14-11a. The variation of belt capacity with

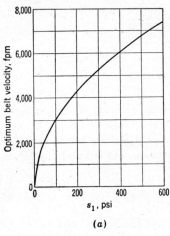

(a)

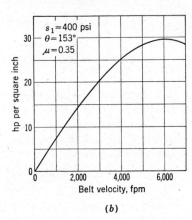

(b)

Fig. 14-11

velocity for the particular case of an allowable stress of $s_1 = 400$ psi, a coefficient of friction of 0.35, and a wrap angle of 153° is shown in Fig. 14-11b.

It is recommended that a belt velocity of 4,000 fpm be used in the absence of other specific requirements. *The increased cost for larger pulleys and a higher noise level may easily outweigh the advantage of the small additional capacity gained by using higher velocities.*

Rotative speeds of the pulleys are usually specified by operating requirements, and the desired pulley diameter can be calculated for a chosen belt velocity. If stock pulleys are to be used, it may be necessary to adjust both the belt velocity and velocity ratio to fit the available pulley diameters. After the pulley diameters and center distance have been selected, Eqs. (14-6) and (14-7) or Eq. (14-8) can be used to calculate the wrap angles. The appropriate values for coefficient of friction can be selected from Table 14-1 and the products $\mu\theta$ calculated for the two pulleys. The smaller value of $\mu\theta$ will then be used in subsequent calculations.

TABLE 14-1. APPROXIMATE COEFFICIENTS OF FRICTION FOR BELT AND PULLEY MATERIAL COMBINATIONS

Belt material	Pulley material		
	Iron-steel	Wood	Paper
Leather.................	0.40	0.45	0.50
Cotton woven...........	0.22	0.25	0.28
Rubber covered..........	0.32	0.35	0.38

Equation (14-27) is next used to determine the required cross-sectional area A. The standard dimensions in Table 14-2 can be used in calculating the width required for one or more of the belt designations. However, the wear life of a belt will be related to the amount of flexing (diameter of small pulley), the frequency of flexing (belt velocity), and the thickness of the belt. Thus, the maximum recommended thickness can be selected from Table 14-3, and it will be necessary to calculate widths for only those belts no thicker than recommended.

TABLE 14-2. BELT DESIGNATIONS AND DIMENSIONS

Ply		Average thickness, in.	Minimum economic width, in.	Maximum width, in.
Symbol	Name			
MS	Medium single	$1\frac{1}{64}$ (0.172)	$1\frac{1}{2}$	8
HS	Heavy single	$1\frac{3}{64}$ (0.203)	2	8
LD	Light double	$1\frac{8}{64}$ (0.281)	3	
MD	Medium double	$2\frac{0}{64}$ (0.313)	$3\frac{1}{2}$	
HD	Heavy double	$2\frac{3}{64}$ (0.359)	4	
MT	Medium triple	$3\frac{0}{64}$ (0.468)	5	
HT	Heavy triple	$3\frac{4}{64}$ (0.531)	6	

TABLE 14-3. MINIMUM PULLEY DIAMETER, IN INCHES, FOR BELT SPEEDS

Belt thickness	0–2,500 fpm	2,500–4,000 fpm	4,000–6,000 fpm
MS	2.5	3	3.5
HS	3	3.5	4
LD	4	4.5	5
MD*	5, 7	6, 8	7, 9
HD*	8, 10	9, 11	10, 12
MT*	16, 20	18, 22	20, 24
HT*	20, 24	22, 26	24, 28

* For belts 8 in. wide and over, use the second figure of the column.

The calculated width should be rounded off to the closest standard belt width listed in Table 14-4.

TABLE 14-4. STANDARD FLAT LEATHER BELT WIDTHS

Range, in.	Increments, in.
½–1	⅛
1–3	¼
3–6	½
6–10	1
10–56	2

The final choice of belt will be based upon the total cost of the drive and upon the availability of components.   Flat-belt-drive problems may also be readily solved by the methods presented in the handbooks and manufacturers' catalogues.   These methods are based upon tables of rated horsepower per inch of width.   The required rated belt horsepower is found by applying service factors to the nominal drive power.   These factors correct for type of motor and starting conditions, size of small pulley (effect of wrap angle), and special operating conditions.

Example 14-1.   The shaft for a centrifugal blower was designed in Example 6-1. The blower was belt-driven at 600 rpm by a 20-hp 1,750-rpm electric motor.   In this example we are asked to design the belt drive with a center distance twice the diam-

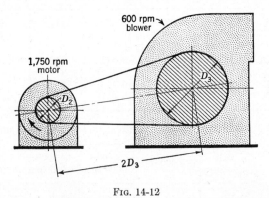

FIG. 14-12

eter of the larger pulley, as shown in Fig. 14-12.   A paper pulley will be used on the motor shaft and a cast-iron pulley on the blower shaft.

We are asked to determine (a) pulley diameters, (b) belt length, (c) widths of flat leather belts, (d) minimum initial tension for operation without slip, and (e) resultant force in the plane of the blower pulley when operating with an initial tension 50 per cent greater than the minimum value.

Solution. (a) *Pulley diameters.* The drive will be designed for a belt velocity of about 4,000 fpm.

$$D_2 = \frac{V}{\pi n_2} = \frac{4,000 \times 12}{\pi \times 1,750} = 8.73 \text{ or } 8\frac{1}{2} \text{ in.}$$

$$D_3 = \frac{n_2}{n_3} D_2 = \frac{1,750}{600} \times 8.5 = 24.8 \text{ or } 25 \text{ in.}$$

Thus, the design belt velocity is

$$V = \pi D_2 n_2 = \pi \times \frac{8.5}{12} \times 1,750 = 3,890 \text{ fpm}$$

and the design blower speed is

$$n_3 = \frac{D_2}{D_3} n_2 = \frac{8.5}{25} \times 1,750 = 595 \text{ rpm}$$

(b) *Belt length.* Equation (14-9) for the length of belt in an open-belt drive is

$$L = 2C + \frac{\pi}{2}(D_3 + D_2) + \frac{(D_3 - D_2)^2}{4C}$$

and

$$C = 2D_3 = 2 \times 25 = 50 \text{ in.}$$

Thus,

$$L = 2 \times 50 + \frac{\pi}{2}(25 + 8.5) + \frac{(25 - 8.5)^2}{4 \times 50} = 154 \text{ in.}$$

(c) *Belt widths.* A design stress between 250 and 400 psi will be expected to give a reasonable life. In lieu of specific requirements, we shall use the average value of 325 psi. The design equation for the cross-sectional area [Eq. (14-27)] is

$$A = \frac{33,000 \times \text{hp}}{V(s_1 - s_c)} \frac{e^{\mu\theta}}{e^{\mu\theta} - 1} \quad \text{in.}^2$$

and hp = 20

$V = 3,890 \text{ fpm}$

$s_1 = 325 \text{ psi}$

$s_c = 3.64 \times 10^{-6} V^2 = 3.64 \times 10^{-6} \times 3,890^2 = 55 \text{ psi}$

Since the angles of wrap and the materials are different for the two pulleys, it is necessary to calculate the products $\mu\theta$ for both pulleys and to use the smaller in determining the belt cross-sectional area.

Motor pulley: $\mu_2 = 0.50$ (leather on paper, Table 14-1). Equation (14-6) for the angle of wrap on the smaller pulley in an open-belt drive is

$$\theta_2 = \pi - 2\alpha = \pi - 2\sin^{-1}\frac{D_3 - D_2}{2C} \quad \text{radians}$$

$$= \pi - 2\sin^{-1}\frac{25 - 8.5}{100} = \pi - 0.332 = 2.81 \text{ radians}$$

Thus

$$\mu_2\theta_2 = 0.50 \times 2.81 = 1.405$$

Blower pulley: $\mu_3 = 0.40$ (leather on cast iron, Table 14-1) and

$$\theta_3 = \pi + 2\alpha = \pi + 0.332 = 3.47 \text{ radians}$$

Thus

$$\mu_3\theta_3 = 0.40 \times 3.47 = 1.388$$

Therefore, the belt will slip first on the blower pulley, and the design must be based on $\mu\theta = 1.388$.

$$e^{\mu\theta} = e^{1.388} = 4.01$$

and

$$A = \frac{33,000 \times 20}{3,890 \times (325 - 55)} \frac{4.01}{4.01 - 1} = 0.837 \text{ in.}^2$$

The usable combinations of widths and thicknesses are given in the following table:

Belt	MS	HS	LD	MD	HD
Thickness, in. (Table 14-2)...	0.172	0.203	0.281	0.313	0.359
Width, in. (calculated).......	4.87	4.12	2.98	2.67	2.33
Width, in. (standard, from Tables 14-4 and 14-2).....	5	4½	3	Not economical	Not economical

While the final choice of belt would be based upon the total cost of pulleys and belts, we shall assume in the remainder of the problem that the belt will be a medium single leather belt 5 in. wide.

It should be noted that the smaller-pulley diameter of 8.5 in. exceeds the minimum values recommended in Table 14-3.

(d) *Minimum initial tension without slip.* Equation (14-25) for the initial tension F_0 is

$$2F_0 = F_1 + F_2$$

$F_1 = s_1A$. For the 5-in. medium single belt,

$$s_1 = 325 \times \frac{4.87}{5} = 317 \text{ psi}$$

$$A = bt = 5 \times 0.172 = 0.860 \text{ in.}^2$$

and

$$F_1 = 317 \times 0.860 = 273 \text{ lb}$$

and

$$\frac{F_1 - F_c}{F_2 - F_c} = e^{\mu\theta}$$

$$F_c = s_cA = 55 \times 0.860 = 47 \text{ lb}$$

Thus

$$\frac{273 - 47}{F_2 - 47} = 4.01$$

from which

$$F_2 = 103 \text{ lb}$$

Therefore,

$$2F_0 = 273 + 103 = 376 \text{ lb}$$

and

$$F_0 = 188 \text{ lb}$$

(e) *Resultant force in plane of pulley when operating with an initial tension 50 per cent greater than the minimum.* The new value of initial tension is

$$F_0' = 1.5F_0 = 1.5 \times 188 = 282 \text{ lb}$$

The resultant force will be the vector sum of $F_1 - F_c$ and $F_2 - F_c$. However, for practical purposes, we may use the algebraic sum, or

$$F_R = F_1 - F_c + F_2 - F_c = F_1 + F_2 - 2F_c$$

and

$$F_1 + F_2 = 2F_0' = 2 \times 282 = 564 \text{ lb}$$

Thus,

$$F_R = 564 - 2 \times 47 = 470 \text{ lb}$$

Summary

Pulleys: 8.5-in.-diameter paper, 25-in.-diameter cast iron.

Belt: Flat leather, medium single, 5 in. wide, 154 in. long, 188 lb minimum initial tension, 470 lb resultant force in plane of blower pulley when operating with initial tension 50 per cent greater than minimum.

14-5. Selection of V Belts. The analysis of forces acting on a V belt is similar to that for the flat belt, except that the pressure between the belt and the sheave is normal to the sides, as shown in Fig. 14-13. The

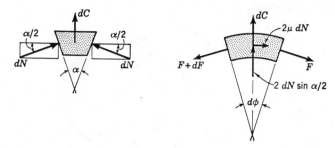

FIG. 14-13. Forces on an element of a V belt.

net result is an increased friction force for a given radial component of pressure, and this is usually expressed in terms of an *effective coefficient of friction,* or

$$\mu' = \frac{\mu}{\sin \alpha/2} \tag{14-31}$$

where μ is the coefficient of friction of the belt material on a flat pulley and α is the included angle of the groove in the sheave. Substitution of μ' for μ in Eq. (14-20) gives

$$\frac{F_1 - F_c}{F_2 - F_c} = e^{\mu'\theta} \tag{14-32}$$

as the basic equation for a V-belt drive. As previously stated, the included angle of the sheave groove varies with the diameter of the sheave from 30 to 38°. If the conservative design value of $\mu = 0.13$ is used, the effective values of μ' range from 0.40 to 0.50. The stiffness of the V belt will result in the actual angle of contact being somewhat less than that given by Eqs. (14-6) and (14-7). Because of the uncertainty of the values of μ' and θ and the fact that the strength and wear properties of V belts vary with the different constructions used, it is advisable to select V belts by following the recommendations of the manufacturer. In general, the catalogue procedure will consider the effect of centrifugal force, angle of wrap, and other operating conditions. An economical size of belt is first determined on the basis of speed and corrected horsepower, and the number of belts required is determined by dividing the corrected horsepower by the horsepower rating for one belt.

V belts are well adapted to use in short-center drives. In many instances, particularly when a flat belt is replaced with V belts, it is economical to use a flat large pulley, and the system is then known as a *V-flat* drive (see Fig. 12-7). In Fig. 14-14 the drive capacity will be limited by the smaller product $\mu'\theta_2$ or $\mu\theta_3$. Usually $\mu\theta_3$ is the smaller, and the large pulley limits the capacity. The V-belt manufacturers' catalogues also consider this type of drive.

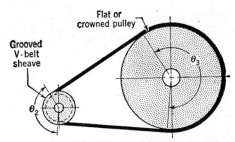

FIG. 14-14. V-flat drive.

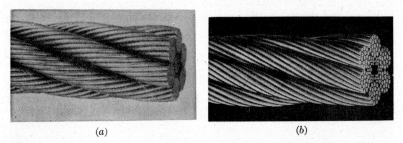

(a)   (b)

FIG. 14-15. Lay of wire rope.   (a) Right regular lay; (b) right lang lay.   (*Courtesy Bethlehem Steel Company.*)

14-6. Selection of Wire Rope.

Wire rope is seldom used today as a power-transmission element in the same sense as flat or V belts.   However, because of its high strength-weight ratio, it has practically replaced chain or hemp rope in hoisting and haulage applications.   Wire rope is also widely used in the form of guys or stays as stationary supporting members for radio-television transmission towers, stacks, etc.

The construction is varied to suit the requirements of the many special applications.   In general, a number of wires, such as 7, 19, or 37, are twisted together to form a strand, and a number of strands, usually six or eight, are then twisted about a core to form the rope.   The term *lay* is used in describing the manner in which the wires and the strands are twisted.   For example, the rope in Fig. 14-15a is called *right regular lay* because the strands are twisted in a right-hand helix and the wires and strands are twisted in opposite directions.   The rope in Fig. 14-15*b* is called *right lang lay* because the strands are twisted in a right-hand helix and the wires and strands are twisted in the same direction.   The core is usually hemp saturated with a lubricant, but wire cores are sometimes

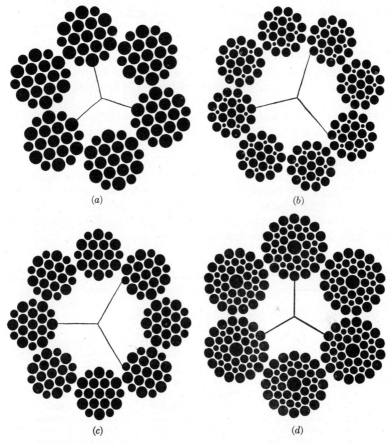

FIG. 14-16. Wire-rope cross sections.   (*a*) 6 × 19 Warrington; (*b*) 8 × 25 filler wire type W; (*c*) 8 × 19 Warrington; (*d*) 6 × 41 Warrington-Seale type M.   (*Courtesy Bethlehem Steel Company.*)

used when strength is more important than flexibility.   The size of wire rope is indicated by the diameter of the circle which encloses all wires, and the type is indicated by two figures giving the number of strands and the nominal number of wires in each strand; e.g., a ½-in. 6 × 19 wire rope has a diameter of ½ in. and is nominally composed of six strands of 19 wires each.   Typical cross sections are shown in Fig. 14-16.   The number of wires per strand is important in relation to flexibility and abrasion resistance. A large number of small wires gives a more flexible rope than a small number of large wires, but at the expense of decreased abrasion resistance. Standard constructions which cover most applications are as follows:

6 × 7 haulage and guy rope. For use under severe operating conditions where abrasion resistance and strength are more important than flexibility.

6 × 19 class wire rope, formerly called standard flexible hoisting rope. Includes 6 × 25 filler wire type W, 6 × 21 filler wire type U, 6 × 21 Seale, 6 × 19 Warrington (Fig. 14-16a), 6 × 19 Seale, and 6 × 17 Seale.

8 × 19 class wire rope, formerly called extra flexible hoisting rope, has somewhat greater flexibility than 6 × 19 rope. Includes 8 × 25 filler wire type W (Fig. 14-16b), 8 × 21 filler wire type U, 8 × 19 Warrington (Fig. 14-16c), and 8 × 19 Seale.

6 × 37 class wire rope, formerly called special flexible hoisting rope. For high-speed use on cranes or where sheaves are small. Includes 6 × 29 filler wire type L, 6 × 36 type L, 6 × 35 type M, 6 × 41 Warrington-Seale type M (Fig. 14-16d), 6 × 41 filler wire type Q, and 6 × 46 filler wire type R.

TABLE 14-5. APPROXIMATE WIRE ROPE AND SHEAVE DATA
d_r = diameter of rope, in.

Type	Diameter of wire d_w, in.	Metallic area of rope A, in.^{2}	Modulus of elasticity of rope E_r, psi	Recommended sheave diameters D, in.	
				Average	Minimum
6 × 7	$0.106d_r$	$0.38d_r{}^2$	14,000,000	$72d_r$	$42d_r$
6 × 19	$0.063d_r$	$0.40d_r{}^2$	12,000,000	$45d_r$	$30d_r$
8 × 19	$0.050d_r$	$0.35d_r{}^2$	10,000,000	$31d_r$	$21d_r$
6 × 37	$0.045d_r$	$0.40d_r{}^2$	11,000,000	$27d_r$	$18d_r$

Type of material	Carbon content, per cent	Breaking strength of wire, s_u, psi*
Iron.......................	0.05–0.15	65,000
Traction steel...............	0.20–0.50	130,000
Mild plow steel..............	0.40–0.70	160,000
Plow steel..................	0.65–0.80	175,000
Improved plow steel..........	0.70–0.85	200,000

* These values are based upon the breaking strength of rope and are 5 to 20 per cent less than for the individual wires.

. The breaking strength of a rope is its most significant mechanical property, and complete tables for the standard rope types and sizes are available from manufacturers, in handbooks, and in other publications.[1] The unusual designations of the materials used in wire ropes, even though peculiar to the industry, are not completely consistent within the industry. The iron or steel is specially heat-treated and then cold-drawn to final wire size. Table 14-5 contains some of the more important properties to be used in selecting wire rope when more complete tables are not available. Common rope diameters vary by $\frac{1}{16}$-in. increments from $\frac{1}{4}$ to $\frac{5}{8}$ in., by $\frac{1}{8}$-in. increments from $\frac{3}{4}$ to $2\frac{1}{4}$ in., and by $\frac{1}{4}$-in. increments from $2\frac{1}{2}$ to $3\frac{1}{2}$ in.

TABLE 14-6. RECOMMENDED MINIMUM FACTORS OF SAFETY FOR WIRE ROPE

Type of service		Type of service*			
		Car speed, fpm	Elevators		Dumb-waiters
			Passenger	Freight	
Track cables.................	3.2				
Guys.......................	3.5				
Mine shafts:					
Depths to 500 ft...........	8	50	7.60	6.65	4.8
500–1,000 ft.............	7	150	8.25	7.30	5.5
1,000–2,000 ft.............	6	300	9.20	8.20	6.6
2,000–3,000 ft.............	5	500	10.25	9.15	8.0
Over 3,000 ft.............	4	800	11.25	10.00	
Miscellaneous hoisting.........	5	1,100	11.70	10.40	
Haulage ropes................	6	1,500	11.90	10.55	
Cranes and derricks...........	6				
Small electric and air hoists....	7				
Hot-ladle cranes..............	8				
Slings......................	8				

* Extracted from Safety Code for Elevators, Dumbwaiters, and Escalators, ASA A17.1-1960, with the permission of the publisher, The American Society of Mechanical Engineers, New York.

The recommended factors of safety given in Table 14-6 are based upon the breaking strength of the wire rope and apply to average conditions. Thus, the factor of safety is defined as

$$\text{f.s.} = \frac{\text{breaking load}}{\text{working load}} = \frac{F_u}{F_s} \qquad (14\text{-}33)$$

The working or service load should properly include the static load and

[1] Simplified Practice Recommendation 198 (OIC 614), U.S. Department of Commerce, 1950.

the acceleration forces, if any, required to start and stop the load. However, the code defines the service load for use with the factors of safety for elevators and dumbwaiters as the "maximum static load imposed on all car ropes with the car and its rated load at any position in the hoistway." The effects of acceleration are included by relating the specified factor of safety to the car speed.

The failure of wire ropes used in hoisting and hauling is generally due to fatigue and wear. The bending and straightening of the wire result in a varying stress and a continual readjustment of wires within the rope. The sliding of the wires on each other and the abrasion between the wire

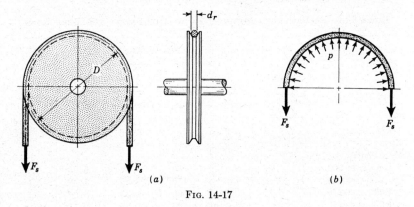

(a) (b)

Fig. 14-17

and the sheave result in wearing away of the load-carrying material. While there have been numerous formulas proposed for the calculation of the bending stress, none is of much use in design, because the simplified assumptions do not correspond to what actually takes place in operation when a wire rope is bent around a sheave or a drum.

The best correlation between calculations and experimental results, good *only for regular-lay ropes*, is that due to Drucker and Tachau,[1] which is based upon the pressure between the rope and the sheave. If the pressure between the rope and the sheave is assumed to be uniformly distributed, as in Fig. 14-17b, it can be shown that

$$p = \frac{2F_s}{d_r D} \qquad \text{psi} \qquad (14\text{-}34)$$

The experimental curves for 6×19 and 6×37 regular-lay wire ropes

[1] D. C. Drucker and H. Tachau, A New Design Criterion for Wire Rope, *Trans. ASME*, vol. 67, p. A-33, 1945.

are given in Fig. 14-18.   It should be noted that the ordinate is a dimensionless number p/s_u, which includes the effect of the class of material selected.   The abscissa, in terms of number of bends to failure, permits the designer to work much more closely to specified operating conditions without using an unnaturally large factor of safety, as must be done with Eq. (14-33).

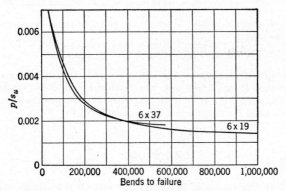

FIG. 14-18. Relationship between p/s_u and number of bends to failure for regular-lay wire ropes. (*From D. C. Drucker and H. Tachau, A New Design Criterion for Wire Rope, Trans. ASME, vol. 67, p. A-33, 1945.*)

For design purposes, Eq. (14-34) can be rewritten as

$$\frac{p}{s_u} = \frac{2(\text{f.s.})F_s}{c s_u d_r{}^2} \qquad (14\text{-}35)$$

where $c =$ the constant from Table 14-5 relating recommended sheave diameters to the type of wire rope being used.

The factor of safety for use with Eq. (14-35) should be selected on the basis of the discussion in Secs. 6-15 to 6-18, rather than from Table 14-6. The design value of p/s_u is taken from the curve corresponding to the type of rope and the number of bends expected during the service life of the rope.   Then either the material or the rope diameter is chosen, and the required value of the other can be calculated.   It is suggested that the rope material be selected first and then the diameter calculated.

Example 14-2.   The brake for a hoist with a rated capacity of 3 tons was designed in Example 13-1.   In this example we are to (*a*) select the proper-size 16 × 19 regular-lay, mild-plow-steel wire rope, (*b*) specify the minimum diameter for sheaves and the cable drum, and (*c*) determine the number of cycles of operation before the rope fails when used on the minimum-recommended-diameter sheave.

Solution.   (*a*) *Wire-rope size.*   The design equation (14-33) is

$$\text{f.s.} = \frac{F_u}{F_s} = \frac{s_u A}{F_s}$$

and f.s. = 5, recommended value for miscellaneous hoists, Table 14-6
  s_u = 160,000 psi, for mild plow steel, from Table 14-5
  A = $0.40d_r^2$, for 6 × 19 wire rope, from Table 14-5
  F_s = 3 tons or 6,000 lb

Thus
$$5 = \frac{160,000 \times 0.40d_r^2}{6,000}$$
from which
$$d_r = 0.685 \text{ or } \tfrac{3}{4} \text{ in.}$$

(b) *Minimum sheave diameter.* From Table 14-5, the minimum sheave diameter for 6 × 19 wire rope is

$$D = 30d_r = 30 \times \tfrac{3}{4} = 22.5 \text{ in.}$$

(c) *Number of cycles to failure.* The cycles, or bends, to failure are shown in Fig. 14-18 as a function of p/s_u, which may be calculated by use of Eq. (14-35). In this case, we shall be analyzing a given situation rather than selecting the rope, and therefore the failure point corresponds to a factor of safety of 1.

$$\frac{p}{s_u} = \frac{2(\text{f.s.})F_s}{cs_u d_r^2}$$

or
$$\frac{p}{s_u} = \frac{2 \times 1 \times 6,000}{30 \times 160,000 \times (\tfrac{3}{4})^2} = 0.00444$$

From Fig. 14-18, the number of bends to failure for a 6 × 19 regular-lay wire rope with $p/s_u = 0.00444$ is found to be about 105,000.

If one cycle of operation is considered to consist of both raising and lowering the rated 3 tons, the cycles of operation to failure would be 105,000/2 = 52,500. However, we expect a hoist to be loaded to capacity in one direction of motion at a time, and therefore the cycles to failure will be about 105,000.

Summary

Wire rope: ¾ in., 6 × 19, regular-lay, mild-plow-steel.
Minimum sheave diameter: 22.5 in.
Cycles to failure: 105,000.

CHAPTER 15

CHAIN DRIVES

A chain drive may be considered to be intermediate between belt and gear drives in that it has features in common with both. Chain drives are used where a positive drive is required, but the action is such that it cannot be used where precise timing is a requirement. Chains are suitable for relatively long or short center-distance drives and give a more compact drive than is possible with belts. The alignment of the shafts must be more accurate than for belts, while the center distance is not as critical as for gear drives. Chain drives are similar to gear drives in that proper lubrication must be provided for a satisfactory service life.

15-1. Types of Power-transmission Chains. The most important types of chains used for transmitting power are the roller chain in Fig. 15-1a and the inverted-tooth, or "silent," chain in Fig. 15-1b. Similar-

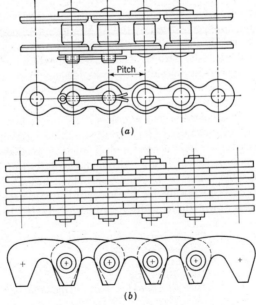

FIG. 15-1. (*a*) Roller and (*b*) inverted-tooth chains.

appearing chains made by different manufacturers usually have some different features of construction. However, most roller and inverted-tooth chains are made to close tolerances with case-hardened and ground rollers, pins, bushings, etc. Alloy steels are extensively used. The construction and nomenclature for roller chains are shown in Fig. 15-2.

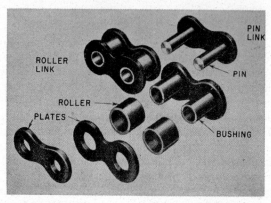

Fig. 15-2. Roller-chain construction. (*Courtesy Morse Chain Company, Division of the Borg-Warner Corporation.*)

Roller chain is available in single- or multiple-strand widths in pitches from $\frac{1}{4}$ to 3 in., as standardized by the ASA[1] and the Association of Roller and Silent-Chain Manufacturers. The capacity increases directly with the number of strands. Roller chains may be obtained with many standard and special attachments or links for use in conveyors, timing devices, etc., as illustrated in Fig. 15-3. Reduced-capacity stainless-steel and bronze chains are made for use in food-processing machinery or in applications where corrosion is a problem. An offset link must be used if conditions require the use of a chain with an odd number of links.

The inverted-tooth chain is commonly known as *silent chain* because it is less noisy than a roller chain. Although, to date, the smallest commercially available standard inverted-tooth chain has a pitch of $\frac{3}{16}$ in., standards[2] have been set up for chains with pitches from $\frac{1}{8}$ to 3 in. Standard chains are made in widths approximately $1\frac{1}{2}$ to 12 times the

---

[1] Transmission Roller Chains and Sprocket Teeth, ASA B29.1-1957, Society of Automotive Engineers, Inc., New York, 1957.

[2] Inverted Tooth (Silent) Chains and Sprocket Teeth, ASA B29.2-1957, Society of Automotive Engineers, Inc., New York, 1957.

Small Pitch Silent Chains and Sprocket Tooth Form, ASA B29.9-1958, Society of Automotive Engineers, Inc., New York, 1958.

pitch, and the chain capacity is directly proportional to its width. Standard inverted-tooth chains have straight-sided teeth in contact with straight-sided sprocket teeth, but several companies now offer chains with tooth profiles so designed that the chain approaches and leaves a sprocket on a line tangent to its pitch circle. This eliminates chordal action (Sec. 15-2) and permits quieter operation at higher speeds. As

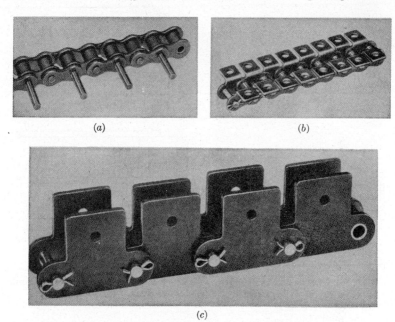

(a)   (b)

(c)

Fig. 15-3. Special attachments for roller chain.   (a) Extended pin, one side; (b) bent-link plate extension, both sides; (c) double-pitch, straight-link plate extension, both sides. (*Courtesy Link-Belt Company.*)

with roller chains, offset links are necessary if the chain has an odd number of links.

There are other, less expensive types of chains which are used to transmit relatively light forces at slow speeds. Four types are shown in Fig. 15-4. Open-link detachable chains are made of unmachined malleable-iron castings, as shown in Fig. 15-4a, or of pressed-steel links. They are best suited for nonabrasive operating conditions and are used in conveyors, elevators, farm machinery, etc. *Pintle* chains (Fig. 15-4b) have cast links coupled with steel pins, which are locked against rotation in the cast side bars. The links are offset, and the chain may be shortened or lengthened by one or more links without requiring a special link. The enclosed joint results in less wear under abrasive conditions than for the open-link detachable chain. The *ladder* chain (Fig. 15-4c) is used in low-power drives for operating models and in small mechanisms such as counters, position-indicating devices, etc.

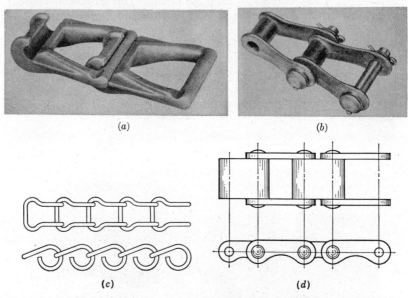

(a)   (b)

(c)   (d)

FIG. 15-4. (a) Ewart detachable chain. (*Courtesy Link-Belt Company.*) (b) Pintle chain. (*Courtesy Link-Belt Company.*) (c) Ladder chain. (d) Block chain.

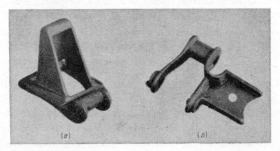

FIG. 15-5. Special attachments (a) for Ewart detachable chain, (b) for pintle chain. (*Courtesy Link-Belt Company.*)

Figure 15-5 shows two of the many types of attachments for Ewart detachable-link and pintle chain.

The Timing belt in Figs. 15-6 and 15-7 is a recent development that, strictly speaking, is neither a belt nor a chain. However, it most closely approximates a rubber inverted-tooth chain in that the teeth on the belt

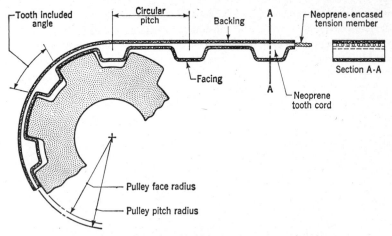

Fig. 15-6. Timing belt and pulley. (*Courtesy United States Rubber Company.*)

Fig. 15-7. Timing-belt drive in a gasoline-dispensing pump. (*Courtesy United States Rubber Company.*)

engage with the teeth on the pulley, or sprocket, to give positive driving. The tension members are either steel cables or cotton cord. The high strength-weight ratio makes the Timing belt ideal for use at high pitch-line velocities, up to 16,000 fpm. The thinness and resulting flexibility of the belt permit its use on small-diameter pulleys without generating excessive heat or rapid wearing. The belt is suitable for use in fixed-center drives because its length does not increase appreciably while in service, particularly when the tension members are steel cables.

15-2. Kinematics of Chain Drives. The most important kinematic consideration of a chain drive is that the chain passes around the sprocket as a series of chordal links, and thus the centerline of the chain is not at a uniform radius, as is the centerline of a belt passing around a pulley. This *chordal action* is illustrated in Fig. 15-8 for the cases of four- and five-tooth sprockets. As can be seen, the chain is approaching and leaving

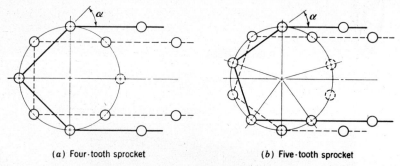

(*a*) Four-tooth sprocket   (*b*) Five-tooth sprocket

FIG. 15-8. Chordal action in a chain drive.

the sprocket at a continually varying radius. Thus, if the driving sprocket rotates at a uniform speed, the driven sprocket rotates at a varying speed that fluctuates above and below some average value. The average speed of the sprocket cannot be determined by use of the pitch diameter, but it will be equal to the length of chain passing around the sprocket in a unit time. Thus,

$$V = \frac{pNn}{12} \quad \text{fpm} \tag{15-1}$$

where V = average chain velocity, fpm
p = pitch of chain, in.
N = number of teeth on sprocket
n = rpm of sprocket

Similarly, the velocity ratio of a chain drive cannot be calculated in terms of the pitch diameters but must be calculated from

$$\frac{n_1}{n_2} = \frac{N_2}{N_1} \tag{15-2}$$

Consideration of the variation in velocity for the chains on the four-tooth and five-tooth sprockets in Fig. 15-8 shows that the amount of variation decreases as the number of teeth increases. It should also be noted that the *angle of articulation* α through which a link rotates during engagement and disengagement is greater for a small number of teeth. The life of a chain is usually limited by wear of the pins and bushings. The greater angle of articulation for a small number of teeth will increase the rate of wear and consequently decrease the life of the chain. Another reason for preferring a large number of teeth is that the larger sprockets give a higher chain velocity and transmit more power for a given chain force. However, dynamic effects, such as centrifugal force and impact between the chain and sprocket teeth, and lubrication difficulties impose a limit to the chain speed. Also, the cost of large sprockets may outweigh the other advantages. The catalogues contain recommendations for the maximum and minimum numbers of teeth on sprockets for use with the various types and sizes of chain.

It has been widely recorded that using odd numbers of teeth on small sprockets would give smoother operation because the vertical motions of the upper and lower parts of the chain are in phase, as in Fig. 15-8b. However, this is no longer considered to be of any great importance. Of more significance is the use of an even number of links to avoid the use of an offset link.

Since alternate links of inverted-tooth chains have odd or even numbers of contact surfaces (5 and 4 in Fig. 15-1b), it is desirable to use odd numbers of teeth on the sprockets, with an even number of links in the chain, to distribute the wear as uniformly as possible over the faces of the sprocket teeth.

For roller chains used in a one-to-one drive, it is particularly important to use a whole number of links in the tangent span. This provides the smoothest possible drive by, in effect, converting the chain and sprockets into a side-rod drive, as shown in Fig. 2-7.

The length of a chain is normally given in terms of the number of links. The equation which gives the approximate length of a chain is

$$L' = \frac{2C}{p} + \frac{N_2 + N_1}{2} + \frac{p(N_2 - N_1)^2}{4\pi^2 C} \qquad \text{links} \qquad (15\text{-}3)$$

where C = center distance, in.
p = pitch of chain, in.
N_2 = number of teeth on large sprocket
N_1 = number of teeth on small sprocket

Obviously, the chain must contain a whole number of links. In general, it is not practical to design a chain drive to fit a specified center distance without provision for some adjustment, unless appreciable sag is permissible.

15-3. Chain Selection. The power transmitted by a chain drive can be found by using the same equation (14-23) as for a belt drive except that the loose-side tension F_2 is considered to be zero.   Thus,

$$hp = \frac{F_1 V}{33,000} \tag{15-4}$$

The allowable value of F_1 for a given type and size of chain is a function of the number of teeth on the small sprocket, the chain velocity, type and degree of lubrication present, unusual operating conditions, etc.   Consequently, it is desirable to use the manufacturers' recommendations and tables to ensure satisfactory service.

A rough estimate of the capacity of a roller-chain drive may be made by considering the following approximations, which apply to all roller chains except No. 41:

Breaking strength
$$F_u = 15,000p^2 \qquad lb \tag{15-5}$$
Endurance load
$$F_e = \tfrac{1}{7}F_u \qquad lb \tag{15-6}$$
Wear load
$$F_w = \tfrac{1}{30}F_u \qquad lb \tag{15-7}$$

It is recommended, for normal chain velocities of 500 to 1,500 fpm, that the velocity ratio be less than 7:1, preferably 5:1, and that the sprockets have a minimum of 16 and a maximum of 125, preferably 20 to 60, teeth.   If the speed is between 10 and 100 rpm, sprockets with only 9 or 10 teeth can be satisfactorily used with velocity ratios as high as 12:1.

Inverted-tooth chains are normally used at velocities below 3,000 fpm, although in recent years commercial applications have been made with velocities over 8,000 fpm.   The breaking strength of inverted-tooth chain is approximately 15,000p per inch of width.   It is recommended that the sprockets have a minimum of 17 and a maximum of 150 teeth.

CHAPTER 16

GEARS

Gears are toothed wheels or multilobed cams used for transmitting motion and power from one shaft to another by means of the positive contact of successively engaging teeth. In comparison with belt and chain drives, gear drives are more compact, can operate at higher speeds, and can be used where precise timing is desired. Gears require more attention to lubrication, cleanliness, shaft alignment, etc., and usually operate in a closed case with provision for proper lubrication.

This chapter will be given to a study of several topics related to the kinematics of tooth profiles and the problem of designing gears for satisfactory strength and wear of the teeth.

16-1. Tooth Profiles. The first gears were made of wood to carry light loads at low speeds, and no particular thought was given to the effects of the shape of the teeth. As both speeds and loads increased, the problems of noise, wear, and rough operation due to improper, as well as inaccurate, tooth profiles became of major importance. Since a pair of mating gear teeth is essentially a direct-contact mechanism, the discussion in Secs. 3-11 to 3-13 is applicable to the selection of the proper curves for gear-tooth profiles.

In almost every application, the major requirement is for the smoothest possible transfer of force and motion. Thus, the basic criterion for gear-tooth profiles is the ability to transmit motion in such a manner that the angular-velocity ratio is constant at all times. As discussed previously in Sec. 3-12, the requirement *for one link to drive another with a constant angular-velocity ratio is that the line of transmission must always intersect the line of centers at the same point.* This is the fundamental law of gear-tooth action. The *pitch point* is the term used in gear nomenclature for the point of intersection of the common normal and the line of centers.

Within practical limits, it is possible to determine a profile which will operate properly, i.e., be *conjugate*, with any other arbitrarily chosen profile. This can be accomplished by utilizing only the statement of the fundamental law of gear-tooth action, but a simpler approach is to consider the two members as a cam and follower, both of which rotate at constant velocities through a small angle, and use the principle of inversion as discussed in Chap. 5. The drawing of the arbitrarily chosen follower profile in its inverted position will require some ingenuity, and the use of templates or overlays may simplify the drawing problem. Aside from the manufacturing problems that would be introduced by the wide choice of shapes, a major disadvantage is that, in general, it would be impossible to determine a third profile that would mesh properly with both of the two previously selected; in other words, the gears would not be interchangeable. *A set of gears is termed interchangeable if all possible combinations of any three gears satisfy the fundamental law of gear-tooth action when meshed.* The two basic curves that have been used for gear profiles are the *cycloid* and the *involute*.

Gears made with cycloidal teeth were used extensively during the period when teeth were usually cast or machined with a form cutter. Since the cycloidal tooth profile is now mainly of academic and historical interest, it will not receive much attention in this book, and the reader is referred to the more detailed discussions in other books.[1]

At the present time almost all gears are made with involute profiles.

16-2. Kinematics of Involute Tooth Profiles.   An *involute* is the curve traced by a point on a line as the line rolls on another curve.   For gear-tooth profiles, the involutes are derived from a circle called the *base circle*.

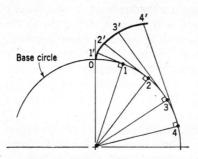

FIG. 16-1. Involute construction.

The fact that an involute can be both generated and checked by using relatively simple tools with straight edges has been a major factor in its almost universal adoption for gear-tooth profiles.

The construction of an involute is shown in Fig. 16-1.   It should be noted that the arc distance from 0 to a point on the base circle is equal to the distance along the straight line from the point to the involute; for example, 02 = 22', 04 = 44', etc.   A basic and obvious characteristic of the involute is that any straight line drawn tangent to the base circle is normal to the involute at its point of intersection.   Thus, by definition, the common normal to any surface in contact with the involute must be tangent to the base circle.   Therefore, if two involutes are in contact,

[1] C. D. Albert and F. S. Rogers, "Kinematics of Machinery," pp. 229–244, John Wiley & Sons, Inc., New York, 1938.

E. Buckingham, "Analytical Mechanics of Gears," pp. 24–32, McGraw-Hill Book Company, Inc., New York, 1949.

C. W. Ham, E. J. Crane, and W. L. Rogers, "Mechanics of Machinery," 4th ed., pp. 113–116, McGraw-Hill Book Company, Inc., New York, 1958.

as in Fig. 16-2, the common normal or line of transmission is the straight line tangent to both base circles. Because contact between two involutes must always lie on this line, it has been given a special descriptive name, the *line of action*. It should be noted that the common normal to the surface in contact will always intersect the line of centers at the same point, the pitch point P, and, therefore, involute profiles satisfy the fundamental law of gear-tooth action.

Since the useful component of the force transmitted along the common normal is that acting perpendicular to the line of centers, the angle ϕ between the line of action and the perpendicular to the line of centers

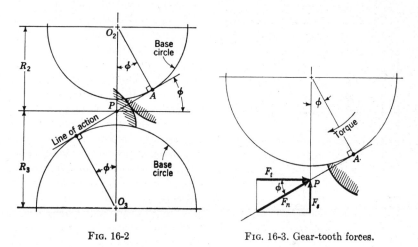

FIG. 16-2   FIG. 16-3. Gear-tooth forces.

is important in relation to the strength and wear of the teeth and in relation to the forces acting on the gear shaft and bearings. The angle ϕ is known as the *pressure angle*. It is also the angle between the line of centers and the radial line from the center to the point at which the line of action is tangent to the base circle. In Fig. 16-3 the *transmitted force* F_t is the useful component, and the *separating force* F_s is the undesirable component of the resultant *normal force* F_n. Since the transmitted force is directly related to the torque on the gear, the other forces may be computed as

$$F_n = \frac{F_t}{\cos \phi} \tag{16-1}$$

and
$$F_s = F_t \tan \phi \tag{16-2}$$

In Fig. 16-2 the distance O_2P is the *pitch radius* R_2 of gear number 2, and the distance O_2A is called the *base radius* R_{b2} of gear number 2.   It is important to note that *the base radius is fundamental and cannot be changed for a given involute, while both the pitch radius and pressure angle exist only when two involute profiles are in contact.* From Fig. 16-2 it can be seen that

$$R_b = R \cos \phi \qquad (16\text{-}3)$$

and the angular-velocity ratio is

$$\frac{\omega_3}{\omega_2} = \frac{R_2}{R_3} = \frac{R_{b2}}{R_{b3}} \qquad (16\text{-}4)$$

In Fig. 16-4 the center distance has been increased by moving O_2 to O_2'. It is evident that increasing

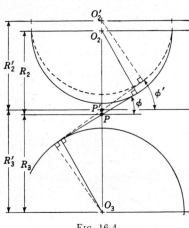

Fig. 16-4

the center distance has increased the pitch radii and the pressure angle for the two gears, but the angular-velocity ratio has not changed. This characteristic of being able to accommodate slight changes in center distance without affecting the angular-velocity ratio is the most important reason that the involute profile has almost completely supplanted the older cycloidal profiles. The cycloidal profile satisfies the fundamental law of gear-tooth action only when a specified center distance is maintained. Minor variations in center distance due to machining tolerances and wear of bearings cause cycloidal gears to transmit a resultant angular velocity consisting of an average value plus an oscillatory component. This can be objectionable, except at very low speeds, because of the increased problems with wear, noise, and vibrations.

In Fig. 16-5 the *pinion* 2 (smaller gear) is driving and rotating in a clockwise direction, as shown. Since meshing teeth can be in contact only along the line of action APD, contact first occurs when the addendum circle (outside radius) of the driven gear 3 intersects the line of action, as shown at B. Similarly, contact is broken when the addendum circle of the driving gear intersects the line of action, as at C.

As the point of contact moves from B to P, the gears rotate through the angles α_2 and α_3, respectively, and B_2' and B_3' move along the pitch circle to P. The arcs $B_2'P$ and $B_3'P$ are equal and known as the *arcs of approach*, and the angles α_2 and α_3 are known as the *angles of approach*. The same reasoning will show why the arcs PC_2' and PC_3' are called the *arcs of recess* and the angles β_2 and β_3 are called the *angles of recess.* The total arcs, $B_2'C_2'$ and $B_3'C_3'$, are the *arcs of action,* and the corresponding angles, $\alpha_2 + \beta_2$ and $\alpha_3 + \beta_3$, are the *angles of action.*

Since the gear-tooth action will be correct only when contact occurs on the line of action and the involute profile extends only outward from the base circle, the maximum radius that can be used on the driven gear 3 is O_3A. Similarly, the maximum permissible radius for gear 2 is O_2D.

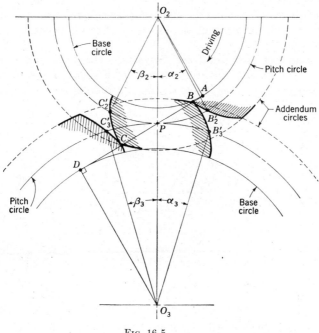

Fig. 16-5

This discussion has been concerned entirely with the action of two involutes while in contact during a fraction of a revolution of the gears. In a practical case, it is necessary to have a number of teeth on each gear so that there is always contact between at least one pair of teeth. The teeth are normally made with involute profiles on both sides so that the gears can be used in either direction of rotation. The pressure angle, the tooth thicknesses, and the length of teeth can be varied to give the combination that will result in optimum performance of the gear drive for any specified operating conditions. However, the cost of designing and manufacturing these gears would be prohibitive in most situations. Thus, standard systems of interchangeable gears have been established so that gears designed and made by one manufacturer will operate properly with gears of the same system designed and made by any other manufacturer. This standardization has also meant that the cutting tools are standard and may be purchased from a number of suppliers.

Certain modifications may be made in the use of standard cutting tools that will readily permit the manufacture of "nonstandard" gears which can give a more compact and quieter-running drive.[1] At the present time, the principal users of nonstandard gears are the aircraft and automotive industries. A detailed discussion of the design procedure for nonstandard gears is beyond the scope of this book, but from time to time comparisons will be made of the characteristics of standard and nonstandard gears.

16-3. Standard Involute Gear Systems. The terminology applied to gears includes so many names peculiar to gears that it seems like another language. Thus, before continuing the discussion on standard

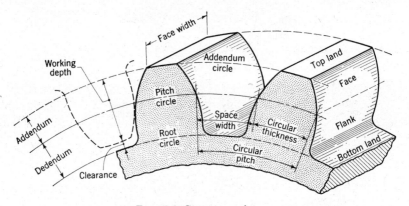

FIG. 16-6. Gear nomenclature.

involute gear systems, it will be helpful to study Fig. 16-6 and the following abbreviated list of terms.[2]

1. *Pitch surfaces* are imaginary planes, cylinders, or cones that roll together without slipping.

2. A *pitch circle* is the curve of intersection of a pitch surface of revolution and a plane of rotation. It is the imaginary circle that rolls without slipping with a pitch circle of a mating gear.

[1] Buckingham, *op. cit.*, chap. 5.

H. H. Mabie and F. W. Ocvirk, "Mechanisms and Dynamics of Machinery," chap. 5, John Wiley & Sons, Inc., New York, 1956.

W. Steeds, "Involute Gears," chap. 7, Longmans, Green & Co., Ltd., London, 1948.

[2] Except for the few cases in which confusion might arise because of conflicts with symbols used in other parts of the book, the nomenclature and symbols correspond to Gear Nomenclature—Terms, Definitions, and Illustrations, AGMA 112.03 or ASA B6.10-1954, American Society of Mechanical Engineers, New York, 1954.

3. *Pitch diameter D* is the diameter of the pitch circle.

4. *Circular pitch p_c* is the distance along the pitch circle between corresponding profiles of adjacent teeth.

$$p_c = \frac{\pi D}{N} \tag{16-5}$$

5. *Base pitch* is the circular pitch measured on the base circle.

6. *Diametral pitch P* is the ratio of the number of teeth to the number of inches in the pitch diameter. When the term *pitch* is used without qualification, it is understood to mean diametral pitch.

$$P = \frac{N}{D} \tag{16-6}$$

There is a fixed relation between diametral pitch and circular pitch, namely,

$$p_c P = \pi \tag{16-7}$$

7. The *addendum circle* coincides with the tops of the teeth in a cross section.

8. *Addendum a* is the height by which a tooth projects beyond the pitch circle or pitch line; also, it is the radial distance between the pitch circle and the addendum circle.

9. The *root circle* is tangent to the bottoms of the tooth spaces in a cross section.

10. *Dedendum d* is the depth of a tooth space below the pitch circle or pitch line; also, it is the radial distance between the pitch circle and the root circle.

11. *Clearance c* is the amount by which the dedendum in a given gear exceeds the addendum of its mating gear.

12. *Working depth h_k* is the depth of engagement of two gears, that is, the sum of their addenda.

13. *Whole depth* is the total depth of a tooth space, equal to addendum plus dedendum; also, it is equal to working depth plus clearance.

14. *Circular thickness t_c* is the length of arc between the two sides of a gear tooth, on the pitch circle unless otherwise specified.

15. *Backlash* is the amount by which the width of a tooth space exceeds the thickness of the engaging tooth on the pitch circles.

Additional terms will be introduced when pertinent to the discussion.

It is evident that interchangeable involute gears must have the following:

1. The same pitch
2. The same pressure angle
3. The same addendum
4. A circular thickness equal to one-half the circular pitch

In view of the relation between the circular and diametral pitches in Eq. (16-7), the tooth thickness becomes

$$t_c = \frac{\pi}{2P} \qquad (16\text{-}8)$$

Since it is logical to expect the addendum to be proportional to the tooth thickness and thus inversely to the diametral pitch, it is specified as

$$a = \frac{k}{P} \qquad (16\text{-}9)$$

where k is a constant.

After the addendum is specified, the working depth is known, and the major remaining unspecified dimension is the clearance. Again, the clearance is denoted as a function of the pitch. Thus, *all the dimensions for a standard system of interchangeable gears can be defined in terms of the pressure angle and pitch.* Additional standard systems have been set up for some of the noninterchangeable gears, such as the *Maag system* for long- and short-addendum spur and helical gears and the *Gleason systems*

TABLE 16-1. STANDARD INTERCHANGEABLE INVOLUTE GEAR SYSTEMS

	$14\frac{1}{2}°$ full depth	$20°$ full depth	$20°$ stub tooth
Pressure angle...............	$14\frac{1}{2}°$	$20°$	$20°$
Addendum...................	$\dfrac{1^*}{P}$	$\dfrac{1}{P}$	$\dfrac{0.80}{P}$
Working depth..............	$\dfrac{2}{P}$	$\dfrac{2}{P}$	$\dfrac{1.60}{P}$
Clearance.............	$\dfrac{0.157}{P}$	$\dfrac{0.157}{P}$	$\dfrac{0.20}{P}$
Dedendum..................	$\dfrac{1.157}{P}$	$\dfrac{1.157}{P}$	$\dfrac{1}{P}$
Whole depth...............	$\dfrac{2.157}{P}$	$\dfrac{2.157}{P}$	$\dfrac{1.80}{P}$
Tooth thickness†	$\dfrac{p_c}{2}$	$\dfrac{p_c}{2}$	$\dfrac{p_c}{2}$
Tooth space†................	$\dfrac{p_c}{2}$	$\dfrac{p_c}{2}$	$\dfrac{p_c}{2}$
Approximate fillet radius‡.....	$\dfrac{0.209}{P}$	$\dfrac{0.239}{P}$	$\dfrac{0.304}{P}$

* All dimensions are in inches and generally should be given to ten-thousandths of an inch.

† With no backlash.

‡ Not standardized—will depend somewhat upon the number of teeth and method of cutting.

for long- and short-addendum bevel gears.   These systems consider the additional factor of velocity ratio.   The Gleason systems will be discussed in more detail in later sections.

16-4. Involute Spur Gears.   A spur gear has a cylindrical pitch surface and has straight teeth.   Some of the standard systems used for interchangeable spur gears are (1) the Brown and Sharpe $14\frac{1}{2}°$ system, (2) the composite $14\frac{1}{2}°$ system, (3) the $14\frac{1}{2}°$ full-depth system, (4) the $20°$ full-depth system, (5) the Fellows stub-tooth system, and (6) the $20°$ stub-tooth system.   At the present time, the three most important systems are the $14\frac{1}{2}°$ full-depth, the $20°$ full-depth, and the $20°$ stub-tooth systems.   The basic dimensions of these systems are given in Table 16-1.

The tooth proportions given in Table 16-1 usually result in a dedendum that is greater than the radial distance from the pitch circle to the base circle.   Since the involute profile cannot exist inside the base circle, it is customary to draw the part of the tooth inside the base circle with radial flanks.

Figure 16-7 is approximately full size and illustrates the variation of size with pitch for full-depth teeth.

FIG. 16-7. Comparative sizes of gear teeth.   (*Courtesy Barber-Colman Company.*)

In most applications, the differences in tooth action and in strength of teeth for the several systems are inconsequential, and the selection is apt to be based on availability of cutting tools or stock gears.   However, the differences become more important as the capacity of the drive increases

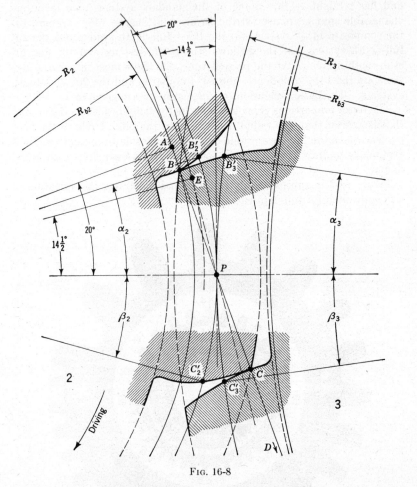

FIG. 16-8

and the size of the gears decreases.   Those factors related to gear-tooth action which result in a given system being most satisfactory for a particular application are illustrated in Figs. 16-8 and 16-9.

Figure 16-8 is a full-sized drawing showing a tooth of an 18-tooth pinion 2 in contact with a tooth of a 40-tooth gear 3.   The pinion is driving in the direction indicated, and the tooth profiles are shown at the first and last points of contact.   The gears have 2-pitch 20° full-depth involute teeth.   The pressure angle ϕ, the base-circle radii R_{b2} and R_{b3}, the pitch-circle radii R_2 and R_3, the line of action APD, the first point of contact B, the last point of contact C, the arcs of approach $B_2'P$ and $B_3'P$, the angles of approach α_2 and α_3, the arcs of recess PC_2' and PC_3', the angles of recess β_2 and β_3, the arcs of action $B_2'C_2'$ and $B_3'C_3'$, and the angles of action $\alpha_2 + \beta_2$ and $\alpha_3 + \beta_3$ have been designated in accordance with the discussion in Sec. 16-2.   However, now that standard tooth dimensions are specified, certain important limitations in relation to number of teeth and pressure angle must be considered.

In order to have continuous transmission of force and motion from one gear to another, it is necessary that a second pair of teeth come into engagement before the first pair breaks contact.   Since the arcs of action are the arcs on the pitch circle through which the gears rotate while a single pair of teeth are in contact, the ratio of the arc of action to the circular pitch may be considered as the average number of pairs of teeth in contact.   This ratio is known as the *contact ratio*.   A contact ratio of 1 would mean that the instant one pair of teeth left engagement another pair would make contact.   Kinematically, this is satisfactory.   Actually, the unavoidable errors in tooth form and spacing introduced during the machining of the gears and the deflection of the teeth under load would result in rough, noisy operation and would probably lead to premature wear or fatigue failure.   This will be considered in more detail in Sec. 16-6, but for the present it is sufficient to say that, all other things being equal, the greater the contact ratio, the smoother the action of the gears. It is usually recommended that the contact ratio be 1.4:1 or greater, although satisfactory service has been given by high-quality gears with contact ratios less than 1.1:1.

Considering that the line of action is also the line which is rolled on the base circle to generate the involute, it should be evident that the *length of contact BC* is the same distance as the arcs on the base circles through which the gears rotate while a single pair of teeth are in contact.   Therefore, the contact ratio is also equal to the length of contact divided by the base pitch.   This definition affords the simplest approach to determining a numerical value of contact ratio, because the length of contact is a distance on a straight line and does not involve the drawing of tooth profiles.   Equations are available[1] for the analytical calculation of contact ratios.

It should be apparent that the use of 20° stub teeth instead of 20° full-

[1] Buckingham, *op. cit.*, p. 22.
    Mabie and Ocvirk, *op. cit.*, pp. 93-94.

depth teeth will result in a smaller contact ratio.   However, the result of substituting a 14½° pressure angle for the 20° pressure angle is not so evident.   The line of action and the base circle for the pinion corresponding to a 14½° pressure angle are shown as dotted lines in Fig. 16-8.   It can be observed that both the length of contact and the base pitch will increase and tend to offset each other.   Actually, the greater increase in the length of contact results in a larger contact ratio for the gears with the smaller pressure angle.

The only difficulty with the comparison just made is that the first point of contact for the 14½° gears occurs to the left of E, the point of tangency of the line of action and the base circle.   As discussed in Sec. 16-2, this means that contact will occur on the pinion tooth inside the base circle.

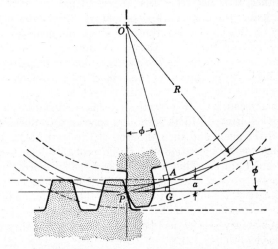

Fig. 16-9. Maximum addendum without involute interference between a rack and a gear.

Since this part of the tooth profile consists of a radial line, not an involute, the tooth action will not only be improper but there will actually be *interference* between the teeth.   The problem of interference is of major importance in relation to cutting and using involute gears with small numbers of teeth.

A gear with an infinite number of teeth will have straight lines for both the pitch and base circles, and the involute profile will be a straight line. This particular gear is known as a *rack* and is illustrated in Fig. 16-9. The addendum has been drawn so that the first point of contact A is at the point of tangency of the line of action and the base circle of the pinion.   This, therefore, represents the maximum possible addendum without interference between the pinion and rack.   From triangle OAP,

$$\sin \phi = \frac{PA}{R} \qquad (16\text{-}10)$$

and from triangle PGA,

$$\sin \phi = \frac{a}{PA} = \frac{k/P}{PA} \qquad (16\text{-}11)$$

The product of Eqs. (16-10) and (16-11) is

$$\sin^2 \phi = \frac{k}{RP} \qquad (16\text{-}12)$$

By definition,

$$P = \frac{N}{D} = \frac{N}{2R} \qquad (16\text{-}13)$$

Substituting from (16-13) for P in (16-12) and solving for N gives

$$N = \frac{2k}{\sin^2 \phi} \qquad (16\text{-}14)$$

where N is the smallest number of teeth that will mesh with a rack without interference.   Applying Eq. (16-14) to the data in Table 16-1 shows that the minimum numbers of teeth for the three systems are 32 for $14\frac{1}{2}°$ full-depth teeth, 18 for 20° full-depth teeth, and 14 for 20° stub teeth. If reference is again made to Fig. 16-9, it can be seen that, if the teeth of the pinion and rack do not interfere, the pinion can be used with any gear having a number of teeth equal to or greater than its own.

When the teeth are cut by one of the processes in which the teeth are generated conjugate to the profile of the cutting edges, the interfering portions will be cut away.   The most important manufacturing processes in this classification are hobbing and shaping.

If a gear is provided with cutting clearance and is hardened, it may be used as a generating tool in a gear shaper.   Figure 16-10 shows the action of the cutter when generating a spur gear.   The cutter reciprocates while it and the gear blank are rotated together at the angular-velocity ratio corresponding to the numbers of teeth on the cutter and the gear. The gear tooth is generated as the envelope of successive positions of the cutter teeth, as shown in Fig. 16-11.   Consequently, if the number of teeth being cut is so small in relation to the number of teeth on the cutter that interference will be a problem, the gear-teeth flanks will be undercut, as illustrated in Fig. 16-12.   Undercut teeth are undesirable because the

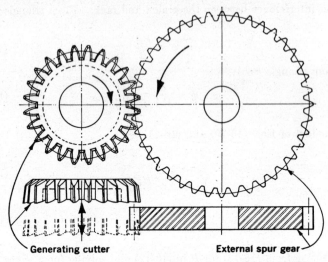

Generating cutter   **External spur gear**

FIG. 16-10. Gear-shaper process for cutting gears.   (*Courtesy Fellows Gear Shaper Company.*)

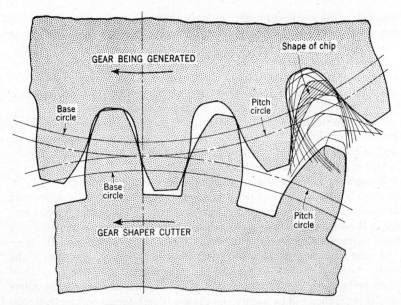

FIG. 16-11. Generating action of a gear-shaper cutter.   (*Courtesy Fellows Gear Shaper Company.*)

loss of material weakens the tooth, and the contact ratio is decreased because of the loss of part of the involute.

It should be observed that the degree of undercutting is dependent upon the pressure angle and the ratio of the number of teeth in the cutter to the number of teeth in the gear. Undercutting will be most severe when hobs (Sec. 16-17) or rack shaper cutters are used.

The strength of gear teeth will be discussed in detail in Sec. 16-7, but it can be noted at this time that a larger pressure angle gives a stronger tooth because of a thicker root section.

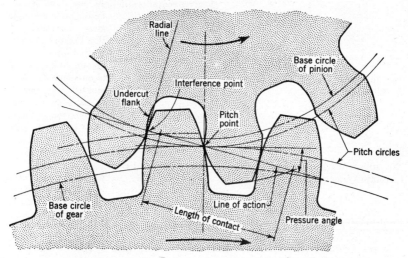

FIG. 16-12. Undercutting of gear teeth. (*Courtesy Fellows Gear Shaper Company.*)

Table 16-2 is a summary of some of the information useful in selecting the standard interchangeable involute gear system that will provide the most satisfactory service under a given set of operating conditions.

Figures 16-8 and 16-9 show that interference takes place between the tips of the teeth of the larger gear and the flanks of the teeth of the pinion. In some cases, the interference may be eliminated by decreasing the addendum of only the gear teeth. Leaving the full addendum on the pinion keeps the contact ratio from decreasing as much as if the interchangeable stub-tooth system were used. Actually, it is usually possible to increase the addendum of the pinion by the amount taken away from the addendum of the gear. The resulting set, known as *long- and short-addendum gears*, can be cut with standard tools and will operate on a standard center distance with a pressure angle corresponding to that of the cutter. An additional advantage of the long- and short-addendum system is that the strengths of the teeth are more nearly balanced, because

TABLE 16-2. GUIDE TO SELECTION OF GEAR-TOOTH SYSTEM

Interchangeable tooth system	Smallest number of teeth that will mesh with rack without interference	Smallest number of teeth in equal pinions that will give continuous driving	Comments
14½° full-depth involute system	32	20 with contact ratio of 1.080 (24 gives contact ratio of 1.469)	Recommended for use when the pinion has 40 or more teeth. Lower pressure angle gives smoother action because of lower normal force and greater contact ratio.
20° full-depth involute system	18	12 with contact ratio of 1.049 (14 gives contact ratio of 1.415)	Recommended for general use. Larger pressure angle gives wider, stronger teeth. Can be used with reasonably small pinions.
20° stub-tooth involute system	14	(12 gives contact ratio of 1.185)	Recommended for use when number of teeth in pinion is too small for satisfactory use of 20° full-depth teeth.

the pinion-tooth thickness is increased and the gear-tooth thickness is decreased. The details of the methods for selecting the best combinations of tooth height, addendum, dedendum, and pressure angle for non-interchangeable gears are beyond the scope of this book, and the reader is referred to more detailed discussions.[1]

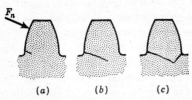

F_n

(a)   (b)   (c)

FIG. 16-13. Fatigue fracture of a gear tooth.

16-5. Gear-tooth Failures. Gears usually fail by fracturing one or more teeth or by excessive wear of the contact surfaces. These types of failure have little in common except that they are fatigue failures due to the repeated loading of the individual teeth.

Figure 16-13 shows several steps in the fracture of a tooth. The tooth fails as a cantilever under fatigue loading. The crack starts at the fillet

[1] Buckingham, op. cit.
  Albert and Rogers, op. cit.
  Steeds, op. cit.

a on the tension side of the tooth, slowly progresses inward and downward *b*, and then, as the tooth becomes much weaker because of the reduced cross section, rapidly breaks through the remainder of the tooth *c*. Section 16-7 will be given to a study of the design of spur gears to prevent this type of failure. The wear of gear teeth may appear as scoring, scratching, pitting, or a general reduction in thickness, with the surfaces remaining smooth (normal wear). These types of failures are also associated with cams, ball and roller bearings, and other direct-contact mechanisms and are considered to be a function of the intensity of the pressure between the surfaces in contact. Section 16-8 will be given to the development of a design procedure based upon the Hertz equation for cylindrical bodies in direct contact.

Gears are often used in positioning devices where the number of cycles of loading is very low and neither fatigue nor impact becomes a problem. In such a case, teeth made from ductile materials could possibly fail by deforming to the extent that improper tooth action results. This problem is not of very great importance, but, even so, the basic consideration is still that of a cantilever in bending, and the discussion in Sec. 16-7 applies.

16-6. Design Load. The useful load on a gear tooth is known as the *transmitted load* F_t. This load is assumed to act at the pitch radius of the gear and may be calculated from

$$F_t = \frac{T}{D/2} = \frac{63,000 \times \text{hp}}{nD/2} = \frac{126,000 \times \text{hp}}{nD} \qquad \text{lb} \qquad (16\text{-}15)$$

For very slow-speed (practically stationary) gears, the nominal transmitted load may be considered to be the actual force present. However, in most cases, the gears are rotating at appreciable speeds, and the impact between the mating teeth results in an additional load. These impacts are due to errors in spacing and profiles of the teeth, as manufactured, and to the slight change in spacing, due to deflection, of the teeth under load relative to the teeth next coming into engagement.

The increased load is considered to be essentially a function of the accuracy of the gears and the velocity with which the teeth come together. Several methods have been proposed for including these factors in determining a *dynamic load* F_V. The simplest approach is to divide the transmitted load by a factor known as a *dynamic factor* K_V to give the dynamic load. Thus,

$$F_V = \frac{F_t}{K_V} \qquad \text{lb} \qquad (16\text{-}16)$$

K_V is based upon experience and includes the effects of both accuracy and velocity. Since higher pitch velocities require more accurate gears for a given operational noise level, it has been found practical to specify the value of K_V in terms of the quality of the gear and the pitch velocity V in feet per minute. Several common values are as follows:

For ordinary commercial-cut gears (normally used with $V < 2,000$ fpm),

$$K_V = \frac{600}{600 + V} \tag{16-17}$$

For carefully cut gears,[1] such as those generated by hobbing or gear

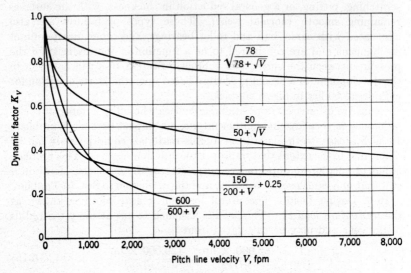

Fig. 16-14. Dynamic factors K_V.

shaping (normally used with $V < 4,000$ fpm),

$$K_V = \frac{50}{50 + \sqrt{V}} \tag{16-18}$$

For highest-accuracy gears,[1] such as shaved, ground, or lapped gears,

$$K_V = \sqrt{\frac{78}{78 + \sqrt{V}}} \tag{16-19}$$

For nonmetallic gears, such as those made of rawhide, micarta, etc.,

$$K_V = \frac{150}{200 + V} + 0.25 \tag{16-20}$$

For convenience in design the values of K_V are given as curves in Fig. 16-14.

The remaining problems in obtaining the design load are (1) the effect of overloads on the strength and wear requirements, (2) the reliability required, and (3) the life required. None of these may be considered in any general rigorous manner, but the AGMA has provided some informa-

[1] Strength of Spur, Helical, Herringbone and Bevel Gear Teeth, AGMA 225.01, American Gear Manufacturers' Association, Washington, 1959.

tion in terms of overload, reliability, and life factors that the designer can use in the absence of more specific data. These factors are presented in Tables 16-3, 16-4, and 16-5.

TABLE 16-3. OVERLOAD FACTORS—K_O*·†

Power source	Load on driven machine		
	Uniform	Moderate shock	Heavy shock
Uniform................	1.00	1.25	1.75
Light shock............	1.25	1.50	2.00
Medium shock..........	1.50	1.75	2.25

* Extracted from AGMA Information Sheet, Strength of Spur, Helical, Herringbone and Bevel Gear Teeth, AGMA 225.01, with permission of the publisher, the American Gear Manufacturers' Association, Washington, 1959.

† Table is for speed-decreasing drives only. For speed-increasing drives of spur and bevel gearing, add 0.01 $(N_G/N_P)^2$ to the factors in the table. N_P = number of teeth in the pinion and N_G = number of teeth in the gear.

TABLE 16-4. RELIABILITY FACTORS—K_R*

Requirements of application	K_R
High reliability....................	1.50–3.00
Fewer than 1 failure in 100...........	1.00–1.25
Fewer than 1 failure in 3.............	0.70–0.80

* Extracted from AGMA Information Sheet, Strength of Spur, Helical, Herringbone and Bevel Gear Teeth, AGMA 225.01, with permission of the publisher, the American Gear Manufacturers' Association, Washington, 1959.

TABLE 16-5. LIFE FACTORS—K_L*·†

Number of cycles	Spur and helical gears	
	250–450 Bhn	Case carburized§
10^3	3.0–4.0‡	2.7
10^4	2.0–2.6‡	2.0
10^5	1.6–1.8‡	1.5
10^6	1.1–1.4‡	1.1
10^7	1.0	1.0
10^8	0.9–1.0	0.9–1.0

* Extracted from AGMA Information Sheet, Strength of Spur, Helical, Herringbone and Bevel Gear Teeth, AGMA 225.01, with permission of the publisher, the American Gear Manufacturers' Association, Washington, D.C., 1959.

† The values in the 250 to 450 Bhn column apply to beam-strength calculations only. Use the values in the case carburized column for all calculations involving wear.

‡ Use the higher values for higher hardness.

§ Case hardness 55-63R_C.

When the factors K_V, K_O, K_R, and K_L are included, the design load F_d can be calculated from

$$F_d = \frac{K_O K_R}{K_V K_L} F_t \quad \text{lb} \tag{16-21}$$

16-7. Beam Strength of Spur Gears.   The simplified design procedure presented here should be adequate for most purposes.   However, if an optimum design is desired or if a large number of gears are to be manufactured, the designer should refer to the AGMA standards for more detailed information and design procedures.   In addition to the factors considered here, the AGMA equations include the effects of face width and misalignment on the distribution of load across the tooth, the maximum height of single-tooth contact, and the effects of size and operating temperature on the mechanical properties of the material.

In Fig. 16-15a the normal force F_n is resolved into components F_b snd F_r at the center of the tooth Q.   The section of maximum bending stress

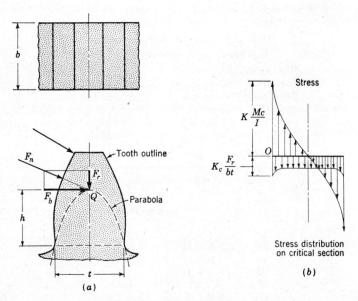

Fig. 16-15. Gear tooth as a cantilever beam.

can be found by drawing a parabola, i.e., a cantilever of uniform strength, through Q and tangent to the tooth profile.   If the slight shift in point of maximum stress due to stress concentration at the fillet is neglected, the point of tangency of the parabola and the tooth profile locates the critical section.   As shown in Fig. 16-15b, the tangential force F_b gives a bending-stress distribution, and the radial force F_r gives a compressive stress on the section. The maximum value of the bending stress is

$$s = K \frac{Mc}{I} = K \frac{F_b h t/2}{bt^3/12} = K \frac{6F_b h}{bt^2} \qquad \text{psi} \qquad (16\text{-}22)$$

and the maximum value of the compressive stress is

$$s = K_c \frac{F_r}{bt} \qquad \text{psi} \qquad (16\text{-}23)$$

Since fatigue failures start on the tension side of the tooth and the relatively small radial compressive stress acts to reduce the net tensile stress, the usual practice is to ignore the effect of the compressive stress.

Except for the unusual case of a contact ratio of 1.0, a single tooth theoretically never carries the full load at the tip of the tooth. Actually, the rigidity of the teeth is such that errors in tooth spacing as small as 0.0005 in. may prevent the sharing of the load. Thus, it is conservative to design the gears on the basis of a single pair of teeth carrying the load through the tips of the teeth.

Substituting the maximum height h_0 and the corresponding thickness t_0 for h and t and rearranging Eq. (16-22) to solve for F_b gives

$$F_b = \frac{sbt_0^2}{K6h_0} \qquad \text{lb} \qquad (16\text{-}24)$$

Multiplying Eq. (16-24) by p_c/p_c gives

$$F_b = \frac{sbt_0^2 p_c}{K6h_0 p_c} = \frac{sbp_c}{K} \frac{t_0^2}{6h_0 p_c} \qquad \text{lb} \qquad (16\text{-}25)$$

The second term, $t_0^2/6h_0 p_c$, is nondimensional, independent of pitch, and is a function only of the tooth system and number of teeth N. The quantity $t_0^2/6h_0 p_c$ is called the *Lewis form factor* and is given the symbol y. Equation (16-25) now becomes the *Lewis equation*

$$F_b = \frac{sbp_c y}{K} \qquad \text{lb} \qquad (16\text{-}26)$$

or, more usefully, in terms of diametral pitch P,

$$F_b = \frac{sb\pi y}{KP} = \frac{sbY}{KP} \qquad \text{lb} \qquad (16\text{-}27)$$

where $Y = \pi y$.

Values of Y are given in Fig. 16-16 for the three most common tooth systems.

The value of s depends upon the operating conditions specified for the gears.   If the use is so infrequent that fatigue is not a problem, s may be based on the yield strength and an appropriate factor of safety.

In most cases, fatigue is a problem, and s must be based on the endurance limit.   The most logical procedure would be based on the discussion in Secs. 6-5 and 6-6.   However, many aspects of manufacturing and use are similar for most gears, and the design can be simplified by using

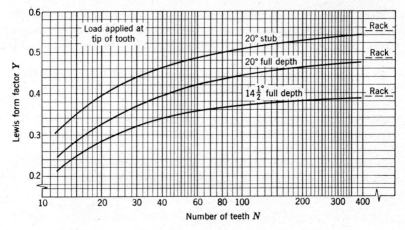

FIG. 16-16. Form factors Y.

allowable stresses, which are based on experience, and the simple beam-strength equation (16-27).

The curve for allowable stresses in Fig. 16-17 is based upon AGMA recommendations and applies to teeth loaded repeatedly in only one direction.   If the gear is to be an idler, such as a planet pinion in a planetary gear train, where the teeth will be subjected to reversed loads, it is recommended that the allowable stress be limited to 70 per cent of that indicated in Fig. 16-17.

There is little information available in relation to the endurance limit of materials other than steel.   Table 16-6 contains allowable stresses that can be used in the absence of better information.

The face width is often dictated by space limitations, heat-treating considerations, and the need for getting the maximum capacity in a minimum radial space.   In the absence of known limitations, the face width is generally chosen as from 3 to $4p_c$.   Thus, a value of $b = 3.5p_c$, or its equivalent $b = 11/P$, may be used to eliminate one variable in the initial solution of a design problem.

If the width of face must be greater than $4p_c$, particular attention must be paid to the alignment of the shafts, the deflection of the shafts, the deflection of the supporting framework, and the warpage due to heat-treatment, to ensure that the load will be reasonably uniformly distributed over the entire width of the tooth.

If the number of teeth is specified, Y can be taken directly from Fig. 16-16. If the number of teeth is not specified, it is necessary to

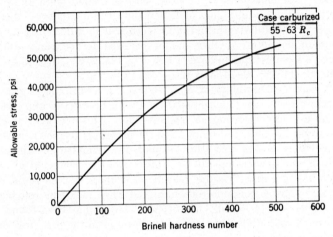

FIG. 16-17. Allowable stresses for steel spur gears. (*After Strength of Spur, Helical, Herringbone and Bevel Gear Teeth, AGMA 225.01, American Gear Manufacturers' Association, Washington, 1959.*)

assume a value for the initial calculations. When the number of teeth is known, it will be desirable to rework the problem with the correct value of form factor. Since the values do not change rapidly and the approximate number of teeth is usually known on the basis of undercutting, interference, or contact ratio, the assumed value will generally be close to the final value.

TABLE 16-6. ALLOWABLE BENDING STRESS FOR GEAR MATERIALS

Material	Allowable stress, psi
Cast iron, ordinary	5,000
Cast iron, medium grade	9,000
Cast iron, highest grade	13,000
Bronze SAE 62	15,000
Phosphor bronze SAE 65	18,000
Meehanite metal, grade GA	18,500
Rawhide, Fabroil	9,000
Micarta, Bakelite, Celoron	13,000
Nylon	6,000

It should be noted that, if the gear and pinion are made of different materials, the smaller product s_2Y_2/K_2 or s_3Y_3/K_3 will designate the critical member.

The stress-concentration factor K for the root fillet of gear teeth is a function of the ratio of fillet radius to tooth thickness and the ratio of tooth thickness to tooth height.[1] The fillet radius for generated gears is, in turn, a function of the cutter-tip radius and the number of teeth in the gear. For most design purposes, with generated teeth and contact at the tip of the tooth, values of $K = 1.6$ for full-depth $14\frac{1}{2}°$ teeth and $K = 1.5$ for 20° full-depth and stub teeth will be sufficiently accurate.

Common diametral pitches should be used for economy and convenience. Common pitches are 1, $1\frac{1}{4}$, $1\frac{1}{2}$, $1\frac{3}{4}$, 2, $2\frac{1}{2}$, 3, $3\frac{1}{2}$, 4, 5, 6, 7, 8, 9, 10, 12, 14, 16, 20, 24, and 32.

16-8. Surface Durability of Spur Gears. The surfaces of gear teeth and other direct-contact mechanisms may fail because of abrasive wear or because of deterioration as pits develop. Abrasive wear is closely related to the lubrication provided and the materials and their treatment. As for clutches, brakes, and other elements involving friction and sliding, the rate of wear is considered to be a function of the product of pV, where p is the pressure and V is the velocity of sliding between the surfaces.

Pitting failure is commonly encountered when members roll together in the presence of a liquid. This type of failure has been found to be related to the materials and their treatment and to the pressure between the bodies in contact.

It should be noted that both abrasive wear and pitting are functions of the pressure and the properties of the materials. Otherwise, they have little in common. In the case of gear teeth, the maximum sliding velocity occurs when contact is made at the farthest distance from the pitch point and pure rolling occurs only when the teeth are in contact at the pitch point.

There is little design information available[2] in relation to any type of surface failure other than that of pitting. Fortunately, in the great majority of cases with enclosed drives and adequate lubrication, gears whose surfaces do not fail by pitting will also not fail by abrasive wear, scoring, etc. Therefore, the only type of surface failure that will be considered further here is that due to pitting. Although this is really a problem in surface durability and not wear, except in very broad terms, the convenience of its short length has led to common use of the expression *wear strength* to describe the load the teeth can carry without destructive pitting.

---

[1] T. J. Dolan and E. L. Broghamer, A Photoelastic Study of Stresses in Gear Tooth Fillets, *Univ. Illinois Eng. Expt. Sta. Bull.* 335, 1942.

[2] For a good practical discussion of the problems of gear-tooth failures see D. W. Dudley, "Practical Gear Design," chap. 7, McGraw-Hill Book Company, Inc., 1954.

As pointed out in the discussion of the beam strength of spur gears, the appropriate AGMA standard[1] should be consulted if the optimum design is required. The AGMA method is basically the same as that presented here, which was formulated by Buckingham[2] on the basis of the Hertz equation for cylindrical bodies in contact. The AGMA method includes, in addition, the effects of nonuniform distribution of load across the face width and the increased stress at the start of single-tooth contact for various numbers of teeth.

The Hertz equation[3] for the maximum pressure occurring at the center of the surface of contact between convex cylinders is

$$q_o = \sqrt{\frac{F'(r_2 + r_3)}{\pi^2(k_2 + k_3)r_2r_3}} \qquad \text{psi} \qquad (16\text{-}28)$$

where $F' =$ load per unit length of contact, lb
r_2 and $r_3 =$ radii of the cylinders, in.
$\qquad k_2 = (1 - \nu_2{}^2)/\pi E_2$
$\qquad k_3 = (1 - \nu_3{}^2)/\pi E_3$
ν_2 and $\nu_3 =$ Poisson's ratios of the respective materials
E_2 and $E_3 =$ moduli of elasticity of the respective materials, psi

In the case of cylinders of the same material and $\nu = 0.3$ (a reasonable assumption for most metals), Eq. (16-28) becomes

$$q_o = 0.418 \sqrt{\frac{F'E(r_2 + r_3)}{r_2r_3}} \qquad \text{psi} \qquad (16\text{-}29)$$

Equations (16-28) and (16-29) can be used also to relate the allowable surface pressure or stress to the dimensions of cams, roller bearings, and gear teeth. It should be emphasized that wear strength is a property of *two* bodies in contact. For *steel gears*, an analysis of Buckingham's data indicates that the allowable contact pressure is directly related to the sum of the hardnesses of the surfaces and is approximately 180 times the sum of the Brinell hardness numbers.

A fundamental property of an involute profile (Sec. 16-2) is that the radius of curvature at any point is equal to the distance along the normal from the profile to its point of tangency with the base circle. Since experience has shown that pitting occurs most often in the vicinity of

[1] Surface Durability of Spur Gears, AGMA 210.01, American Gear Manufacturers' Association, Washington, 1946.

[2] E. Buckingham, The Relation of Load to Wear of Gear Teeth, paper presented at the Annual Meeting of the American Gear Manufacturers' Association, Detroit, Mich., May 14, 1926.

Buckingham, *op. cit.*, chap. 23.

[3] S. Timoshenko, "Strength of Materials," 3d ed., pt. II, pp. 339–345, D. Van Nostrand Company, Inc., Princeton, N.J., 1956.

the pitch circle, the radii of curvature become

$$r_2 = \frac{D_2}{2} \sin \phi \quad \text{and} \quad r_3 = \frac{D_3}{2} \sin \phi \qquad (16\text{-}30)$$

Assuming $\nu = 0.3$, substituting from (16-30) for r_2 and r_3, and rearranging to solve for F', Eq. (16-28) becomes

$$F' = 1.42q_o{}^2 \sin \phi \left(\frac{1}{E_2} + \frac{1}{E_3}\right) \frac{D_2 D_3}{D_2 + D_3} \qquad \text{lb/in.} \qquad (16\text{-}31)$$

If Eq. (16-31) is multiplied by the face width b and certain terms are grouped together, the *Buckingham wear equation* appears as

$$F_w = D_2 b K_w Q \qquad \text{lb} \qquad (16\text{-}32)$$

where D_2 = diameter of the pinion, in.
   b = face width, in.
   $K_w = 0.71q_o{}^2 \sin \phi (1/E_2 + 1/E_3)$ psi
   $Q = 2D_3/(D_2 + D_3) = 2N_3/(N_2 + N_3)$

The value of K_w, the *load-stress factor*, can be readily calculated for any desirable combination of materials and tooth system. However, most design problems can be solved by choosing the proper hardnesses and value of K_w from the recommended combinations in Table 16-7.

16-9. Application of the Beam- and Wear-strength Equations. In many specialized applications such as marine gear and aircraft power-plant designs, the materials, heat-treatment, pitches, and gear ratios in use are comparable for all cases. The beam-strength equation does not need to be used, and the wear-strength equation can be simplified to

$$F_w = D_p b K_w' \qquad (16\text{-}33)$$

Here, K_w' is the allowable tooth load per inch of pinion diameter per inch of face width, and its value is based upon the accumulated experience of the company in a relatively narrow field of application.

In general, it is not desirable to specify a series of steps in designing a gear train, because the approach will have to be altered as the initial specifications are varied. Basically, the procedure consists of the following:

1. Calculating a design load. It may be either static or dynamic and will include factors for overloads, reliability, and life.

2. Using the Lewis, or beam-strength, equation (16-27) to determine the proper combination of pitch, face width, and material properties of the teeth to ensure that $F_b \geqq F_d$.

3. Using the Buckingham wear equation (16-32) to determine the required material property (Bhn) of the surfaces of the teeth to ensure that $F_w \geqq F_d$.

TABLE 16-7. VALUES OF LOAD-STRESS FACTOR K_w

| Pinion | | Gear | | Allowable q_o, psi | K_w, psi | |
Material	Bhn	Material	Bhn		$\phi = 14\frac{1}{2}°$	$\phi = 20°$
Steel.........	150	Steel	150	50,000	30	41
Steel.........	200	Steel	150	60,000	43	58
Steel.........	250	Steel	150	70,000	58	79
Steel.........	200	Steel	200	70,000	58	79
Steel.........	250	Steel	200	80,000	76	103
Steel.........	300	Steel	200	90,000	96	131
Steel.........	250	Steel	250	90,000	96	131
Steel.........	300	Steel	250	100,000	119	162
Steel.........	350	Steel	250	110,000	144	196
Steel.........	300	Steel	300	110,000	144	196
Steel.........	350	Steel	300	120,000	171	233
Steel.........	400	Steel	300	125,000	186	254
Steel.........	350	Steel	350	130,000	201	275
Steel.........	400	Steel	350	140,000	233	318
Steel.........	450	Steel	350	145,000	250	342
Steel.........	400	Steel	400	150,000	268	366
Steel.........	500	Steel	400	155,000	286	391
Steel.........	600	Steel	400	160,000	305	417
Steel.........	550	Steel	450	170,000	344	470
Steel.........	600	Steel	450	175,000	364	497
Steel.........	500	Steel	500	190,000	430	588
Steel.........	600	Steel	600	230,000	630	861
Steel.........	150	Cast iron	...	50,000	44	60
Steel.........	200	Cast iron	...	70,000	87	119
Steel.........	250 and higher	Cast iron	...	90,000	144	196
Steel.........	150	Phosphor bronze	...	50,000	46	62
Steel.........	200	Phosphor bronze	...	70,000	87	119
Steel.........	250 and higher	Phosphor bronze	...	85,000	135	204
Cast iron.....	...	Cast iron	...	90,000	193	284

4. Reviewing steps 2 and 3 to adjust design parameters so that the beam and wear strengths are equal and both equal or exceed the design load. In many instances, it is neither practical nor desirable to go to the limits that might be required to equalize the beam and wear strengths of the gears. In such a situation, the lower of the two strengths must equal or exceed the design load.

Example 16-1. The brake for a hoist was designed in Example 13-1, and the wire rope was selected in Example 14-2. In this example we are to select an electric motor and make a preliminary design of a spur-gear reduction drive. The lifting speed of the 3-ton load was specified as 200 fpm and the cable drum diameter as 24 in. The motor will be a crane and hoist motor with a synchronous (no-load) speed of 1,800 rpm. Crane and hoist motors are similar to Design D motors with high starting torque (about 275 per cent of rated torque), low starting current, and relatively high full-load slip (about 20 per cent).[1]

Solution. (*a*) *Motor selection.* The power required to raise 6,000 lb at the rate of 200 fpm is

$$\text{hp} = \frac{FV}{33,000} = \frac{6,000 \times 200}{33,000} = 36.4$$

The mechanical efficiency of the hoist is not known and cannot be readily calculated at this time. However, 90 per cent is a reasonable value, and thus the motor should be rated at about $36.4/0.90 = 40.4$ hp. The closest standard motor ratings are 40 and 50 hp (Appendix G), and a 40-hp motor will be specified.

(*b*) *Gear-train velocity ratios.* Velocity ratios for gear trains in general are considered in Chap. 17. In this case, the cable-drum speed will be

$$n_{CD} = \frac{V}{\pi D} = \frac{200}{\pi \times 24/12} = 31.8 \text{ rpm}$$

and the motor full-load speed will be

$$n_M = 1,800(1 - S) = 1,800(1 - 0.20) = 1,440 \text{ rpm}$$

The reduction ratio will be

$$\frac{n_M}{n_{CD}} = \frac{1,440}{31.8} = 45.3$$

It is not practical to achieve such a large reduction in a single step with spur gears, and a compound gear train must be used (Sec. 17-1). If two reductions are used, the simplest procedure is to make each equal to $(45.3)^{1/2}$:1 or 6.73:1. Even this ratio is large for spur gears, but since the hoist is to be used in the construction industry, where neither space requirements nor quietness of operation is critical, the double reduction drive will be used because of its relative simplicity. Alternatives include using a more expensive lower speed motor or using a triple reduction with its additional parts and complexity.

(*c*) *Gear-tooth system and numbers of teeth.* The large reduction ratio of 6.73:1 in each step requires pinions with maximum strength and a minimum number of teeth. We shall therefore use 20° stub-tooth pinions with 14 teeth (Table 16-2). For a velocity ratio of 6.73:1, the gears should have $14 \times 6.73 = 94.2$ or 94 teeth, with a ratio of $94/14 = 6.71$.

It should be noted that rounding off the tooth number to a whole number of teeth has slightly changed the over-all reduction ratio and thus the hoist speed. The increase, from 200 to 201 fpm, is negligible, particularly when it is considered that the high-slip motor speed will vary over a range of several per cent as the load changes.

---

[1] C. C. Libby, "Motor Selection and Application," p. 46, McGraw-Hill Book Company, Inc., New York, 1960.

The proposed drive for the hoist is shown in Fig. 16-18. It should be noted that the gear train, brake, and cable drum are designed as an integral unit, with the gears enclosed in an oil- and dust-tight case.

(d) *Gears for first reduction.* Since the dynamic factor K_V is a function of both the accuracy of the gears and the pitch line velocity and nothing can be specified about either at this moment, we shall defer consideration of it until a more appropriate time.

In operation, the hoist will be subjected to frequent starts. The load on the driven machine is uniform, and we shall consider the starting to be equivalent to a light shock for the power source. Thus, from Table 16-3, $K_O = 1.25$.

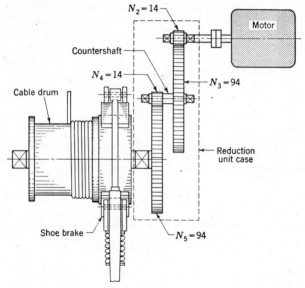

Fig. 16-18

Safety considerations indicate a need for high reliability, but since the hoist will not be used to carry people, we shall choose $K_R = 1.50$, the lowest recommended value for high reliability in Table 16-4.

At 1,440 rpm, 10 million cycles will be completed in

$$10 \times 10^6/1{,}440 \times 60 = 115.8 \text{ hr}$$

of use under full load. To get an idea of the significance of this number, let us assume that each lifting operation requires 1 min and that the hoist is used ten times each hour during an 8-hr workday. This operation will take $(115.8 \times 60)/(1 \times 10 \times 8) = 86.8$ days, which is not unreasonable. We shall assume that $K_L = 1.0$, from Table 16-5, will be appropriate.

The final choice of materials and material treatments will be based upon a number of factors. In order to gain an idea as to a practical maximum size of drive, we are asked to make our calculations on the basis of hobbed pinions made from heat-treated steel with a Brinell hardness number of 250 and hobbed gears made from a medium-grade cast iron.

Beam strength. The pitch will be selected on the basis of adequate beam strength by use of the Lewis equation (16-27):

$$F_b = \frac{sbY}{KP}$$

Since the materials and tooth system have been specified and the numbers of teeth are known, the products $s_2 Y_2/K_2$ and $s_3 Y_3/K_3$ can be calculated to determine whether the pinion or the gear will be the weaker.

Pinion:

$$s_2 = 35,000 \text{ psi} \qquad \text{from Fig. 16-17}$$
$$Y_2 = 0.34 \qquad \text{from Fig. 16-16}$$
$$K_2 = 1.5$$

and
$$\frac{s_2 Y_2}{K_2} = \frac{35,000 \times 0.34}{1.5} = 7,940$$

Gear:

$$s_3 = 9,000 \text{ psi} \qquad \text{from Table 16-6 for medium-grade cast iron}$$
$$Y_3 = 0.504 \qquad \text{from Fig. 16-16}$$
$$K_3 = 1.5$$

and
$$\frac{s_3 Y_3}{K_3} = \frac{9,000 \times 0.504}{1.5} = 3,020$$

Therefore, the beam-strength design will consider only gear 3, the weaker of the pair.

The beam strength F_b will be made equal to the design load F_d when the hoist is operating at its rated capacity. The design-load equation (16-21) is

$$F_d = \frac{K_O K_R}{K_V K_L} F_t$$

where
$$F_t = \frac{126,000 \times \text{hp}}{n_2 D_2} = \frac{126,000 \times 40}{1,440 D_2} = \frac{3,500}{D_2}$$

$$K_V = \frac{50}{50 + \sqrt{V}} \qquad \text{for hobbed gears}$$

or, in terms of D_2,

$$K_V = \frac{50}{50 + \sqrt{\pi D_2 n}} = \frac{50}{50 + \sqrt{\pi D_2\, 1,440/12}} = \frac{50}{50 + \sqrt{377 D_2}}$$

and K_O, K_R, and K_L have been determined previously as 1.25, 1.50, and 1.0, respectively. Neither F_t nor K_V can be calculated until D_2 is determined. However, D_2 can be expressed as N_2/P which, in this case, equals $14/P$. This will be convenient for F_t, which becomes

$$F_t = \frac{3,500}{14/P} = 250P$$

The expression for K_V in terms of P becomes cumbersome to use, in that the Lewis equation must be solved by trial and error. It is usually more practical to estimate the velocity, calculate K_V, and determine the required pitch. The calculated value of pitch is then used to determine better values for D, V, and K_V, after which a second value for the required pitch should be calculated. Usually the second trial will be adequate, but sometimes a third solution is necessary.

We shall estimate the pinion diameter as 5.0000 in.

$$V = \pi D n = \pi \times \tfrac{5}{12} \times 1,440 = 1,880 \text{ fpm}$$

and, from Fig. 16-14, $K_V = 0.54$. Thus,

$$F_d = \frac{1.25 \times 1.50}{0.54 \times 1.0} \, 250P = 868P$$

Therefore, substituting in the Lewis equation,

$$868P = \frac{9,000 \times 11/P \times 0.504}{1.5 \times P}$$

from which $P = 3.39$. Rounding off to a pitch of 3, which is the nearest common standard pitch with stronger teeth, we find for our second trial that

$$D_2 = \frac{N_2}{P} = \frac{14}{3} = 4.6667 \text{ in.}$$

$$V = \frac{4.6667}{5.0000} \times 1,880 = 1,760 \text{ fpm}$$

and $K_V \approx 0.54$, which is the value we had guessed. Therefore, our estimate of velocity for the first trial was most fortunate. However, it should be noted that, since K_V does not change rapidly and since the pitch is determined as the cube root, a less fortunate estimate for V will not greatly affect the amount of work required to reach a final answer.

Rounding off the pitch from 3.29 to 3 results in a gear that is stronger than necessary. In the interest of saving material and machining time, it will be worthwhile to adjust the face width to make $F_b = F_d$. Therefore,

$$\frac{9,000 \times b \times 0.504}{1.5 \times 3} = 868 \times 3$$

from which $b = 2.60$, or $2\frac{5}{8}$ in. This is appreciably narrower than the assumed value of $11/P = 1\frac{1}{3} = 3.667$ in.

Wear strength. The 3-pitch pinion and gear will now be checked to see whether $F_w \geqq F_d$ or not.

$$F_w = D_2 b K_w Q \tag{16-32}$$

and

$$D_2 = 4.6667 \text{ in.}$$
$$b = 2\frac{5}{8} \text{ in.}$$
$$K_w = 196 \quad \text{from Table 16-7}$$
$$Q = \frac{2N_3}{N_2 + N_3} = \frac{2 \times 94}{14 + 94} = 1.74$$

Thus,

$$F_w = 4.6667 \times 2.625 \times 196 \times 1.74 = 4,180 \text{ lb}$$

From the beam-strength solution,

$$F_d = 868P = 868 \times 3 = 2,600 \text{ lb}$$

Therefore, since $4,180 > 2,600$, the 3-pitch pinion and gear have more than adequate wear strength. (Although one example does not really justify a sweeping generalization, it will be found that, except under most unusual circumstances, the wear strength will exceed the beam strength whenever a cast-iron gear is in contact with a steel gear or another cast-iron gear.)

(e) *Gears for second reduction.* The procedure for designing the second-reduction gears will be similar to that for the first-reduction gears. We know that the countershaft torque will be 6.71 times the torque on the input (motor) shaft and that the countershaft speed will be $1,440/6.71 = 215$ rpm. It is apparent also that, since the materials are the same, the pitch, and thus the diameters, will be larger for the second-reduction gears. In terms of pitch, the transmitted force will be

$$F_t = 6.71 \times 250P = 1,680P$$

If the pinions 2 and 4 were the same diameter, the pitch-line velocity for the second reduction would be $1,760/6.71$, or 262 fpm. However, the increased size of 4 required to carry the greater force will result in a velocity somewhat higher than 262 fpm. We shall estimate $V = 500$ fpm for the first trial.

Beam strength

$$K_V = 0.69$$
$$F_d = \frac{1.25 \times 1.5}{0.69 \times 1.0} \, 1,680P = 4,570P$$
$$4,570P = \frac{9,000 \times 11/P \times 0.504}{1.5 \times P}$$

from which $P = 1.94$. We now have the choice of dropping down to $P = 1\frac{3}{4}$ and cutting down the face width or going up to $P = 2$ and increasing the face width. Since the gears are going to be quite large in diameter in any case, we shall use a pitch of 2 and find the value of face width that will make $F_b = F_d$.

$$D_4 = \frac{N_4}{P} = \frac{14}{2} = 7.0000 \text{ in.}$$
$$V = \pi \times \frac{7.0000}{12} \times 215 = 394 \text{ fpm}$$
$$K_V = 0.71$$
$$F_d = \frac{0.69}{0.71} \times 4,570P = 4,440 \times 2 = 8,880 \text{ lb}$$

Thus, for $F_b = F_d$

$$\frac{9,000 \times b \times 0.504}{1.5 \times 2} = 8,880 \text{ lb}$$

from which $b = 5.87$ or $5\frac{7}{8}$ in.

Wear strength. Although it really is not necessary to check the wear strength in this case, it is

$$F_w = 7.0000 \times 5.875 \times 196 \times 1.74 = 14,000 \text{ lb}$$

which, as was expected, is greater than the design load of 8,880 lb and is therefore adequate.

Summary

First reduction:
   Pinion: steel, 250 Bhn, 14 teeth, 3-pitch, 20° stub-tooth system, 4.6667-in. pitch
      diameter, 5.2000-in. blank diameter $(D + 2a)$, $2\frac{5}{8}$-in. face width
   Gear: medium-grade cast iron, 94 teeth, 3-pitch, 20° stub-tooth system, 31.3333-in.
      pitch diameter, 31.8667-in. blank diameter, $2\frac{5}{8}$-in. face width
Second reduction:
   Pinion: steel, 250 Bhn, 14 teeth, 2-pitch, 20° stub-tooth system, 7.0000-in. pitch
      diameter, 7.8000-in. blank diameter, $5\frac{7}{8}$-in. face width
   Gear: Medium-grade cast iron, 94 teeth, 20° stub-tooth system, 47.0000-in. pitch
      diameter, 47.8000-in. blank diameter, $5\frac{7}{8}$-in. face width

Closure. At this point it is worthwhile to review the design and to outline the course to be followed in arriving at the final solution. The design feature of greatest interest is the large size of the gears. This was not unexpected. The design was purposely based on a cast-iron gear in order to give us an idea as to the maximum space required. The gear teeth were found to be less than half as strong as the pinion teeth, with $s_3 Y_3 / K_3 = 3{,}020$ as compared with $s_2 Y_2 / K_2 = 7{,}940$. Thus the over-all size can be decreased somewhat by using steel instead of cast iron. The problem would become somewhat different because, with a steel pinion and gear, wear rather than fracture is likely to become the critical factor. An additional decrease in over-all size would be gained by using steels that are heat-treated, either through- or case-hardened, for increased mechanical properties.

However, probably the best place to gain in compactness would be to substitute a triple-reduction, with ratios of $(45.3)^{1/3}:1$ or $3.57:1$, for the double-reduction gear train. It should be noted that a triple-reduction drive would require manufacturing or purchasing additional parts, e.g., two more gears, another shaft, and two more bearings, and would also increase the complexity of the reduction-unit frame and case.

The final choice of materials and number of reductions will be based upon both economic and space considerations.

16-10. Involute Helical Gears on Parallel Shafts.

One problem inherent in the operation of spur gears is that a tooth on one gear makes instantaneous full-length contact with a tooth on the mating gear. This sudden contact means that there must be a sudden deformation of the tooth to compensate for the deflection under load of the teeth just previously engaged. Also, the effects of errors in tooth form and spacing are most pronounced. The staggered teeth of the stepped spur gear in Fig. 16-19a is one approach toward securing more gradual contact. The

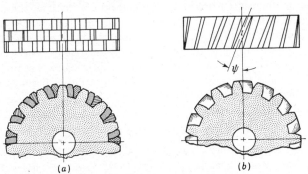

Fig. 16-19. (a) Stepped spur gear; (b) helical gear.

stepped gear is not practical because it would have to be made as three separate spur gears and then fastened together, with the laminations displaced in increments of one-third of the circular pitch.

A better approach to gradual contact is the *helical gear* in Fig. 16-19b. A helical gear is, in effect, a stepped gear with an infinite number of steps.

As the name implies, the teeth of helical gears form true helices which make a constant angle, the *helix angle* ψ, with the axis of rotation.

Another term important in the discussion of gears with curved teeth is the *hand* of the gear.   The gear in Fig. 16-19b is a *right-hand* helical gear. The designation of hand is the same as that used for screw threads.   It should be noted that a pair of helical gears on parallel shafts have the same helix angle but are of opposite hand.   Since the force must be transmitted along the normal between the mating teeth, the plane perpendicular to the teeth, known as the *normal plane*, has an important place in the study of helical gears.   The relation between tooth dimensions and angles for the normal plane and the plane of rotation can be determined by considering Fig. 16-20, from which

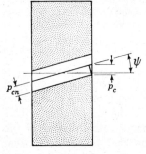

FIG. 16-20

$$p_{cn} = p_c \cos \psi \qquad (16\text{-}34)$$

Recalling that $p_c P = \pi$, Eq. (16-34) may be rewritten as

$$P_n = \frac{P}{\cos \psi} \qquad (16\text{-}35)$$

where p_{cn} and P_n are the circular and diametral pitches, respectively, in the normal plane.

These relations are important in that the dimensions of the gear depend upon the method of cutting; i.e., if the gears are hobbed, the pitch and pressure angle of the hob will apply in the normal plane, while if the gears are cut by the Fellows gear-shaper method, the pitch and pressure angle of the cutter will apply to the plane of rotation.   Thus, since the dimensions most useful in design, such as those for locating axes of rotation and calculating tooth forces, are in the plane of rotation, it becomes essential to know how the gears are to be made.

From Fig. 16-21,

$$\tan \phi_n = \tan \phi \cos \psi \qquad (16\text{-}36)$$

and

$$F_n = \frac{F_t}{\cos \phi_n \cos \psi} \qquad (16\text{-}37)$$

$$F_s = F_t \tan \phi \qquad (16\text{-}38)$$

$$F_a = F_t \tan \psi \qquad (16\text{-}39)$$

where F_a is the component of force in the axial direction.

The contact between mating teeth is gradual, starting at one end and moving along the teeth so that at any instant the line of contact runs diagonally across the teeth.   Also, the contact ratio is greater than for spur gears because the total arc of contact is the sum of the arc of contact of a section in the plane of rotation plus the additional arc given by the twist of the teeth.   If the helix angle and the face width are such that in Fig. 16-22

$$b \tan \psi = p_c \qquad (16\text{-}40)$$

there will always be a pair of teeth in contact because of the twist of the helix, without regard to the nominal arc of action.   It is usually recommended that the overlap be about 15 per cent of the circular pitch, or

$$b = 1.15 \frac{p_c}{\tan \psi} \qquad (16\text{-}41)$$

Both the gradual contact and greater contact ratio contribute materially to the smoother running of a pair of helical gears, as compared with similar spur gears.

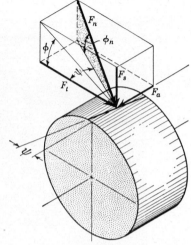

Fig. 16-21. Helical-gear-tooth forces.

The axial component of force introduced by the helix angle results in an additional bending moment on the shaft and requires at least one bearing to carry a thrust as well as a radial load.   In a double-reduction gear unit, such as in Fig. 16-23, the helix angles of the gears on the countershaft are usually so related to the diameters that the net axial force on

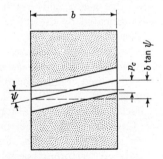

Fig. 16-22. Overlap of helical-gear teeth.

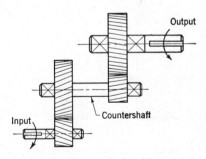

Fig. 16-23. Countershaft with zero net axial force.

the countershaft is zero.   Thus, only a simple thrust bearing or washer is required to carry the small axial load introduced by machining errors, acceleration forces, gravity forces, etc.

Figure 16-24 shows this principle as applied to a Falk Motoreducer, which combines a motor and a gear reduction unit into a self-contained drive. The figure shows a double-reduction unit in which the input and output shafts are on the same axis. The torque on the countershaft is

FIG. 16-24. Double-reduction Falk Motoreducer. (*Courtesy The Falk Corporation.*)

due to the gears only, and the transmitted force is greater for the pinion than for the gear. Thus, to balance the axial force, the helix angle is greater for the gear, and to balance tooth strengths, the face width is greater for the pinion. The case is a weldment fabricated from steel plate, and the gears and bearings are splash-lubricated by the action of the countershaft gear which dips into the oil in the bottom of the case.

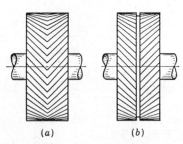

(a)   (b)

FIG. 16-25. (a) Continuous-tooth herringbone gear; (b) double-helical gear.

Herringbone gears and *double-helical gears* are widely used because the axial thrust forces are balanced within the gear itself and the net axial force is zero. Continuous-tooth herringbone gears, as in Fig. 16-25a, must be cut with a reciprocating cutter, while the double-helical gears in Fig. 16-25b can be either hobbed, if space between halves is enough for hob clearance, or cut with a reciprocating cutter. Since the balancing of the axial forces is dependent upon the equal distribution of load between the right- and left-hand parts of the gear, it is necessary for one gear of a pair to be axially located relative to the other gear with a high degree of precision.   The simplest and most common method is to use thrust bearings for one shaft only and to allow the other shafts to float axially.   The gear-tooth forces will then automatically position the other gears so that no external thrust forces are present.

Figure 16-26 shows a double-reduction speed reducer with the top half of the cast-iron housing removed.   The double-helical-pinion teeth are

Fig. 16-26. Reduction gear unit with double-helical gears. (*Courtesy The Falk Corporation.*)

machined integral with the shafts, and the gears are made from steel castings.   Tapered roller bearings[1] on the low-speed shaft carry both radial and thrust loads and provide for axial positioning of the gear train. Cylindrical roller bearings, with no axial restraint, are used on the high- and intermediate-speed shafts.

Helix angles for single-helical gears are generally between 15 and 25°, while helix angles for herringbone gears vary from 20 to 45°.

Helical gears may also be used on nonparallel or crossed shafts, but the tooth action and design considerations are quite different from those for helical gears on parallel shafts.   Since helical gears on nonparallel shafts are closely related to worm and worm-gear drives, this particular

[1] See Sec. 18-18 for discussion of roller bearings.

application will be considered in Sec. 16-17 as an introduction to the discussion of worm-gear drives in Sec. 16-18.

16-11. Beam Strength of Helical Gears. The appropriate AGMA standards[1] should be consulted for a more rigorous treatment, but for many purposes a suitable pitch for helical gears can be selected by using the Lewis equation (16-27) for spur gears. The pitch and the form factor will correspond to the plane of rotation. Any errors introduced will be on the safe side.

16-12. Wear Strength of Helical Gears. The AGMA standard for determining the surface strength of helical and herringbone gears[2] considers the effects of face width on load distribution, material properties, tooth form, gear ratio, dynamic factor, and the revolutions per minute and pitch diameter of the pinion.

The Buckingham wear equation for helical gears becomes

$$F_w = \frac{D_2 b K_w Q}{\cos^2 \psi} \tag{16-42}$$

where the numerator is the same as in Eq. (16-32) for the wear strength of spur-gear teeth.

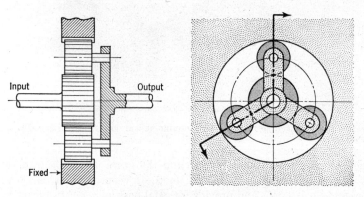

Input   Output

Fixed →

FIG. 16-27. Planetary gear train.

16-13. Internal Involute Gears. An internal involute gear has either spur or helical teeth cut on the inside of a ring. Its most common use is in a planetary gear train, as in Fig. 16-27, but it can be used in a more

---

[1] Strength of Spur, Helical, Herringbone and Bevel Gear Teeth, AGMA 225.01, 1959, and Strength of Helical and Herringbone Gear Teeth, AGMA 221.01, 1948, American Gear Manufacturers' Association, Washington.

[2] Surface Durability of Helical and Herringbone Gears, AGMA 211.01, American Gear Manufacturers' Association, Washington, 1944.

compact ordinary reduction unit, as in Fig. 16-28.   The calculation of velocity ratios for planetary gear trains will be considered in Secs. 17-3 to 17-5.   As illustrated, a number of intermediate, or planet, gears may be used to share the load so that this type of drive will give the maximum

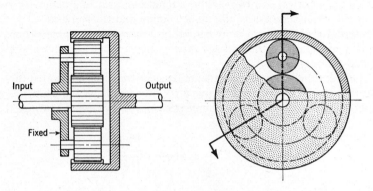

Fig. 16-28

capacity in a given space.   Because of unavoidable errors in manufacturing, the total load is not divided equally among the sets of gears.   A reasonable assumption for commercial units is that each set carries 0.9 of the load that it could carry as a simple pair of gears.

Figure 16-29 shows a pinion in mesh with the internal or ring gear.   It can be seen that the tooth of the internal gear corresponds to the tooth space of a similar external gear and that the tooth space of the internal gear corresponds to the tooth of an external gear.   This results in a relatively thick and strong internal-gear tooth and in the contact of a convex with a concave surface, instead of convex on convex as for external gears.

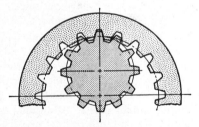

Fig. 16-29

The stronger internal-gear tooth means that it is ordinarily not necessary to calculate its beam strength.

The contact between the convex and concave surfaces results in increased wear strength.   This can be seen by noting that in Eq. (16-28) the $r_2 + r_3$ term applies to contact between convex surfaces and that, when one surface is convex and the other concave, the term becomes either $r_2 - r_3$ or $r_3 - r_2$, depending on whether r_2 is greater or less than r_3. The net result is that, when using Eq. (16-32) or (16-42) for an internal and an external gear,

$$Q = \frac{2D_3}{D_3 - D_2} = \frac{2N_3}{N_3 - N_2} \qquad (16\text{-}43)$$

Thus, the critical pair of gears for strength or wear in a planetary gear train will almost always be the planet pinion and the sun gear.

Internal gears also run more smoothly because of the convex-concave contact and because the contact ratio is higher than that for a similar pair of external gears.

The use of helical gears in a planetary gear train is complicated by the fact that the axial tooth forces give a rocking couple on the planet gears, as shown in Fig. 16-30. This couple necessitates the use of bearings that

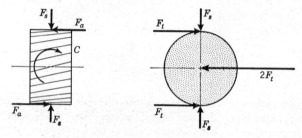

FIG. 16-30. Tooth forces on helical idler gears.

can keep the gears in correct alignment without undue wear. Consequently, the smaller planetary gear trains are usually made with spur gears and simple bronze bushings.

If the difference in numbers of teeth between the internal and external gears is less than seven for stub teeth or less than 12 for full-depth teeth, the teeth will interfere or foul well away from the line of action. This fouling is not the same as the involute interference previously discussed. It is generally not a problem, because the difference in size between the internal gear and the planets of a planetary gear train is so great that fouling cannot exist. Additional considerations peculiar to gear trains with multiple-tooth contact are discussed in Sec. 17-4.

16-14. Bevel Gears. *Bevel gearing* properly describes those gears based upon pitch surfaces which are frustums of cones. Figure 16-31 shows two pairs of cones in contact. The elements of the cones in Fig. 16-31a intersect at the point of intersection of the axes of rotation. It can be seen that, since the radii of both gears are proportional to their distance from the apex, the cones may roll together without sliding. In Fig. 16-31b the elements of both cones do not intersect at the point of shaft intersection, and consequently there can be pure rolling at only one point of contact and there must be tangential sliding at all other points of contact.   Therefore, these cones cannot be used as pitch surfaces because

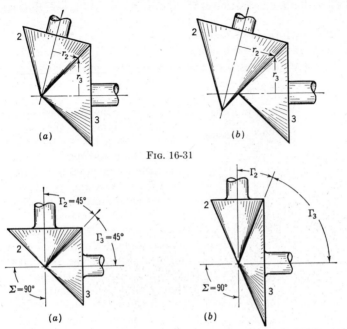

(a)   (b)

Fɪɢ. 16-31

Fɪɢ. 16-32. Pitch cones for bevel gears with 90° shaft angles.

it is impossible to have positive driving and sliding in the *same direction* at the same time.   Thus, the elements of bevel-gear pitch cones and the shafts must intersect at the same point.

In Fig. 16-32 the shafts intersect at right angles.   In Fig. 16-32a the cones are identical with *pitch angles* $\Gamma = 45°$.   Gears based on these pitch cones are known as *miter gears*. While bevel gears are commonly used to connect shafts intersecting at 90°, they often find use where the angle of intersection is other than 90°. Figure 16-33 shows the pitch cones for the particular case where one of

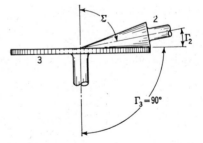

Fɪɢ. 16-33. Pitch cones for a bevel pinion and a crown gear.

the bevel gears has a pitch angle of 90° and is known as a *crown gear*. The crown gear corresponds to a rack in spur gearing.

The involute teeth for a spur gear can be generated by the edge of a plane as it rolls on a base cylinder. A similar analysis for a bevel gear will show that a true section of the resulting involute lies on the surface of a sphere and thus cannot be accurately represented on a plane surface. Actually, the form is so complex that it has not been possible to build a machine that will cut true spherical involute surfaces. In practice, most bevel gears are generated so that they are conjugate to crown-gear teeth with flat sides.

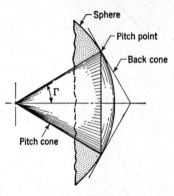

FIG. 16-34

It is not possible to represent on a plane surface the exact profile of a bevel-gear tooth lying on the surface of a sphere. Since the strength and kinematic characteristics are dependent upon the tooth profiles, it is important to approximate the bevel-gear tooth profiles as accurately as possible. The method used is known as *Tredgold's approximation*. This approximation is based upon the fact that a cone tangent to the sphere at the pitch point will closely approximate the surface of the sphere for a short distance either side of the pitch point, as shown in Fig. 16-34. The cone, known as the *back cone*, can then be developed as a plane surface, and spur-gear teeth corresponding to the pitch and pressure angle of the bevel gear and the radius of the developed cone can be drawn. This procedure is better illustrated in Fig. 16-35, where Γ is the pitch angle and R is the pitch radius of the bevel gear. The form of the teeth does not depend upon R but rather upon the back-cone distance or *equivalent pitch radius R_e*. From Fig. 16-35 it can be seen that

$$R_e = \frac{R}{\cos \Gamma} \qquad (16\text{-}44)$$

and that the *equivalent number of teeth N_e* is given by

$$N_e = 2PR_e = \frac{2PR}{\cos \Gamma} = \frac{N}{\cos \Gamma} \qquad (16\text{-}45)$$

The action of the bevel gears will be the same as that of the equivalent spur gears. Since the equivalent number of teeth is always greater than the actual number of teeth, a given pair of bevel gears will have a larger contact ratio and will run more smoothly than will a pair of spur gears with the same numbers of teeth.

Bevel gears can be, and have been, made with proportions corresponding to many of the standard gear-tooth systems. However, since bevel gears are inherently noninterchangeable and usually operate as matched pairs, there is much to be gained by using unequal-addendum teeth. Thus, most bevel gears are designed in accordance with the standards developed by the Gleason Works, Rochester, N.Y. These standards are known as the *Gleason systems* and were adopted by the AGMA in 1922 as recommended standards. Since the design and cutting of bevel gears are too highly specialized for discussion in detail here, only some of the

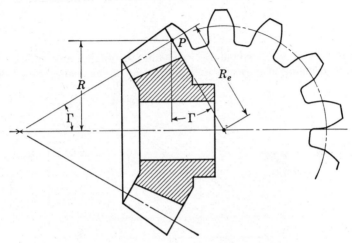

Fig. 16-35. Equivalent spur gear.

more practical aspects in relation to the different types of bevel gears will be covered. The reader is referred to the publications of the Gleason Works and the AGMA for additional information.

Figure 16-36 is a section view of two bevel gears in mesh, illustrating the dimensions and terms most important in the study of bevel gears. The pitch diameters are measured at the large, or heel, ends of the teeth. It should be noted that the elements of the pitch and root cones intersect at the pitch apex or shaft intersection, while the face cone does not. The face cone of each gear is turned parallel to the root cone of the other. This gives a constant clearance which eliminates possible fillet interference at the small ends of the teeth. The sum of the addenda of the pinion and gear is a function of the diametral pitch and is constant, but the actual addenda are selected on the basis of the velocity ratio of the drive. For a velocity ratio of 1:1, the gear and pinion addenda are equal, while for ratios greater than about 7:1, the gear addendum is only 37 per cent of the pinion addendum.

The basic pressure angle is 20°, but other angles from 14½ to 25° may be used when desired.

By the use of similar triangles, it can be shown that the pitch angles Γ for the pinion and gear can be found from

$$\tan \Gamma_2 = \frac{\sin \Sigma}{N_3/N_2 + \cos \Sigma} \qquad (16\text{-}46)$$

and

$$\Gamma_3 = \Sigma - \Gamma_2 \qquad (16\text{-}47)$$

For the usual case, with $\Sigma = 90°$, Eq. (16-46) becomes

$$\tan \Gamma_2 = \frac{N_2}{N_3} \qquad (16\text{-}48)$$

Other angles, e.g., dedendum, root, and face angles, are important in the dimensioning of drawings and the cutting of the gears. Since these

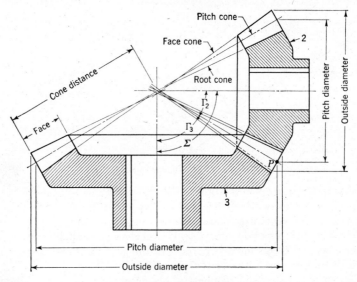

Fig. 16-36. Bevel gears.

angles depend upon the addenda of the gears, the publications of the Gleason Works or the AGMA should be consulted for the appropriate equations.

It is recommended that the face width not exceed 0.3 of the cone distance or $10/P$, whichever is smaller. The slight increase in strength gained by increasing the width of the face at the small end of the tooth is outweighed by the increased difficulties in manufacturing and in maintaining uniform contact over the entire face. The alignment of bevel gears is much more critical than that of spur or helical gears. Machining errors, deflections of supporting members, thermal expansion of gears and shafts, and play in bearings may combine to cause the small end of the teeth to carry most of the load. While keeping the face width fairly short helps to alleviate this problem, the modern approach is to manufacture the gears with slightly curved teeth so that the ends are relieved.

The *straight bevel gear* system has the simplest form, with straight teeth. It may be used with pinions containing at least 13 teeth. The calculations and the setting up of the gear cutting machines are the least complicated, and consequently the straight bevel gear system is the most economical for small-lot production. For the purpose of calculating shaft bending moments and bearing loads, the transmitted force F_t is assumed to act at the mean pitch diameter. As shown in Fig. 16-37, the pressure angle and the pitch angle result in the transmitted force being accompanied by radial and axial separating forces. These may be calculated from

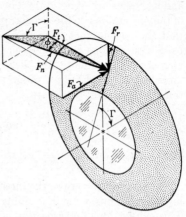

Fig. 16-37. Tooth forces on a bevel gear.

$$F_r = F_t \tan \phi \cos \Gamma \qquad (16\text{-}49)$$
and
$$F_a = F_t \tan \phi \sin \Gamma \qquad (16\text{-}50)$$

Straight bevel gears are not recommended for use with peripheral speeds

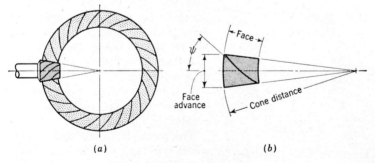

(a)   (b)

Fig. 16-38. Spiral bevel gears.

greater than 1,000 fpm or 1,000 rpm, the main reasons being that the teeth make instantaneous line contact along the entire width of face.

The *spiral bevel gear* in Fig. 16-38a is related to a straight bevel gear in the same manner as a helical gear is related to a spur gear. The teeth are circular arcs, and the *spiral angle* ψ (Fig. 16-38b) is measured at the mid-point of the face.

The spiral angle improves the tooth action by giving gradual contact and a greater arc of action. The *face-contact ratio* is defined as the ratio of the *face advance* to the circular pitch. The Gleason Works recommends that the face-contact ratio be at least 1.25. Spiral bevel gears may be used with pinions containing as few as six teeth. The gear-tooth-force analysis becomes more complicated for spiral bevel gears because of the introduction of additional radial and axial separating forces. The equations for the total forces are

$$F_r = \frac{F_t}{\cos \psi} (\tan \phi \cos \Gamma \pm \psi \sin \Gamma) \tag{16-51}$$

and

$$F_a = \frac{F_t}{\cos \psi} (\tan \phi \sin \Gamma \mp \sin \psi \cos \Gamma) \tag{16-52}$$

The hand of the gears, the direction of rotation, and which gear is driving must be considered in deciding whether the plus or minus is to be used. This can be done by inspection. The hand of spiral bevel gears is determined in the same manner as for helical gears, but since the pinion and gear will be of opposite hand, the hand of the pair is denoted by that of the pinion. The choice of hand is important because under certain conditions the axial force may, with one direction of rotation, draw the gears tightly together, whereas with the other direction of rotation the axial force pushes the gears apart. In general, it is desirable to have the axial forces on both the gear and pinion, or at least on the pinion, tending to move the gears out of mesh. Thus, the only ill effect of excessive end play will be to increase backlash. The necessity for maintaining precise alignment with large axial and radial loads has resulted in the almost universal use of ball and roller bearings in spiral-bevel-gear mountings.

It is recommended that ground gears be used when the peripheral speed is greater than 8,000 fpm.

The particular spiral bevel gear with a zero spiral angle in Fig. 16-39 is known as a *Zerol bevel gear*. Zerol bevel gears

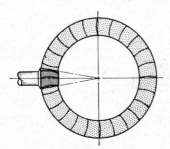

FIG. 16-39. A Zerol bevel gear.

have the same tooth action and the same tooth forces and are recommended for use with the same peripheral speed (under 1,000 fpm or 1,000 rpm) as straight bevel gears. Zerol bevel gears theoretically give somewhat more gradual contact and a slightly larger contact ratio. However, this increase in contact ratio is too small to be of much practical value.

The principal advantage of the Zerol bevel gear is the fact that it can be manufactured on the same cutting and grinding equipment used for spiral bevel gears. The teeth must be ground if the gears are to be hardened or carburized, and the highest accuracy is required.

The appropriate AGMA standards[1] or the publications of the Gleason Works should be consulted when designing bevel gears for satisfactory fatigue and wear life.

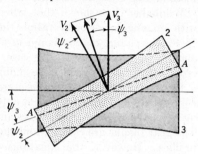

16-15. Hypoid Gears. In many applications, it is advantageous or necessary to use gears which operate on nonparallel and nonintersecting shafts. One of the most important types is that developed by the Gleason Works as a replacement for spiral bevel gears in automotive differentials to permit the lowering of the drive shaft and the floor of the automobile. These gears are known as *hypoid gears* and are based upon pitch surfaces which are hyperboloids of revolution. The hyperboloids of revolution in Fig. 16-40 may be generated by rotating a straight line about an axis which it does not intersect and to which it is not parallel. If two hyperboloids are correctly aligned or designed, they will be tangent along a common generating straight line A-A.[2]

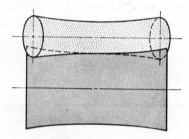

Fig. 16-40. Hyperboloids of revolution.

When the two hyperboloids are rotated together as shown, the resultant motion is a combination of rolling and sliding. The sliding is along the line of tangency.[3] It should be evident that, for hyperboloids in contact,

---

[1] Strength of Spur, Helical, Herringbone and Bevel Gear Teeth, AGMA 225.01, 1959, and Surface Durability of Straight Bevel and Spiral Bevel Gears, AGMA 212.01, 1944, American Gear Manufacturers' Association, Washington.

[2] Buckingham, *op. cit.*, chap. 17.

Albert and Rogers, *op. cit.*, pp. 207–210.

[3] It should be noted that the accepted definition of pitch surfaces (Sec. 16-3) considers only those surfaces which may roll together without slipping, and, therefore, hyperboloids, with their combination of rolling in the direction of transmission of motion and sliding along the line of tangency, should not be classed as pitch surfaces. Nevertheless, since the rolling component of motion for the hyperboloids is the same as for other pitch surfaces and since there is no other term available, this book will consider hyperboloids to be the pitch surfaces for hypoid gears.

as shown in Fig. 16-40, the gear teeth must be straight and parallel to the line of tangency. In the past, hypoid gears with straight teeth have been called *skew bevel gears*. Current practice, however, is to use the term *skew hypoid gears*, reserving the name *skew bevel* for bevel gears with teeth having straight elements not intersecting the axis.

Wildhaber[1] has shown that when hyperboloids are rotated together with an angular-velocity ratio other than that corresponding to Fig. 16-40, the sliding motion is not directed along a straight line but rather along a curved path. In this case, the teeth must follow the curves in order to permit the lengthwise sliding.

In practice, the term *hypoid gears* almost always refers to gears with curved teeth. Hypoid gears not only look like spiral bevel gears but have many of the same characteristics and are made on the same equipment. The gear blanks are conical, and the teeth are generated with conjugate profiles. However, the equations for diameter, equivalent number of teeth, etc., presented in Sec. 16-14 for bevel gears are not valid for hypoid gears, and publications of the Gleason Works must be consulted for details. Hypoid-gear teeth are unsymmetrical, and, in order to have equal arcs of action for the two sides, each side has a different pressure

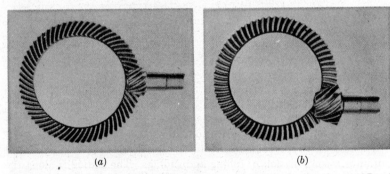

(a)   (b)

Fig. 16-41. Comparison of bevel and hypoid gear sets with the same ratio.   (a) Bevel; (b) hypoid. (*Courtesy Gleason Works.*)

angle.   Figure 16-41 compares a hypoid gear and pinion with an equivalent spiral bevel gear and pinion. The fact that the hypoid pinion is considerably larger and stronger than the spiral bevel pinion is responsible for the increasing use of hypoid gears in heavily loaded drives where the offset-shaft feature is of relatively little importance, e.g., heavy-duty truck and locomotive drives. When used in heavily loaded drives, the pinion offset is usually less than 20 per cent of the cone distance of the equivalent bevel-gear combinations.   The offset may be up to 40 per cent

[1] Ernest Wildhaber, Basic Relationship of Hypoid Gears: I, *Am. Machinist*, vol. 90, no. 4, pp. 108–111, February, 1946.

for normal power drives and slightly over 40 per cent for very lightly loaded drives.

Another important difference between hypoid and spiral bevel gears is that the hypoid teeth are subjected to sliding along the teeth. This additional sliding results in smoother operation but at the expense of decreased efficiency and an increased difficulty in providing adequate lubrication. The efficiency of bevel-gear drives is in the order of 98 to 99 per cent, whereas that of hypoid-gear drives, with ratios less than 10:1, is in the order of 96 to 98 per cent. The lengthwise sliding under heavy load tends to weld the contacting tooth surfaces together, and a special lubricating oil is required. This oil contains an antiflux and is known as an *extreme-pressure lubricant*.

16-16. Face Gears. The Fellows Gear Shaper Company, Springfield, Vt., has developed a type of gearing which utilizes a spur gear pinion meshing with conjugate teeth cut into the face of a ring gear.[1] If, in a plan view of the face gear, the axes intersect (as for bevel gears), the face gear is described as being *on center*, while if the axes do not intersect the face gear is described as being *off center* (as for hypoid gears). An on-center face gear is illustrated in Fig. 16-42.

On-center face gears may be considered as special bevel gears which, instead of having teeth whose dimensions are proportional to the

FIG. 16-42. Face gear and spur pinion. (*Courtesy Fellows Gear Shaper Company.*)

distance from the point of intersection of the axes, have teeth whose dimensions vary in such a manner that the pinion becomes a spur gear.

The advantages of face gears are that spur-gear cutters and gear shapers can be used to cut the face gears and that the axial position of the pinion is not critical, as it is for bevel- and hypoid-gear units.

The major disadvantage is that the width of tooth face on the face gear is limited by the trimming of the flank of the tooth at the inner end and by the top land coming to a point at the outer end.

16-17. Involute Helical Gears on Crossed Axes. As discussed in Sec. 16-10, helical gears have cylindrical pitch surfaces. If the axes of rotation are parallel, as in Fig. 16-43a, the pitch cylinders are in contact along a line parallel to the axes, and the cylinders can roll together without one

[1] B. Bloomfield, Designing Face Gears, *Machine Design*, vol. 19, pp. 129–134, April, 1947.

cylinder sliding on the other. However, if the axes are crossed, as in Fig. 16-43b, the pitch cylinders are in contact at only one point, and rotation will be accompanied by sliding of one cylinder on the other. Figure 16-44a illustrates the case where cylinders 2 and 3 are so rotating that the tangential velocity of the points of contact are V_2 and V_3, respectively. Since, as discussed in Sec. 3-11, the velocity of sliding is the vector difference between the absolute velocities of the points in contact, it must be directed along a line parallel to MN. Therefore, if teeth are to be machined on these pitch surfaces, the teeth must permit this sliding to

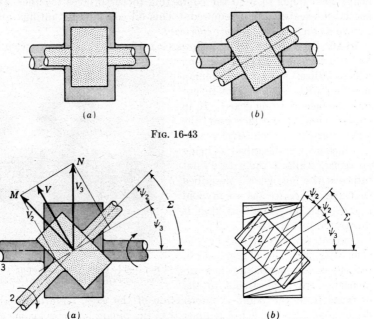

FIG. 16-43

FIG. 16-44. Helical gears on crossed axes.

take place and thus must also be parallel to MN at the point of contact. Figure 16-44b shows the correct orientation of the teeth for this particular situation.

It should be noted that in Fig. 16-44b both gears are left-hand helical gears and that the angle between shafts is the sum of the helix angles; that is, $\Sigma = \psi_2 + \psi_3$. It can be shown that, when the gears have the same hand, the shaft angle equals the sum of the helix angles and that, when the gears have opposite hands, the shaft angle equals the difference between the helix angles. For the particular, and most common, case, with the shafts at 90°, the gears must have the same hand.

It is apparent that the angular-velocity ratio of *any* pair of gears is

$$\frac{\omega_2}{\omega_3} = \frac{N_3}{N_2} \tag{16-53}$$

It is often desirable to calculate the velocity ratio in terms of the pitch diameters or to calculate the pitch diameters required to give a specified velocity ratio with a specified center distance. The gears studied so far, with the exception of hypoid and off-center face gears, have pitch diameters which are directly proportional to the numbers of teeth. However, when helical gears on crossed axes or hypoid gears or a worm and worm gear are used, the relation between velocity ratio and pitch diameter is not so simple. From Fig. 16-44a,

$$\omega_2 = \frac{V_2}{D_2/2} \quad \text{and} \quad \omega_3 = \frac{V_3}{D_3/2} \tag{16-54}$$

or

$$\frac{\omega_2}{\omega_3} = \frac{V_2}{D_2} \frac{D_3}{V_3} \tag{16-55}$$

but

$$V_2 = \frac{V}{\cos \psi_2} \quad \text{and} \quad V_3 = \frac{V}{\cos \psi_3} \tag{16-56}$$

Thus

$$\frac{\omega_2}{\omega_3} = \frac{D_3 \cos \psi_3}{D_2 \cos \psi_2} \tag{16-57}$$

For the teeth to mesh properly, the two gears must have the same pitch and pressure angle in the normal plane. *It should be noted that these are fundamental requirements for any pair of mating involute gears.*

Since helical gears on crossed axes have point loading and sliding along the teeth, they are useful only in relatively light-duty service. Two common automotive applications are the speedometer cable drive and the oil-pump and distributor drive.

The sliding along the teeth for helical gears on crossed axes is utilized in the generation process of *gear hobbing* and the finishing process of *gear shaving.*

Fig. 16-45. Hobbing of a helical gear. (*Courtesy Barber-Colman Company.*)

Figure 16-45 shows a hob in position for cutting the teeth on helical gears. The hob is essentially a helical gear with a small number of teeth that have been gashed to provide cutting edges. For greatest accuracy, a single-thread (one-tooth) hob is used, but multiple-thread hobs are often used for roughing operations because of the increased rate of production.

One hob of given pitch can be used to cut any size of spur or helical gear of either hand with helix angles up to 30°. For helix angles greater than 30°, the hob and gear should be of the same hand.

During the cutting operation, the hob is fed parallel to the gear-blank axis while the hob and the blank are rotated together at the correct speed ratio. When spur gears are being cut, the ratio is simply the inverse ratio of the numbers of teeth, whereas when helical gears are cut, the ratio must also consider the combined effects of the helix angle of the gear and the rate of feed of the hob.

The rotary shaving cutter in Fig. 16-46 is a high-speed-steel helical gear with hardened and ground teeth containing a number of radial

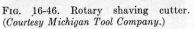

FIG. 16-46. Rotary shaving cutter. *(Courtesy Michigan Tool Company.)*   FIG. 16-47. Shaving a helical gear. *(Courtesy Michigan Tool Company.)*

grooves that form the cutting edges of the tool. Figure 16-47 shows the cutter as used in shaving a helical gear. Gear shaving is a finishing operation and is applied to gears after the teeth have been generated by shaping or hobbing. The gear is mounted on live centers and is driven by the cutter as it rotates and is fed slowly parallel to the gear axis. The slow feed, about 0.010 in. per revolution of the gear blank, and the large number of cutting edges combine with the sliding velocity to remove the "scallops" left by the initial generation of the teeth (Fig. 16-11) and to correct minor errors in tooth profile.

The depth of cut during the hobbing or shaping operation is greater than standard for gears which are to be finished by shaving. This is to avoid fouling the tips of the cutter teeth with the fillet at the base of the gear teeth. The depth of cut is usually $2.35/P$ as compared with $2.157/P$ for standard full-depth teeth.

The most efficient shear angle, i.e., shaft angle, is between 12 and 15°, but angles as low as 3° may be used.

16-18. Worms and Worm Gears. Two types of worms are commonly used, the *cylindrical worm* and the *double-enveloping*, or hourglass, worm. The cylindrical worm in Fig. 16-48a is a true involute helical gear with

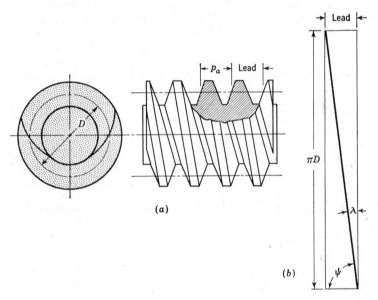

FIG. 16-48. Single-thread cylindrical worm.

one tooth. The combination of a relatively large helix angle and a small diameter results in the gear having an appearance similar to a screw thread. Actually, the two have much in common, and the same terminology applies to both. Figure 16-48b shows the trace of a tooth on the developed pitch surface of the worm in Fig. 16-48a. The *axial pitch p_a* corresponds to the pitch of a screw thread. The *lead* is the distance a point on one tooth or thread advances during one revolution. As for screw threads, the lead equals the axial pitch for a single-thread worm, the lead is twice the axial pitch for a double-thread worm, etc. The *lead angle λ* is the angle, at the pitch diameter, between a thread and the plane of rotation.

The sum of the helix and lead angles is 90°. The equations for determining the tooth-force components and relating the dimensions and angles in the normal plane to the plane of rotation were presented in Sec. 16-10.

If the worm is to be used with involute spur or helical gears, it must have true involute profiles in the plane of rotation. Worm teeth which have straight sides in an axial section are close enough to the true involute for most practical purposes. When a worm meshes with a spur gear or a helical gear, the pair are actually helical gears on crossed axes, and the discussion and equations in Sec. 16-17 should be applied. The major advantage of a worm drive is that it can be used with much larger velocity ratios than any other type of gearing. The disadvantage is that, when a worm is used with a helical gear, there is kinematic point contact, and only light loads can be carried.

One practical answer to this problem of point contact is the use of a *worm gear* or *worm wheel* in place of a helical gear. The worm gear in Fig. 16-49 is machined by a hob which is identical, except for clearance, to

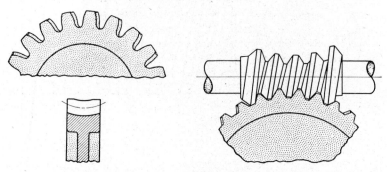

FIG. 16-49. Worm gear for use with a cylindrical worm.

FIG. 16-50. Double-enveloping worm and worm-gear set.

the worm. Thus, even though the worm gear no longer has involute teeth, the mating teeth will be conjugate, and the enveloping of the worm by the worm gear results in line instead of point contact between the teeth.[1]

A later development is the hourglass, or double-enveloping, worm and worm gear in Fig. 16-50.   In this case, the worm envelops the worm gear, and the worm gear envelops the worm; the result is area contact in place of line contact. Thus, the contact pressure between teeth will be lower, and a given-capacity drive will occupy less space for a double-enveloping worm-gear set than for the other types. The alignment of these gears is more critical than for a cylindrical worm and worm gear.

[1] See Figs. 18-63 and 18-64 for cutaway views of typical worm-gear speed reducers.

The effects of the high sliding velocity between teeth are the same as for hypoid gears (Sec. 16-15), only more pronounced. If the relatively minor losses due to friction in bearings are neglected, the efficiency of a worm drive with shafts at 90° is

$$\eta = \tan \lambda \frac{\cos \phi_n - \mu \tan \lambda}{\cos \phi_n \tan \lambda + \mu} \tag{16-58}$$

The coefficient of friction μ is a function of materials, lubricant, workmanship, and rubbing velocity. The worm is usually made of heat-treated steel and the worm gear of bronze, and the coefficient of friction will normally vary between 0.03 and 0.15. It should be noted that Eq. (16-58) is the same as Eq. (10-10) for translation screws, except that the pressure angle in the normal plane ϕ_n has replaced its equivalent α'. The curves for the efficiency of the Acme screw thread with $\alpha = 14.5°$ in Fig. 10-5 are also applicable, if bearing friction is neglected, to a worm-gear unit with a $14\frac{1}{2}°$ involute worm.

The capacity of a worm-gear reduction unit is usually limited by the rate of heat dissipation or wear of the members. High-capacity commercial worm-gear reduction units are often equipped with built-in blowers for forced convection cooling or are provided with coils for circulating cooling water through the lubricating oil.

The quiet operation at high speeds with a relatively large change in speed has resulted in the worm and worm-gear drive being used in speed-increasing as well as speed-reducing applications. Typical applications have been to automotive supercharger drives and centrifugal cream separators. In speed-increasing applications, it is desirable to use a multiple-thread worm with a relatively large lead angle to give a reasonable efficiency; e.g., an automotive supercharger used a six-threaded worm with a lead angle of about 45°.

A worm-gear set is self-locking if the worm gear cannot drive the worm. As for screw threads, the criterion for self-locking is a function of the lead angle, the pressure angle, and the coefficient of friction. As an approximation, a drive will be self-locking if the coefficient of friction is greater than the tangent of the lead angle. The AGMA has a number of standards[1] related to the designing of worms, worm gears, and reduction units.

16-19. Miscellaneous Gears. In recent years numerous special gears have been developed for transmitting power or motion between nonparallel shafts. In general, they represent variations of the types dis-

---

[1] Surface Durability of Cylindrical-Wormgearing, AGMA 213.02, 1952; Surface Durability of Double Enveloping Wormgearing, AGMA 214.02, 1954; Practice for Cylindrical Wormgear Speed Reducers, AGMA 440.03, 1959; Practice for Double Enveloping Wormgear Speed Reducers, AGMA 441.02, 1954, American Gear Manufacturers' Association, Washington.

cussed above.   However, for certain applications, some have additional
features such as greater economy of manufacturing, less critical mounting
requirements, and the possibility of controlling or eliminating backlash.
Some of the trade names are Beveloid, Spiroid, Helicon, and Planoid.
The reader should consult the literature[1] for more information.

[1] R. C. Bryant and D. W. Dudley, Which Right-angle Gear System?   *Prod. Eng.*,
vol. 31, no. 46, pp. 55–66, Nov. 7, 1960.

CHAPTER 17

GEAR TRAINS, VARIABLE-SPEED DRIVES, AND DIFFERENTIAL MECHANISMS

The transmission of angular motion between shafts by belts, chains, clutches, couplings, and gears has been discussed in earlier chapters where the major emphasis was given to the special problems of kinematics, strength, and wear in relation to their design and use. This chapter will be devoted mainly to the study of combinations of these and other mechanisms that are used to change the output characteristics, e.g., speed, direction, type of motion, force, torque, etc., in relation to the characteristics of the prime mover.

17-1. Gear Trains. The gear trains illustrated in Fig. 17-1 are known as *simple gear trains* because in each case only two gears are involved.

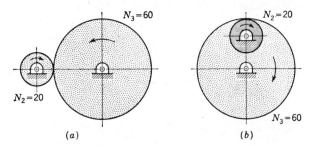

FIG. 17-1. Simple gear trains.

The *velocity ratio* of a gear train is defined as the ratio of the angular velocity of the driving gear to that of the driven gear. In both Fig. 17-1a and b, with gear 2 driving, the velocity ratio is

$$\frac{\omega_2}{\omega_3} = \frac{N_3}{N_2} = {}^{60}\!/_{20} = 3 \qquad (17\text{-}1)$$

It should be noted that the shafts of the two external gears a rotate in opposite directions, whereas the shafts of the external and internal gears b rotate in the same direction.

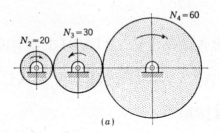

(a)

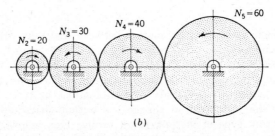

(b)

FIG. 17-2. Idler gear trains.

In the gear train illustrated in Fig. 17-2, each parallel shaft carries only one gear.   The velocity ratio of the train in Fig. 17-2a is

$$\frac{\omega_2}{\omega_4} = \frac{N_3}{N_2}\frac{N_4}{N_3} = {}^{30}\!/_{20} \times {}^{60}\!/_{30} = 3 \qquad (17\text{-}2)$$

Similarly, the velocity ratio of the train in Fig. 17-2b is

$$\frac{\omega_2}{\omega_5} = \frac{N_3}{N_2}\frac{N_4}{N_3}\frac{N_5}{N_4} = {}^{30}\!/_{20} \times {}^{40}\!/_{30} \times {}^{60}\!/_{40} = 3 \qquad (17\text{-}3)$$

In both cases, the velocity ratio depends only upon the numbers of teeth in the first and last gears of the train.   The remaining gears are called *idler gears*, and the train is known as an *idler gear train*.   Idlers are used to fill the space between the first and last shafts and/or to change the direction of rotation of the last shaft relative to the first.   It should be noted that, if an odd number of idlers is used, the first and last shafts rotate in the same direction; if an even (or zero) number of idlers is used, the first

and last shafts rotate in opposite directions.   Figure 17-3 shows the use
of idler gears in a simple reversing mechanism, as used on lathes for
reversing the rotation of the feed screw and the direction of the tool feed
relative to the rotation of the work piece.

The obvious disadvantage of a simple gear train is that, for large
velocity ratios, the size of the last gear and, consequently, the over-all
size of the train become large.   In addition to the problem of size, the
strength and wear capacities become unbalanced, and inefficient use is

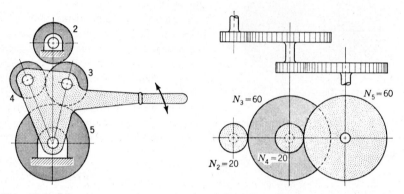

FIG. 17-3. Idler-gear reversing mecha-   FIG. 17-4. Compound gear train.
nism.

made of the materials.   Even when helical or herringbone gears are used,
single-reduction commercial gear reducers are usually limited to a maximum ratio of 10:1.

In order to use large velocity ratios economically with spur or helical
gears, it is desirable to use a *compound gear train,* as shown in Figs.
16-18, 16-24, 16-26, and 17-4.   In Fig. 17-4, gears 3 and 4 are known as
a *compound gear* and are either manufactured as a single piece or are keyed
to the countershaft so that they must rotate together.   With gear 2
driving, the velocity ratio is

$$\frac{\omega_2}{\omega_5} = \frac{N_3}{N_2}\frac{N_5}{N_4} = {}^{60}\!/_{20} \times {}^{60}\!/_{20} = 9 \qquad (17\text{-}4)$$

In this case, the train is, in effect, two simple trains in series, and the
intermediate gears must be considered when the velocity ratio is
calculated.

Study of Eqs. (17-1) to (17-4) will show that the velocity ratio is equal
to the ratio of the continued product of the numbers of teeth on the driven
gears to the continued product of the numbers of teeth on the driving
gears.

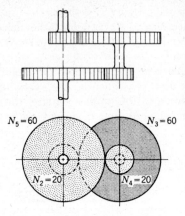

FIG. 17-5. Reverted gear train.

If the numbers of teeth on the several gears are so chosen that the center distance for gear 2 and gears 3 and 4 equals the center distance for gears 3 and 4 and gear 5, a compound gear train may have the input and output shafts located on the same centerline, as in Figs. 16-24 and 17-5. This arrangement is known as a *reverted gear train*. It is useful in mechanisms such as clocks, altimeters, air-speed indicators, etc., where it is desirable to have two pointers on concentric shafts moving with a specified velocity ratio. Reverted gear trains are also particularly useful for securing the maximum drive capacity in a minimum space. In this case, two or more idler shafts or countershafts are used to divide the load between two or more pairs of gears, as in Fig. 16-28.

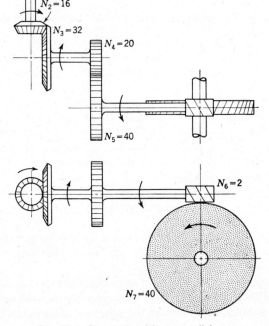

FIG. 17-6. Gear train with nonparallel axes.

17-2. Gear Trains with Nonparallel Axes.

The angular-velocity ratio for a gear train containing bevel gears, worms and worm wheels, etc., can be determined in the same manner as described in the preceding section for gears with parallel axes. However, the problem of determining the directions of rotation of the several shafts requires some additional thought. In Fig. 17-6, gear 2 is the driver and rotates as shown. The angular-velocity ratio is

$$\frac{\omega_2}{\omega_7} = \frac{N_3\,N_5\,N_7}{N_2\,N_4\,N_6} = {}^{32}\!/_{16} \times {}^{40}\!/_{20} \times {}^{40}\!/_2 = 80 \qquad (17\text{-}5)$$

The directions of rotation of the bevel gears are observed to be as indicated in Fig. 17-6. There are a number of methods available for determining the direction of rotation imparted to the worm wheel by the worm. One method is to consider the worm wheel as a nut on a screw thread and to note the direction the nut must move when the screw rotates while remaining in the same axial position. The motion of the nut corresponds to the motion of the worm wheel at the point of contact. Another useful method is to consider the worm teeth as an inclined plane moving perpendicular to its axis of rotation and to note the direction along the axis of the worm in which a point on the worm wheel is driven by the inclined plane.

17-3. Epicyclic or Planetary Gear Trains.

An *epicyclic gear train* is basically a gear train with one gear fixed so that the meshing gear or gears have a motion composed of a rotation about their own axes and a rotation about the axis of the fixed gear. The name comes from the fact that points on the revolving gears trace out epicycloidal curves. Popularly, such a gear train is known as a *planetary gear train*, the fixed gear as the *sun gear*, and the revolving gears as *planet gears*.

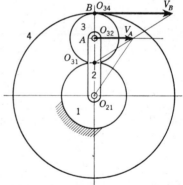

Fig. 17-7. Epicyclic gear train.

Calculation of velocity ratios for planetary gear trains is complicated by the double rotation of the planets. There are numerous methods available for solving any problem concerning a planetary gear train, but only three methods will be discussed here.

Instant-center Method. A simple method is that using instant centers (Secs. 4-3 to 4-10). In Fig. 17-7, gear 1 is fixed, and it is desired to determine the ratio of the angular velocity of the ring gear 4 to the planet carrier 2. The gears must necessarily be of the same pitch; therefore, the diameters are found by dividing the numbers of teeth N by the pitch P. The velocity of A as a point in 2 is

$$V_A = \omega_2 \frac{D_3 + D_1}{2} = \omega_2 \frac{N_3 + N_1}{2P} \tag{17-6}$$

Since A is a point in both 3 and 2, a gage line for all points in link 3 is drawn from O_{31} through the tip of the arrow of the vector V_A. Since B is a point in 4, the angular velocity of gear 4 is

$$\omega_4 = \frac{2V_A}{D_4/2} = \frac{4V_A}{D_4} = \frac{4V_A}{N_4/P} \tag{17-7}$$

Substituting for V_A from (17-6), (17-7) becomes

$$\omega_4 = 2\omega_2 \frac{N_3 + N_1}{N_4} \tag{17-8}$$

Then
$$\frac{\omega_4}{\omega_2} = 2 \frac{N_3 + N_1}{N_4} \tag{17-9}$$

The solution may be left as Eq. (17-9), or it may be rearranged by substituting $(N_4 - N_1)/2$ for N_3 and then simplifying to

$$\frac{\omega_4}{\omega_2} = 1 + \frac{N_1}{N_4} \tag{17-10}$$

Analytical Method. Another basic approach is that of using the equations for relative angular velocities:

$$\omega_{4/1} = \omega_{4/2} + \omega_{2/1} \tag{17-11}$$

Dividing both sides by $\omega_{2/1}$ gives

$$\frac{\omega_{4/1}}{\omega_{2/1}} = \frac{\omega_{4/2}}{\omega_{2/1}} + 1 \tag{17-12}$$

Equation (17-12) is not very useful because the ratio $\omega_{4/2}/\omega_{2/1}$ cannot be readily solved. However, if 2 is considered to be the frame, i.e., fixed, the gear train becomes a simple idler train, and the velocity ratios for any gears can be calculated. Therefore, it is desirable to substitute $-\omega_{1/2}$ for

$\omega_{2/1}$ in the right-hand side of the equation.   Thus,

$$\frac{\omega_{4/1}}{\omega_{2/1}} = 1 - \frac{\omega_{4/2}}{\omega_{1/2}} \tag{17-13}$$

and

$$\frac{\omega_{4/2}}{\omega_{1/2}} = -\frac{N_1}{N_4} \tag{17-14}$$

where the minus sign indicates opposite directions of rotation.   Therefore,

$$\frac{\omega_{4/1}}{\omega_{2/1}} \quad \text{or} \quad \frac{\omega_4}{\omega_2} = 1 + \frac{N_1}{N_4} \tag{17-15}$$

Superposition or Tabular Method.   The last method presented here is basically the same as the analytical method except that the effects of two operations are superposed to give the final solution.   As previously mentioned, the resultant motion of the planet gears is the sum of a rotation with the carrier about the axis of the fixed gear and a rotation about their own axes on the carrier.   Thus the procedure is as follows:

1. To consider the train locked and rotated a given number of revolutions with the carrier.

2. To consider the carrier fixed and to rotate the gear, which has a known absolute motion, the required number of revolutions in the required direction so that the sum of its motions in steps 1 and 2 equals its known absolute motion.

3. To add the motions in steps 1 and 2.   The sums will be related correctly for all members so that angular velocities or velocity ratios may be easily calculated.

The procedure as applied to the gears in Fig. 17-7 is illustrated in Table 17-1.

TABLE 17-1

Operation	Number of revolutions of each member			
	1	2	3	4
With carrier.............	$+1$	$+1$	$+1$	$+1$
Relative to carrier........	-1	0	$+\dfrac{N_1}{N_3}$	$+\dfrac{N_1}{N_3}\dfrac{N_3}{N_4}$
Resultant...............	0	$+1$	$1+\dfrac{N_1}{N_3}$	$1+\dfrac{N_1}{N_4}$

From the resultant motions,

$$\frac{\omega_4}{\omega_2} = 1 + \frac{N_1}{N_4} \tag{17-16}$$

It is important to remember that in step 2 (relative to the carrier) the carrier *must be held fixed* and 0 must be entered in the corresponding box in the table.

The compound epicyclic gear train in Fig. 17-8 illustrates how a very large velocity ratio can be achieved in a small space with spur gears. The velocity ratio ω_5/ω_2 is desired.

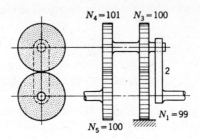

$N_4 = 101 \quad N_3 = 100$

2

$N_1 = 99$

$N_5 = 100$

Fig. 17-8. Compound epicyclic gear train.

By the analytical method,

$$\omega_{5/1} = \omega_{5/2} + \omega_{2/1} \tag{17-17}$$

Thus

$$\frac{\omega_{5/1}}{\omega_{2/1}} = \frac{\omega_{5/2}}{\omega_{2/1}} + 1 = 1 - \frac{\omega_{5/2}}{\omega_{1/2}} \tag{17-18}$$

Therefore

$$\frac{\omega_{5/1}}{\omega_{2/1}} = 1 - \frac{N_1}{N_3} \frac{N_4}{N_5} = 1 - \frac{99}{100} \times \frac{101}{100} \tag{17-19}$$

or

$$\frac{\omega_5}{\omega_2} = \frac{1}{10{,}000} \tag{17-20}$$

For the tabular method, see Table 17-2.

TABLE 17-2

Operation	Number of revolutions of each member				
	1	2	3	4	5
With carrier..........	$+1$	$+1$	$+1$	$+1$	$+1$
Relative to carrier.....	-1	0	$+\dfrac{N_1}{N_3}$	$+\dfrac{N_1}{N_3}$	$-\dfrac{N_1}{N_3}\dfrac{N_4}{N_5}$
Resultant.............	0	$+1$	$1+\dfrac{N_1}{N_3}$	$1+\dfrac{N_1}{N_3}$	$1-\dfrac{N_1}{N_3}\dfrac{N_4}{N_5}$

and

$$\frac{\omega_5}{\omega_2} = 1 - \frac{N_1}{N_3} \frac{N_4}{N_5} = \frac{1}{10{,}000} \tag{17-21}$$

It should be noted that, since $N_1 + N_3 = 199$ and $N_4 + N_5 = 201$, it will be necessary to use nonstandard gears so that the pairs of gears may operate on the same center distance.   If standard gears are used, the maximum velocity ratio will be when $N_1 = 99$, $N_3 = 101$, $N_4 = 100$, and $N_5 = 100$.   The velocity ratio with these numbers of teeth is

$$\frac{\omega_5}{\omega_2} = 1 - {}^{99}\!/_{101} \times {}^{100}\!/_{100} = {}^{2}\!/_{101} \qquad (17\text{-}22)$$

17-4. Design Considerations Peculiar to Gear Trains with Multiple-tooth Contact.   Planetary and other reverted gear trains are usually designed so that two or more pairs of gears are in contact at all times. This is advantageous from the viewpoint of load capacity, but several limitations due to interference must be considered.

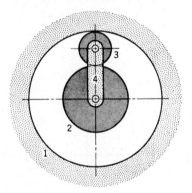

FIG 17-9

In Fig. 17-9, the ring gear 1 is stationary, and the velocity ratio is

$$\frac{\omega_2}{\omega_4} = 1 + \frac{N_1}{N_2} \qquad (17\text{-}23)$$

If equally spaced planet gears are to be used, the following relation must be true:

$$\frac{\text{Number of sun-gear teeth} + \text{number of ring-gear teeth}}{\text{Number of planet gears}} = \text{an integer} \qquad (17\text{-}24)$$

It is possible to use unequally spaced planet gears if Eq. (17-24) cannot be satisfied.   Aircraft engines have used over 20 planet gears in order to achieve a high-capacity drive in a small space with relatively small teeth.

If more than two planet gears are used, the teeth on adjacent gears may interfere if the velocity ratio is very large. Tables are available[1] relating the maximum velocity ratio, as defined by Eq. (17-23), permissible with standard full-depth teeth to the number of teeth on the sun gear, the number of teeth on a planet gear, and the number of planet gears. The maximum ratio may be calculated as

$$\left(\frac{\omega_2}{\omega_4}\right)_{max} = \frac{2N_2 - 4}{N_2(1 - \sin 180/n_p)} \tag{17-25}$$

where n_p = number of planet gears.

17-5. Gear-train Applications. *Lathe Thread-turning Train.* The gear train which is used in turning threads on a lathe is also employed in wrapping machinery, electron-tube grid winding, and spring coiling, as well as in many other similar applications where it is desired to achieve a helical motion by combining a translatory motion with a rotation. Figure 17-10 shows a simplified screw-threading train. The lead screw is

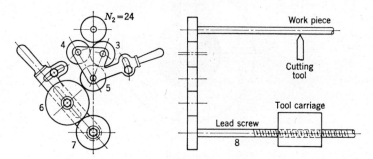

FIG. 17-10. Thread-turning gear train for a lathe.

a single four-threads-per-inch screw and thus has a lead of $\frac{1}{4}$ in. Since all gears except 2 and 7 are idlers, the velocity ratio of the spindle to the lead screw is

$$\frac{\omega_2}{\omega_7} = \frac{N_7}{N_2} \tag{17-26}$$

The tool carriage moves $\frac{1}{4}$ in. for every revolution of the lead screw 8. Therefore, the lead of the threads being turned is

$$l = \frac{1}{4}\frac{N_2}{N_7} \tag{17-27}$$

[1] C. Carmichael (ed.), "Kent's Mechanical Engineers' Handbook," "Design and Production" volume, 12th ed., pp. 14–52 and 14–53, John Wiley & Sons, Inc., New York, 1950.

As the lead for a single thread is $1/N$, where N is the number of threads per inch, (17-27) may be rewritten as

$$N = 4\,\frac{N_7}{N_2} \tag{17-28}$$

The number of teeth on gear 2 cannot be changed, and all adjustment must be made by changing gear 7. The idler gears 3, 4, and 5 are provided for reversing the direction of feed in relation to spindle rotation, as discussed in Sec. 17-1. The position of the axis of idler gear 6 can be adjusted to accommodate any desired number of teeth on gear 7. Involute gears should be used for change gears because it is not practical to provide for maintaining exact, i.e., theoretically correct, center distances for all the necessary ratios and only involute gears give a velocity ratio that is insensitive to errors in center distance. The practical arrangement of Eq. (17-28) is that which designates the number of teeth required on gear 7 to give the specified threads per inch on the work piece. After substituting $N_2 = 24$ and rearranging, (17-28) becomes

$$N_7 = 6N \tag{17-29}$$

Modern lathes and other machine tools utilize a transmission so that a lever may be used to change the velocity ratio without actually replacing any gears. One of the types of transmissions often used is the *tumbler-and-cone-gear* arrangement in Fig. 17-11, which permits the connecting of

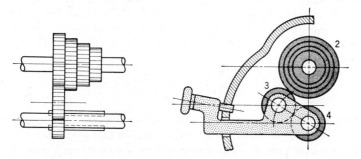

Fig. 17-11. Tumbler-and-cone-gear transmission.

shaft 4 with any step on the cone gear 2. Gear 4 is connected to its shaft through feather keys so that it will transmit a torque in any of its axial positions. The housing has four holes so located that, when the latch pin is inserted, the idler gear 3 is held in engagement with the desired step on the cone gear 2.

Conventional Four-speed Automotive Transmission.   The conventional automotive transmission in Fig. 17-12 has three forward speeds and one

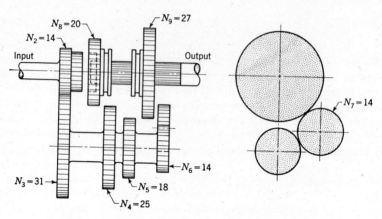

FIG. 17-12. Conventional four-speed automotive transmission.

reverse speed.   The synchromesh clutches (Fig. 12-8) have been omitted in the interest of simplification.   Gear 2 is connected to the engine through the clutch.   Gears 3, 4, 5, and 6 are a compound gear, with gear 3 being in mesh with gear 2 at all times.   Gear 7 is the reverse idler gear and is in mesh with gear 6 at all times.   Gears 8 and 9 are splined to the output shaft and are shifted axially into engagement by forks riding in the shifting-ring grooves.   As drawn, gears 8 and 9 are not engaged, and the transmission is in neutral.   When it is desired to start in low gear (first), gear 9 is moved to the left where it engages with gear 5, and the flow of power is through gears 2, 3, 5, and 9.   The velocity ratio is

$$\frac{\omega_2}{\omega_9} = \frac{N_3}{N_2}\frac{N_9}{N_5} = {}^{31}\!/_{14} \times {}^{27}\!/_{18} = 3.32 \qquad (17\text{-}30)$$

When shifting into second gear, the shifting levers are so operated that gear 9 moves to the right to its neutral position and gear 8 moves to the right to engage gear 4.   The flow of power is now through gears 2, 3, 4, and 8, and the velocity ratio is

$$\frac{\omega_2}{\omega_8} = \frac{N_3}{N_2}\frac{N_8}{N_4} = {}^{31}\!/_{14} \times {}^{20}\!/_{25} = 1.77 \qquad (17\text{-}31)$$

The highest speed is when gear 8 moves to the left, where its internal teeth engage the external teeth on 2 as a splined connection, and the velocity ratio is

$$\frac{\omega_2}{\omega_8} = 1 \qquad (17\text{-}32)$$

To reverse the motion of the car, gear 8 is moved to its neutral position, and gear 9 is shifted to the right into engagement with the reverse idler 7. The velocity ratio is

$$\frac{\omega_2}{\omega_9} = -\frac{N_3}{N_2}\frac{N_9}{N_6} = -\,{}^{31}\!/_{14} \times {}^{27}\!/_{14} = -4.27 \qquad (17\text{-}33)$$

Planetary Automotive Transmissions. One of the early and better-known planetary transmissions used in an automobile was the Ford Model T transmission illustrated in Fig. 17-13. The input shaft is the

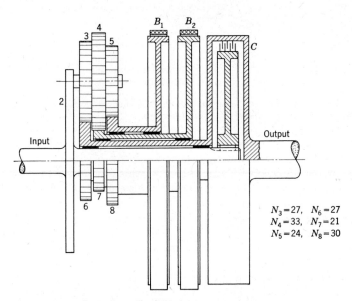

$N_3 = 27, \quad N_6 = 27$
$N_4 = 33, \quad N_7 = 21$
$N_5 = 24, \quad N_8 = 30$

Fig. 17-13. Ford Model T transmission.

planet carrier 2, and gear 6 is rigidly connected to the output shaft. The lower forward speed is given when the band brake B_2 is engaged and gear 7 is locked to the frame with a reduction ratio of 2.75:1. High is given by engaging the clutch C, which connects the input directly to the output, and the entire train turns as a unit with no reduction in speed. For reverse, the band brake B_1 is engaged, locking gear 8 to the frame with a reduction ratio of 4:1.

A modern automatic transmission is the Buick Dynaflow transmission.

A major part of the torque increase is given by the torque converter, but a planetary gear train, as shown in Fig. 17-14, is used to give low-range forward and reverse drives.

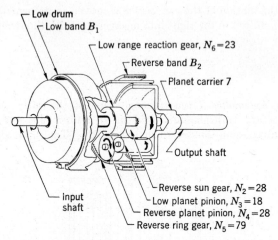

Fig. 17-14. Planetary gear train in the Buick Dynaflow transmission. (*Courtesy Buick Motor Division, General Motors Corporation.*)

In low-range drive, the band brake B_1 is engaged, and the sun gear 6 is locked to the frame. The input gear 2 is in mesh with the idler planet gear 3 which meshes with planet gear 4, which in turn is in mesh with the locked sun gear 6. The ring gear 5 is also in mesh with 4, but, since it is free to rotate, it has no effect on the motion of the train. In low range, the speed reduction, or torque multiplication, in the planetary gear train is 1.82:1. The torque converter gives a maximum torque multiplication of 2.25:1. Thus the maximum total torque multiplication is $1.82 \times 2.25 = 4.09$.

In high-range drive, a multiple-disk clutch is engaged, locking gear 6 to the input shaft. There can be no relative rotation of the several gears, and the torque is transmitted through static engagement of the gears with no multiplication.

In reverse drive, band brake B_2 is engaged, locking the ring gear 5 to the frame. As for low range, the torque multiplication in the planetary gear train is 1.82:1, and the total maximum torque multiplication is 4.09:1.

Automotive Bevel-gear Differential. The automotive differential shown in Fig. 17-15 is technically not a differential mechanism (Sec. 17-7) but is more properly called a *cyclic equalizing gear.* The correct name comes from its basic function of dividing the input torque equally between the two rear wheels, irrespective of their relative speeds of rotation. The tabular method of computing velocity ratios for epicyclic gear trains is the most useful method in analyzing the operation of the differential. Table 17-3 illustrates the solution for the case of a car going around a corner with the right wheel 4 turning at 48 rpm and the left wheel 2 turning at 50 rpm.

The column for gear 3 is left blank because gears 3 rotate about an axis perpendicular to that of the gears of interest and their direction of rotation cannot be denoted by plus or minus. Since $X + Y = 50$ rpm and $X - Y = 48$ rpm, X, the speed of the hypoid ring gear 5, must be 49 rpm, the average of the two wheel speeds.

The same result can be deduced on a logical basis by considering the case where the rear end of the automobile is jacked up and the hypoid ring gear is held fixed. Then, if the right wheel is given one plus revolution, the left wheel will make one minus revolution, and the average of the two will be zero, the motion of the ring gear.

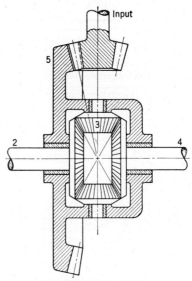

Fig. 17-15. Bevel-gear differential.

Spur-gear Differential. As shown in relation to a bevel-gear differential, the motion of the carrier is directly proportional to the sum of the motions of the two bevel-gear shafts. Because of this, the differential is

TABLE 17-3.

Operation	Number of revolutions of each member			
	2	3	4	5
With carrier..............	$+X$	...	$+X$	$+X$
Relative to carrier.........	$+Y$	...	$-Y$	0
Resultant................	50	...	48	$+X$

an important part of many computing devices where it is necessary to add the effects of several independent variables.

As discussed in Sec. 16-14, the proper mounting of bevel gears is more difficult than that of spur gears. Consequently, the spur-gear differential, as shown in Fig. 17-16, has been widely used in computers. It should be noted that two planet pinions are required to act as two idlers so that, when the carrier is considered fixed, the input shafts will rotate in opposite directions.

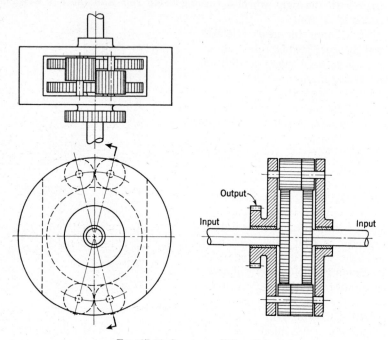

FIG. 17-16. Spur-gear differential.

Another important use of a spur- or bevel-gear differential is to change the angular phase relationship between two shafts. In this application, the carrier is normally fixed, and the input motion is transmitted without change, except in the direction of rotation; e.g., in Fig. 17-16 the carrier labeled "output" is fixed, and one of the input shafts becomes the output shaft. Then, if the carrier is rotated through an angle, the output shaft will be given an additional rotation that results in moving the output shaft ahead or behind the input shaft by an amount depending upon the gear ratio and the amount and direction of carrier rotation.

Reversing Gear Train. Previous discussions have shown how idler and planetary gear trains can be used to provide reversible outputs for unidirectional input rotation. The bevel reversing gear train in Fig. 17-17 is similar to that used in applications such as outboard motors. The gears rotate at all times, and the output shaft is engaged to either of the bevel pinions by shifting the clutch to the right or to the left.   The complexity and cost of the clutch may range from that of the square-jaw clutch, as illustrated, to that of a multiple-disk friction clutch (Fig. 12-5), depending upon the relative speeds of the shafts being connected together.

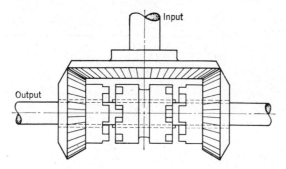

FIG. 17-17. Bevel reversing gear train.

17-6. Variable-speed Drives.   Gears and chains must be used if a change in angular velocity is required while maintaining a definite phase relationship between the input and output shafts.   Chain reduction units have been used in relatively large-capacity slow-speed oil-field applications.   If more than two different speeds are required, the size and cost of chain drives become relatively high.   Gear units may be designed for any desired capacity, but if more than several speed ratios are specified, the cost may become excessive.

Flat-belt drives with cone or stepped pulleys have been used to give several speed ratios.   Unless some method is used to maintain the proper belt tension, it is necessary to determine accurately the pulley diameters so that the belt will fit on all pairs with the same tightness.   For a crossed-belt drive, the requirement is that the sum of the pulley diameters be the same for all steps.   However, since the length of belt on an open-belt drive is a function of both the sum and the difference of the diameters (Eq. 14-9), a direct solution is not possible.   Handbooks[1] should be consulted for detailed instructions.

In many applications it is desirable or necessary to have a very large number, or even an infinite number, of velocity ratios available.   *The mechanisms that provide for varying the velocity ratio continuously* (infinitely variable) *rather than in steps are known as variable-speed drives.*   Several of the more common types will be discussed below.

V-belt Variable-speed Drives.   V belts or modifications are widely used in connection with adjustable split sheaves to give variable-speed ratios.

[1] L. S. Marks (ed.), revised by T. Baumeister, "Mechanical Engineers' Handbook," 6th ed., p. 8-67, McGraw-Hill Book Company, Inc., New York, 1958.

(a)

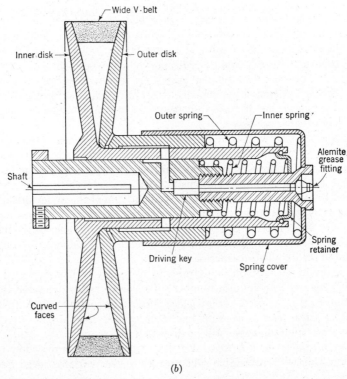

(b)

FIG. 17-18. (a) V-belt variable-speed drive with adjustable motor base; (b) spring-loaded variable-pitch sheave. (*Courtesy Lovejoy Flexible Coupling Company.*)

One or both of the sheaves are so made that the two halves may be moved axially to change the effective pitch diameters. The simplest type of drive, shown in Fig. 17-18a, consists of a wide V belt, a spring-loaded sheave, a fixed-diameter sheave, and an adjustable motor base. The spring-loaded sheave (Fig. 17-18b) automatically adjusts both its diameter and axial position as the motor is moved toward or away from the output shaft.

The U.S. Varidrive motor in Fig. 17-19 uses one spring-loaded self-adjusting sheave and one positively adjusted sheave to give a wider

FIG. 17-19. V-belt variable-speed drive with two variable-pitch sheaves. (*Courtesy U.S. Electrical Motors, Inc.*)

range of speed ratios than is possible with a single variable-pitch sheave. Some drives with two variable-pitch sheaves utilize a mechanical linkage to adjust both sheaves simultaneously so that the belt remains under correct tension at all times. An additional advantage of the spring-loaded adjustable sheave is that it can also be used as a load-limiting device, i.e., a slip clutch. Heavy-duty dual-belt drives are available with capacities up to 60 hp.

The Select-O-Speed transmission in Fig. 17-20 is, in effect, a compound gear (gears 3 and 4 in Fig. 17-4) with variable-diameter gears. The positions of the driving and driven units and the diameters of their sheaves remain fixed at all times. The position of the control lever determines the ratio of the distances from the compound-sheave axis to the driving-

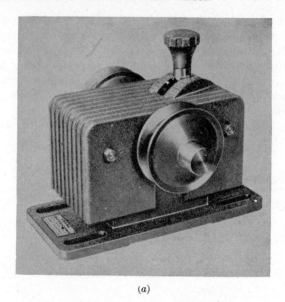

(a)

Control lever

Balanced
sheaves

Heavy duty
ball bearings

Alemite
grease
fitting

Precision ground
shaft

All metal construction

(b)

FIG. 17-20. V-belt variable-speed drive with variable-pitch compound sheave.   (*Courtesy Lovejoy Flexible Coupling Company.*)

and driven-member axes. The belt lengths are constant, and the compound-sheave diameters adjust automatically for the new center distances. It should be noted that, since the axial position of the inside sheave halves is fixed by the ball-bearing mounting, the control-lever pivot axis must be at an angle with the sheave-rotation axis so that, as the diameters change, the planes of the belts will remain in the planes of the sheaves on the driving and driven units.

Mechanical-linkage Variable-speed Drives. The Zero-Max transmission in Fig. 17-21a is a combination of four-bar linkages that convert

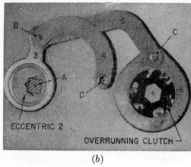

(a)   (b)

FIG. 17-21. (a) The Zero-Max transmission; (b) operating linkage. (*Courtesy Revco Incorporated.*)

rotation of the cranks into oscillation of the followers and of overrunning clutches that convert the oscillation of the followers into one-direction rotation of the output shaft. When four or eight linkages are used with uniformly spaced cranks, resultant output motion is made up of the maximum-velocity part of the motion of each linkage and approaches uniform rotation—within several per cent, depending upon the number of linkages and the inertia of the load.

One of the linkages is shown in Fig. 17-21b. Link 4 is a secondary or control link used to change the reduction ratio of the transmission by changing the position of point D which controls the path of point B which in turn determines the motion transmitted by link 5 and the angle through which link 6 oscillates. Note the use of a square driving shaft and an eight-point hole in the eccentric to provide quick and accurate indexing of the eccentrics as well as positive driving.

Standard Zero-Max transmissions are available with torque ratings up to 450 lb-in. and with output speed ranges of two-thirds input speed to zero and from two-thirds input speed through zero to two-thirds input speed in reverse.

Rolling-contact Variable-speed Drives.   Several methods by which the friction between rolling surfaces in contact is used to give a variable-speed drive are shown in Fig. 17-22.

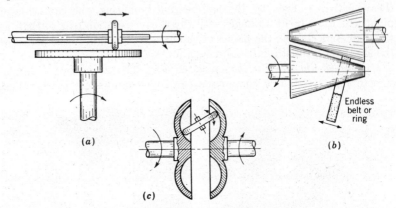

FIG. 17-22. Rolling-contact variable-speed drives.

The Metron miniature variable-speed changer in Fig. 17-23 is a commercial application of the principle illustrated in Fig. 17-22c.   The unit is slightly over 4 in. long, and its torque capacity is rated in ounce-inches. The velocity ratio of the unit shown may be varied from 1:5 to 5:1, a total range of 25:1.   A relatively light spring 4 keeps the wheels 2 and the toroidal disks 1 in contact at all times.   The relative angular position of cams 3 is a function of the applied torque, and the design is such that the contact force is proportional to the torque.

The Graham variable-speed transmission in Fig. 17-24 is a compound planetary drive in which the positive contact between the ring gear and one end of the compound planet gear has been replaced by rolling contact between a reaction ring and a tapered planetary roller.   The axial position of the reaction ring determines the effective diameter of the planetary roller and, thus, the speed ratio of the drive.   The planet carrier is driven by a constant-speed motor, and the centrifugal force on the planetary rollers supplies the traction pressure between the rollers and the ring.   Two types are available.   The standard (nonreversing) types give output speeds ranging from a maximum of approximately one-third input speed in the same direction, through zero, to approximately one one-hundredth input speed in reverse.   The B (reversing) types give equal speeds of approximately one-fifth input speeds both sides of zero. Built-in reduction or step-up gear trains are normally used to give a wider range of maximum output speeds.   Capacities range from $\frac{1}{15}$ to 5 hp at 1,750-rpm input speeds, and drive efficiencies range from 65 to 85 per cent, depending upon the size of the unit.

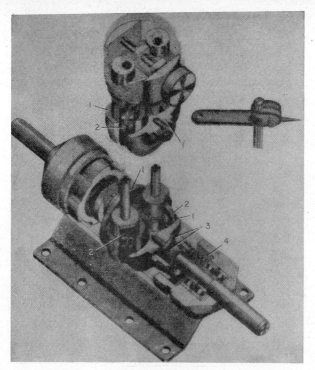

Fig. 17-23. Metron miniature variable-speed changer. (*Courtesy Metron Instrument Company.*)

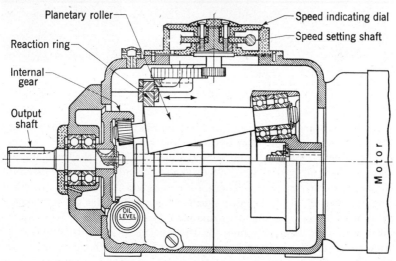

Fig. 17-24. Graham variable-speed transmission. (*Courtesy Graham Transmissions, Inc.*)

Variable-pitch Chain Drives. The unique drive in Fig. 17-25 is known as the *P.I.V.* (positive, infinitely variable) drive and uses a self-tooth-forming chain running on the grooved faces of opposed conical wheels. The sheaves are mechanically positioned axially to change the pitch diameters in a manner similar to that for a V-belt drive. The operating parts are splash-lubricated and operate in a closed case. The maximum standard rating is 25 hp at 500-rpm input speed, and the mechanical efficiency is in the order of 90 per cent.

(a)   (b)

Fig. 17-25. P.I.V. variable-speed drive. (*Courtesy Link-Belt Company.*)

Hydraulic Variable-speed Drives. Hydraulic variable-speed drives may usually be classified as either hydrodynamic or hydrostatic.

The principal hydrodynamic type is a modification of the fluid coupling (Sec. 11-7), as shown in Fig. 17-26. The inner and outer casing assembly rotates with the impeller. There are a number of calibrated orifices in the inner casing which permit a steady flow of oil, under the influence of centrifugal force, from the torus ring into the outer casing. The rotating oil in the outer casing impinges on the stationary scoop tube and is led back to the reservoir, from which it is pumped through the cooler to the coupling. The radial position of the scoop controls the amount of oil in the outer casing and, consequently, the amount of oil in the torus. The

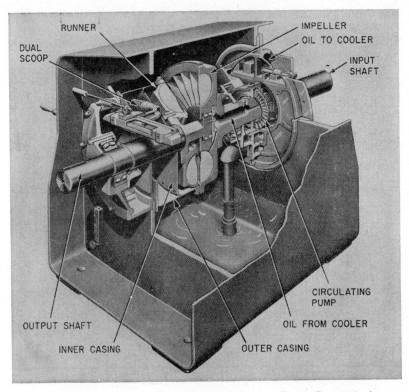

DUAL SCOOP

RUNNER

IMPELLER

OIL TO COOLER

INPUT SHAFT

CIRCULATING PUMP

OIL FROM COOLER

OUTPUT SHAFT

INNER CASING

OUTER CASING

FIG. 17-26. Gyrol fluid drive.   (*Courtesy American Blower Corporation.*)

amount of oil in the torus determines the minimum effective radius of the impeller, and this controls the transfer of energy from the impeller to the runner.   The volume of oil in the torus can be regulated to suit the output load and the speed requirements.

As previously discussed, there is no change in torque through the coupling, and the efficiency is proportional to slip.   Consequently, this type of drive is best suited for applications such as fans, centrifugal compressors, and centrifugal pumps, where the horsepower required varies as the cube of the speed.   Figure 17-27 shows typical power-speed curves for a hydraulic coupling when used as a variable-speed transmission driving a fan.   It should be noted that, while the drive is inefficient at low speeds, the actual power loss is not large.

The hydrostatic types of variable-speed drives use a variable-displacement pump and either a constant- or variable-displacement hydraulic motor.   Generally both pump and motor are of the positive-displacement piston type, and power is transmitted by a positive flow of oil under high pressure. The major advantage of this type of variable-speed drive is the high torque-inertia ratio and its suitability to applications where rapid response is essential.

The axial-piston type of hydraulic transmission is illustrated in Fig.

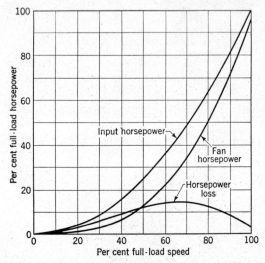

Fig. 17-27. Horsepower-speed requirements for variable-speed fluid coupling driving a fan.

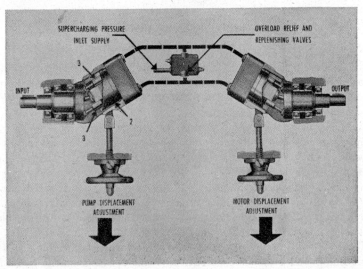

Fig. 17-28. Axial-piston hydraulic transmission. (*Courtesy Vickers Incorporated, Division of Sperry Rand Corporation.*)

17-28.   The input shaft is driven by an electric motor or another constant-speed prime mover.   The input shaft and the cylinder barrel 2 rotate together.   The angularity results in the pistons 3 being driven in and out of the cylinder barrel during the rotation.   The amount of piston motion, and thus the displacement of the pump, is a function of the angle between the input-shaft and cylinder-barrel axes.   The adjustment of the pump displacement is accomplished by use of a handwheel or by a suitable mechanical, electrical, or hydraulic device.   The hydraulic motor is basically the same as the pump.   The pump and motor may be separated by a considerable distance or may be mounted in the same housing to give a compact transmission.   The overload relief valve is designed to limit the pressure to a safe value so that the mechanism will not be damaged and the driving motor will not be overloaded in case the driven load becomes too large or is stalled.   Standard transmissions are available with capacities ranging from about 12 hp at 3,600 rpm to about 190 hp at 850 rpm. The transmission mechanical efficiency is up to 85 per cent.

Another common type of variable-displacement pump is that using radial pistons, as illustrated in Fig. 17-29.   The horizontal position of the slide block determines the eccentricity of the rotating cylinder and thus

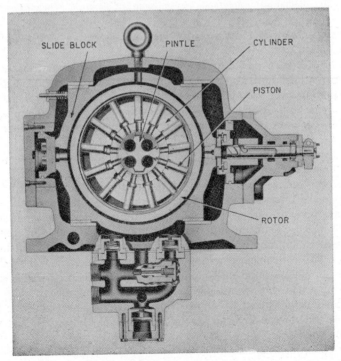

FIG. 17-29. Radial-piston hydraulic pump.   (*Courtesy The Oilgear Company.*)

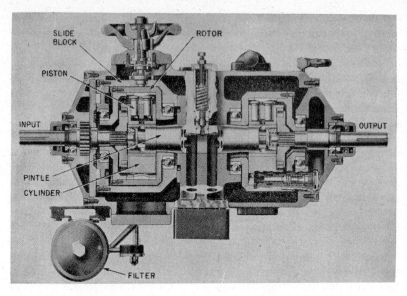

Fig. 17-30.   Radial-piston hydraulic transmission.   (*Courtesy The Oilgear Company.*)

determines the displacement of the reciprocating pistons and the output of the pump.

Figure 17-30 is a section view of a radial-piston hydraulic transmission with handwheel control. The unit illustrated has a variable-displacement pump and a constant-displacement motor and is known as a *constant-torque transmission*. When a constant-displacement pump and a variable-displacement motor are used, it becomes a *constant-power transmission*. It should be noted that the system in Fig. 17-28 may be used for either, or a combination of, a constant-torque or a constant-power transmission.

Electric Variable-speed Drives. A d-c motor in combination with a motor-generator set and the appropriate control circuit is one of the more efficient types of electric variable-speed drives. Cost, size, and complexity usually limit its use to high-power applications and applications where, for relatively long periods, high torque is required at low speed. The convenience and rapid response of electric control have resulted in the development of a number of types of electric variable-speed drives in which a constant-speed motor is the prime mover. The output speed is controlled by a relatively small d-c signal which in turn controls the torque-slip characteristics and thus the output speed for a given load.

The operating characteristics are quite similar to those of hydraulic-coupling variable-speed drives; there is no multiplication of torque; the efficiency, if the relatively small power loss in the electrical circuit is neglected, is 100 minus the per cent slip; and it is best adapted for driving loads whose torque requirements decrease as the speed decreases.

The torque transmission in eddy-current couplings is due to the magnetic force from the current induced when the driving and driven members have relative motion in the presence of a magnetic field. Eddy-current couplings are available in a wide range of capacities, up to 18,000 hp at 500 rpm. Air-cooled units may be used up to 900 hp, but water cooling is necessary for the largest units and is often desirable even for low-capacity drives. The normal slip at rated torque varies from 1½ to 5 per cent. Additional losses at rated speed and torque due to windage, friction, magnetic drag, and excitation power are in the order of 2 per cent of rated power.

The Dynaspede coupling in Fig. 17-31 has a stationary-field coil and is water-cooled. Units are available with normal capacities up to 1,500 hp at 1,800 rpm. The cooling systems can absorb approximately 200 per

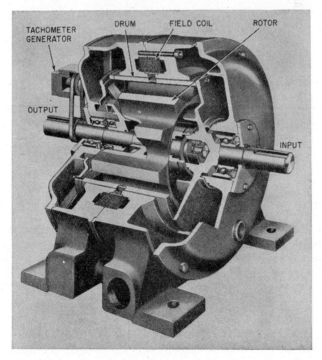

FIG. 17-31. Dynaspede stationary-field, liquid-cooled, eddy-current coupling. (*Courtesy Dynamatic Division, Eaton Manufacturing Company.*)

cent of rated power.   The tachometer generator supplies the output-speed signal for the control circuit.

Figure 17-32 shows a self-contained unit, combining a squirrel-cage motor with a rotating-field eddy-current coupling, and the electronic control system.   Adjusto-Spede drives are available in sizes up to 75 hp at 1,700 rpm.   The units are air-cooled.   The smaller sizes dissipate rated power at zero output speed, but the percentage of rated power that can be

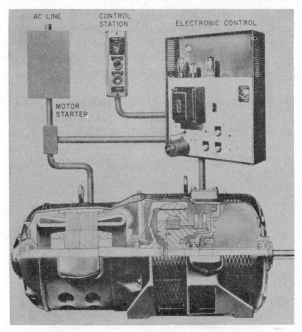

FIG. 17-32.   Adjusto-Spede drive.   (*Courtesy Dynamatic Division, Eaton Manufacturing Company.*)

dissipated decreases with increasing size.   The speed ratio for rated torque operation is limited to a maximum of $2\frac{1}{2}:1$ for the largest units. The empty space in the right end of the unit provides room for mounting an eddy-current brake for more rapid deceleration.   It should be noted that, like hydrodynamic brakes, eddy-current brakes cannot develop braking torque without some slip and consequently cannot be used to stop or hold a load.

The electromagnetic coupling (Fig. 11-17) and the magnetic-particle clutch (Fig. 12-16) also find limited use in variable-speed drives.

17-7. Differential Mechanisms.   A differential mechanism is one characterized by similar parts moving simultaneously at different rates.

One type is the *differential screw* in Fig. 17-33.   The two threaded portions have the same hand but different pitches.   For the case shown, one revolution of the screw moves the screw axially $\frac{1}{28}$ in. to the left relative to the frame.   The slider, which is restrained from rotating, moves $\frac{1}{32}$ in. to the right relative to the screw, and resultant motion relative to the frame becomes

$$p_1 - p_2 = \frac{1}{28} - \frac{1}{32} = \frac{4}{896} = 0.00446 \text{ in./revolution} \quad (17\text{-}34)$$

It can be seen that, even with standard threads, as listed in Table 8-1, the differential action can be used to give very minute and precise adjustment.

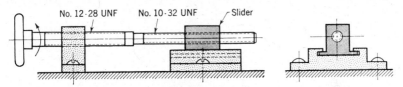

No. 12-28 UNF   No. 10-32 UNF   Slider

FIG. 17-33. Differential screw.

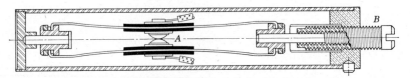

FIG. 17-34. Thermoswitch.   (*Courtesy Fenwal Incorporated.*)

The Fenwal Thermoswitch illustrated in Fig. 17-34 is a good example of the application of a differential screw.   The contact points A are located on flat struts made from a material with a low coefficient of thermal expansion, such as Invar, and the case is made from a material with a relatively high coefficient of thermal expansion, such as stainless steel.   As the temperature of the case rises, its length increases, the struts are stretched, and the contact points separate.   The temperature at which the contact points open is dependent upon the initial axial position of the right end of the struts relative to the case.   Since the magnitude of the difference in increased lengths of the tube and struts is in the order of only 6 to 30 $\times$ 10^{-6} in./°F, it is important to be able to make a very fine adjustment of the position of the right end of the struts relative to the case.   A differential screw B is commonly used as the adjusting mechanism.

If the threads are of opposite hand, the motion becomes the sum of the pitches, and the mechanism is known as a *compound screw*.   The compound screw is not very important, because a multiple-thread screw can readily be made to give rapid motion.

The *differential chain hoist* in Fig. 17-35a is another common example of a differential mechanism. In this case, the compound sprocket wheel

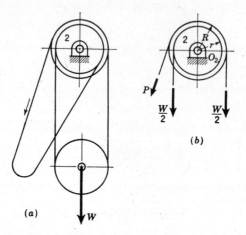

FIG. 17-35. Differential chain hoist.

2 rotates, and the endless chain operates at two different radii, R and r. From b, neglecting friction,

$$\Sigma M_{O_2} = 0$$

from which

$$P = \frac{W(R - r)}{2R} \qquad (17\text{-}35)$$

and, in terms of a mechanical advantage,

$$\text{M.A.} = \frac{W}{P} = \frac{2R}{R - r} \qquad (17\text{-}36)$$

CHAPTER 18

BEARINGS AND LUBRICATION

Bearings are machine elements that permit relative motion of two parts in one or more directions with a minimum of friction while preventing motion in the direction of the applied load. In general, bearings may be divided into two groups:

1. Bearings in which the surfaces either make sliding contact or are separated by a film of lubricant

2. Bearings which have rolling contact between surfaces

This chapter will be devoted to a discussion of the important basic principles underlying the operation of several types of each group of bearings. The major emphasis will be given to those factors influencing the choice of a particular type of bearing for a given application. The subject of bearings is too extensive and specialized to be adequately covered in this chapter. The literature should be consulted for more detailed information.[1]

18-1. Sliding Bearings. The term *sliding bearings* is not really general enough, but for convenience, we shall use it to describe all bearings that do not use rollers or balls as load-supporting elements.

It is often convenient to classify sliding bearings into (1) thick-film bearings in which the surfaces are completely separated from each other by the lubricant; (2) thin-film or boundary-lubricated bearings in which, although lubricant is present, the surfaces partially contact each other at least part of the time; and (3) zero-film bearings which operate without any lubricant present.

Furthermore, sliding bearings are usually designed for a single purpose, either (1) to carry radial loads or (2) to carry axial or thrust loads. Most sliding bearings are not suitable for carrying a combination of radial and

[1] M. C. Shaw and F. Macks, "Analysis and Lubrication of Bearings," McGraw-Hill Book Company, Inc., New York, 1949.

D. D. Fuller, "Theory and Practice of Lubrication for Engineers," John Wiley & Sons, Inc., New York, 1956.

R. R. Slaymaker, "Bearing Lubrication Analysis," John Wiley & Sons, Inc., New York, 1955.

O. Pinkus and B. Sternlicht, "Theory of Hydrodynamic Lubrication," McGraw-Hill Book Company, Inc., New York, 1961.

thrust loads unless the magnitude of one component is negligible in relation to the other; therefore, further classification into radial and thrust bearings is both convenient and useful.

From the viewpoint of general usefulness in machines, the most important sliding bearing is the thick-film radial bearing, and as such, it will receive a more thorough treatment than the other types.

18-2. Journal or Sleeve Bearings.   Sliding bearings designed to carry a radial load while permitting relative rotation or oscillation are known as *journal* or *sleeve bearings*.   The full journal bearing in Fig. 18-1a and

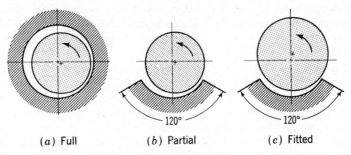

(a) Full   (b) Partial   (c) Fitted

Fɪɢ. 18-1. Journal bearings.

the 120° partial journal bearing in Fig. 18-1b are known as *clearance bearings* because the diameter of the journal is less than that of the bearing bore.   The 120° partial journal bearing in Fig. 18-1c has no clearance, i.e., the diameters of the journal and bushing are equal, and is known as a *fitted bearing*.   A partial journal bearing has less friction than a full journal bearing, but it can be used only where the load is always in one direction.   The most common application of a partial journal bearing is to railroad-car axles.

The inner member is known as the *journal* and the outer member as the *bearing*.

The lubricant used to separate the journal and bearing is usually a mineral oil refined from petroleum, but vegetable oils, silicone oils, greases, graphite, and compounds such as molybdenum disulfide find widespread use.   Gases, such as air, and even liquid metals are used in special applications.

As discussed in books in the field of fluid mechanics and lubrication, Tower, in 1883, discovered that, when the bearing is supplied with sufficient oil, a pressure is built up in the clearance space when a journal is rotating about an axis that is eccentric with the bearing axis.   He also showed that the load can be supported by this fluid pressure without any actual contact between the two members.   Osborne Reynolds learned of the investigation and proceeded to develop the basic differential equation for hydrodynamic lubrication.

The load-carrying ability of a hydrodynamic bearing arises simply because a viscous fluid resists being pushed around. Under the proper conditions, this resistance to motion will develop a pressure distribution in the film that can support a useful load. There are three mechanisms that are mainly responsible, two of which are hydrodynamic and the third is hydrostatic.

The load-supporting pressure in hydrodynamic bearings arises from either (1) the flow of a viscous fluid in a converging channel, the *wedge film*, or (2) the resistance of a viscous fluid to being squeezed out from between approaching surfaces, the *squeeze film*.

The hydrostatic bearing derives its load-carrying ability from the resistance of a viscous fluid to flow through a restricted passage. This type of bearing will be discussed in Sec. 18-10 and will not be considered further at this time.

Many bearings must operate under conditions requiring consideration of both the wedge film and the squeeze film simultaneously. Unfortunately, at the present time, there is no satisfactory solution or design procedure for the most general case, and the designer is often forced to rely upon experience gained through many years of trial-and-error development in a narrow field of special application, e.g., the main bearings in automobile engines. Since most of the information from these sources cannot be extrapolated to new situations with any degree of accuracy, our time will be more profitably spent in discussing the design of bearings for use when it is fairly obvious that the main support comes from only one of the hydrodynamic films.

18-3. Wedge-film Journal Bearings. The load-carrying ability of a wedge-film journal bearing results when the journal and/or the bearing rotates relative to the load. The most common case is that of a steady load, a fixed (nonrotating) bearing, and a rotating journal. Figure 18-2*a* shows a statically loaded journal at rest. In Fig. 18-2*b* the journal has

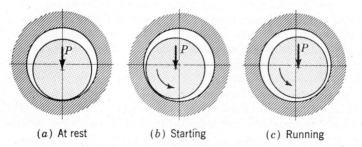

(*a*) At rest   (*b*) Starting   (*c*) Running

Fig. 18-2. Journal positions.

started to rotate under load without any oil present. As shown, friction causes the journal to climb the side of the bearing. Obviously, this type of operation would result in very rapid wear of the parts. However, the oil adheres to both the journal and bearing and rotation of the journal literally pulls oil between the journal and the bearing. This forces the two members to separate, and the journal moves to a position, such as in Fig. 18-2c, where the resultant force from the pressure in the oil film exactly counterbalances the load.

The reader should apply the above reasoning to show that, if the journal were to be fixed and the bearing were to rotate in the direction shown, the position of the journal relative to the bearing would be the same as shown in Fig. 18-2c when running equilibrium had been reached, although the starting position would not be the same as in Fig. 18-2b. Since there is no metal-to-metal contact, there will be no wear of the parts, and the only friction present is that due to the shearing of the viscous oil. However, there is ordinarily little wear during the starting of a machine. Many machines start under light loads, and in any case, if sufficient oil is present, the oil film is built up almost instantly; thus, normally there is little tendency for a journal to crawl up the bearing as shown in Fig. 18-2b.

18-4. Load Capacity of Wedge-film Journal Bearings. The best known solution of Reynolds' equation is that by Sommerfeld in which the integration was performed for an infinitely long bearing in which the oil film extended all around the bearing and both positive and negative oil pressures contributed to support of the load. Comparison of theoretical and experimental results shows that the axial flow of oil (end leakage) from a finite-length bearing and the fact that an oil film does not develop appreciable negative pressures cannot be neglected if reasonable accuracy is required. Consequently, experimental data have been plotted as a function of different variables, and end-leakage factors have been used to correct for the lack of agreement between the theoretical assumptions and actual bearing practice. The pressure distribution in the hydrodynamic oil film is complicated. As illustrated in Fig. 18-3, the pressure distribution varies in both the circumferential and axial directions. Thus, in addition to the circumferential oil flow due to the relative motion of the bearing and journal, there is both a circumferential and an axial flow caused by the variation of pressure in these directions. Ocvirk[1] has presented a solution for Reynolds' equation that is at the other extreme from the Sommerfeld solution, in that it is more nearly correct for bearings whose lengths are very small in relation to their diameters. This solution, known as the *short-bearing approximation*, includes the effect of pressure on the axial flow and neglects the effect of pressure on the circum-

[1] F. W. Ocvirk, Short Bearing Approximation for Full Journal Bearings, *Natl. Advisory Comm. Aeronaut. Tech. Note* 2808, 1952.

ferential flow.   The major usefulness of Ocvirk's work is that it led to the consideration of a new nondimensional group, the *load number* N_L, as a basis for design.   Further investigations[1] showed that experimental results for bearings with length-to-diameter ratios of 1 or less are in close agreement when plotted as functions of the load number, without using any factors to adjust for end leakage.   Thus, the design procedure for short bearings has been greatly simplified.   Actually, a slight modification, which will be discussed below, permits the extension of the use of the

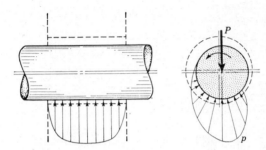

FIG. 18-3. Oil-film pressure distribution.

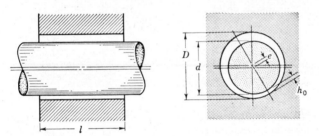

FIG. 18-4. Bearing dimensions.

experimental curves to the design of bearings with length-to-diameter ratios somewhat greater than 1.

Since the majority of modern journal bearings can be classed as short bearings, the test data are of direct use in most cases.   Figure 18-4 shows

---

[1] G. B. DuBois and F. W. Ocvirk, Analytical Derivation and Experimental Evaluation of Short-bearing Approximation for Full Journal Bearings, *Natl. Advisory Comm. Aeronaut. Rept.* 1157, 1953.

G. B. Dubois and F. W. Ocvirk, The Short Bearing Approximation for Plain Journal Bearings, *ASME Paper* 54-*Lub*-5, 1954.

G. B. DuBois, F. W. Ocvirk, and R. L. Wehe, Experimental Investigation of Eccentricity Ratio, Friction, and Oil Flow of Long and Short Journal Bearings with Load-number Charts, *Natl. Advisory Comm. Aeronaut. Tech. Note* 3491, 1955.

the dimensions important in the performance of a journal bearing. The practical criterion for determining whether or not a bearing will operate satisfactorily is the value of minimum film thickness h_0. The limiting thickness is dependent upon the surface finish of the parts, deflections of the parts, and cleanliness of the lubricant. Carefully made bearings have given satisfactory service with oil films as thin as 0.000025 in. However, in practical design, it is recommended that the minimum film thickness be at least 0.00015 in./in. of bearing diameter. Unfortunately, at the present time it is not possible to design a bearing directly upon the basis

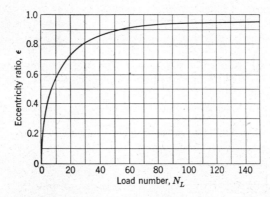

Fig. 18-5. Eccentricity ratio as a function of load number. (*After G. B. DuBois, F. W. Ocvirk, and R. L. Wehe, Experimental Investigation of Eccentricity Ratio, Friction, and Oil Flow of Long and Short Journal Bearings with Load-number Charts, Natl. Advisory Comm. Aeronaut. Tech. Note 3491, 1955.*)

of a selected minimum film thickness. Instead, it is necessary to make use of a nondimensional quantity called the *eccentricity ratio* ϵ, which is defined as

$$\epsilon = \frac{2e}{c_d} \tag{18-1}$$

where e = eccentricity of the journal center relative to the bearing center, in., and c_d = diametral clearance = $D - d$, in. The minimum film thickness is related directly to the eccentricity ratio by

$$h_0 = \frac{c_d}{2}(1 - \epsilon) \tag{18-2}$$

For design purposes, it is desirable to rewrite (18-2) as

$$\epsilon = 1 - \frac{2h_0}{c_d} \tag{18-3}$$

Since most bearings are designed with a diametral clearance of 0.001 to 0.002 in./in. of diameter, a desirable maximum value of ϵ can be readily calculated. The eccentricity ratio is a function of the load number, which is defined as

$$N_L = \frac{60p}{\mu n} \left(\frac{c_d}{d}\right)^2 \left(\frac{d}{l}\right)^2 \qquad (18\text{-}4)$$

where p = pressure on projected area = P/ld, psi
$\quad \mu$ = viscosity in reyns, lb-sec/in.^{2}
$\quad n$ = rotative speed, rpm

The experimentally determined relationship between the eccentricity ratio and the load number is shown in Fig. 18-5. As previously pointed out, this applies best to bearings with $l/d \le 1$. However, there is evidence that only minor error, on the safe side, is introduced for bearings with $l/d = 1$ to 2 if the bearings are considered to have $l/d = 1$. That is, let $(d/l)^2 = 1$ when calculating the load number.

The viscosity μ is the only variable in Eq. (18-4) that cannot be explicitly determined for most bearing problems and consequently must be considered in some detail.

18-5. Viscosity. *Viscosity* is the internal friction that resists relative motion in liquids. If, as in Fig. 18-6, an unloaded surface moves parallel

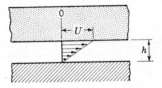

FIG. 18-6

to a stationary surface and the space between is filled with fluid, the velocity gradient will be a straight line. The fluid shear stress for most fluids is proportional to the rate of shear, or

$$\tau = \mu \frac{U}{h} \qquad \text{psi} \qquad (18\text{-}5)$$

where μ = coefficient of absolute viscosity (or simply viscosity), lb-sec/in.^{2}.

The force required to move one surface relative to the other is

$$F = \tau A = \mu \frac{U}{h} A \qquad \text{lb} \qquad (18\text{-}6)$$

where A = area upon which the shear stress acts, in.^{2}.

The engineering unit of viscosity is the reyn (named after Reynolds), and the chart in Fig. 18-7 shows the average viscosity-temperature curves of several oils, as classified by the Society of Automotive Engineers. As illustrated by the curves, the viscosity varies considerably with temperature. This variation of viscosity with temperature is more pronounced for some oils than for others. The term *viscosity index* is used to denote the degree of variation of viscosity with temperature. A viscosity index has been established in terms of the viscosity at 100 and 210°F. An oil with a high viscosity index shows a smaller decrease in

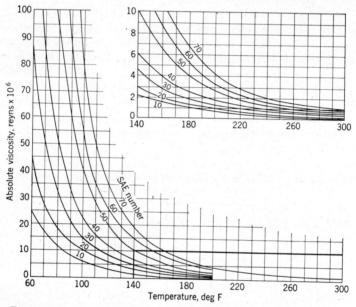

Fig. 18-7. Average viscosity-temperature curves for SAE classified oils.

viscosity with an increase in temperature than does an oil with a low viscosity index.

The viscosity has also been observed to be increased at high pressures. This variation can be neglected in most lubrication problems.

18-6. Friction. As discussed in the previous section, the friction force [Eq. (18-6)] is a function of the viscosity of the oil, the shear rate of the oil, and the area on which the stress acts. For the case of an unloaded journal bearing, the eccentricity is zero, and the friction torque on the stationary element equals that on the rotating element. In this case, the friction torque can be calculated by use of the *Petroff* equation

$$T_0 = \frac{\pi^2 \mu n l d^2}{60(c_a/d)} \qquad \text{lb-in.} \tag{18-7}$$

The Petroff equation also gives a close approximation for lightly loaded bearings with low eccentricity ratios. As the eccentricity ratio increases, the friction torques increase. The friction torque on the rotating element increases more rapidly than that on the stationary element, by the product of the bearing load and the component of eccentricity perpendicular to the load direction. Since the energy lost in the bearing is a function of the friction torque on and the rotative speed of the rotating element, only the rotating element need be considered.

The curves in Fig. 18-8 show the ratio of the torque T_r on a loaded rotating member to the torque T_0 on an unloaded rotating member as a function of the l/d ratio and the load number of the bearing.

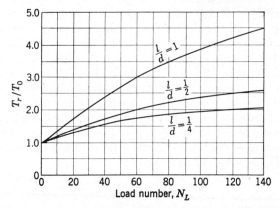

Fig. 18-8. (*After G. B. DuBois, F. W. Ocvirk, and R. L. Wehe, Experimental Investigation of Eccentricity Ratio, Friction, and Oil Flow of Long and Short Journal Bearings with Load-number Charts, Natl. Advisory Comm. Aeronaut. Tech. Note 3491, 1955.*)

18-7. Heat Balance in Journal Bearings.   The rate at which energy is generated by the viscous friction in a journal bearing is

$$H_\theta = \frac{T\omega}{778} = \frac{T_r \times 2\pi n}{12 \times 778} = \frac{T_r n}{1{,}486} \qquad \text{Btu/min} \qquad (18\text{-}8)$$

where T_r = friction torque on rotating element, lb-in., and n = speed of rotating element, rpm.

In operation, the temperature of the bearing rises until a balance is reached between the rates at which the heat is generated and dissipated. Heat is carried away by the oil flow and by conduction to other parts of the machine where it is finally dissipated by convection and radiation.

In many applications, oil is supplied under pressure to the bearing and is drained to a sump from which it is again pumped to the bearing. Study of Eqs. (18-7) and (18-8) will show that, for unloaded journal bearings, the rate of heat generation varies with the square of the speed.   This applies approximately for all journal bearings; consequently, high-speed bearings often require a heat exchanger or radiator for cooling the oil. When the oil flow is sufficient to carry away the heat and the temperature of the oil entering the bearing is known, the temperature rise of the oil when passing through the bearing and the heat absorbed by the oil are related by

$$H_d = C_p Q(t_2 - t_1) \qquad \text{Btu/min} \tag{18-9}$$

where C_p = specific heat, Btu/(lb)(°F) (about 0.49 for petroleum oils)
 Q = oil flow rate, lb/min
 t_2 = oil leaving temperature, °F
 t_1 = oil entering temperature, °F

The temperature t_2 of the leaving oil is used when calculating the load number [Eq. (18-4)] and the heat generation rate [Eq. (18-8)].

If the oil flow cannot be depended upon to remove the heat generated in the bearing, the problem of determining the equilibrium temperature becomes complicated. An expert in the field of heat transfer may arrive at a reasonable answer, but the number of variables and assumptions almost precludes determining an accurate answer. For most purposes, the work of Lasche[1] in 1902 is still the best basis for calculating the rate of heat dissipation from journal bearings. Lasche's equation assumes that the external area of the bearing is a function of the projected area of the bearing. The rate of heat dissipation is

$$H_d = Cld(t_b - t_a) \qquad \text{Btu/min} \tag{18-10}$$

where C = heat-dissipation coefficient, Btu/(min) (in.[2] projected bearing area)(°F)
 t_b = temperature of external surface of bearing, °F
 t_a = temperature of ambient air, °F

Since C includes the effects of both convection and radiation, it is not independent of the temperature; consequently, it is most convenient to use the curve in Fig. 18-9, which gives the product $C(t_b - t_a)$ as a function of $t_b - t_a$ for normal operating conditions.

Although it is probably overly conservative in many cases, a reasonable assumption relating the oil-film temperature t_o and the external bearing-surface temperature t_b is that t_b is midway between the oil-film and ambient temperatures,[2] or

$$t_b - t_a = \tfrac{1}{2}(t_o - t_a) \tag{18-11}$$

Example 18-1. The shaft for the centrifugal blower in Examples 6-1, 6-2, and 14-1 is to be mounted on ring-oiled babbitt bearings (Fig. 18-25). The blower speed is

[1] O. Lasche, Bearings for High Speed, *Traction and Transmission*, vol. 6, pp. 33–64, 1903.

[2] M. D. Hersey, "Theory of Lubrication," John Wiley & Sons, Inc., New York, 1938.

600 rpm, and the minimum journal diameter, based on strength calculations in Example 6-1, is $1\frac{7}{8}$ in.   To minimize the number of different parts that must be manufactured and kept in stock for replacements, identical bearings will be used at each end of the shaft.   The bearing loads due to the weight of the rotor and the belt forces have been determined to be 125 and 630 lb.   The blower must operate continuously, without the oil temperature exceeding 180°F when the ambient temperature is between 60 and 100°F.   We are asked to design the bearing, specify the SAE number for the oil, and determine the maximum power loss in the bearings.

Solution.   The limiting conditions are a satisfactory minimum film thickness and a maximum oil temperature of 180°F at equilibrium conditions.   Since the film thickness will be less for the most heavily loaded bearing, the design load will be 630 lb, and, since the equilibrium temperature increases with ambient temperature, the design ambient temperature will be 100°F.   The power loss will be maximum at the lowest operating temperature, or when the ambient temperature is 60°F.

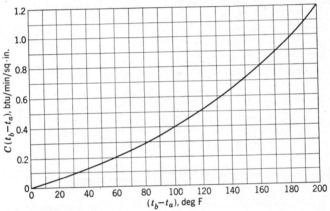

FIG. 18-9. Heat-dissipation coefficients for journal bearings. (*After O. Lasche, Bearings for High Speed, Traction and Transmission, vol. 6, pp. 33–64, 1903.*)

We shall use the average recommended clearance ratio of 0.0015 in./in. of diameter and the recommended value for minimum film thickness of 0.00015 in./in. of diameter.   Thus, from Eq. (18-3), the design eccentricity ratio is

$$\epsilon = 1 - \frac{2h_0}{c_d} = 1 - \frac{2 \times 0.00015d}{0.0015d} = 0.80$$

Therefore, from Fig. 18-5, the load number N_L must be equal to or less than 28. However, N_L cannot be calculated until the operating oil temperature is known. Thus the first step is to select an oil and determine the equilibrium temperature of the bearing.

The first trial will be made with a light oil (SAE 10) and the minimum journal diameter ($1\frac{7}{8}$ in.) because, if satisfactory, this combination will give the minimum power loss in the bearings.   An l/d ratio of 1:1 will also be assumed in the first trial.

The design procedure will be to assume oil temperatures and then calculate N_L, T_o, T_r, the rate of heat generation, and the rate of heat dissipation for each assumed temperature.   Curves will then be plotted for heat generated and heat dissipated as a function of oil temperature, the intersections being the equilibrium temperature. Since each value must be calculated several times, we shall first set up the equations for N_L and T_0 as function of μ only.

Equilibrium temperature. The equation for load number [Eq. (18-4)] is

$$N_L = \frac{60p}{\mu n} \left(\frac{c_d}{d}\right)^2 \left(\frac{d}{l}\right)^2$$

and $p = P/ld = 630/(1.875 \times 1.875) = 179.2$ psi
$\mu =$ (to be taken from Fig. 18-7) reyn
$n = 600$ rpm
$c_d/d = 0.0015$
$d/l = 1$

Thus
$$N_L = \frac{60 \times 179.2}{\mu \times 600} (0.0015)^2 (1)^2 = \frac{40.3 \times 10^{-6}}{\mu}$$

The equation (18-7) for the torque on an unloaded rotating member is

$$T_0 = \frac{\pi^2 \mu n l d^2}{60 \times c_d/d} = \frac{\pi^2 \mu \times 600 \times 1.875 \times 1.875^2}{60 \times 0.0015} = 0.434\mu \times 10^6 \qquad \text{lb-in.}$$

Heat generated, 120° oil temperature

$$\mu = 3 \times 10^{-6} \text{ reyn} \qquad \text{from Fig. 18-7}$$
$$N_L = \frac{40.3 \times 10^{-6}}{3 \times 10^{-6}} = 13.4$$
$$T_0 = 0.434 \times 3 \times 10^{-6} \times 10^6 = 1.30 \text{ lb-in.}$$
$$T_r/T_0 = 1.5 \qquad \text{from Fig. 18-8 for } N_L = 13.4$$
$$T_r = 1.5 \times 1.30 = 1.95 \text{ lb-in.}$$

From Eq. (18-8),

$$H_g = \frac{T_r n}{1{,}486} = \frac{1.95 \times 600}{1{,}486} = 0.79 \text{ Btu/min}$$

Heat dissipated, 120° oil temperature and 100° ambient temperature. From Eq. (18-10),

$$H_d = Cld(t_b - t_a)$$
and
$$t_b - t_a = \tfrac{1}{2}(t_o - t_a) = \tfrac{1}{2}(120 - 100) = 10°F$$
$$C(t_b - t_a) = 0.02 \text{ Btu/(min)(in.}^2) \qquad \text{from Fig. 18-9 for } t_b - t_a = 10°F$$

Thus,
$$H_d = 0.02 \times 1.875 \times 1.875 = 0.070 \text{ Btu/min}$$

The results of calculations for oil temperatures from 120 to 200°F are tabulated below, and the values of H_g and H_d are plotted as functions of oil temperature in

t_o, °F	120	140	160	180	200
N_L	13.4	18.4	26.3	38.3	50.4
H_g, Btu/min	0.79	0.66	0.50	0.43	0.38
H_d, Btu/min	0.07	0.18	0.28	0.42	0.56

Fig. 18-10. It should be noted that the curves do not pass through all the calculated points. This is because of the coarseness of the values read from the curves.

The operating oil temperature will be about 181°F for a 100°F ambient temperature. N_L will be somewhat greater than the value of 38.3 for 180°F oil temperature. Since 38.3 > 28, the minimum film thickness will be less than the specified value of 0.00015 in./in. of diameter, and the design must be reconsidered. Since the difference of 1°F between the calculted equilibrium temperature and the specified maximum oil temperature is less than the errors in our assumptions and in reading the curves, it is·apparent that we should place emphasis on means for decreasing N_L.

Examination of Eq. (18-4) indicates that the N_L may be decreased by decreasing p, c_d/d, and d/l. However, it should be recalled that $l/d = 1$ is to be used for values of $1 < l/d < 2$, or $1 > d/l > \frac{1}{2}$. Thus, we actually may consider decreasing only p and c_d/d. Since our operating temperature is already at the maximum specified value of 180°F, there is no point in decreasing c_d/d or increasing d to decrease p, because both steps would result in a higher operating temperature. It can be further noted that the ratio T_r/T_0 decreases as N_L decreases. Thus, while both H_g and H_d increase directly with l, H_g decreases relative to H_d as N_L decreases. Therefore, the equilibrium temperature for the longer bearing will be less than 181°F if $N_L < 38.3$.

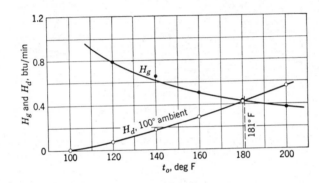

FIG. 18-10

Hence, keeping everything the same except for increasing l to decrease N_L by decreasing p, we find

$$l_2 = l_1 \times \frac{38.3}{28} = 1.875 \times \frac{38.3}{28} = 2.565 \text{ or } 2\frac{9}{16} \text{ in.}$$

and $$\frac{l}{d} = \frac{2.5625}{1.875} = 1.37$$

Also $$N_L = \frac{1.875}{2.5625} \times \frac{40.3}{\mu} \times 10^{-6} = \frac{29.5}{\mu} \times 10^{-6}$$

and $$T_0 = \frac{2.5625}{1.875} \times 0.434\mu \times 10^6 = 0.593\mu \times 10^6 \qquad \text{lb-in.}$$

Calculated values of N_L, H_g, and H_d for oil temperatures from 80 to 200°F and for ambient temperatures of 60 and 100°F are tabulated below. The values of H_g and

t_o, °F	80	100	120	140	160	180	200
N_L	2.6	4.9	9.8	13.4	18.4	28.1	35.9
H_g, Btu/min	3.03	1.74	0.97	0.79	0.65	0.52	0.44
$H_{d,60°}$, Btu/min	0.10	0.25	0.38	0.58	0.77	0.96	
$H_{d,100°}$, Btu/min			0.10	0.25	0.38	0.58	0.77

H_d are plotted as functions of t_o in Fig. 18-11. As noted, the equilibrium temperature will be about 175°F when the ambient temperature is 100°F. The value of N_L will be less than the value of 28.1 for 180°F oil temperature, and the minimum film thickness will be greater than the specified minimum value. Therefore, both the load-capacity and temperature characteristics of this bearing will be satisfactory.

When the ambient temperature is 60°F, the equilibrium temperature will be about 152°F, and the rate of heat generation will be 0.70 Btu/min. The power loss will actually be 0.70 Btu/min for the heavily loaded bearing and something less for the

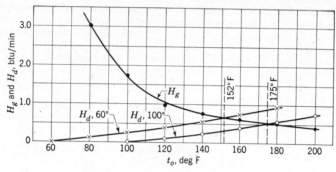

FIG. 18-11

more lightly loaded bearing. However, the accuracy of any bearing calculation is questionable, and for practical purposes we may consider the power loss to be the same for each bearing. Thus, the total power loss is

$$2 \times \frac{0.70 \times 778}{33,000} = 0.033 \text{ hp}$$

Summary

Bearing dimensions: $1\frac{7}{8}$-in.-diameter bore, $2\frac{9}{16}$ in. long, 0.0028-in. diametral clearance.

Oil: SAE 10.

Power loss: 0.033 hp for two bearings at 60°F ambient temperature.

18-8. Wedge-film Journal Bearings with Rotating Journals, Bearings, and Loads.

Many journal bearings support rotating members that are not perfectly balanced. The unbalance results in a centrifugal force, which is then a rotating load. In other bearings, e.g., in the transmissions in Figs. 17-13 and 17-14, both the journal and bearing rotate, sometimes in the same direction and sometimes in opposite directions.

If the magnitudes of the speeds of rotation and of the load are constant or nearly so, the performance of these bearings may be handled exactly as in Sec. 18-4 provided the speed of rotation n is replaced in the load number calculation, Eq. (18-4), by an *equivalent speed* n_e[*] which is

[*] A. B. Miller, discussion of A. F. Underwood and J. M. Stone, Load-carrying Capacity of Journal Bearings, *Trans. SAE*, vol. 1, no. 1, p. 69, January, 1947.

defined as

$$n_e = |n_{J/P} + n_{B/P}| \qquad (18\text{-}12)$$

where the vertical lines signify "the absolute value of" and

$$n_{J/P} = n_J - n_P$$
$$n_{B/P} = n_B - n_P$$

Also, n_J, n_B, and n_P are the algebraic, e.g., ccw is positive, values of the speeds of rotation of the journal, bearing, and load, respectively.

It is interesting to note that, when the bearing is stationary and the load rotates at one-half the speed of and in the same direction as the journal, the equivalent speed is zero and the bearing has zero load capacity.

18-9. Squeeze-film Journal Bearings. From a literal viewpoint a journal bearing in which neither the journal nor the bearing rotates is just as trivial as a bearing with either zero or infinite length.   However, many bearings must oscillate or rotate so slowly that the wedge film cannot provide a satisfactory film thickness.   If the load is uniform or even varying in magnitude while acting in a constant direction, this becomes a thin-film or, possibly, a zero-film problem.   But if the load reverses its direction, the squeeze film may develop sufficient capacity to carry the dynamic loads without contact between the journal and the bearing.

Although not very well understood, the difference between operating with a static and a dynamic load has long been recognized.   Before data and methods were available for designing a bearing to provide a specified minimum film thickness, it was necessary to determine the dimensions on the basis of allowable pressures selected from tables.   These allowable pressures were based upon experience in specific applications, and seldom would any values greater than 300 psi be given for bearings that carry a static load.   However, where the squeeze film is important, much higher pressures were commonly permitted, e.g., up to 1,800 psi for wristpin bearings in even slow-speed steam engines, up to 5,000 psi for wristpin bearings in aircraft engines, and up to 8,000 psi for the crankpin bearings in punching and shearing machinery.

Until recently, the designer had no other realistic basis for design. However, papers by Hays[1] and Phelan[2] present methods that now permit at least a reasonable prediction of the performance of bearings when the squeeze film is the predominant factor and oil is supplied under pressure to the bearing.   Hays presents the results of a digitial-computer solution

[1] D. F. Hays, Squeeze Films: A Finite Journal Bearing with a Fluctuating Load, *ASME Paper* 61-LubS-7, 1961.

[2] R. M. Phelan, Non-rotating Journal Bearings under Sinusoidal Loads, *ASME Paper* 61-LubS-6, 1961.

for the case of a half (or π) bearing, while Phelan presents experimental results and a semiempirical equation that correlates experimental data with a simple mathematical model that is based upon the analogy between approaching flat, circular plates and the journal approaching the bearing. Both methods are in reasonable agreement with experimental results. However, only the latter method will be discussed further here because its relative simplicity permits a direct solution in many cases and involves only simple graphical or numerical integrations in any case.

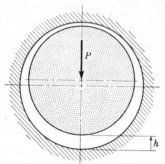

FIG. 18-12. Nonrotating journal bearing.

Figure 18-12 shows a load acting downward on a nonrotating journal in a stationary bearing. The journal moves along a radial line in the direction of the applied load. Thus, the film thickness h decreases and the oil is forced to flow up around the journal and out the ends of the bearing. The relationship between the instantaneous value of the bearing pressure and the film thickness is

$$p = 3.86\mu \left(\frac{d}{c_d}\right)^2 \left(\frac{l}{d}\right)^2 \left(\frac{c_d}{h}\right)^{0.28} \frac{dh/dt}{h} \quad \text{psi} \quad (18\text{-}13)$$

where, as before, $p = P/ld$ and we use $(l/d) = 1$ when $l > d$. Also, note that P and thus p will be negative numbers, since they act in a direction opposite to positive h. The eccentricity ratio no longer enters the problem, and we are now dealing directly with the film thickness. For a given bearing and film thickness, it can be seen that the load capacity of the squeeze film is directly proportional to the velocity. The squeeze film acts as a dashpot or viscous damper and resists the load only when the film thickness is decreasing.

When the direction of the load reverses, the reference for film thickness must immediately move to a point diametrically opposite its original position.

In the general case, we are interested in preventing the film thickness from becoming less than a specified value when the load varies with time in a given manner. Thus, to solve for film thicknesses, Eq. (18-13) is rewritten as

$$\int_{h_1}^{h_2} \frac{1}{h^{1.28}} \, dh = \frac{(c_d/d)^2 (d/l)^2}{3.86\mu c_d^{0.28}} \int_{t_1}^{t_2} p \, dt \quad (18\text{-}14)$$

which becomes, upon integration and substitution of the limits into the left-hand side of the equation,

$$-\frac{1}{0.28h_2{}^{0.28}} + \frac{1}{0.28h_1{}^{0.28}} = \frac{(c_d/d)^2(d/l)^2}{3.86\mu c_d{}^{0.28}} \int_{t_1}^{t_2} p\,dt \qquad (18\text{-}15)$$

or, more usefully,

$$\frac{1}{h_2{}^{0.28}} = \frac{1}{h_1{}^{0.28}} - \frac{0.0725(c_d/d)^2(d/l)^2}{\mu c_d{}^{0.28}} \int_{t_1}^{t_2} p\,dt \qquad (18\text{-}16)$$

Equation (18-16) can be applied directly to the *analysis* of a given bearing *provided* the load does not reverse its direction and the initial film thickness is known. In the general case, the designer will be dealing with a periodic load and will not know the value of film thickness at any time. It will be necessary to divide the load-time diagram into sections

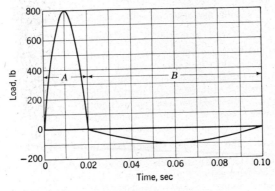

Fig. 18-13

corresponding to the times at which the load reverses direction, assume an initial thickness at the beginning of one section, and then work through the complete cycle until the starting point is reached again. If the calculated film thickness at the end of the cycle agrees with the initially assumed value, the problem is solved. If not, the process must be repeated, using the final calculated position to determine the new initial film thickness. The procedure to follow will be explained in Example 18-2.

Example 18-2. The load-time curve for a bearing with an oscillating journal may be approximated by the half-sine curves shown in Fig. 18-13. We are asked to determine whether or not the following bearing specifications will be satisfactory:

$$l = d = 1\tfrac{1}{2} \text{ in.}$$

$$\frac{c_d}{d} = 0.0015$$

$$\mu = 1.5 \times 10^{-6} \text{ reyn}$$

Solution. The load-time curve is divided into sections A and B, as indicated. The general sinusoidal load will be described by

$$P = -P_0 \sin \omega t \qquad (a)$$

where P_0 = load amplitude, lb, and ω = frequency, radians/sec.
For section A, Eq. (a) becomes

$$P_A = -800 \sin \frac{\pi}{0.02} t \qquad \text{lb}$$

or, in terms of pressure on the projected area,

$$p_A = \frac{P_A}{ld} = \frac{-800}{1.5 \times 1.5} \sin \frac{\pi}{0.02} t$$

$$= -355 \sin \frac{\pi}{0.02} t \qquad \text{psi}$$

Similarly, for section B

$$p_B = -44.4 \sin \frac{\pi}{0.08} t \qquad \text{psi}$$

The maximum possible film thickness will be equal to the clearance, which is

$$c_d = 0.0015d = 0.00225 \text{ in.}$$

The minimum film thickness will occur at the end of one of the sections. From Eq. (18-16), it can be seen that, for a given value of h_1, the film thickness decreases as the area under the load-time curve increases. Since the loading is periodic, the motion must also be periodic, and therefore, the journal will move through the same distance in each direction. It can also be seen that an infinite time will be required for any finite load to make h_2 equal zero. The not so obvious conclusion is that the minimum film thickness for the cycle will occur at the end of the section having the greatest value of $\int_{t_1}^{t_2} p \, dt$. This is the best starting point in the analysis because an error in the assumed initial thickness at this point will have the least effect on the calculations. Thus, we shall consider section B first; i.e., we shall start our analysis at $t_1 = 0.02$ sec.

For section B, let us assume that $h_{1,B} = 0.00200$ in. and start measuring time from $t = 0.02$ sec. The appropriate equation (18-16) is

$$\frac{1}{h_2^{0.28}} = \frac{1}{h_1^{0.28}} + \frac{0.0725(c_d/d)^2(d/l)^2}{\mu c_d^{0.28}} \int_{t_1}^{t_2} p_0 \sin \omega t \, dt$$

$$= \frac{1}{0.00200^{0.28}} + \frac{0.0725(0.0015)^2(1)^2}{1.5 \times 10^{-6} \times 0.00225^{0.28}} \int_0^{0.08} 44.4 \sin \frac{\pi}{0.08} t \, dt$$

$$= 5.68 + 0.601 \int_0^{0.08} 44.4 \sin \frac{\pi}{0.08} t \, dt$$

$$= 5.68 + 1.36 = 7.04$$

and $\qquad h_{2,B} = 0.00095$ in.

Now, when the load reverses direction at the beginning of section A, the reference for film thickness moves to the other side of the bearing and $h_{1,A}$ becomes the difference between the clearance and the value of h_2 for section B. Thus,

$$h_{1,A} = 0.00225 - 0.00095 = 0.00130 \text{ in.}$$

and $\qquad \dfrac{1}{h_2^{0.28}} = \dfrac{1}{0.00130^{0.28}} + 0.601 \int_0^{0.02} 355 \sin \dfrac{\pi}{0.02} t \, dt$

$$= 6.41 + 2.72 = 9.13$$

and $\qquad h_{2,A} = 0.00037$ in.

This completes one cycle. Comparing the new value of $h_{1,B} = 0.00225 - 0.00037$ $= 0.00188$ in. with the estimated value of 0.00200 in., we see that the motion is almost, but not quite, periodic and at least one more cycle should be considered. Noting that the calculated value is less than the estimated leads to estimating $h_{1,B}$ as 0.00187 in.

The iterative process is actually not so tedious and time-consuming as it may appear, because the terms involving the integrals remain as before. Thus,

$$\frac{1}{h_{2,B}{}^{0.28}} = \frac{1}{0.00187^{0.28}} + 1.36 = 7.17$$

and
$$h_{2,B} = 0.00088 \text{ in.}$$

Then,
$$h_{1,A} = 0.00225 - 0.00088 = 0.00137 \text{ in.}$$
$$\frac{1}{h_{2,A}{}^{0.28}} = \frac{1}{0.00137^{0.28}} + 2.72 = 9.06$$

and
$$h_{2,A} = 0.00038 \text{ in.}$$

At the completion of the cycle

$$h_{1,B} = 0.00225 - 0.00038 = 0.00187 \text{ in.}$$

which agrees exactly with the estimated value.

Thus, our calculations indicate that the minimum film thickness will be approximately 0.00038 in. No reliable design criterion has been established for this type of bearing. However, a breakdown of the film will not be quite so serious with a reversing load and an oscillating journal as with a steady load and a constant speed of rotation. For example, if the film breaks down at the instant the journal speed is zero there will be no wear and, thus, no difficulty. Therefore, if the film-thickness ratio h_m/d is greater than 0.00015 in./in. as recommended in Sec. 18-4 for wedge-film bearings, the squeeze-film bearing may be expected to operate satisfactorily. In this example $h_m/d = 0.00025$ in./in. and the proposed bearing specifications will be satisfactory.

Figure 18-14 shows complete curves for the variation with respect to time of film thickness and load for the bearing considered in Example 18-2. The observation of major interest is that thick films offer little resistance to motion. Thus, even a relatively small load will result in a high velocity and a considerable displacement when it acts upon a relatively thick film, and we find that the squeeze film can actually support a periodic load with an average or steady component. As shown in Fig. 18-14, the average load in this case is 51 lb.

The major practical consideration for squeeze-film bearings is the necessity of filling up the unloaded side of the bearing with oil in the short time available. This requires careful attention to the location of oil-supply holes and grooves and to the oil-supply pressure. The experimental results show that with small clearance ratios, for example, $c_d/d < 0.001$ in./in., it becomes very difficult to pump oil through the narrow passage at a rate sufficient to ensure that the clearance space will always be full when the load reverses direction. However, it seems likely that this is not so serious as it may first seem, at least for heavily loaded bearings.

For example, as discussed above, little support is given by the thick film, and the presence of a void in this region should have only a minor effect on the final answer.

The reader may have observed that no mention has been made of a heat balance for squeeze-film bearings. The reason is simply that very little heat is generated in the squeeze film. In fact, experience in the laboratory indicates that, if the temperature of the oil supply is several degrees above ambient, the temperature of the oil will actually drop when

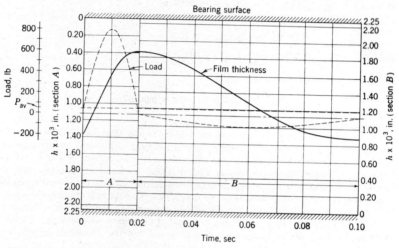

FIG. 18-14. Film-thickness and load-time curves for the bearing in Example 18-2.

passing through the bearing. The reason for this behavior becomes apparent when one considers that the work done on the oil is

$$W = \int P \, dh$$

and the total distance the force moves per cycle is not very great.

18-10. Hydrostatic Journal Bearings. The major advantage of hydrostatic, or externally pressurized, bearings is that they can support steady loads without any relative motion between members. An additional advantage is that this type of bearing will have less friction than any other type of bearing—except possibly a magnetic bearing, i.e., a bearing in which the support is due to magnetic forces—at very low speeds. The reason is that the friction is due solely to the shear of a viscous fluid [Eq. (18-6)] and the shear rate is practically zero at very low speeds.

Both liquids and gases are used. In some respects the principles of operation are similar, but gases are compressible while, for practical purposes, the liquids are incompressible.

Air is being used in many high-speed applications, where its low viscosity and the resulting low friction torque are important, e.g., in small, high-speed (100,000 rpm) grinders and in gyroscopes. Although many air bearings are not externally pressurized, the problem of starting from rest with metal-to-metal contact and the "spring" action due to the compressibility give rise to problems that are less critical when the support is hydrostatic rather than hydrodynamic. The subject of air bearings is too specialized to be considered further here, and the reader is referred to the rapidly increasing volume of literature[1] on this subject.

Figure 18-15 is a schematic representation of a hydrostatic journal bearing suitable for use with an incompressible fluid such as oil. When

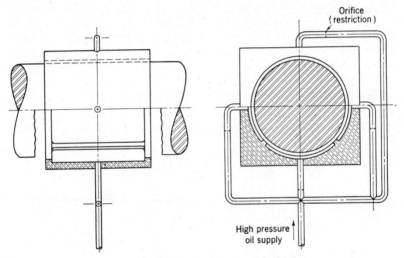

FIG. 18-15. Hydrostatic journal bearing.

the journal is centered in the bearing, the area for axial flow out the ends is the same for each recessed region. Thus, if the orifices are identical, the flow rates for all regions will be the same and the pressure drops across the orifices will be identical. Consequently, the force exerted on the journal will be the same for each region. Now, if a load is applied to the journal, it will move off center and the flow area will increase on one side and decrease on the other. There will be an increased drop in pressure across the orifice(s) to the region(s) with the increased flow area and a decreased drop in pressure across the orifice(s) to the region(s) with

[1] For example, Fuller, *op. cit.*, chap. 9.

Pinkus and Sternlicht, *op. cit.*, chap. 6.

C. B. Adams, J. Dworski, and E. M. Shoemaker, Externally Pressurized Step Journal Bearings, *ASME Paper* 61-LubS-8, 1961.

the decreased flow area. The result is a decreased force from one side and an increased force from the other. The net force acts to oppose the load, and the equilibrium position of the journal will depend upon the relative magnitudes of the load and the supply pressure.

An example of the application of this type of bearing is found in the 200-in. Mount Palomar telescope, where the 1,245,000-lb dead weight is supported on hydrostatic bearings and a fractional-horsepower motor supplies the turning effort for positioning the telescope.[1]

18-11. Thin-film and Zero-film Journal Bearings. The design considerations for and performance of bearings that must operate under such conditions that a thick film cannot be provided or where no lubricant can be used become almost entirely functions of the materials involved.

For thin-film or boundary lubrication, the most important factor is the ability of the lubricant to form an adsorbed film on the bearing surface, so that metal-to-metal contact between the journal and bearing is prevented. This characteristic is often a joint property of the bearing material and lubricant known as oiliness.   Further discussion will be left for Sec. 18-14.

18-12. Sliding Thrust Bearings. The function of a thrust bearing is to restrain a shaft against an axial load. As discussed in previous sections, a converging film is automatically provided in a journal or sleeve

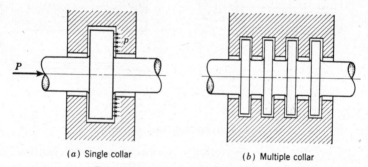

(a) Single collar   (b) Multiple collar

Fig. 18-16. Collar-type thrust bearings.

bearing whenever the axes of the journal and bearing do not coincide. However, this is not the case for the simple thrust bearing in Fig. 18-16a, where both the rotating and stationary elements are plane surfaces. The operation of such a thrust bearing is essentially a case of boundary lubrication, because the hydrodynamic effects of radial grooves, the variation of viscosity due to the temperature gradient within the film, etc., cannot ordinarily be depended upon to provide an oil film thick enough to separate the surfaces. Before the invention of hydrodynamic thrust bear-

[1] M. B. Karelitz, Oil-pad Bearings and Driving Gears of 200-inch Telescope, *Mech. Eng.*, vol. 60, pp. 541–544, 1938.

ings, it was necessary to use multiple collars, as in Fig. 18-16b, to make the average pressure low enough so that wear would not be too rapid.   Even with the most exacting workmanship, it is impossible to achieve a uniform distribution of the load among the several collars and bearing surfaces.[1]

Several types of hydrodynamic thrust bearings are illustrated in Fig. 18-17.   The *tapered-land thrust bearing a* depends upon a machined inclination to provide the necessary wedge-shaped film.   The flat portion carries the thrust load at zero or low rotative speeds.   The major difficulty with this type of thrust bearing is that the total taper per pad should be in the order of 0.001 in. and the required accuracy of machining is close to the limit of practical work.

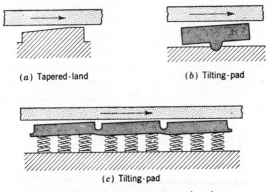

(a) Tapered-land   (b) Tilting-pad

(c) Tilting-pad

FIG. 18-17. Hydrodynamic thrust bearings.

The two types illustrated in Fig. 18-17b and c depend upon the tilting of the bearing surfaces under the hydrodynamic pressure of the oil film to provide automatically the converging oil film.   The *tilting-pad thrust bearing* in Fig. 18-17b was independently developed in the 1890s by Kingsbury in this country and by Michell in Great Britain.

Hydrodynamic thrust bearings have been used extensively to carry the thrust load from marine propellers, steam turbines, and centrifugal pumps, to support the weight of hydroelectric turbine rotors, and in many other similar applications.   Large tilting-pad thrust bearings have been used to carry loads as high as 3,000,000 lb.

The design and operational characteristics of hydrostatic thrust bearings are based upon the principles that were discussed in Sec. 18-10 in relation to journal bearings.   The use of spherical seats in combination with the inherent self-aligning feature makes hydrostatic thrust bearings

---

[1] Refer to the discussion on translation screws (Sec. 10-4 and Table 10-2) for values of coefficient of friction for collar-type thrust bearings.

ideally suited for use in experimental apparatus, such as bearing test machines,[1] where large forces must be transmitted and the friction must be as small as possible.

18-13. Special-purpose Bearings.   The tilting pad commonly used in thrust bearings has also been applied to the journal bearing, as shown in Fig. 18-18.   The added expense of such bearings cannot be justified unless an ordinary journal bearing fails to function satisfactorily.

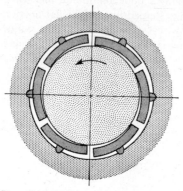

FIG. 18-18. Tilting-pad journal bearing.

A limitation of the sleeve bearing is that, under a combination of light loads and high speeds, instability, *known as oil-film whirl*, often develops. Oil-film whirl is a self-excited vibration in which the journal moves within the clearance space along a closed path in the direction of rotation.   This disturbance is of particular importance in high-speed grinding spindles.   The tilting-pad journal bearing is not subject to oil-film whirl.   Another advantage of tilting-pad bearings is that the pads may be adjusted radially so that the shoes are preloaded; i.e., all shoes carry a load even when the journal is at rest.   This preloading decreases the radial motion of the journal as the load varies in magnitude.

18-14. Materials for Sliding Bearings.   If a journal and bearing are always separated by a film of clean, noncorrosive lubricant, the only requirement of the materials is that they have sufficient strength and rigidity.   However, the conditions under which bearings must operate in service are generally far from ideal, and certain other properties of the materials become important.

In a majority of applications, the journal is made of steel because of the need for strength and rigidity; thus the following discussion will apply to bearing materials when used with steel journals.

Since the localized peak pressure on the bearing may be many times greater than that calculated by dividing the load by the projected area, it is important that the bearing material have a *compressive strength* high enough to prevent extrusion or other permanent deformation of the bearing.

The ability of the bearing material to withstand repeated loads without developing surface fatigue cracks is called *fatigue strength* and is of major importance in aircraft and automotive engines.

[1] G. B. DuBois and F. W. Ocvirk, Experimental Investigation of Eccentricity Ratio, Friction, and Oil Flow of Short Journal Bearings, *Natl. Advisory Comm. Aeronaut. Tech. Note* 2809, 1952.

The *conformability* of a bearing material is its ability to accommodate shaft deflections and bearing inaccuracies by plastic deformation or creep without excessive wear and heating.

Lubrication systems should be designed to keep the oil clean and free from foreign particles. However, it is difficult to prevent completely the entry of particles of dust, grit, etc. If the bearing material is hard, the particles act as an abrasive and the bearing and/or journal may be subjected to rapid wear. This may not be a serious problem if the bearing material has sufficient *embeddability*, i.e., is soft enough for the particles to become embedded without pressing against the journal.

The ability of a material to operate with boundary lubrication without undue wear, scoring, or bearing seizure is of primary importance in the cases of bearings which must start from rest under full load and when inadequate lubrication is expected during service. As discussed in Sec. 18-11, this may be related to the joint property of the bearing material and lubricant known as oiliness. It may also be a joint property of the journal and bearing materials. One of the primary considerations is the use of dissimilar metals or similar metals with special surface treatments for the journal and bearing.

Corrosion resistance of a bearing material is of particular importance in internal-combustion engines where the same oil is used to lubricate the cylinder walls and the bearings. The inevitable entry of products of combustion into the lubricating oil may result in a rapid deterioration of the bearings due to corrosion. Certain additives are combined with modern oils to inhibit the formation of corrosive products.

A high *thermal conductivity* is desirable to permit the rapid removal of the heat generated by friction.

Many high-capacity bearings are made by bonding one or more thin layers of a bearing material to a high-strength steel shell. In this case, the strength of the bond, or *bondability*, is an important consideration.

The coefficient of *thermal expansion* and its effect on the clearance must be considered if the operating temperature is relatively high or if the bearing must operate over a wide range of temperatures.

Unfortunately, no single bearing material possesses all the attributes to a desirable degree. Many different alloys and methods of manufacturing have been used in the attempt to secure the best possible combination of properties for a given application. While the manufacturers' catalogues should be consulted for more specific information, there are a number of references in periodicals[1] and books[2] related to the materials used in

[1] Plain Sleeve Bearings, Materials and Design, *Prod. Eng.*, vol. 18, pp. 130–160, October, 1948.

[2] C. Carmichael (ed.), "Kent's Mechanical Engineers' Handbook," "Design and Production" volume, 12th ed., pp. 4-73 to 4-78, John Wiley & Sons, Inc., New York, 1950.

Shaw and Macks, *op. cit.*, pp. 463–475.

bearings.   Table 18-1 is a comparative compilation of some of the properties of the more common metallic bearing materials.

Babbitts are recommended for most applications where the maximum bearing pressure (on projected area) is not over 1,000 to 2,000 psi. When applied in automobile engines, the babbitt is generally used as a thin layer, 0.002 to 0.006 in. thick, bonded to an insert or steel shell, as illustrated in Fig. 18-19.

TABLE 18-1. COMPARATIVE PROPERTIES OF BEARING METALS

Material	Fatigue strength	Conformability	Embeddability	Antiscoring	Corrosion resistance	Thermal conductivity
Tin-base babbitt[a].	Poor	Good	Excellent	Excellent	Excellent	Poor
High-lead babbitt[b]	Poor to fair	Good	Good	Good to excellent	Fair to good	Poor
Lead bronze[c].....	Fair	Poor	Poor	Poor	Good	Fair
Copper lead[d].....	Fair	Poor	Poor to fair	Poor to fair	Poor to fair	Fair to good
Aluminum[e].......	Good	Poor to fair	Poor	Good	Excellent	Fair
Silver...........	Excellent	Almost none	Poor	Poor	Excellent	Excellent
Silver lead indium[f]........	Excellent	Excellent	Poor	Fair to good	Excellent	Excellent

[a] Typical analysis: SAE 10—nominally 90 per cent tin, 4.5 per cent antimony, 0.5 per cent lead, 4.5 per cent copper, 0.1 per cent arsenic.

[b] Typical analysis: SAE 13—nominally 6 per cent tin, 10 per cent antimony, 84 per cent lead, 0.5 per cent copper, 0.25 per cent arsenic.

[c] Typical analysis: SAE 64—nominally 80 per cent copper, 10 per cent tin, 9.5 per cent lead, 0.75 per cent zinc, 0.5 per cent antimony, 0.5 per cent nickel, 0.25 per cent phosphorus.

[d] Typical analysis: SAE 48—nominally 70.5 per cent copper, 28.5 per cent lead, 1.5 per cent silver, 0.1 per cent zinc, 0.35 per cent iron, 0.25 per cent tin.

[e] Typical analysis: 91.5 per cent aluminum, 6.5 per cent tin, 1 per cent nickel, 1 per cent copper.

[f] Typical procedure: Layer of silver 0.02 to 0.03 in. thick, lead overlay 0.001 to 0.003 in. thick, and 4 to 5 per cent indium, based on the weight of lead, electrolytically deposited and thermally diffused into the layer of lead.

The silver and silver-lead-indium bearings are used almost exclusively in aircraft engines where the fatigue strength is the most important consideration.

Cast-iron bearings have been used with steel journals.   Lubrication must be adequate, and usually the pressure is limited to 500 psi and the speed to 130 fpm.

The *porous bronze* bearings in Fig. 18-20 are designed for use with little or no attention to lubrication.   These bearings are made by pressing a mixture of powdered copper and tin into the desired shape and then sintering.   The bearing is porous and may be impregnated with oil to the extent of 20 per cent of the total bearing volume. The selection of porous bronze bearings is usually based on the fundamental friction-wear criterion of an allowable value of pV. Typical design values are in the order of $pV = 35,000$ to $50,000$ for journal bearings and $pV = 10,000$

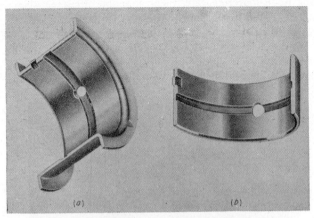

Fig. 18-19. Automotive sleeve-type half bearings. (a) Flanged; (b) plain. (*Courtesy Federal-Mogul Division, Federal-Mogul-Bower Bearings, Inc.*)

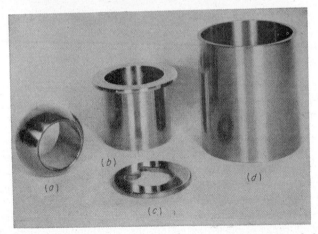

Fig. 18-20. Oilite porous bronze bearings. (a) Self-aligning; (b) flanged sleeve; (c) thrust; and (d) sleeve. (*Courtesy Amplex Division, Chrysler Corporation.*)

for thrust bearings, where p is the unit pressure on the projected area in pounds per square inch and V is the sliding velocity in feet per minute. The allowable values of pV decrease at very low speeds and at sliding velocities greater than 1,200 fpm. In Fig. 2-4, the motor shaft is mounted in self-aligning and the cutter-follower shaft in plain porous bronze bearings to ensure a satisfactory life without additional lubrication. An additional limitation is that the static-load capacity must not be exceeded. The static-load capacity varies from about 8,500 to 30,000 psi, depending upon materials, etc. The manufacturers' catalogues should be consulted for additional information.

A number of nonmetallic materials are used in bearings. In general, these materials have not been as widely used as those previously discussed, but, because of special properties, their use is increasing rapidly, particularly for unusual or extreme conditions.

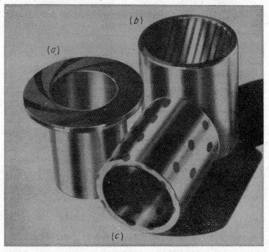

FIG. 18-21. Graphited bronze bearings. (a) Thrust; (b) serrated; (c) plug. (Courtesy Johnson Bronze Company.)

One important class of bearings is that depending upon the low shear strength of graphite to give a low coefficient of friction and prevent rapid wear. The bearing may consist entirely of a carbon-graphite mixture or a carbon-graphite-metal mixture, or it may be a simple bushing with graphite inlaid in grooves or holes.

The *carbon-graphite* types are self-lubricating, are dimensionally stable over a wide range of operating conditions, are chemically inert, and can operate at higher temperatures than other bearings. Because of these properties, carbon-graphite bearings are used in food-processing and other equipment where contamination by oil or grease must be prohibited, in pumps handling corrosive liquids, in bearings that must operate at high temperature, in bearings inaccessible for ordinary lubrication service, and in applications where the shaft speed is too low to maintain a hydrodynamic oil film.   The nonmetallic carbon-graphite bearings are limited to a maximum pressure of 20 psi but may be used at a temperature of 900°F in an oxidizing atmosphere or up to 5400°F in a nonoxidizing atmosphere.   The carbon-graphite-metal composition may be used with bearing pressures of 350 to 600 psi, depending upon whether the bearing is operated dry or wet.   The limiting temperature depends upon the melting point of the metal and atmospheric conditions; e.g., if babbitt metal is used, the maximum temperature is 464°F.   Both types must be used with hard, smooth journals.

The *graphited bronze* bearings in Fig. 18-21 have graphite compressed into dovetailed grooves, or serrations, or into drilled holes.   They are particularly useful at low speeds, $V < 30$ fpm, where it is difficult to maintain an oil film thick enough to prevent metal-to-metal contact.   The bearings are rated for continuous rotation without supplementary lubrication on a basis of $pV = 1,500$ with p being limited to a maximum of 1,500 psi.   For oscillating shafts, $pV = 5,000$ is permissible; with supplementary lubrication and rotating shafts, $pV = 10,000$, with p not exceeding 350 psi.   Standard serrated types may be used at temperatures up to 275°F, and standard plug types

may be used at temperatures up to 400°F.   Special plug-type bearings may be used at temperatures up to 1000°F.

Soft rubber bearings, as shown in Fig. 18-22, are used with water or other low-viscosity lubricants, particularly where sand or other large particles are present.   In addition to the high degree of embeddability and conformability, the rubber bearings are excellent for absorbing shock loads and vibrations.   The load capacity is

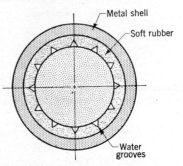

Fig. 18-22. Water-lubricated rubber bearing.

relatively low, being limited to 50 psi, but there is no apparent limit to the surface velocity.   Rubber bearings are used mainly on marine propeller shafts, rudder pintles, hydraulic turbines, and pumps.

Wood bearings have been used in numerous applications where low cost, cleanliness, inattention to lubrication, and antiseizing are important. The most commonly used wood is a dense (sp gr 1.33) tropical wood known as *lignum vitae*. Lignum vitae contains an oily resinous gum which acts as a natural lubricant, and it is generally used without additional lubrication.   *Rock maple* and *oak* are also used, but the bearings must be impregnated with oil.   Wood bearings change dimensions with a change in moisture content and usually have larger clearance-diameter ratios than other bearings. The allowable bearing pressure varies from 75 to 300 psi, depending upon the direction of the grain of the wood. Wood bearings will not seize the journal and are good for carrying shock loads. Though largely replaced by laminated plastic bearings, lignum vitae has been used with adequate lubrication to carry loads up to 2,000 psi in steel-mill roll-neck bearings.

Hard-maple bearings may operate at slightly higher temperatures, but lignum vitae bearings should not be used at temperatures above 120°F.

The use of plastics for bearing materials is a fairly recent development; the most widely used are nylon and Teflon. Both have many characteristics desirable in a bearing material, and both can be used dry, i.e., as zero film bearings. Nylon is stronger, harder, and more resistant to abrasive wear and is better suited for applications in which these properties are important, e.g., elevator gibs or bearings, cams in telephone dials, drawer rollers, and the connecting rod in the Sunbeam Shavemaster electric shaver in Fig. 2-4. However, Teflon is rapidly replacing nylon

TABLE 18-2. DESIGN DATA FOR TEFLON BEARINGS, DRY OPERATION

Conditions	pV^*	Maximum p, psi
Unmodified resin.................	1,000	500
Modified and reinforced†	80,000	40,000

* Where p is in pounds per square inch, based on projected area, and V is in feet per minute.

† Used in a thin layer, not exceeding 0.04 in., on a strong, heat-conducting backing.

as a wear surface or liner for journal and other sliding bearings because (1) it has a lower coefficient of friction, about 0.04 (dry) as compared with about 0.15; (2) it can be used at higher temperatures, up to about 600°F as compared with about 250°F; (3) it is dimensionally stable because it does not absorb moisture; and (4) it is practically chemically inert.

When used with an adequate supply of lubricant and under conditions leading to hydrodynamic or thick-film operation, Teflon is no better than many other materials except for those situations in which the journal must start up under a heavy load and a minimum starting friction and rate of wear are required.

The real advantages of Teflon show up when the bearing must run dry, e.g., as in food machinery where cleanliness is important, or at elevated temperatures where conventional lubricants tend to evaporate or break down. When the bearing is running dry, the important factors are the rates of wear and heat generation. Both rates are proportional to the product of the coefficient of friction, the pressure, and the sliding velocity.

For a given material, the basis for design is, again, an allowable value of pV. For Teflon, the allowable pV is not a constant but depends upon whether the Teflon resin is used in its unmodified form, filled with inorganic fibers, e.g., glass, graphite, asbestos and certain metals, or reinforced with a backing.[1] Although the variables are too numerous to permit the presentation here of adequate design data, the information in Table 18-2 will provide an introduction to the possibilities.

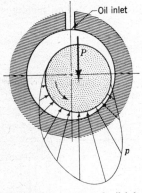

FIG. 18-23. Preferred oil-inlet location.

18-15. Mechanical Design Details. A major problem in most applications is that of ensuring an adequate supply of the proper lubricant in the right place at all times. The simplest case is that where the bearing is to be supplied by oil under pressure. In this situation, the satisfactory operation of the bearing may depend upon the location of the oil supply hole and grooves.

If the load is always in one direction, the oil hole should be located on the unloaded side of the bearing, as shown in Fig. 18-23. Circumferential and diagonal grooves should be avoided if possible. As illustrated in

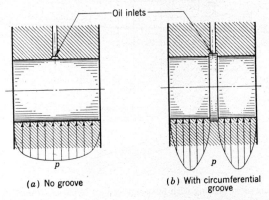

(a) No groove   (b) With circumferential groove

FIG. 18-24. Axial pressure distribution for bearings with single-hole oil supply and with circumferential-groove oil supply.

Fig. 18-24, the circumferential groove has the same effect as actually cutting the bearing in two, and the peak pressures must be much higher to give the same average pressure. Reference to Eq. (18-4) will show

[1] Bearing Design with Teflon Flourocarbon Resins, *J. Teflon*, Reprint 2, E. I. du Pont de Nemours and Company, Wilmington, 1960.

that, for the same total load, two bearings with $l/d = 0.5$ will each operate at a load number four times that for a single bearing with $l/d = 1$; consequently, the minimum film thickness will be less than for the single bearing. If oil must be transmitted through a hole in the rotating journal to drilled passages in the shaft in order to supply other bearings,

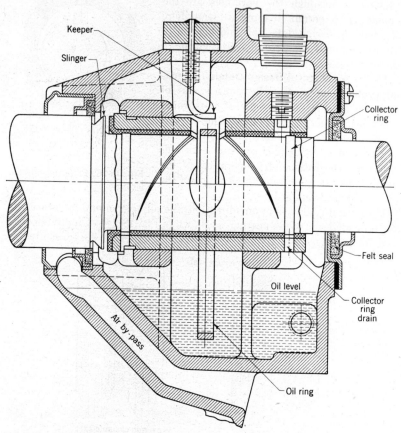

Keeper

Slinger

Collector ring

Felt seal

Oil level

Collector ring drain

Air by-pass

Oil ring

FIG. 18-25. Ring-oiled bearing with felt seals and air bypass. (*Courtesy Westinghouse Electric Corporation.*)

e.g., connecting-rod bearings in an internal-combustion engine, it will be necessary to use circumferential grooves, and the best one can do is to consider this fact in the design.

Large bearings often have an axial groove on the unloaded side of the bearing to help distribute the oil more evenly and to increase the oil flow rate to assist in carrying away heat.

Figure 18-25 shows a typical ring-oiled electric-motor bearing mounting. The friction between the journal and the ring causes the ring to rotate and thus raise oil up from the reservoir to the bearing. The use of oil collector rings, slingers, and seals is required to ensure that the oil returns to the reservoir and that dirt is excluded. Waste packing is

FIG. 18-26. Lubrication system for a high-speed gear unit. (*Courtesy Westinghouse Electric Corporation.*)

used in motors, railroad journal boxes, etc., to transfer oil from the reservoir to the bearing by capillary action. Chains and rotating disks are also commonly used to lift the oil from the reservoir.

Figure 18-26 is a cutaway view of a high-speed gear unit designed specifically for use as a speed increaser. The pinion shaft may rotate at 6,000 rpm or more, and the gear pitch-line velocities may exceed 12,000 fpm. The lubrication requirements cannot be met by a simple splash system and ring-oiled bearings, as for lower-speed operation.   As indicated in Fig. 18-26, a force-feed circulating system, including an oil filter and an oil cooler, supplies oil to the bearings and the gears.

Many light-duty bearings must depend upon intermittent lubrication. Generally an oil cup or well is filled from an oil can, and the oil drips or is fed by a wick to the bearing.   Aside from the lack of continuous lubrication, some of the faults of these systems are loss of oil after leaving the bearing, necessity for frequent attention, and the possibility of inadequate or too much oiling.

The small clearance used in sleeve bearings and the resulting need for accurate alignment of the journal and bearing axes are sources of many operating difficulties.   Problems may arise because of machining errors, thermal differential expansion, and too much flexibility on the part of

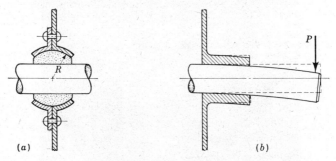

(a)   (b)

FIG. 18-27. (a) Self-aligning bearing; (b) elastic matching.

the shaft or the supporting frame.   As discussed in Sec. 18-14, the selection of the proper material may help to alleviate this problem.   However, if the use or economics of the design results in more misalignment than can be tolerated by the bearing material, it is necessary to resort to a *self-aligning bearing* or to what is called *elastic matching*.   Small self-aligning bearings are usually made (Fig. 18-20a) with a spherical external surface which is held in a spherical seat in the housing, as shown in Fig. 18-27a.   It should be noted that even self-aligning bearings do not help much when three or more bearings are used.

Elastic matching is the term given to the designing of the bearing support so that an eccentric load gives a deflection of the bearing that will match the slope of the deflection curve of the shaft.   An example of this is shown in Fig. 18-27b.

18-16. Rolling-contact Bearings.   The second major classification of bearings consists of those depending upon the contact between rolling members to give the desired restraint with a minimum of friction.

Probably the outstanding advantage of a rolling-contact bearing over a sliding bearing is its low starting friction.   As previously discussed, the ordinary sliding bearing starts from rest with practically metal-to-metal contact and has a high coefficient of friction as compared with that between rolling members.   This feature is of particular importance in the case of bearings which must carry the same load at rest as when running; for example, less than one-thirtieth as much force is required to start a railroad freight car equipped with roller bearings as with plain journal bearings.   However, most bearings carry relatively light loads while starting and do not become heavily loaded until the speed is high enough for a hydrodynamic film to be built up.   At this time the friction is that in the lubricant, and in a properly designed journal bearing the viscous friction will be in the same order of magnitude as that for a rolling-contact bearing.

Another important feature of rolling-contact bearings is that many types of ball and roller bearings can carry combined radial and axial loads without any additional complications.

Rolling-contact bearings inherently have much smaller l/d ratios than comparable journal bearings.   This is often advantageous in that a more compact design may be possible with smaller-diameter shafts for the same levels of rigidity and stress.

Another major advantage is that, in general, it is easier to provide adequate lubrication for rolling bearings than for sliding bearings.   This is particularly true for slow-speed bearings where grease may be used. The grease is retained by simple and inexpensive seals, and the initial supply of lubricant often lasts for the life of the machine.

It is difficult to position a shaft precisely by use of a journal bearing because the center of the rotating element moves relative to the center of the fixed element with changes in load, speed, or lubricant viscosity. Rolling bearings may be mounted so that there is very little motion of the parts, and relatively accurate alignment can be maintained.

Rolling-contact bearings are made by many different manufacturers in many countries according to the same standards.   The feature of universal availability of standard, easily replaced bearings is an important factor when service of equipment is considered.

There are many variations, but all rolling-contact bearings may be classed, according to the shape of the intermediate rolling member, as either ball or roller bearings.   The classification of bearings into those designed to carry radial loads and those designed to carry thrust loads becomes less important for rolling-contact bearings than for sliding bearings, because most ball and roller bearings can carry both types of loads. Some of the particular types and their characteristics will be discussed in the following sections.

18-17. Ball Bearings.   A ball bearing consists of grooved inner and outer races separated by a number of balls.   As shown in Fig. 18-28, the radius of the ball is slightly less than the radii of curvature of the grooves in the races.   Kinematically, this gives point contact, and the balls and races may roll freely without any sliding.   How-

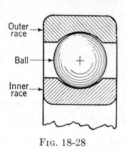

ever, elastic and plastic deformation of the balls and races under load results in contact occurring over a small area and a resulting deviation from pure rolling motion.   Ball bearings have ground balls and may have either ground or unground races.   The unground and low-precision-ground bearings are relatively inexpensive and perform quite satisfactorily in a number of applications. These bearings will be discussed in more detail in Sec. 18-19, and the remainder of this discussion

Fig. 18-28

will be related to the precision-ground bearings, whose dimensions and tolerances are specified in tenths of thousandths of an inch.

The most common type of ball bearing is the *deep-groove bearing* in Fig. 18-29.   During assembly, the races are offset, and the maximum number

Fig. 18-29. Deep-groove ball bearing. (*Courtesy New Departure Division, General Motors Corporation.*)

Fig. 18-30. Filling-notch ball bearing. (*Courtesy New Departure Division, General Motors Corporation.*)

of balls are placed between the races.   The races are then centered, and the balls are symmetrically located by use of a retainer or cage.

This bearing is commonly known as a *radial ball bearing*.   However, since all balls share a thrust load and only several balls carry the radial load, the thrust capacity will equal or exceed the radial capacity. The load capacity of a ball bearing is related to the size and number of the balls.

The *filling-notch bearing* in Fig. 18-30 derives its name from the notches in the inner and outer races which permit more balls to be inserted than in a comparable deep-groove ball bearing. The notches do not extend to the bottom of the raceway, and this type of bearing can carry a considerable thrust load, though not to the same extent as a deep-groove bearing.

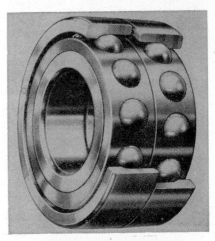

FIG. 18-31. Angular-contact ball bearing. (*Courtesy New Departure Division, General Motors Corporation.*)

FIG. 18-32. Duplex ball bearings. (*Courtesy New Departure Division, General Motors Corporation.*)

The *angular-contact bearing* in Fig. 18-31 has one side of the outer-race groove cut away to permit the insertion of more balls than in a deep-groove bearing but without having a notch cut into both races. This permits the bearing to carry a relatively large axial load *in one direction* while also carrying a relatively large radial load. Angular-contact bearings are usually used in pairs so that thrust loads may be carried in either direction. If the maximum rigidity is desired, matched pairs of angular-contact bearings known as *duplex bearings* (Fig. 18-32) are often specified. The bearings may be obtained with the ring widths ground with the correct amount of offset so that, when they are clamped together, all the clearance is taken up and the bearings are given an initial load. The bearings are then *preloaded*.

Double-row ball bearings are made with either radial or angular contact between the balls and races. The angular-contact double-row ball bearing in Fig. 18-33 is assembled with a specified preload.   The double-row bearing is appreciably narrower than two single-row bearings, and its capacity is slightly less than twice that of a single-row bearing.

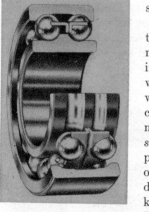

The normal clearances in a ball bearing are too small to accommodate any appreciable misalignment of the shaft relative to the housing.   Consequently, if the unit is assembled with shaft misalignment present, the bearing will be subjected to a load that may be in excess of the design value, and premature failure may occur.   Figure 18-34 shows two types of *self-aligning bearings* designed for use where the presence of precise alignment may be doubtful or when it is impractical to provide for the desired precision.   The type in Fig. 18-34*a* is known as an *externally self-aligning bearing.* The outer race has a spherical surface which fits in a mating spherical surface in an aligning ring (as shown) or in another housing.   This bearing is designed to adjust its position infre-

FIG. 18-33. Double-row ball bearing. (*Courtesy New Departure Division, General Motors Corporation.*)

quently, e.g., during assembly and disassembly.   The type in Fig. 18-34*b* is known as an *internally self-aligning bearing.*   The inner surface of the outer race is spherical.   Consequently, the outer race may be displaced

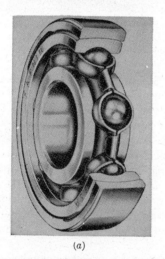

(a)   (b)

FIG. 18-34. Self-aligning ball bearings.   (a) Externally; (b) internally.   (*Courtesy The Fafnir Bearing Company.*)

through a small angle without interfering with the normal operation of the bearing. The internally self-aligning ball bearing is interchangeable with other ball bearings. However, it is more efficiently used with predominantly radial loads because of its limited thrust capacity. It also has the lowest friction of any rolling-contact bearing, e.g., a coefficient of friction, referred to the bore diameter, of about 0.0010, as compared with about 0.0015 for the deep-groove ball bearing.[1]

Ball thrust bearings, illustrated in Fig. 18-35, are useful for carrying pure thrust loads. The use of a large number of balls results in a relatively high capacity in a small space. The effects of centrifugal force limit the maximum speed to about 60 per cent of that recommended for a similar-size radial ball bearing. Thus, at high speeds, it is recommended that angular-contact ball bearings be used in place of ball thrust bearings.

FIG. 18-35. Ball thrust bearing. (*Courtesy Marlin-Rockwell Corporation.*)

18-18. Roller Bearings. When the maximum load capacity is required in a given space, it becomes desirable to substitute the line contact between a roller and the races of a roller bearing for the point contact in ball bearings. The rollers may be in the form of cylinders, of convex or concave surfaces of revolution, or frustums of cones. The discussion in this section will be concerned with the manner in which the form of the roller influences the characteristics of several of the more common types of roller bearings.

Figure 18-36 illustrates two types of *needle*, or *quill*, *bearings*. The rollers are relatively slender and completely fill the space so that neither a cage nor a retainer is needed. The outer race of the bearing in Fig. 18-36a is a drawn, case-hardened steel shell which depends upon the support from the housing bore into which it is pressed to maintain its cylindrical shape. Both bearings may run directly on the shaft. However, as

---

[1] A. Palmgren, "Ball and Roller Bearing Engineering," p. 39, SKF Industries, Inc., Philadelphia, 1945.

(a)   (b)

FIG. 18-36. Needle bearings.   (*Courtesy The Torrington Company.*)

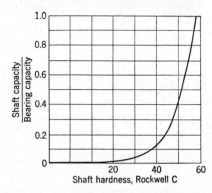

FIG. 18-37. Relationship of load capacity to shaft hardness for needle bearings running directly on a shaft.   (*Courtesy The Torrington Company.*)

shown in Fig. 18-37, the load capacity decreases rapidly if the shaft hardness falls below the recommended value of Rockwell C 58.   When a hardened and ground shaft is not practical or where radial space is not at a premium, a standard inner race, as illustrated in Fig. 18-36b, should be used.

The advantage of this type of bearing is the large ratio of load capacity to size.   For example, a standard needle bearing with an outside diameter of 1⅜ in. is rated to carry a load of 3,630 lb for 500 hr at 100 rpm, while a filling-notch ball bearing with a comparable outside diameter (1.3780 in.) can carry only 1,180 lb under the same conditions.   However, completely filling up the space with long rollers results not only in the much higher load capacity but also in more friction (a coefficient of friction of about 0.0025)* and an increased tendency for the rollers to skew, particularly in the presence of an eccentric load.   Needle bearings are ideally suited for use when heavy loads are carried with an oscillatory motion, e.g., piston-pin bearings in heavy-duty diesel engines, where the reversal of motion tends to keep the rollers in correct alignment.

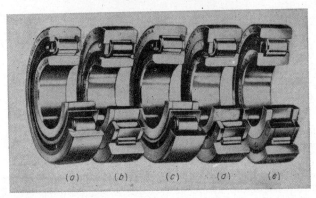

FIG. 18-38. Cylindrical roller bearings.   (*Courtesy Hyatt Bearings Division, General Motors Corporation.*)

Needle bearings are often used to replace sleeve bearings.   The small external diameter permits the change in type of bearing with little or, in many cases, no change in design of the parts.

Several types of roller bearings used to carry radial loads without thrust are shown in Fig. 18-38.   These bearings with relatively short rollers that are positioned and guided by a cage are commonly referred to as *cylindrical roller bearings*.   These bearings are relatively rigid against radial motion and have the lowest friction of any form of heavy-duty rolling-contact bearings (a coefficient of friction of about 0.0011).†   Consequently, they are extensively used in high-speed service.   The bearings *a*, *b*, and *e* permit complete axial freedom of one race relative to the other. It is often desirable to use one of these bearings on one end of a shaft

* Personal communication from J. R. Hull, The Torrington Company, January, 1962.
† Palmgren, *loc. cit.*

when a duplex or other type of ball bearing is used at the other end. In this case, the ball bearing carries all the thrust load, and the roller bearing does not restrain the axial motion of the shaft relative to the housing.

The *wound roller bearing* in Fig. 18-39 has rollers made from helically wound strips of spring steel. The rollers have alternately right- and left-hand helices to assist in distributing the lubricant. The bearings are not

Fig. 18-39. Wound roller bearing. (*Courtesy Hyatt Bearings Division, General Motors Corporation.*)

recommended for high-speed operation, but the flexibility of the rollers permits their use under severe shock loads. Wound roller bearings may be used with inner and outer races, as shown, or the rollers may run directly on the shaft. For best results, the shaft should be hardened and ground with a hardness of Rockwell C 54.

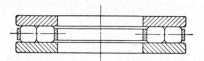

Fig. 18-40. Cylindrical roller thrust bearing.

The *cylindrical roller thrust bearing* in Fig. 18-40 can carry heavy thrust loads but is not recommended for general use because of the sliding between the rollers and the races.

The *spherical roller bearings* in Figs. 18-41a and 18-46 are so named because one of the raceways is ground in the form of a sphere. The internally self-aligning feature of these bearings eliminates the sensitivity to eccentric loading and misalignment inherent in all other types of roller bearings. The bearings can normally tolerate angular misalignment in the order of $\pm 1\frac{1}{2}°$ and, when used with a double row of rollers, as shown, can carry thrust loads in either direction.

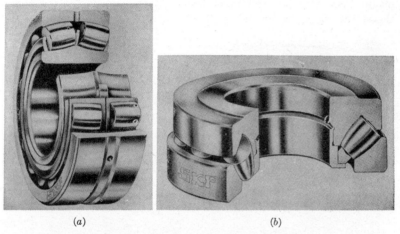

(a)   (b)

FIG. 18-41. Spherical roller bearings. (a) Double-row; (b) thrust. (*Courtesy SKF Industries, Inc.*)

If the thrust load is high in comparison with the radial load, it is more efficient to use a *spherical roller thrust bearing*, illustrated in Fig. 18-41b.

The *tapered roller bearing* in Fig. 18-42 is designed so that the rollers and raceways are frustums of cones whose elements intersect at a common

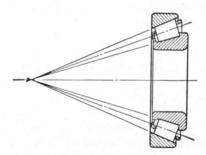

FIG. 18-42. Tapered roller bearing.

point. As discussed in Sec. 16-14 in relation to bevel gears, this is the basic requirement for pure rolling between conical surfaces. This type of bearing can carry both radial and thrust loads. In fact, the presence of either component results in the other acting on the bearing. Thus, tapered roller bearings are almost always used as opposed pairs. The axial position of the bearings can be controlled to preload the bearings as desired. The bearings are available in various combinations as double-row bearings and with different cone angles for use with different relative magnitudes of radial and thrust loads. A steep-angle bearing and a thrust bearing are shown in Fig. 18-43.

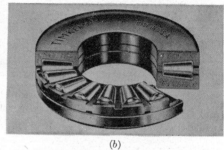

(a)   (b)

Fig. 18-43. Tapered roller bearings. (a) Steep-angle; (b) thrust. (*Courtesy The Timken Roller Bearing Company.*)

18-19. Special Features of Commercially Available Rolling-contact Bearings. In addition to the fundamental differences between types of rolling-contact bearings discussed in the previous sections, there are almost innumerable minor variations that have been made in order to simplify the use, installation, and service of the bearing.

One of the major features in relation to convenience of use is the addition of *shields* and *seals* designed to prevent the lubricant from escaping and to prevent dirt and moisture from entering. Several types are illustrated in Fig. 18-44. The early types of sealed bearings used felt seals *a* or labyrinth seals *b* and were wider than standard single-row bearings. Shielded bearings *c* were developed to be interchangeable with regular bearings, but the sealing is not very effective, and the lubricant must be replaced often. A fairly recent innovation is the plastic contact seal *d* which is thin enough to permit the use of standard single-row dimensions and yet effective enough to retain the initial supply of grease for a relatively long time, often for the life of the bearing or machine. Various combinations, such as one shield and one seal, etc., can also be obtained

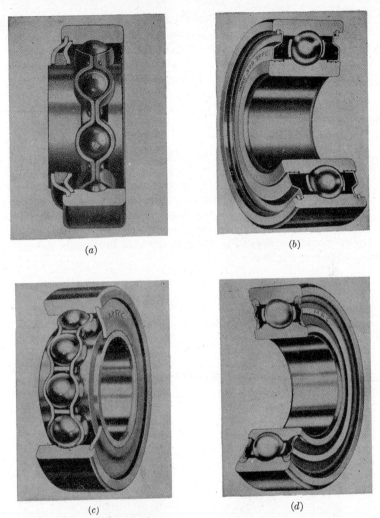

(a)   (b)

(c)   (d)

FIG. 18-44. Ball bearings with integral seals and shields.   (*Courtesy Marlin-Rockwell Corporation.*)

as stock items.   The bearings that depend upon rubbing contact for sealing have more friction than the labyrinth-sealed and clearance-shielded bearings and are not as satisfactory for use at high speeds.

Many types of bearings are available with tapered bores that can be used with adapter rings for more convenient mounting when the thrust load is relatively low.

Wide-inner-race bearings (Fig. 18-45) are seldom used in general design but are commonly used in complete units such as *pillow blocks* (Fig. 18-46)

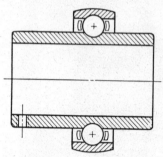

and *flanged housings* (Fig. 18-47). Pillow blocks and flanged housings are usually self-aligning, and heavy-duty units are made with spherical roller bearings, as shown in Fig. 18-46. Most types are designed to be self-lubricating, with shaft seals and an adequate lubricant-storage capacity. Simplicity of mounting and ability to operate for long periods without service have been responsible for the almost universal use of pillow blocks and cartridges in conveying and other materials-handling equipment.

FIG. 18-45. Wide-inner-race ball bearing.

The *snap-ring bearing* in Fig. 18-48 is provided to simplify the mounting, as shown in Fig. 8-27e.

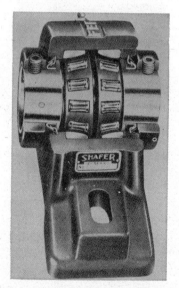

FIG. 18-46. Spherical-roller-bearing pillow block. (*Courtesy Chain Belt Company.*)

FIG. 18-47. Ball-bearing flanged housing. (*Courtesy SKF Industries, Inc.*)

One type of precision ball bearing that is special only in relation to its size is the miniature ball bearing (Fig. 18-49). Standard miniature full-race ball bearings are made in many sizes, the smallest being 0.0250-in.

bore and 0.1000-in. OD. The smallest standard pivot bearing has an outside diameter of 0.0590 in. and can be used with shafts with diameters as small as 0.020 in. The load capacities are measured in ounces or a few pounds.

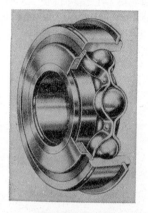

Fig. 18-48. Snap-ring bearing. (*Courtesy New Departure Division, General Motors Corporation.*)

Fig. 18-49. Miniature ball bearings, (*Courtesy Miniature Precision Bearings. Inc.*)

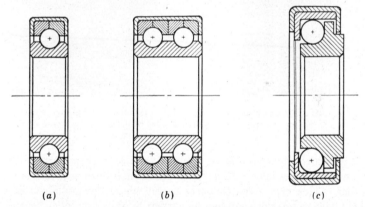

(*a*)   (*b*)   (*c*)

Fig. 18-50. Low-precision ball bearings. (*a*) Radial; (*b*) double-row; (*c*) angular-contact.

The low-precision bearings in Fig. 18-50 have either turned or ground and hardened races. The divided construction permits filling the entire space between the races with balls so that a cage is not necessary. The bearings are relatively economical and can be used effectively in many places where precision bearings have historically been used.

The specific needs of the aircraft industry have resulted in the design and manufacture of a wide variety of special bearings that may be of considerable use in other applications. The names of several types are bell cranks, track rollers, guide rollers, pulley bearings, torque tube bearings, and rod ends. Self-aligning rod-end bearings (Fig. 18-51) are particularly useful in linkages. The shank may have internal or external threads, and either right- or left-hand threads are available. The use of a right- and a left-hand threaded member at the ends of a rod forms a turn-buckle and permits an accurate setting of the center distance. A similar

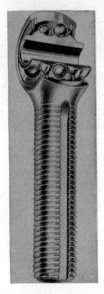

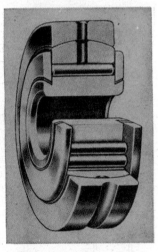

FIG. 18-51. Self-aligning rod-end bearing. (*Courtesy The Fafnir Bearing Company.*)

FIG. 18-52. Self-aligning aircraft-type needle bearing. (*Courtesy The Torrington Company.*)

bearing, which does not have rolling contact, uses a hardened-steel spherical inner member resting on a Teflon or bronze seat in the outer member. This type is particularly useful where the motion is limited and the load is heavier than can be carried by a ball or roller bearing.

The self-aligning *aircraft-type needle bearing* in Fig. 18-52 is designed for use when the maximum capacity is desired with oscillating motion and where it is difficult to obtain alignment during assembly or where deflections make a self-aligning bearing desirable.

Standard ball bearings are often used as *cam rollers*, even though the outer race is too thin and flexible for best service. Commercial cam followers using needle bearings are available, as shown in Fig. 18-53.

Three recent applications of rolling contact to reduce the friction of normally high-friction devices are the ball-bearing spline (Fig. 8-25), the ball-bearing screw (Fig. 10-7), and the *ball bushing* in Fig. 18-54. As for the ball-bearing spline (Sec. 8-7), a passage is provided so that the

Fig. 18-53. Cam-follower bearing. (*Courtesy McGill Manufacturing Company, Inc.*)

Fig. 18-54. Ball bushing. (*Courtesy Thomson Industries, Inc.*)

balls which have used up their rolling distance are free to return to their initial starting place, where they again take up their share of the load.

18-20. Standard Dimensions and Designations of Ball Bearings.   The dimensions that have been standardized on an international basis are indicated in Fig. 18-55. These dimensions are a function of the bearing bore and the series of bearing. Some bearings are available with the bore given in fractions of an inch. However, the majority of ball and roller bearings are based upon the metric system, and the dimensions are expressed in whole millimeters. For engineering purposes, these dimensions are given in inches. This is not as inconvenient as it first appears. The required accuracy of the shaft and housing bore diameters is higher than can be normally achieved by means other than grinding or careful turning or boring. Thus, since stock shafting and standard reamers are not sufficiently accurate, the fact that the nominal dimension is not a fraction of an inch becomes inconsequential. Another factor is that most shafts need to have shoulders to locate the axial position of the bearing and to carry the thrust load. Thus, it is often convenient to start with inch-dimensioned rolled stock and then decrease the diameter to that required for the bearing.

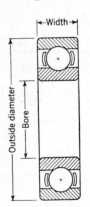

Fig. 18-55. Standardized dimensions for ball bearings.

Bearings are designated by a number. There are some exceptions, but in general the number consists of at least three digits. Additional digits or letters are used to indicate special features, e.g., deep groove, filling notch, snap ring, shields, seals, etc. The last three digits give the series and the bore of the bearing. The last two digits, from 04 on up, when multiplied by 5, give the bore diameter in millimeters.

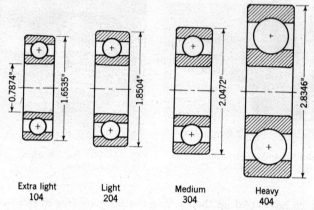

Extra light   **Light**   **Medium**   **Heavy**
104   204   304   404

FIG. 18-56. Variation of outside diameter with series for bearings with the same bore diameter.

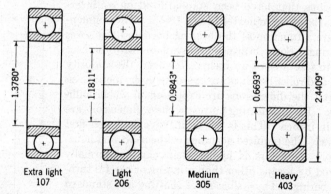

Extra light   **Light**   **Medium**   **Heavy**
107   206   305   403

FIG. 18-57. Variation of bore diameter with series for bearings with the same outside diameter.

The third from the last digit designates the series of the bearing. The three most common are the 200, or light, series; the 300, or medium, series; and the 400, or heavy, series. An extra-light series is also used, but its designation has not been standardized.

The outside diameters (OD) have been standardized in relation to the bore and series so that there is often a bearing in each series with the same outside diameter. This is particularly convenient when space requirements are more important than load-carrying capacity. Figure 18-56 shows the variation of outside diameter with series for bearings with the same bore diameter, and Fig. 18-57 shows the variation of bore diameter for bearings with the same outside diameter.

18-21. Load Capacity of Rolling-contact Bearings. Some general-purpose rolling-contact bearings are made from casehardened steel, but the majority are made from a high-carbon chrome-alloy through-hardening steel. The alloy used corresponds closely to SAE 52100, which contains about 1 per cent carbon and 1.5 per cent chromium and can be through-hardened to 62 to 65 Rockwell C. These materials are suitable for use at temperatures up to about 350°F, but their hardness and thus their load capacities decrease rapidly at higher temperatures.

For medium-high temperature service (up to about 600°F) AISI 440C stainless steel is usually preferred. For still higher temperatures, some tool steels are usable at 1000°F and materials such as Pyroceram, alumina, and titanium-carbide cermet appear to be useful at temperatures exceeding 1500°F.

Special bearings are made from beryllium copper for use when a non-magnetic bearing is required. The load capacity of a beryllium-copper bearing is about 25 per cent that of a similar SAE 52100 bearing.

Since there are so many types of rolling-contact bearings and each has its own special considerations, such as the ability to carry both radial and thrust loads, speed, and other service restrictions, it is not practical to cover these details here. The reader should refer to manufacturers' catalogues for specific information about a particular type of bearing. However, some factors of major importance are common to all rolling-contact bearings, and these will be considered in some detail below.

The calculation of the load capacity of rolling-contact bearings, as for other direct-contact mechanisms such as gear teeth (Sec. 16-8) and cams, is based upon the Hertz equations for bodies in direct contact. However, the actual useful rating must be determined on the basis of experimental tests and correlation of the test data with the theoretical results.

Bearings are often required to carry loads while at rest and while rotating. The load with no rotation is spoken of as a static load, and the static-load rating is limited by the requirement that the permanent deformation of the balls and races be less than that which would give rough and noisy operation when rotating. Palmgren[1] has found that this requirement is met if the combined permanent deformation does not exceed 0.0001 in./in. of diameter of the ball or roller. It is fortunate that bearings will operate satisfactorily with permanent deformation. Experi-

[1] Palmgren, *op. cit.*, p. 78.

ments[1] have shown that a 19.7-lb load on a ⅜-in. steel ball will cause a permanent deformation of 0.00001 in. in a flat plate. This same ball may be required to carry a load in excess of 1,800 lb when used in a bearing. It is interesting to note that a bearing rotating at a low speed may carry a considerably higher load than if the speed were zero. This is because of the even distribution of the plastic deformation of the races.

The fact that any practical load will be sufficient to cause plastic deformation of the balls and races means that rolling-contact bearings operate at such high stress levels that they have no endurance limit in the usual sense. In other words, ball and roller bearings have finite service life, and the problem becomes that of determining the load corresponding to the desired service life.

As discussed in Sec. 6-2 the problem of finite life under fatigue loads becomes a problem in statistics or, as far as design is concerned, a question

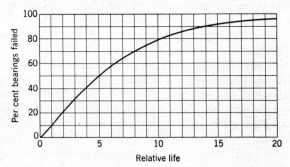

Fig. 18-58. Cumulative failure curve for rolling-contact bearings. (*Courtesy SKF Industries, Inc.*)

of reliability. Analyses of test data for all types of rolling-contact bearings result in cumulative failure curves similar to that in Fig. 18-58. While not strictly correct terminology, the life of the 50 per cent bearing is commonly termed the *average life* and the 50 per cent bearing is called the *average bearing*. It has been found that the longest life rarely exceeds four times the average life and that 90 per cent of the bearings have a life longer than one-fifth of the average life.[2]

In the past, the catalogue values of load ratings were specified either as the load the bearing could carry for a specified number of hours (500, 3,500, 3,800, etc.) at a given speed or as the load it could carry for a specified number of revolutions (usually 10^6). To complicate the problem further some manufacturers used the load corresponding to 50 per

[1] After A. B. Jones, "New Departure Analysis of Stresses and Deflections," vol. I, p. 173, New Departure Division of General Motors Corporation, Bristol, Conn., 1946.

[2] Palmgren, *op. cit.*, pp. 68–69.

cent failures while others used the load corresponding to only 10 per cent failures. The designer must still read the catalogues with care, but almost all manufacturers now base their ratings on 10 per cent failures in 10^6 revolutions. If the load that a given size and type of bearing can carry for 10^6 revolutions with 10 per cent failures is called the *basic dynamic load rating* C and endurance tests are run for many bearings, the results would appear as shown in Fig. 18-59.

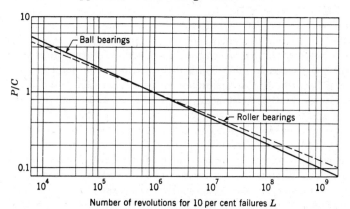

FIG. 18-59. Typical fatigue test curves for rolling-contact bearings.

The number of revolutions giving 10 per cent failures under a given load is sometimes called the L_{10} or B_{10} life; however, we shall use simply L, with the understanding that it refers to 10 per cent failures unless otherwise indicated.

From Fig. 18-59, we can write, for ball bearings,

$$\frac{P}{C} = \left(\frac{10^6}{L}\right)^{1/3} \tag{18-17}$$

and for roller bearings,

$$\frac{P}{C} = \left(\frac{10^6}{L}\right)^{0.3} \tag{18-18}$$

In the selection of bearings to meet the requirements of a particular problem, the designer is interested primarily in finding the value of C so that he can enter the tables in a catalogue and pick out a bearing or bearings. Thus, Eqs. (18-17) and (18-18) are more useful when rewritten as, respectively,

$$C = P\left(\frac{L}{10^6}\right)^{1/3} \tag{18-19}$$

and

$$C = P\left(\frac{L}{10^6}\right)^{0.3} \tag{18-20}$$

The remaining problems are due to the fact that the basic dynamic load rating C applies to the bearing when (1) it is carrying only the type load it was primarily designed to carry, e.g., radial or thrust, and (2) the outer ring is fixed and the inner ring is rotating. The former is important because the relationship between the resultant of a radial and a thrust load and the contact pressure is a problem in geometry of the bearings, i.e., radii of curvature of the elements in contact, and is therefore different for each type of bearing. The latter is significant because it is not really the number of revolutions that is important but, rather, the number of times a particular spot must carry the load. For a statically loaded bearing, the relationship between radii of curvature is such that, except for internally self-aligning ball bearings, the point of maximum stress is between the balls or rollers and the inner race. When the inner race rotates relative to the load, every point on the circumference of the race shares equally in carrying the load, but when the inner race is fixed and the outer race rotates, one—and only one—point on the inner race is subjected to a peak stress every time a ball or roller passes it. Also, if one considers the rolling elements as pitch circles and the retainer as the carrier of a planetary gear train, it can be shown that, for a given number of revolutions of the input, the carrier will make the greater number of revolutions when the inner race is fixed and the outer race is the input. Thus, for a specified life, bearings can carry a greater load when the inner race rotates relative to the load. For most ball and roller bearings carrying radial loads, the ratio of capacities is 1.2:1.

The effects of combined radial and axial loads and which ring is rotating relative to the load are considered by using factors to convert the actual load into an equivalent load, which is then used to find the basic dynamic load rating. The calculation procedure varies from manufacturer to manufacturer, but all provide tables, charts, nomographs, etc., so that the selection of a bearing with sufficient capacity for a specified service becomes largely a routine operation.

It should be noted that, since the permissible load varies inversely as the cube root of the desired life, the life will vary inversely as the cube of the load. Thus, one should never select a bearing with even a slightly lower rating unless he is willing to accept a considerably shorter service life.

18-22. Rolling-contact Bearings under Variable Loads. As for journal bearings, many rolling-contact bearings must carry loads that vary in magnitude and/or direction. There is not much information available for use by the designer when the load varies in direction. Actually, if anything, the changing direction would be beneficial in that it would result in a more uniform distribution of wear or surface damage on the fixed member.

For the situation in which the magnitude of the load varies with time, we are faced with the problem of cumulative damage, a problem that is not fully understood in relation to general design.[1]   However, the facts that, as shown in Fig. 18-59, endurance test data plot as a straight line over a wide range of loads and numbers of revolutions and the relationship between load and life can be expressed as in Eqs. (18-17) and (18-18) indicate that the bearing fails when the cumulation of damage reaches some definite point.   This point, for operation at a given load, can be determined by noting from Eq. (18-17) that failure for ball bearings occurs when

$$LP^3 = 10^6 C^3 = K_C \qquad (18\text{-}21)$$

where K_C = cumulative damage constant.

If we extend this idea to assuming that the damage occurring at a given load is directly proportional to the product of the number of cycles at that load and the cube of the load, we find that failure would be expected to occur when

$$L_1 P_1{}^3 + L_2 P_2{}^3 + \cdots = K_C$$

or
$$\sum_{i=1,2,\ldots} L_i P_i{}^3 = K_C \qquad (18\text{-}22)$$

This has been found to agree well with experience.[2]   For a continuously varying load Eq. (18-22) becomes

$$\int_0^{L_F} P^3 \, dL = K_C \qquad (18\text{-}23)$$

where L_F = life in revolutions.

For practical use where time and speeds are more convenient than revolutions,

$$L = nt \qquad \text{and} \qquad dL = n \, dt \qquad (18\text{-}24)$$

and Eq. (18-23) becomes

$$\int_0^{t_F} P^3 n \, dt = K_C \qquad (18\text{-}25)$$

where t_F = time to failure, min, and n = speed of rotation, rpm.

Equation (18-25) can often be solved analytically if the equations for $P = f(t)$ and $n = f(t)$ are known.   Otherwise, numerical or graphical techniques must be used.

For design purposes Eqs. (18-22), (18-23), and (18-25) become,

[1] G. Sines and J. L. Waisman (eds.), "Metal Fatigue," chap. 12, McGraw-Hill Book Company, Inc., New York, 1959.

[2] Palmgren, op. cit., p. 82.

respectively,

$$C = \left(\frac{\Sigma L_i P_i^3}{10^6}\right)^{1/3} \tag{18-26}$$

$$C = \left(\frac{\int_0^{L_F} P^3 \, dL}{10^6}\right)^{1/3} \tag{18-27}$$

and

$$C = \left(\frac{\int_0^{t_F} P^3 n \, dt}{10^6}\right)^{1/3} \tag{18-28}$$

Similar equations can be developed for roller bearings by using the proper exponent.

Example 18-3   Ring-oiled journal bearings were designed in Example 18-1 to support the shaft of a centrifugal blower. The blower speed was 600 rpm, and the bearing loads were 125 and 630 lb. We are now in the process of making an alternative design which will use rolling-contact bearings. As specified in Example 6-2, the bearings must be self-aligning, and for simplicity we shall plan to use ball- or roller-bearing pillow blocks (Fig. 18-46) with internally self-aligning bearings.

The journal bearings were designed for hydrodynamic lubrication and, for practical purposes, infinite life. Ball and roller bearings must, however, be selected for a finite life, and, in this case, it has been decided that the blower should run continuously for two years without requiring replacement of more than 10 per cent of the bearings.

As in Example 18-1, we shall plan to decrease the number of parts to be purchased and stocked by using identical bearings at both ends of the shaft. However, since the difference in loads is relatively large, the increased cost of an unnecessarily large bearing for the lightly loaded end may outweigh the advantages of using only one size. For lack of information, we cannot determine here whether one or two sizes will actually be more economical in the long run. Our analysis will be limited to the more heavily loaded bearing.

Specifically, we are asked to determine the equivalent load to be used in selecting the bearings from catalogue A, in which the ratings are based upon a 90 per cent life expectancy of 1,000,000 revolutions, and catalogue B, in which the ratings are based upon an average life of 3,800 h at 1,000 rpm.

Solution.   Catalogue A. Since the catalogue ratings are based upon the same life expectancy (90 per cent) specified for our design, we need only relate the desired load and life to the basic dynamic load ratings by using Eqs. (18-19) and (18-20):

$$C = P\left(\frac{L}{10^6}\right)^{1/3} \quad \text{and} \quad C = P\left(\frac{L}{10^6}\right)^{0.3}$$

where $P = 630$ lb, bearing load, and $L = $ desired life $= 600 \times 60 \times 24 \times 2 \times 365$ $= 631 \times 10^6$ revolutions for two years of continuous duty at 600 rpm. Thus, for ball bearings,

$$C = 630 \left(\frac{631 \times 10^6}{10^6}\right)^{1/3} = 5,400 \text{ lb}$$

and for roller bearings,

$$C = 630 \left(\frac{631 \times 10^6}{10^6}\right)^{0.3} = 4,350 \text{ lb}$$

Catalogue B.   Since the average life will be five times the 90 per cent life, the total revolutions for which the bearings will be selected on an average-life basis will be

$$5 \times 631 \times 10^6 = 3,160 \times 10^6 \text{ revolutions}$$

The rating life is

$$3,800 \times 60 \times 1,000 = 228 \times 10^6 \text{ revolutions}$$

To avoid confusion with the standard ratings based on 10 per cent failures, Eqs. (18-19) and (18-20) are rewritten to be, respectively,

$$P_R = P\left(\frac{L}{L_R}\right)^{1/3} \quad \text{and} \quad P_R = P\left(\frac{L}{L_R}\right)^{0.3}$$

where P_R = catalogue rated load and L_R = catalogue rating life.   Thus, for ball bearings,

$$P_R = 630\left(\frac{3,160 \times 10^6}{228 \times 10^6}\right)^{1/3} = 1,510 \text{ lb}$$

and, for roller bearings,

$$P_R = 630\left(\frac{3,160 \times 10^6}{228 \times 10^6}\right)^{0.3} = 1,390 \text{ lb}$$

Example 18-4.   Ball bearings are to be selected for an application in which the radial load is 400 lb during 90 per cent of the time and 1,600 lb during the remaining 10 per cent.   The shaft is to rotate at 150 rpm, and we are asked to determine the minimum value of the basic dynamic load rating to be used in selecting bearings to give 5,000 hr of operation with not more than 10 per cent failures.

Solution.   The appropriate equation (18-26) is

$$C = \left(\frac{\Sigma L_i P_i^3}{10^6}\right)^{1/3} = \left(\frac{L_1 P_1^3 + L_2 P_2^3}{10^6}\right)^{1/3}$$

where L_1 = 0.90 × 150 × 60 × 5,000 = 40.5 × 10^6 revolutions
$\quad\quad P_1$ = 400 lb
$\quad\quad L_2$ = 0.10 × 150 × 60 × 5,000 = 4.5 × 10^6 revolutions
$\quad\quad P_2$ = 1,600 lb

Thus,
$$C = \left(\frac{40.5 \times 10^6 \times 400^3 + 4.5 \times 10^6 \times 1,600^3}{10^6}\right)^{1/3}$$
$$= (2.59 \times 10^9 + 18.45 \times 10^9)^{1/3} = 2,760 \text{ lb}$$

18-23. Mounting and Lubrication of Rolling-contact Bearings.   The mounting of rolling-contact bearings should provide for all the restraint required without introducing any additional restraint, and it should permit disassembly without any unnecessary complications.   Figure 18-60 illustrates a typical mounting when two deep-groove ball bearings are used on a rotating shaft.   It should be noted that both inner races are positioned axially by positive means, while the outer race of only one bearing is positioned positively against axial motion and the other is relatively free to move to accommodate any differential expansion or contraction that might occur with temperature changes.

The fit between the inner race and the shaft is an interference fit that prevents relative rotation and corresponding wear between the race and

shaft.   The outer race may also be an interference fit in the housing l ut
to a lesser degree than for the inner race.   If the load is always in the
same direction, it is considered beneficial to permit the stationary race to
creep slowly so that the wear on the race will be more uniformly distributed.   The correct fit is difficult to achieve.   The Annular Bearing
Engineers' Committee (ABEC) has standardized the tolerances and

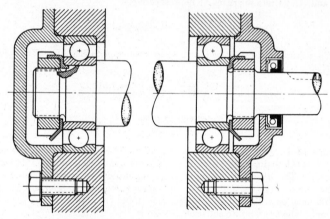

FIG. 18-60. Single-row deep-groove ball-bearing mounting.

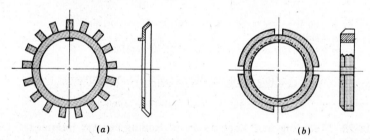

(a)   (b)

FIG. 18-61. (a) Lock washer; (b) lock nut.

allowances for bearing mounting, and complete tables are included in
most bearing catalogues.

The shoulder heights, both on the shaft and in the housing, are important.   The height must be sufficient so that adequate contact is made
between the races and shoulders for transmitting the thrust load, and,
at the same time, there must be enough of the bearing race exposed so
that the removing force can be applied to the race in question.   The
relatively large force involved in pressing a bearing off a shaft may easily
ruin the bearing if it must be transmitted through the balls from one race

to the other.   Tables of recommended shoulder heights are found in the catalogues.

The shaft assembly may be held together by a lock washer and lock nut made for this purpose.   The lock washer has a tongue that fits in a shallow keyway in the shaft and prevents the washer from turning.   The lock washer in Fig. 18-61a has a number of radial tabs, and the lock nut b has a number of wrench slots.   The numbers of tabs and slots are chosen so that the ratio is not factorable.   Thus, a fine adjustment is provided, and the tab that lines up with a slot is bent down to lock the nut against backing off in service.   In some cases, particularly with larger shafts, it is more convenient and economical to drill and tap the shaft so that a cap screw may hold an end plate against the bearing, as shown in Fig. 18-62.   If two or more cap screws are used, the heads may be drilled and safety wires used to prevent loosening of the screws.

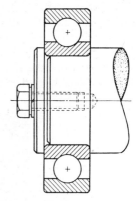

FIG. 18-62

Many small bearings, e.g., in low-power electric motors, are held on the shaft by the friction of the press fit.   Other bearings carrying relatively low thrust loads are held in place by snap rings, as in Fig. 8-27d.

The double-reduction worm-gear speed reducer in Fig. 18-63 illustrates two common types of rolling-contact-bearing mountings.   The large

FIG. 18-63. Ball- and tapered-roller-bearing mounting in a worm-gear speed reducer. (*Courtesy Cleveland Worm and Gear Company.*)

thrust loads from the worm are carried by the preloaded duplex bearings at the left end of the high- and intermediate-speed shafts. The deep-groove ball bearings at the right ends of the shafts are mounted with axial clearance for the outer races and carry only radial loads.

The output shaft is mounted on tapered roller bearings. The bearings are initially positioned with zero axial clearance, and consequently there is no opportunity for axial adjustment with temperature changes. This is not important when the bearing span is small, as in this case, because the tapered construction keeps the load from increasing too rapidly with the small axial interference due to differential expansion.

FIG. 18-64. Mounting for worm-gear shaft with extra-large overhung load.   (*Courtesy Cleveland Worm and Gear Company.*)

Figure 18-64 shows the type of mounting used when a worm-gear speed reducer is to be used with extra-large overhung loads. Under this condition, it is desirable to use cylindrical roller bearings to carry the radial and to use ball thrust bearings to carry the axial components of the load and the worm-gear tooth forces.

In many situations, similar to that in Fig. 18-60, where heavy radial loads combined with heavy thrust loads must be carried with a minimum of axial flexibility, it is convenient to use a double-row spherical roller bearing or a double-row tapered roller bearing on one end of the shaft to carry the thrust load and provide the axial rigidity and to use a cylindrical roller bearing on the other end of the shaft to carry the radial load, thus permitting thermal expansion and contraction of the shaft without adding to the bearing loads.

Rolling-contact bearings require lubrication to (1) prevent corrosion, (2) aid in excluding dirt, water, etc., (3) reduce friction and wear between the sliding parts of the bearing, and (4) dissipate heat.

The critical point is the sliding contact between the retainer and the balls or rollers.   Providing adequate lubrication in this region becomes increasingly more difficult with higher operating temperatures.

Oil is the preferred lubricant, and special attention must be given to ensuring a steady supply to the bearings and to sealing against leakage. When the operating speed is high, it may be necessary to use a circulating system so that a minimum amount of oil will be in contact with the rotating parts of the bearing and the heat may be dissipated in a heat exchanger.   For normal speeds of operation, it is satisfactory to have the lower part of the bearing run in the oil.   However, too much oil may be as bad as too little because of the rapid heating of the oil by the churning action of the rotating parts of the bearing.   It is recommended that the oil level be at the center of the lowest ball or roller.   It is often desirable to provide an inspection-drain plug, pet cocks (Figs. 18-63 and 18-64), or a sight glass with an inscribed line indicating the desired oil level (Fig. 18-26).

For operation at high speeds, when external cooling is not required, the oil level may be kept below the rotating parts with a slinger used to splash-lubricate the bearing.

Very-high-speed bearings are often lubricated by a mist of oil and air sprayed through jets directly onto the bearing.

The lubrication of high-speed vertical-axis bearings is doubly complicated by the problem of churning and oil retention.   The bearing manufacturers' catalogues contain numerous examples of satisfactory designs.

A seal must prevent the entrance of contaminants and must prevent the loss of lubricant.   Rubbing seals, as shown in Figs. 16-24, 17-30, 18-60, 18-63, and 18-64, are available as stock items and are widely used. The friction introduced by the seal may exceed that of the bearing itself and may cause rapid heating and deterioration of the seal at high speeds. If the seal must operate under a static head of oil or where there must be practically zero leakage, it is necessary to use a packing gland or a face seal, as for the fluid coupling in Fig. 11-15 and the high-speed worm shaft in Fig. 18-63.   Several types of nonrubbing labyrinth seals, with slingers, grooved shafts, or grooved housings, that are suitable for high-speed operation are shown in Fig. 18-65.

Grease is used for lubricating many rolling-contact bearings because it offers the simplest method of obtaining reliable lubrication.   An addi-

tional advantage is that grease is effective in preventing the entry of dirt
Grease is essentially a suspension of a metallic soap in a lubricating oil.
The consistency depends upon the type and amount of the constituents.
The greases most satisfactory for use with rolling-contact bearings are a

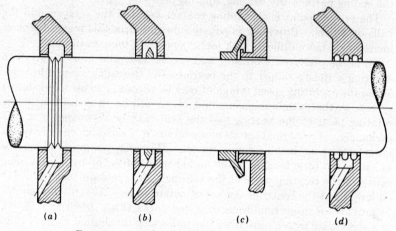

FIG. 18-65. Nonrubbing seals for high-speed service.

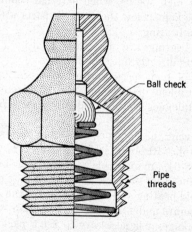

FIG. 18-66. Alemite lubrication fitting.   (*Courtesy Stewart-Warner Corporation.*)

combination of a lime-base soap and mineral oil and a soda-base soap and
mineral oil.

The lime-soap grease is limited to use between 32 and 115°F.   At
higher temperatures, the mineral oil separates permanently from the soap,

and a harmful residue is left. It is also nonemulsifying and does not afford the best protection against corrosion.

The soda-base grease can be used continuously between −5 and +160°F and for short periods up to 210°F. If the grease melts, it does not separate but will return to its original state upon cooling. The soda-base grease is comparatively stiff and consequently helps the sealing of the housing. It also can absorb some water and thus offers excellent protection against rust.

As with oil lubrication, it is important to avoid using too much lubricant. Generally a housing should not be over half full of grease. This permits the grease to "expand" and to channel so that working or churning is kept to a minimum. For proper lubrication, the grease and the oil in the grease gradually flow into the channel and contact the rotating parts.

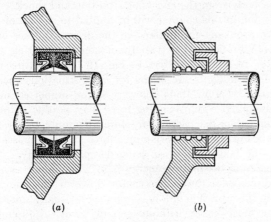

(a) (b)

Fig. 18-67. Double-purpose seals. (a) Rubbing; (b) nonrubbing.

Most grease-lubricated bearings require attention only once or twice a year. Large high-speed bearings require more frequent attention, and lubrication fittings, as shown in Fig. 18-66, are provided so that a grease gun can be used. It is important, in any case, to be sure that the housing is not overfilled with grease.

Sealing a grease-lubricated bearing is relatively simple because it is not necessary to return the lubricant to a sump or reservoir. In general, the methods illustrated in Figs. 18-60 and 18-65d will be satisfactory. In particularly dusty or damp environments, it may be desirable to use a double seal, as in Fig. 18-67a, where one lip is turned in to keep in the grease and the other is turned out to prevent the entrance of dirt and water. Figure 18-67b shows a nonrubbing seal with an external slinger for use at higher speeds than is permissible with rubbing seals.

CHAPTER 19

VIBRATIONS IN MACHINERY

Many books and papers have been written on the subject of vibrations. The books by Thomson,[1] Den Hartog,[2] and Timoshenko and Young[3] are excellent references on the general theory of vibrations, while the book by Crede[4] is of particular importance in the application of theory to the isolation of a body from the effects of vibration and shock.

The discussion in this chapter will be limited to several of the simpler commonplace problems encountered in the design and use of machines.

19-1. Free Vibration. The two fundamental vibrating systems are shown in Fig. 19-1. The system in Fig. 19-1a, consisting of a simple spring-supported mass, is the model for most situations encountered in machine design. Figure 19-1b shows a spring-supported mass accompanied by damping. The damping force may be due to internal hysteresis, to coulomb or dry friction, or, more commonly, to viscous friction, as indicated in Fig. 19-1. The damping force always opposes the motion of the vibrating body and in so doing dissipates energy. This fundamental property of dissipating energy is of great importance in certain cases, to be discussed in the next section. However, the amount of damping present in the vibration of a machine part is usually small and can be neglected in most cases without serious error. Consequently, the simplified discussion in this chapter will neglect damping except in relation to critical situations.

Figure 19-2a shows the system of Fig. 19-1a when the mass has been deflected a distance x from its equilibrium (at-rest) position. The free-body diagram of the forces acting on the mass when it is released is shown in Fig. 19-2b.[5] The spring and inertia forces must be equal in magnitude

[1] W. T. Thomson, "Mechanical Vibrations," 2d ed., Prentice-Hall, Inc., Englewood Cliffs, N.J., 1953.

[2] J. P. Den Hartog, "Mechanical Vibrations," 4th ed., McGraw-Hill Book Company, Inc., New York, 1956.

[3] S. Timoshenko and D. H. Young, "Vibration Problems in Engineering," 2d ed., D. Van Nostrand Company, Inc., Princeton, N.J., 1954.

[4] C. E. Crede, "Vibration and Shock Isolation," John Wiley & Sons, Inc., New York, 1951.

[5] Since the deflection is measured from the equilibrium position, the gravity force and its equal and opposite spring force need not be included.

and opposite in direction and

$$\Sigma F_x = 0$$
$$m\ddot{x} + kx = 0 \tag{19-1}$$

where m = mass of body, lb-sec²/in.

$\ddot{x} = d^2x/dt^2$, acceleration in x direction, in./sec²

k = spring rate, lb/in.

x = displacement, in.

The solution of Eq. (19-1), as given in any calculus or vibrations book, is

$$x = A \sin \omega_n t + B \cos \omega_n t \tag{19-2}$$

where A and B = const depending upon initial conditions and $\omega_n = \sqrt{k/m}$, natural frequency of free vibration, radians/sec. The most important

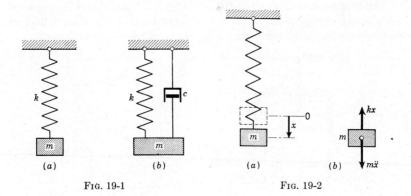

FIG. 19-1   FIG. 19-2

point here is that the *natural frequency of vibration* ω_n is a function of the mass and the spring rate only. Thus, the equation

$$\omega_n = \sqrt{\frac{k}{m}} \qquad \text{radians/sec} \tag{19-3}$$

or its equivalent
$$f_n = \frac{1}{2\pi} \sqrt{\frac{k}{m}} \qquad \text{cps} \tag{19-4}$$

merits some additional study. By definition,

$$m = \frac{W}{g} \qquad \text{lb-sec}^2/\text{in.} \tag{19-5}$$

where W = weight, lb, g = acceleration of gravity, in./sec², and

$$k = \frac{W}{\delta_{st}} \qquad \text{lb/in.} \tag{19-6}$$

where δ_{st} = static deflection of the spring, in. Substituting Eqs. (19-5) and (19-6) into (19-3) and (19-4) results in

$$\omega_n = \sqrt{\frac{g}{\delta_{st}}} \qquad \text{radians/sec} \qquad (19\text{-}7)$$

and

$$f_n = \frac{1}{2\pi}\sqrt{\frac{g}{\delta_{st}}} \qquad \text{cps} \qquad (19\text{-}8)$$

Equation (19-8) can be put into a more convenient form by substituting 386 in./sec² for g, which gives

$$f_n = 3.13 \sqrt{\frac{1}{\delta_{st}}} \qquad \text{cps} \qquad (19\text{-}9)$$

or

$$f_n = 188 \sqrt{\frac{1}{\delta_{st}}} \qquad \text{cpm} \qquad (19\text{-}10)$$

The relationship in Eq. (19-10) is particularly useful and is expressed graphically in Fig. 19-3. It should be noted that a large static deflection is necessary if a suspension with a low natural frequency is desired.

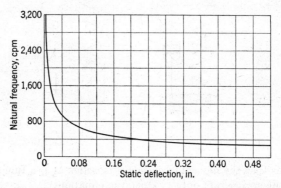

FIG. 19-3. Relationship of natural frequency to static deflection.

19-2. Forced Vibration and Transmissibility. If an unbalanced rotating mass or a mass reciprocating with simple harmonic motion is mounted on the spring-supported mass, the system in Fig. 19-1a becomes that in Fig. 19-4. The component of the disturbing force in the x direction is

$$F = F_0 \sin \omega t \qquad (19\text{-}11)$$

where F_0 = maximum force, lb, and ω = frequency of disturbing force, radians/sec. The differential equation becomes

$$m\ddot{x} + kx = F_0 \sin \omega t \qquad (19\text{-}12)$$

The general solution of Eq. (19-12) is

$$x = A \sin \omega_n t + B \cos \omega_n t + \frac{F_0}{k - m\omega^2} \sin \omega t \qquad (19\text{-}13)$$

The first two terms make up the complementary solution, and the last term is the particular solution.   The complementary solution is the same as Eq. (19-2) for the undamped free vibration.   Since it is impossible to have a vibrating system without damping, the free vibration, or transient, will soon die out, leaving only the particular solution.   Thus, for steady-state conditions after the free vibration has damped out,

$$x = \frac{F_0}{k - m\omega^2} \sin \omega t \tag{19-14}$$

After rearranging and substituting $\omega_n{}^2$ for k/m, Eq. (19-14) becomes

$$x = \frac{F_0/k}{1 - (\omega/\omega_n)^2} \sin \omega t \tag{19-15}$$

As indicated in Eq. (19-15), the system must vibrate at the frequency

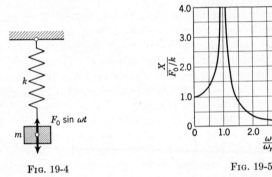

FIG. 19-4   FIG. 19-5

of the forcing function, and the maximum value of x is

$$X = \frac{F_0/k}{1 - (\omega/\omega_n)^2} \tag{19-16}$$

The numerator F_0/k is the deflection that would be given by F_0 if it were a static force.   Therefore, Eq. (19-16) may be written in non-dimensional form as

$$\frac{X}{F_0/k} = \frac{1}{1 - (\omega/\omega_n)^2} \tag{19-17}$$

It can be seen that the ratio is positive for all values of $\omega < \omega_n$ and negative for all values of $\omega > \omega_n$.   The significance of the sign is that, when it is positive, the mass is moving in the same direction (in phase) as the force and, when it is negative, the mass is moving in the direction opposite (out of phase) to the force.   Since the mass is moving with simple harmonic motion, the phase relationship is ordinarily inconsequential, and the relation expressed in Eq. (19-17) is usually shown, as in Fig. 19-5, in terms of magnitude only. Thus, Eq. (19-17) is more usefully written as

$$\frac{X}{F_0/k} = \frac{1}{|1 - (\omega/\omega_n)^2|} \tag{19-18}$$

where the vertical lines mean "the absolute value of."[1]

The important points illustrated by the curve in Fig. 19-5 are (1) when $\omega = \omega_n$, the vibration amplitude becomes infinite; (2) when $\omega \gg \omega_n$, the motion of the mass becomes very small; and (3) when $\omega < \omega_n$, the motion of the mass is magnified in comparison with the deflection of the spring due to the force alone.

Obviously, because of the physical limitations of the parts, an infinite amplitude cannot be reached. However, the amplitude can easily become so large that the noise and vibration transmitted through the air and floor may be objectionable to personnel, and it is possible that the machine may be damaged.

The feature that permits a machine to run through a critical speed to reach the smooth-operating region where $\omega \gg \omega_n$ is that energy must be supplied to the system in order to increase the deflection of the spring and to replace that dissipated by damping. Thus, if the machine accelerates to its operating speed in a short period of time, there will not be enough time or cycles of vibration in the critical region for the amplitude to build up to a destructive level.

The effect of damping on the system is shown in Fig. 19-6. The amount of damping is shown as a ratio of the actual damping present to the critical damping, c/c_c. *Critical damping* is that value of damping which permits the quickest free return of a deflected body to its equilibrium position without overshooting. This type of motion is often called "dead-beat" motion. The concept of critical damping is important in the study of instruments, servomechanisms, artillery recoil systems, automobile suspensions, and many similar applications where it is desired to have the effect of a sudden change in position die out in the shortest possible time.

As shown in Fig. 19-6, the presence of a relatively small amount of damping has an appreciable effect toward decreasing the magnitude of the amplitude of vibration in the region near $\omega/\omega_n = 1$. It can also be noted that the presence of damping results in the maximum amplitude occurring at a lower frequency than for the undamped system.

---

[1] To avoid confusion with the absolute-value term, Eq. (19-18) can be considered as two equations:

$$\frac{X}{F_0/k} = \frac{1}{1 - (\omega/\omega_n)^2} \qquad \text{when } \frac{\omega}{\omega_n} < 1$$

and

$$\frac{X}{F_0/k} = \frac{1}{(\omega/\omega_n)^2 - 1} \qquad \text{when } \frac{\omega}{\omega_n} > 1$$

The preceding discussion has been concerned almost entirely with the effect of the forcing function on the motion of the spring-supported mass. When it is desired to consider the effect of the forcing function and vibration of the mass on the floor, etc., the problem becomes that of determining the force transmitted from the machine to the foundation.   If no

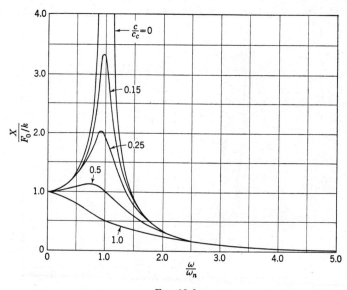

FIG. 19-6

damping is present, the entire transmitted force is that due to the deflection of the spring.   Thus, the maximum value of transmitted force is

$$F_{TR} = kX \qquad \text{lb} \tag{19-19}$$

The ratio of the transmitted force to the disturbing force is called *transmissibility* TR.   From Eqs. (19-18) and (19-19),

$$TR = \frac{F_{TR}}{F_0} = \frac{1}{|1 - (\omega/\omega_n)^2|} \tag{19-20}$$

A comparison of Eqs. (19-20) and (19-18) shows that, for zero damping, the right-hand terms are identical.   However, the complete equations, including damping, are not the same.   Figure 19-7 shows the effect of frequency ratio and damping on transmissibility.   The most important observation here is that damping is beneficial for values of $\omega/\omega_n < \sqrt{2}$ and detrimental for values of $\omega/\omega_n > \sqrt{2}$.   The reason that damping increases the transmissibility when $\omega/\omega_n > \sqrt{2}$ is that part of the transmitted force is due to the viscous friction, which increases with velocity. Therefore, a spring suspension used to isolate a machine from its surroundings should have negligible damping and should have a natural frequency in the order of one-quarter or less of the frequency of the disturbing force. If the machine accelerates slowly up to its operating speed, it may be necessary to provide buffers or enough damping to limit the amplitude at resonance to a safe value.

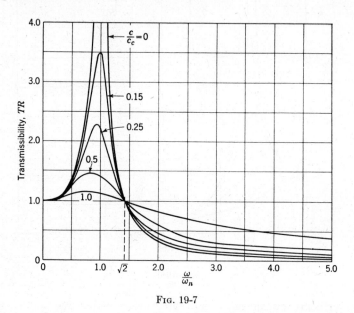

FIG. 19-7

Another class of problems includes the isolation of an automobile body from the effects of bumps in the road surface and of isolating electronic and other relatively fragile equipment from any undesirable motion of the supporting structure. In Fig. 19-8 the support has a motion of $y = Y \sin \omega t$. The force due to the difference between the motion of one end of the spring and the other acts directly on the mass. The force equation is

$$m\ddot{x} = -k(x - y) \qquad (19\text{-}21)$$

Hence,
$$m\ddot{x} + kx = kY \sin \omega t \qquad (19\text{-}22)$$

After integrating and letting the transient free vibration die out, the resulting equation of motion is

$$x = \frac{Y}{1 - (\omega/\omega_n)^2} \sin \omega t \qquad (19\text{-}23)$$

Solving Eq. (19-23) for the ratio of maximum values gives

$$\frac{X}{Y} = TR = \frac{1}{|1 - (\omega/\omega_n)^2|} \qquad (19\text{-}24)$$

Undamped motion transmissibility is seen to be the same as undamped force transmissibility. While not readily apparent, it can also be shown that for the case of damped forced vibration the curves in Fig. 19-7 apply equally well to both motion and force transmissibility.

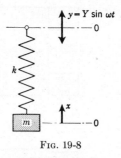

FIG. 19-8

Additional problems arise because of the necessity of mounting the machine or equipment on more than one spring. The usual case is that shown in Fig. 19-9, where four springs are used to support the machine. This example now becomes a vibration problem with *six degrees of freedom*, in that six numbers or coordinates, for example, x, y, z, ϕ, θ, and ψ, are required to specify the position of the body in space. The resulting vibration is the sum of the three translations and the three rotations. If, as in Fig. 19-9, the four springs are identical and the center of gravity

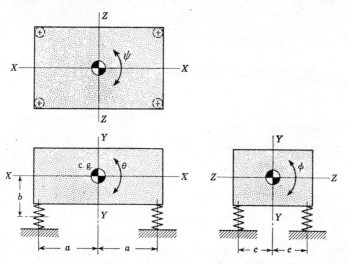

FIG. 19-9. Suspension system with two planes of symmetry.

is midway between the springs but at a distance b above the center of the spring, the spring forces due to a vertical deflection y are symmetrical with respect to the center of gravity, and the system is considered to have two planes of symmetry. For this case, a vertical force through the center of gravity will cause only a vertical motion y, and a couple about the Y-Y axis will cause only a rotation ψ about the Y-Y axis. However, a horizontal force through the center of gravity will cause both a translation (x or z) and a rotation (θ or ϕ). Since the inertia force acts through the center of gravity, the translatory and oscillatory *modes* of vibration

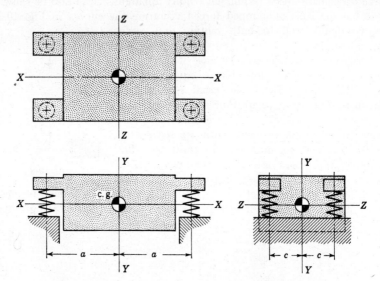

FIG. 19-10. Suspension system with three planes of symmetry.

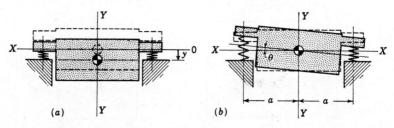

FIG. 19-11

will exist simultaneously, and the modes are said to be *coupled*. Since the vertical translation and the oscillation about the Y-Y axis can exist independently, these two modes are said to be *decoupled*.

If the distance b is made equal to zero, three planes of symmetry will exist, and all modes are decoupled. This case is shown in Fig. 19-10. Usually the two most important modes of vibration are the vertical translation y and the oscillation about the Z-Z axis, as shown in Fig. 19-11.

The force equation for the vertical motion is

$$\Sigma F_y = 0$$

$$m\ddot{y} + 4ky = 0 \tag{19-25}$$

and

$$\omega_{n,y} = \sqrt{\frac{4k}{m}} \quad \text{radians/sec} \tag{19-26}$$

where k = spring rate of each isolator, lb/in.   The torque equation for small oscillations about the Z-Z axis is

$$\Sigma T_{z\text{-}z} = 0$$

$$I_{z\text{-}z}\ddot{\theta} + 4ka^2\theta = 0 \tag{19-27}$$

and

$$\omega_{n,z\text{-}z} = \sqrt{\frac{4ka^2}{I_{z\text{-}z}}} \quad \text{radians/sec} \tag{19-28}$$

where $I_{z\text{-}z}$ = moment of inertia about the Z-Z axis, lb-in.-sec^2.

In many situations, the distance b is small compared with the distance a, and an approximate solution made by considering b equal to zero will give a usable answer.

In the general case of a forced vibration of a spring-mounted machine, the disturbing force will have different components in the X and Y directions, and neither will act through the center of gravity of the body. The problem becomes somewhat more complicated, and the reader is referred to the previously mentioned book by Crede for a more thorough and rigorous discussion of this problem.

19-3. Types of Vibration and Shock Mounts.   One of the most practical vibration or shock mounts is the simple metallic helical compression or tension spring.   Its outstanding features are high strength, very low internal damping, a constant spring rate (unless otherwise desired), and relative freedom from effects of changes in temperature, humidity, etc. When additional damping is required, it is usually desirable to use commercially available units, such as shown in Fig. 19-12.   Both mounts have nonlinear spring rates with stiffness increasing with deflection so that the natural frequency of the system will be constant for all loads within the application range of the mount; that is, $\sqrt{k/m}$ = const.   They provide for snubbing when the force or deflection becomes greater than can be carried by the springs without bottoming, and they utilize dry friction to give maximum stability in damping effectiveness over a wide range of operating conditions.

Rubber is one of the most effective materials for use in vibration and shock isolation because of its elasticity and inherent damping properties. Natural rubber has the least set, or creep, with time under a static load and is superior to most synthetic rubbers at low temperature because its stiffness has not significantly increased at $-50°$F.   Neoprene and buna

N can be used at temperatures up to 250°F; have much better resistance to oil, gasoline, etc.; and are less affected by oxidation and sunlight than is natural rubber.

Silicone rubbers have relatively low strength and a high drift rate but can be used at temperatures between −70 and +400°F.

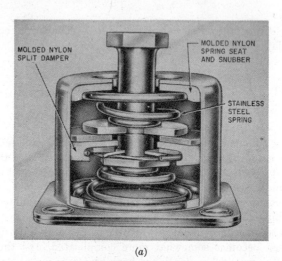

(a)

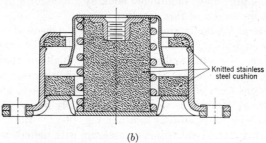

(b)

Fig. 19-12. Commercial vibration isolators and shock mounts.   (a) All-Angl mount. (*Courtesy Barry Controls Incorporated.*)   (b) Met-L-Flex unit mount. (*Courtesy Robinson Vibrashock Division, Robinson Technical Products, Inc.*)

Rubber may be loaded in compression, tension, shear, or bending. However, rubber is seldom used in tension because of the problem of obtaining satisfactory end connections and the sensitiveness of rubber to surface scratches when subject to a high tensile stress.

Rubber is practically incompressible, and a compression deflection in one direction will be accompanied by a considerable expansion in the direction perpendicular to the compression.   Thus, when a rubber pad bounded to two parallel plates is compressed from an initial free position, as indicated by the dotted lines in Fig. 19-13, the final shape is as shown in solid lines.   This lateral constraint results in a stiffer member than if the entire height had been permitted to expand.   The stiffness of a rubber pad increases rapidly with an increasing ratio of area under compression to the lateral area for expansion.   Consequently, there is little to be gained from using large sheets of solid rubber to isolate vibrations.

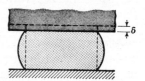

One way in which sheets of rubber can be used efficiently for vibration isolation is illustrated in Fig. 19-14, which shows an Isomode pad under the foot of a machine.   The grooves permit the necessary lateral expansion with a minimum restraint.   A single-thickness pad (as shown) is normally used with a static deflection of $\frac{1}{16}$ in., corresponding to a loading of 50 psi on the pad.

Fig. 19-13.   Rubber pad under compression loading.

Another problem of using rubber in compression is that it is relatively stiff unless very thick sheets or pads are used.   Thus, rubber pads are not

FIG. 19-14. Isomode-pad application.   (*Courtesy MB Electronics, Division of Textron Electronics, Inc.*)

very useful in isolating low-frequency vibrations, though they are quite effective for isolating the higher-frequency vibrations that create objectionable sound or noise.

When rubber is loaded in shear, the problem of lateral expansion no longer exists.   The shear modulus of elasticity is also less than that for compression.   These two factors combine to give greater flexibility and more efficient use of material.   Consequently, most commercial vibration or shock mounts use rubber under shear stress.   Several typical kinds of construction of shear mounts are shown in Fig. 19-15.   Manufacturers' catalogues should be consulted for details, but these mounts may be readily used to isolate frequencies above 1,000 cpm and in special cases at somewhat lower frequencies.

One particular application of rubber in shear is that illustrated in Fig. 19-16, where the motor hubs on an electric motor are mounted in rubber bushings.   The rubber resists any lateral motion because of its stiffness in compression but has relatively high torsional flexibility.   This is especially useful in providing quieter operation by isolating the torsional

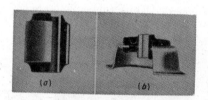

FIG. 19-15. Commercial vibration isolators using rubber in shear.   (a) Tube form mounting; (b) plate form mounting; and (c) Chan-L-Mount.   (*Courtesy Lord Manufacturing Company.*)

vibration due to the 120 torque impulses per second on the rotor of a single-phase 60-cycle a-c motor.

When a system, such as shown in Fig. 19-10, is subjected to disturbances in all planes, it is usually advantageous to use mounts with equal spring rates in all directions.   The All-Angl mount in Fig. 19-12a and the Isomode units in Fig. 19-17 are examples of typical commercial isolators with approximately equal spring rates in all directions.

In some applications where metallic springs are used to isolate a major low-frequency disturbance, a high-frequency noise is transmitted along the wire of the spring.   In this situation, it is often necessary to use a pad of rubber between the spring and the machine or frame to isolate the high-frequency component of the vibration.

Sponge rubber is a cellular rubber containing many minute pockets of air or gas.   The pad is considerably more flexible than if the rubber were solid.   Sponge rubber may be obtained in a wide range of stiffnesses and is successfully used in slab form in many applications.

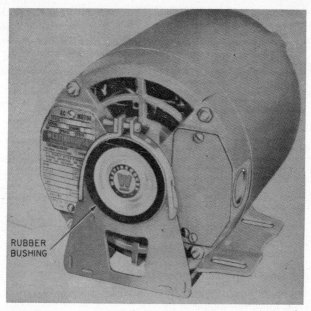

Fig. 19-16. Rubber-mounted single-phase a-c motor.   (*Courtesy Westinghouse Electric Corporation.*)

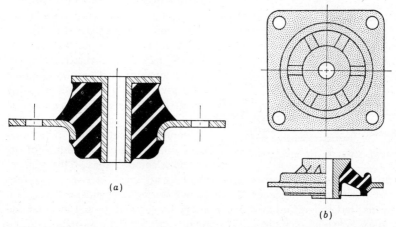

(*a*)

(*b*)

Fig. 19-17. Vibration isolators with equal spring rates in all directions.   (*a*) Isomode unit, type 5; (*b*) plate Isomode unit.   (*Courtesy MB Electronics, Division of Textron Electronics, Inc.*)

The vibration of large machines is often isolated from the building by use of cork slabs.   Simple pads, protected from oil, water, chemicals, etc., may be used under the feet of a machine, or the entire foundation of the machine, as shown in Fig. 19-18, may rest on cork.   Since cork is relatively stiff, it is useful principally for sound isolation.   Machines, such

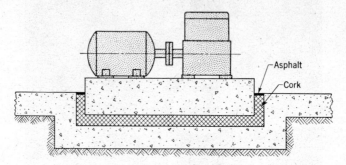

FIG. 19-18. Vibration isolation by use of cork slabs.

as elevator drives, blowers, compressors, and motor-generator sets, commonly used in office buildings and hotels where quiet operation is a major requirement, are often mounted on cork.

Felt has properties that are somewhat similar to those of sponge rubber and cork and is used in similar applications.

When vibrations with very low frequencies, i.e., in the order of several hundred cycles per minute or less, must be isolated, large static deflections are required.   It is difficult to provide a practical metallic spring suspension with adequate stability and almost impossible to use rubber, cork, felt, etc.   A relatively recent development that can be designed to give almost any desired spring rate is the *air spring* shown in Fig. 19-19.   The natural frequency is found[1] to be a function of the area of the piston and volume of air under compression.   This type of spring has been applied to motor buses with great success.   The fact that the natural frequency of the suspension is independent of the load means that the optimum riding comfort will be maintained whether the vehicle is loaded or empty.   The air pressure must be varied according to the load.   An additional feature is that damping can be introduced

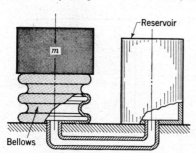

FIG. 19-19. Air spring.

[1] Thomson, *op. cit.*, pp. 80–81.

and controlled by installing an orifice or throttling valve in the line between the cylinder and reservoir.

Example 19-1. A light weight (33-lb) portable air compressor with a built-in direct-drive motor has been on the market for a short time. The compressor was designed for the light-duty service of spray painting in home workshops, commercial sign and photography studios, and in classrooms. The units have been supplied with four rubber buttons (feet) to prevent scratching of the table or work surface and to minimize vibration and noise due mainly to the inertia forces from the reciprocating parts in the compressor. Evidently the latter function of the rubber buttons has not been adequately performed because a number of complaints about noisy operation have been turned in by people using the compressors in commercial studios and in classrooms.

The motor is a $\frac{1}{4}$-hp 1,750-rpm 115-V single-phase a-c electric motor.

The proposed remedy is to redesign the base to incorporate vibration isolators. We are asked to specify the spring rate and working deflection for isolators that will permit only 5 per cent of the disturbing forces to be transmitted to the table.

Solution. The suspension will be designed so that each of the four identical isolators will carry one-quarter of the 33-lb weight, or 8.25 lb. The frequency for the disturbance is that corresponding to the motor speed, or 1,750 rpm.

The compressor will accelerate to running speed in a fraction of a second. Therefore, the natural frequency will be passed so quickly that there will be little opportunity for the vibration amplitude to build up to a harmful level, even without damping. Therefore, since zero damping results in a minimum transmissibility, when $TR < 1$ and $\omega/\omega_n > \sqrt{2}$, we shall want to have as little damping as possible and will probably use metallic springs as isolators.

To determine the natural frequency of the system for a transmissibility of 5 per cent, we shall use Eq. (19-20):

$$TR = \frac{1}{|1 - (\omega/\omega_n)^2|} = \frac{1}{|1 - (f/f_n)^2|}$$

and $TR = 0.05$ and $f = 1,750$ rpm. Thus

$$0.05 = \frac{1}{|1 - (1,750/f_n)^2|}$$

from which $f_n = 382$ cpm. We can now use Eq. (19-10) to determine the static deflection:

$$f_n = 188 \sqrt{\frac{1}{\delta_{st}}} \quad \text{cpm}$$

or

$$382 = 188 \sqrt{\frac{1}{\delta_{st}}}$$

from which $\delta_{st} = 0.242$ in. (see also Fig. 19-3). The spring rate is

$$k = \frac{W}{\delta_{st}} = \frac{8.25}{0.242} = 34.1 \text{ lb/in.}$$

Closure. It should be noted that the required static deflection of 0.242 in. is too large to be achieved conveniently by any means other than metallic springs. However, it is important to realize that, in an actual situation, an experimental investigation should be made to determine what level of transmissibility could really be tolerated. If the compressor is well balanced so that the shaking forces are relatively small, satisfactory results may be obtained with transmissibilities of 20 per cent or more. In this example, 20 per cent transmissibility would require a static deflection of only 0.069 in., which is well within the practical range for isolators using rubber elastic members.

19-4. Torsional Vibrations of Shafts.

Practically all machines have one or more shafts. The shaft is essentially a torsion spring, and the shaft and the masses mounted on a shaft make up a vibrating system. The basic torsional system is shown in Fig. 19-20a, where one end of the

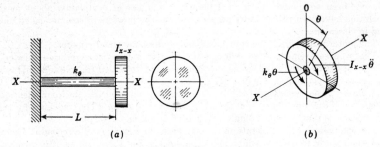

(a)   (b)

FIG. 19-20. Torsional vibration.

shaft is fixed and the other end carries a mass. If the disk is given an angular deflection and is released, the torques acting on the disk are as shown in Fig. 19-20b, and the equilibrium equation is

$$I_{x\text{-}x}\ddot{\theta} + k_\theta\theta = 0 \qquad (19\text{-}29)$$

where $I_{x\text{-}x}$ = moment of inertia of the disk about the X-X axis, lb-in.-sec² and k_θ = angular spring rate T/θ, lb-in./radian.

Comparing Eq. (19-29) with Eq. (19-1) will show that the two are similar. In fact, all the equations and discussion in the previous sections are directly applicable to angular vibrations after each lineal term is replaced by the corresponding angular term. Therefore,

$$\omega_n = \sqrt{\frac{k_\theta}{I_{x\text{-}x}}} \qquad \text{radians/sec} \qquad (19\text{-}30)$$

The simplest practical problem is that shown in Fig. 19-21a, where the shaft with a disk on each end is free to rotate. It can be seen that, if there is to be a vibration, one plane section of the shaft will remain stationary and the disks must oscillate out of phase; i.e., when disk 2 is rotating in a clockwise direction, disk 3 must be rotating in a counterclockwise direction. The natural frequency can be calculated in several ways. The simplest approach is to consider the system as two simple systems (Fig. 19-20) where the shaft is divided into two parts at the plane of zero motion or *node*, as indicated in Fig. 19-21b. The condition that must be fulfilled is that the two simple systems have the same natural frequency. Therefore,

$$\omega_n = \sqrt{\frac{k_{\theta,2}}{I_2}} = \sqrt{\frac{k_{\theta,3}}{I_3}} \qquad \text{radians/sec} \qquad (19\text{-}31)$$

For a solid shaft with a uniform diameter,

$$k_\theta = \frac{T}{\theta} = \frac{JG}{L} \qquad \text{lb-in./radian} \qquad (19\text{-}32)$$

where J = polar moment of inertia of shaft cross section, in.[4]

G = shear modulus of elasticity, psi (about 11,500,000 psi for steel)

L = length, in.

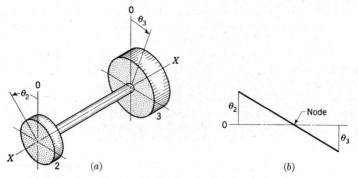

FIG. 19-21. Torsional vibration.

Substituting from Eq. (19-32) for k_θ in (19-31) gives

$$\omega_n = \sqrt{\frac{JG}{L_2 I_2}} = \sqrt{\frac{JG}{L_3 I_3}} \qquad \text{radians/sec} \qquad (19\text{-}33)$$

Thus $\qquad\qquad\qquad\qquad L_2 I_2 = L_3 I_3 \qquad\qquad\qquad\qquad\qquad (19\text{-}34)$

Solving Eq. (19-34) for L_2 or L_3 in terms of L and substituting into (19-33) gives

$$\omega_n = \sqrt{\frac{JG}{L} \frac{I_2 + I_3}{I_2 I_3}} = \sqrt{k_\theta \frac{I_2 + I_3}{I_2 I_3}} \qquad \text{radians/sec} \qquad (19\text{-}35)$$

It should be noted that, if I_2 is infinite, disk 2 does not move and Eq. (19-35) becomes the same as Eq. (19-30).

When more than two masses or a gear train with a speed reduction or increase are involved, the problem becomes somewhat more complicated and the reader should refer to one of the books mentioned at the beginning of this chapter.

19-5. Shaft Whirl. Rotating shafts must carry the functional load and an additional load due to the fact that the center of gravity of the system does not coincide with the center of rotation. The magnitude of the additional force is

$$F = me\omega^2 \quad \text{lb} \tag{19-36}$$

where e = eccentricity of the center of gravity measured from the axis of rotation, in.

This force is known as an *unbalanced force*, and, because it rotates with the shaft, it is the disturbing force of a forced vibration. Since the magnitude of the force increases with the square of the speed, it becomes necessary to balance accurately parts which must rotate at high speeds. As a matter of economics, it is not practical to balance any part more accurately than required for a given set of operating conditions. In the last analysis, it is impossible to achieve perfect balance, except by chance, and consequently there will be a rotating unbalanced force whenever there is a rotating mass. In many cases, the force is so small that its effect on the bearings and the vibration of the machine is nominal and easily isolated. However, under certain conditions, the deflection of the shaft and the forces involved may be so high as to impair the usefulness of the machine. The term used to describe this phenomenon is *shaft whirl*.

Figure 19-22 shows the forces acting on a mass that is mounted eccentrically on a rotating shaft. The rotor is assumed to be located at the

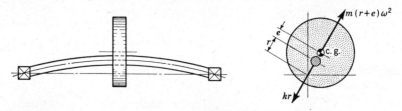

FIG. 19-22. Shaft whirl.

point of zero slope (to eliminate gyroscopic effects), and the mass of the shaft is considered to be small compared with that of the rotor or disk and can, therefore, be neglected without introducing serious error. Summing up the radial forces gives

$$kr - m(r + e)\omega^2 = 0 \tag{19-37}$$

where k = spring constant of the shaft, lb-in., and r = deflection of the shaft from the at-rest equilibrium position, in. The spring constant for a uniform shaft can be calculated by using the appropriate equation in Appendix A and for a nonuniform shaft by the graphical-numerical procedure in Sec. 6-9.

Solving Eq. (19-37) for r gives

$$r = e\,\frac{\omega^2}{k/m - \omega^2}\qquad \text{in.}\qquad (19\text{-}38)$$

Noting that the natural frequency of the shaft for lateral vibrations is

$$\omega_n = \sqrt{\frac{k}{m}}\qquad \text{radians/sec}$$

Eq. (19-38) may be written as

$$r = e\,\frac{\omega^2}{\omega_n{}^2 - \omega^2} = e\,\frac{(\omega/\omega_n)^2}{1 - (\omega/\omega_n)^2}\qquad \text{in.}\qquad (19\text{-}39)$$

The relationship expressed by Eq. (19-39) is shown in Fig. 19-23. It should be noted that, when $\omega/\omega_n < 1$, r and e are in the same direction and that, when $\omega/\omega_n > 1$, r and e are out of phase. The significance of these phase relationships is that (1) below the critical frequency the unbalanced force is greater than that due only to the eccentricity of the mass and (2), when ω/ω_n is very large, the unbalanced force approaches zero. In the latter case, $r \cong -e$, and the shaft assumes a deflected position such that the mass rotates about its gravity axis, not the shaft axis.

This concept is very useful at high rotative speeds and has been applied to small high-speed (10,000-rpm) steam turbines, cream separators, centrifuges, supercharger drives, etc.,

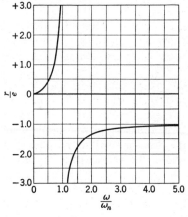

Fig. 19-23

where a large rigid shaft and very accurate balance would otherwise be required to minimize the disturbing force when operating below the critical speed.

When more than one mass is mounted on the shaft, the number of critical speeds is determined by the number of masses. The most general treatment of this case is beyond the scope of this book, and the reader is again referred to the literature for more information. However, as pointed out earlier in this section, the natural frequencies of whirling and lateral vibrations are identical, and thus, the discussion in Sec. 19-6 can be used to find the lowest natural frequency of whirl for any system consisting of one or more masses mounted on a shaft.

19-6. Lateral Vibrations with Lumped Masses. The principle of conservation of energy is one of the most powerful tools available for the solution of vibration problems involving complex systems. This is due to the fact that mechanical elements normally have very little damping and the error introduced in natural frequency calculations by considering damping as zero is negligible for most practical purposes.

We may consider that, with no energy being dissipated, the sum of the kinetic and potential energies must be a constant or, more usefully in this case, that the maximum value of kinetic energy in a cycle must equal the maximum value of potential energy.

The procedure presented here is known as the Rayleigh method and is based upon the observation that for a system such as shown in Fig. 19-24,

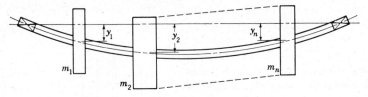

FIG. 19-24. Shaft or beam with many masses.

the lowest natural frequency will occur when all points reach their positions of maximum deflection simultaneously.

Thus, if we assume that the motion is simple harmonic, as was found to be the case for the simple mass-spring system in Sec. 19-1 and will be the case for small amplitudes of vibration of a shaft or beam, the deflection of the several masses will be

$$y_1 = Y_1 \sin \omega t$$
$$y_2 = Y_2 \sin \omega t$$
$$\cdot \ \cdot \ \cdot \ \cdot \ \cdot \ \cdot \ \cdot \ \cdot$$
$$y_n = Y_n \sin \omega t \tag{19-40}$$

The kinetic energy is, by definition,

$$\text{KE} = \tfrac{1}{2} m V^2$$

and in terms of the maximum values of $V = dy/dt$

$$\begin{aligned} \text{KE}_{\max} &= \tfrac{1}{2} m_1 (Y_1 \omega)^2 + \tfrac{1}{2} m_2 (Y_2 \omega)^2 + \cdots + \tfrac{1}{2} m_n (Y_n \omega)^2 \\ &= \tfrac{1}{2} \omega^2 \sum_{i=1 \text{ to } n} m_i Y_i^2 \end{aligned} \tag{19-41}$$

Since there are no losses, the maximum potential energy in the system will be equal to the work done in moving the masses to their positions of maximum deflection. Thus, if the shaft is considered as a linear spring and the forces acting when the masses are at their maximum deflection positions are F_1, F_2, . . . , F_n, the maximum potential energy becomes

$$\text{PE}_{\text{max}} = \tfrac{1}{2}F_1Y_1 + \tfrac{1}{2}F_2Y_2 + \cdots + \tfrac{1}{2}F_nY_n$$
$$= \tfrac{1}{2}\sum_{i=1 \text{ to } n} F_iY_i \qquad (19\text{-}42)$$

The problem now is that we know neither the values of Y_i nor the values of F_i, and we must assume a deflection curve before continuing. Rayleigh proposed assuming that the deflection curve during the vibration is similar to the static deflection curve. That is,

$$Y_i = C\delta_{st,i} \qquad (19\text{-}43)$$

where $C =$ a constant of proportionality. Therefore,

$$\frac{F_i}{Y_i} = \frac{W_i}{\delta_{st,i}} \qquad (19\text{-}44)$$

Rearranging Eq. (19-44) gives

$$F_i = \frac{W_i}{\delta_{st,i}} Y_i \qquad (19\text{-}45)$$

Using the relations in Eqs. (19-43) and (19-45), we can write Eqs. (19-41) and (19-42) as, respectively,

$$\text{KE}_{\text{max}} = \frac{1}{2} \omega^2 \Sigma m_i (C\delta_{st,i})^2 = \frac{C^2\omega^2}{2g} \Sigma W_i\delta_{st,i}^2 \qquad (19\text{-}46)$$

and
$$\text{PE}_{\text{max}} = \frac{C^2}{2} \Sigma W_i\delta_{st,i} \qquad (19\text{-}47)$$

Setting Eq. (19-46) equal to Eq. (19-47) and solving for ω gives

$$\omega = \sqrt{g \frac{\Sigma W_i\delta_{st,i}}{\Sigma W_i\delta_{st,i}^2}} \qquad \text{radians/sec} \qquad (19\text{-}48)$$

Substituting 386 in./sec² for g in Eq. (19-48) and dividing by 2π gives

$$f_n = 3.13 \sqrt{\frac{\Sigma W_i\delta_{st,i}}{\Sigma W_i\delta_{st,i}^2}} \qquad \text{cps} \qquad (19\text{-}49)$$

Equations (19-48) and (19-49) are not exact, because the actual deflection curve is not exactly similar to the static deflection curve as assumed. However, the calculated frequency will be within several per cent of the correct value, always on the high side,[1] and the effect of the incorrect deflection curve is probably minor in relation to that of unavoidable errors in other assumptions.

In most practical problems, the shaft is stepped and the graphical-numerical method in Sec. 6-9 may be useful in determining the static deflection curve. The mass of the shaft can be included by breaking it down into sections and considering each as a lumped mass.

[1] Thompson, *op. cit.*, p. 33.

APPENDIX A

BENDING FORMULAS

L = length of beam, in.
x = distance along beam, in.
y = deflection, in.
M = bending moment, lb-in.
θ = slope of beam, radians
E = modulus of elasticity, psi
I = moment of inertia of section, in.^{4}
w = uniformly distributed load, lb/in.

Beam loading and support	M	y	θ
	$-Px$	$-\dfrac{P}{6EI}\,(x^3 - 3L^2x + 2L^3)$	$-\dfrac{PL^2}{2EI}$
	$-\dfrac{wx^2}{2}$	$-\dfrac{w}{24EI}\,(x^4 - 4L^3x + 3L^4)$	$-\dfrac{wL^3}{6EI}$

In length a $+\dfrac{Pbx}{L}$ In length b $+\dfrac{Pa}{L}(L-x)$	In length a $-\dfrac{Pbx}{6EIL}(L^2-b^2-x^2)$ In length b $-\dfrac{Pa(L-x)}{6EIL}[2Lb-b^2-(L-x)^2]$	$\theta_1 = -\dfrac{P}{6EI}\left(bL-\dfrac{b^3}{L}\right)$ $\theta_2 = \dfrac{P}{6EI}\left(2bL+\dfrac{b^3}{L}-3b^2\right)$
$\dfrac{wL}{2}\left(x-\dfrac{x^2}{L}\right)$	$-\dfrac{wx}{24EI}(L^3-2Lx^2+x^3)$	$\theta_1 = -\dfrac{wL^3}{24EI}$ $\theta_2 = -\theta_1$
$M_1 = -\dfrac{Pab^2}{L^2}$ $M_2 = -\dfrac{Pa^2b}{L^2}$	In length a $-\dfrac{Pb^2x^2}{6EIL^3}(3aL-3ax-bx)$ In length b $-\dfrac{Pa^2(L-x)^2}{6EIL^3}[3bL-(3b+a)(L-x)]$	
$M_1 = -\dfrac{wL^2}{12}$ $M_2 = M_1$	$y = -\dfrac{wx^2}{24EI}(L^2+x^2-2Lx)$	

471

PROPERTIES OF CROSS SECTIONS

Section	Rectangular		Polar	
	Moment of inertia I	$\dfrac{I}{c}$	Moment of inertia J	$\dfrac{J}{r}$
	$\dfrac{bh^3}{12}$	$\dfrac{bh^2}{6}$		
	$\dfrac{\pi D^4}{64}$	$\dfrac{\pi D^3}{32}$	$\dfrac{\pi D^4}{32}$	$\dfrac{\pi D^3}{16}$
	$\dfrac{\pi(D^4 - d^4)}{64}$	$\dfrac{\pi(D^4 - d^4)}{32D}$	$\dfrac{\pi(D^4 - d^4)}{32}$	$\dfrac{\pi(D^4 - d^4)}{16D}$

APPENDIX C: VALUES OF $\pi d^3/16$ AND $\pi d^3/32$

Values of $\pi d^3/16$ (d = diameter of shaft)

d	0	1/16	1/8	3/16	1/4	5/16	3/8	7/16	1/2	9/16	5/8	11/16	3/4	13/16	7/8	15/16
0	0	0.000048	0.00038	0.0013	0.0031	0.006	0.0104	0.0164	0.0245	0.0349	0.0479	0.0638	0.0828	0.1053	0.1315	0.1618
1	0.196	0.236	0.280	0.329	0.384	0.444	0.510	0.583	0.663	0.749	0.843	0.944	1.052	1.169	1.294	1.428
2	1.571	1.723	1.884	2.055	2.236	2.428	2.630	2.843	3.068	3.304	3.551	3.811	4.083	4.368	4.666	4.977
3	5.301	5.639	5.992	6.359	6.740	7.136	7.548	7.975	8.416	8.877	9.352	9.845	10.35	10.88	11.42	11.99
4	12.57	13.16	13.78	14.42	15.07	15.75	16.44	17.16	17.89	18.65	19.42	20.22	21.04	21.88	22.75	23.63
5	24.54	25.47	26.43	27.41	28.41	29.44	30.49	31.56	32.66	33.79	34.94	36.12	37.33	38.56	39.82	41.10
6	42.41	43.75	45.12	46.51	47.93	49.39	50.87	52.38	53.92	55.49	57.09	58.72	60.38	62.08	63.80	65.56
7	67.35		71.02		74.82		78.76		82.83		87.04		91.39		95.89	
8	100.5		105.3		110.3		115.3		120.6		126.0		131.5		137.3	
9	143.1		149.2		155.4		161.8		168.3		175.1		182.0		189.1	
10	196.4		203.8		211.4		219.3		227.3		235.5		243.9		252.5	
11	261.3		270.3		279.6		289.0		298.6		308.5		318.5		328.8	
12	339.3		350.0		361.0		372.1		383.5		395.1		407.0		419.1	
13	431.4		443.9		456.8		469.8		483.1		496.6		510.4		524.5	
14	538.8		553.3		568.2		583.3		598.6		614.2		630.1		646.5	
15	662.7		679.4		696.4		713.6		731.2		749.0		767.1		785.6	

Values of $\pi d^3/32$ (d = diameter of shaft)

d	0	1/16	1/8	3/16	1/4	5/16	3/8	7/16	1/2	9/16	5/8	11/16	3/4	13/16	7/8	15/16
0	0	0.000024	0.00019	0.00065	0.00154	0.003	0.0052	0.0082	0.0123	0.0175	0.0239	0.0319	0.0414	0.0527	0.0658	0.0809
1	0.098	0.118	0.14	0.164	0.192	0.222	0.255	0.292	0.331	0.375	0.421	0.472	0.526	0.585	0.647	0.714
2	0.785	0.862	0.942	1.028	1.118	1.214	1.315	1.422	1.534	1.652	1.776	1.906	2.042	2.184	2.333	2.489
3	2.651	2.82	2.996	3.18	3.37	3.568	3.774	3.988	4.208	4.439	4.676	4.923	5.176	5.44	5.712	5.995
4	6.283	6.58	6.892	7.21	7.535	7.876	8.220	8.580	8.946	9.326	9.712	10.11	10.52	10.94	11.38	11.82
5	12.27	12.74	13.22	13.71	14.20	14.72	15.24	15.73	16.33	16.89	17.42	18.06	18.66	19.28	19.91	20.55
6	21.21	21.88	22.56	23.26	23.97	24.70	25.44	26.19	26.96	27.75	28.55	29.36	30.19	31.04	31.90	32.78
7	33.68		35.51		37.41		39.38		41.42		43.52		45.70		47.95	
8	50.27		52.66		55.13		57.67		60.29		62.99		65.77		68.63	
9	71.57		74.59		77.70		80.90		84.17		87.54		90.99		94.53	
10	98.18		101.9		105.7		109.6		113.6		117.8		122.0		126.3	
11	130.7		135.2		139.8		144.5		149.3		154.2		159.3		164.4	
12	169.7		175.0		180.5		186.0		191.8		197.6		203.5		209.5	
13	215.7		222.0		228.4		234.9		241.6		248.3		255.2		262.2	
14	269.4		276.7		284.1		291.6		299.3		307.1		315.1		323.1	
15	331.3		339.7		348.2		356.8		365.6		374.5		383.6		392.8	

APPENDIX D

MECHANICAL PROPERTIES OF MATERIALS

Material	Designation number	Description	Tensile ultimate strength, 10³ psi	Tensile yield strength, 10³ psi	Shear strength, 10³ psi	Compressive strength, 10³ psi	Endurance limit, 10³ psi	Modulus of elasticity in tension, 10⁶ psi	Elongation in 2 in., %
Cast iron..........	ASTM 20	....	20	...	32	95	10	12	
	ASTM 30	....	30	...	44	115	14.5	15	
	ASTM 40	....	40	...	57	143	21	17	
	ASTM 60	....	60	...	...	170	....	19.5	
Malleable cast iron.	ASTM 47-33	Grade 32510	50	32	...	...	....	25	10
Meehanite........	....	....	55	35	...	...	....	....	10
Carbon steels*.....	AISI-C1015	Hot-rolled	60	35	...	...		30	35
		Cold-drawn	75	65	...	...		30	20
	AISI-C1020	Hot-rolled	65	43	...	...		30	35
		Cold-drawn	78	66	...	...	See Sec. 6-5	30	20
	AISI-C1030	Hot-rolled	80	47	...	...		30	30
		Cold-drawn	93	78	...	...		30	17
	AISI-C1040	Hot-rolled	84	53	...	...		30	28
		Cold-drawn	100	92	...	...		30	15
Stainless steel.....	302 (18-8)	Annealed	90	35	...	...	42	28	55
		Cold-drawn bars	115	85	...	...	....	28	24
Aluminum alloys..	2024-O	Annealed	26	11	18	...	12	10.6	19
	2024-T4	Heat-treated	64	42	41	...	18	10.6	19
	113	As cast	24	15	20	...	10	....	1.5
	195-T4	Heat-treated casting	32	16	24	...	6	....	8.5
Magnesium alloy..	A 10	As sand-cast	22	12	18	...	10	6.5	2
		Heat-treated and aged casting	40	19	22	...	10	6.5	1
		Extruded bar	51	38	23	...	18	6.5	9
Bronze...........	Leaded tin	As cast	38	16	35	...	....	13	35

* See Appendix E for heat-treated steels.

474

MECHANICAL PROPERTIES OF HEAT-TREATED STEELS

AISI-C 1030, Fine Grain

(Water Quenched)

PROPERTIES CHART
(Average Values)

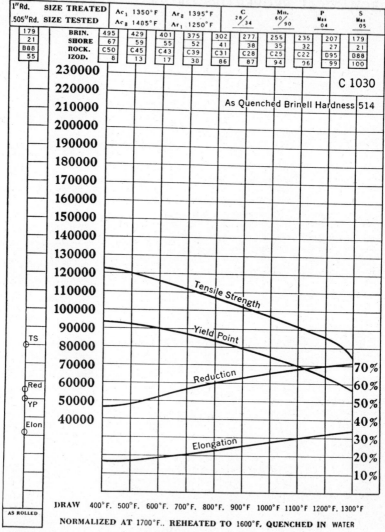

C 1030

As Quenched Brinell Hardness 514

NORMALIZED AT 1700°F.. REHEATED TO 1600°F. QUENCHED IN WATER

Reproduced with permission of the publisher from "Modern Steels and Their Properties," 3d ed., Bethlehem Steel Company, Bethlehem, Pa., 1955.

AISI-C1030
MASS EFFECT DATA
(Single Heat Results)

C	Mn	P	S	Si	Ni	Cr	Mo	Grain Size
.31	.65	.023	.026	.14	....	....	....	5-7

Annealed—(Heated to 1550 F, furnace-cooled 20 F per hour to 1200 F, air-cooled)

Size	T.S.	Y.P.	El. (2 in.)	Red.	Brinell	Izod
1 in. Rd	67,250	49,500	31.2	57.9	126	51.2

Normalized—(Heated to 1700 F, air-cooled)

Size	T.S.	Y.P.	El. (2 in.)	Red.	Brinell	Izod
½ in. Rd	77,500	50,000	32.1	61.1	156	69.7
1 in. Rd	75,500	50,000	32.0	60.8	149	69.0
2 in. Rd	74,000	49,500	29.5	58.9	137	63.0
4 in. Rd	72,500	47,250	29.7	56.2	137	61.2

As-Quenched Rockwell$_C$ Hardness (Water)

	½ in. Rd	1 in. Rd	2 in. Rd	4 in. Rd
Surface	50	46	30	R_B97
½ Radius	50	23	R_B93	R_B88
Center	23	21	R_B90	R_B85

Water-quenched from 1600 F, tempered at 1000 F.

Size	T.S.	Y.P.	El. (2 in.)	Red.	Brinell	Izod
½ in. Rd	91,500	75,000	28.2	58.0	187	88.5
1 in. Rd	88,000	68,500	28.0	68.6	179	92.2
2 in. Rd	86,500	63,750	28.2	65.8	170	93.5
4 in. Rd	80,750	54,750	32.0	68.2	163	95.5

Water-quenched from 1600 F, tempered at 1100 F.

½ in. Rd	88,500	64,000	28.9	69.7	179	94.5
1 in. Rd	85,250	63,000	29.0	70.8	170	98.7
2 in. Rd	83,750	57,250	29.0	69.1	167	100.7
4 in. Rd	80,500	54,500	32.0	68.5	163	96.5

Water-quenched from 1600 F, tempered at 1200 F.

½ in. Rd	85,500	62,000	29.9	70.5	174	99.5
1 in. Rd	84,500	61,500	28.5	71.4	170	102.7
2 in. Rd	80,000	56,750	30.2	70.9	156	105.0
4 in. Rd	74,500	49,500	34.2	71.0	149	100.7

Reproduced with permission of the publisher from "Modern Steels and Their Properties," 3d ed., Bethlehem Steel Company, Bethlehem, Pa., 1955.

AISI-C 1040, Fine Grain
(Oil Quenched)
PROPERTIES CHART
(Average Values)

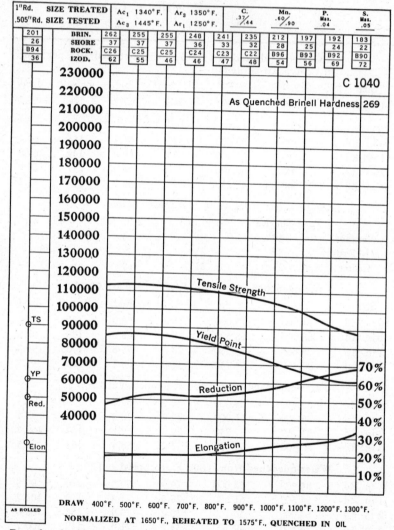

1"Rd.	SIZE TREATED	Ac₁ 1340°F.		Ar₃ 1350°F.		C. .3/.44		Mn. .60/.90		P. Max. .04		S. Max. .05
.505"Rd.	SIZE TESTED	Ac₃ 1445°F.		Ar₁ 1250°F.								
201	BRIN.	262	255	255	248	241	235	212	197	192	183	
26	SHORE	37	37	37	36	33	32	28	25	24	22	
B94	ROCK.	C26	C25	C25	C24	C23	C22	B96	B93	B92	B90	
36	IZOD.	62	55	46	46	47	48	54	56	69	72	

C 1040

As Quenched Brinell Hardness 269

230000
220000
210000
200000
190000
180000
170000
160000
150000
140000
130000
120000
110000
100000
90000 — TS
80000
70000
60000 — YP
50000 — Red.
40000

70%
60%
50%
40%
30%
20%
10%

Elon.

Tensile Strength

Yield Point

Reduction

Elongation

AS ROLLED

DRAW   400°F. 500°F. 600°F. 700°F. 800°F. 900°F. 1000°F. 1100°F. 1200°F. 1300°F.

NORMALIZED AT 1650°F., REHEATED TO 1575°F., QUENCHED IN OIL

Reproduced with permission of the publisher from "Modern Steels and Their Properties," 3d ed., Bethlehem Steel Company, Bethlehem, Pa., 1955.

AISI-C 1040

MASS EFFECT DATA
(Single Heat Results)

C	Mn	P	S	Si	Ni	Cr	Mo	Grain Size
.39	.71	.019	.036	.15	....	....	....	5-7

Annealed—(Heated to 1450 F, furnace-cooled 20 F per hour to 1200 F, air-cooled)

Size	T.S.	Y.P.	El. (2 in.)	Red.	Brinell	Izod
1 in. Rd	75,250	51,250	30.2	57.2	149	32.7

Normalized—(Heated to 1650 F, air-cooled)

Size	T.S.	Y.P.	El. (2 in.)	Red.	Brinell	Izod
½ in. Rd	88,250	58,500	30.0	56.5	183	51.7
1 in. Rd	85,500	54,250	28.0	54.9	170	48.0
2 in. Rd	84,250	53,000	28.0	53.3	167	51.5
4 in. Rd	83,500	49,250	27.0	51.8	167	39.0

As-Quenched Rockwell$_C$ Hardness (Oil)

	½ in. Rd	1 in. Rd	2 in. Rd	4 in. Rd
Surface	28	23	$R_B 93$	$R_B 91$
½ Radius	22	21	$R_B 92$	$R_B 91$
Center	21	18	$R_B 91$	$R_B 89$

Oil-quenched from 1575 F, tempered at 1000 F.

Size	T.S.	Y.P.	El. (2 in.)	Red.	Brinell	Izod
½ in. Rd	104,750	72,500	27.0	62.0	217	66.5
1 in. Rd	96,250	68,000	26.5	61.1	197	68.0
2 in. Rd	92,250	59,750	27.0	59.7	187	75.2
4 in. Rd	90,000	57,500	27.0	60.3	179	61.0

Oil-quenched from 1575 F, tempered at 1100 F.

½ in. Rd	100,500	69,500	27.0	65.2	207	76.0
1 in. Rd	91,500	64,250	28.2	63.5	187	80.7
2 in. Rd	86,750	56,875	28.0	62.5	174	91.5
4 in. Rd	82,750	52,250	30.0	61.6	170	81.0

Oil-quenched from 1575 F, tempered at 1200 F.

½ in. Rd	95,000	66,625	28.9	65.4	197	86.0
1 in. Rd	85,250	60,250	30.0	67.4	170	88.2
2 in. Rd	82,500	54,500	31.0	66.4	167	93.7
4 in. Rd	78,750	50,000	31.2	64.5	156	85.5

Reproduced with permission of the publisher from "Modern Steels and Their Properties," 3d ed., Bethlehem Steel Company, Bethlehem, Pa., 1955.

AISI - 3140

(Oil Quenched)

PROPERTIES CHART
(Single Heat Results)

530"Rd. SIZE TREATED 505"Rd. SIZE TESTED	Ac₁ 1360°F Ac₃ 1420°F	Ar₃ 1265°F Ar₁ 1200°F	C. 38/43	Mn. 70/90	P. Max. 04	S. Max. 04	Si. 20/35	Ni. 1 10/1 40	Cr. 55/75	Mo. Nil	Grain Size
HEAT TESTED			39	76	013	026	25	1 20	65	.08	6-8

BRIN.	555	514	477	461	388	352	311	285	262	223	
SHORE	75	70	65	63	54	49	44	40	37	32	
ROCK.	C55	C52	C49	C47	C41	C37	C33	C30	C26	C20	
IZOD.	35	22	22	24	40	51	66	80	97	105	

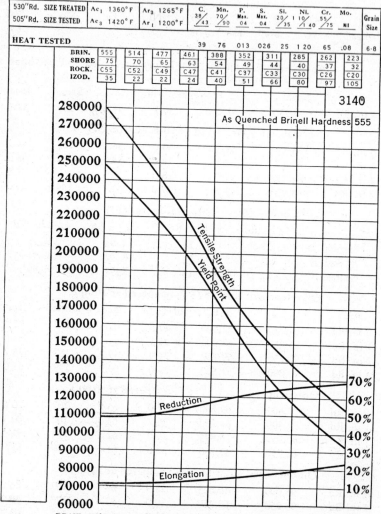

3140

As Quenched Brinell Hardness 555

Tensile Strength
Yield Point
Reduction
Elongation

280000
270000
260000
250000
240000
230000
220000
210000
200000
190000
180000
170000
160000
150000
140000
130000   70%
120000   60%
110000   50%
100000   40%
90000   30%
80000   20%
70000   10%
60000

DRAW 400°F. 500 F. 600°F. 700°F. 800°F. 900°F. 1000°F. 1100°F. 1200°F. 1300°F.

NORMALIZED AT 1600°F., REHEATED TO 1525°F., QUENCHED IN AGITATED OIL

AISI-3140

MASS EFFECT DATA
(Single Heat Results)

C	Mn	P	S	Si	Ni	Cr	Mo	Grain Size
.40	.90	.025	.018	.27	1.21	.62	.02	5-8

Annealed—(Heated to 1500 F, furnace-cooled 20 F per hour to 1150 F, air-cooled)

Size	T.S.	Y.P.	El. (2 in.)	Red.	Brinell	Izod
1 in. Rd	100,000	61,250	24.5	50.8	197	34.2

Normalized—(Heated to 1600 F, air-cooled)

Size	T.S.	Y.P.	El. (2 in.)	Red.	Brinell	Izod
½ in. Rd	148,500	94,750	16.4	43.0	302	21.0
1 in. Rd	129,250	87,000	19.7	57.5	262	39.5
2 in. Rd	119,250	73,500	21.7	59.0	248	44.7
4 in. Rd	118,000	71,000	21.0	56.4	241	30.0

As-Quenched Rockwell$_C$ Hardness (Oil)

	½ in. Rd	1 in. Rd	2 in. Rd	4 in. Rd
Surface	57	55	46	34
½ Radius	57	55	40	33.5
Center	57	55	40	33.5

Oil-quenched from 1525 F, tempered at 1000 F.

Size	T.S.	Y.P.	El. (2 in.)	Red.	Brinell	Izod
½ in. Rd	150,500	140,500	17.4	54.8	302	46.2
1 in. Rd	146,750	132,000	17.5	57.0	293	47.5
2 in. Rd	130,250	102,750	19.7	60.7	269	64.5
4 in. Rd	121,250	91,250	20.0	56.9	248	50.2

Oil-quenched from 1525 F, tempered at 1100 F.

½ in. Rd	138,000	125,750	20.0	59.2	277	57.0
1 in. Rd	131,750	118,875	21.2	61.0	269	65.0
2 in. Rd	123,250	95,500	21.5	63.3	248	69.5
4 in. Rd	109,250	78,250	23.7	65.0	217	85.5

Oil-quenched from 1525 F, tempered at 1200 F.

½ in. Rd	128,250	117,250	21.7	62.0	255	72.5
1 in. Rd	124,500	109,000	21.5	60.2	248	66.5
2 in. Rd	114,500	85,000	23.5	66.7	229	92.0
4 in. Rd	108,000	79,000	24.2	65.6	217	91.0

Reproduced with permission of the publisher from "Modern Steels and Their Properties," 3d ed., Bethlehem Steel Company, Bethlehem, Pa., 1955.

APPENDIX F

DECIMAL EQUIVALENTS OF FRACTIONS

Advancing by sixty-fourths

$\frac{1}{64}$			= 0.015625	$\frac{33}{64}$			= 0.515625
	$\frac{1}{32}$		= 0.03125		$\frac{17}{32}$		= 0.53125
$\frac{3}{64}$			= 0.046875	$\frac{35}{64}$			= 0.546875
		$\frac{1}{16}$	= 0.0625			$\frac{9}{16}$	= 0.5625
$\frac{5}{64}$			= 0.078125	$\frac{37}{64}$			= 0.578125
	$\frac{3}{32}$		= 0.09375		$\frac{19}{32}$		= 0.59375
$\frac{7}{64}$			= 0.109375	$\frac{39}{64}$			= 0.609375
		$\frac{1}{8}$	= 0.125			$\frac{5}{8}$	= 0.625
$\frac{9}{64}$			= 0.140625	$\frac{41}{64}$			= 0.640625
	$\frac{5}{32}$		= 0.15625		$\frac{21}{32}$		= 0.65625
$\frac{11}{64}$			= 0.171875	$\frac{43}{64}$			= 0.671875
		$\frac{3}{16}$	= 0.1875			$\frac{11}{16}$	= 0.6875
$\frac{13}{64}$			= 0.203125	$\frac{45}{64}$			= 0.703125
	$\frac{7}{32}$		= 0.21875		$\frac{23}{32}$		= 0.71875
$\frac{15}{64}$			= 0.234375	$\frac{47}{64}$			= 0.734375
		$\frac{1}{4}$	= 0.25			$\frac{3}{4}$	= 0.75
$\frac{17}{64}$			= 0.265625	$\frac{49}{64}$			= 0.765625
	$\frac{9}{32}$		= 0.28125		$\frac{25}{32}$		= 0.78125
$\frac{19}{64}$			= 0.296875	$\frac{51}{64}$			= 0.796875
		$\frac{5}{16}$	= 0.3125			$\frac{13}{16}$	= 0.8125
$\frac{21}{64}$			= 0.328125	$\frac{53}{64}$			= 0.828125
	$\frac{11}{32}$		= 0.34375		$\frac{27}{32}$		= 0.84375
$\frac{23}{64}$			= 0.359375	$\frac{55}{64}$			= 0.859375
		$\frac{3}{8}$	= 0.375			$\frac{7}{8}$	= 0.875
$\frac{25}{64}$			= 0.390625	$\frac{57}{64}$			= 0.890625
	$\frac{13}{32}$		= 0.40625		$\frac{29}{32}$		= 0.90625
$\frac{27}{64}$			= 0.421875	$\frac{59}{64}$			= 0.921875
		$\frac{7}{16}$	= 0.4375			$\frac{15}{16}$	= 0.9375
$\frac{29}{64}$			= 0.453125	$\frac{61}{64}$			= 0.953125
	$\frac{15}{32}$		= 0.46875		$\frac{31}{32}$		= 0.96875
$\frac{31}{64}$			= 0.484375	$\frac{63}{64}$			= 0.984375
		$\frac{1}{2}$	= 0.50				

APPENDIX G

PARTIAL LIST OF NEMA STANDARDS FOR HORSE-POWERS, SPEEDS, FRAME NUMBERS, AND DIMENSIONS FOR OPEN-TYPE, POLY-PHASE, SQUIRREL-CAGE, GENERAL-PURPOSE INDUCTION MOTORS

Hp	Frame numbers				Frame size	Dimensions	
	Rpm					A*	U†
	3,600	1,800	1,200	900			
½	...	...	...	182	182	9	⅞
¾	...	...	182	184	184	9	⅞
1	...	182	184	213	213	10½	1⅛
1½	182	184	184	213	215	10½	1⅛
2	184	184	213	215	254U	12½	1⅜
3	184	213	215	254U	256U	12½	1⅜
5	213	215	254U	256U	284U	14	1⅝
7½	215	254U	256U	284U	286U	14	1⅝
10	254U	256U	284U	286U	324S	16	1⅝
15	256U	284U	324U	326U	324U	16	1⅞
20	284U	286U	326U	404	326S	16	1⅝
25	286U	324U	404	405	326U	16	1⅞
30	324S	326U	405	444	404S	20	1⅞
40	326S	404	444	445	404	20	2⅛
50	404S	405S	445	504U	405S	20	1⅞
60	405S	444S	504U	505	405	20	2⅛
75	444S	445S	505	...	444S	22	2⅛
100	445S	504S	...	...	444	22	2⅜
					445S	22	2⅛
					445	22	2⅜
					504S	25	2⅛
					504U	25	2⅞
					505	25	2⅞

* A = outside diameter of motor, in.

† U = shaft extension diameter, in.

PROBLEMS

Chapter 2.   Linkages

2-1. For the case of link 2 driving in Fig. P 1, (a) determine whether the driver and the follower rotate or oscillate, and (b) locate all dead points and limiting positions of travel.

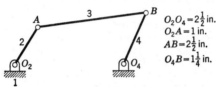

$O_2O_4 = 2\frac{1}{2}$ in.
$O_2A = 1$ in.
$AB = 2\frac{1}{2}$ in.
$O_4B = 1\frac{1}{4}$ in.

Fig. P 1

2-2. Same as Prob. 2-1 except that link 4, instead of link 2, is the driver.

2-3. Link 2, the driver in Fig. P 1, is rotating counterclockwise with uniform angular velocity. (a) Considering the driver to be in its zero position when the follower is at its farthest position to the right, divide the driver crank circle (A) into 24 equal portions. (b) Determine the corresponding positions of point B on the follower. (c) Plot the displacement (arc distance in inches) of B as a function of the crank position.

NOTE: Since the crank is rotating at a uniform rate, the crank positions are directly proportional to time, and the resulting graph is known as a *displacement-time diagram*.

2-4. Same as Prob. 2-3 except use the mechanism in Fig. P 2 instead of that in Fig. P 1.

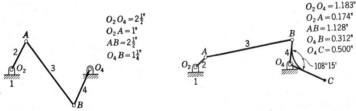

$O_2O_4 = 2\frac{1}{2}''$
$O_2A = 1''$
$AB = 2\frac{1}{2}''$
$O_4B = 1\frac{1}{4}''$

Fig. P 2

$O_2O_4 = 1.183''$
$O_2A = 0.174''$
$AB = 1.128''$
$O_4B = 0.312''$
$O_4C = 0.500''$

$108°15'$

Fig. P 3. (*Courtesy Sunbeam Corporation.*)

2-5. The dimensions of the Sunbeam Shavemaster mechanism (Fig. 2-4) are given in Fig. P 3. Crank 2 rotates ccw at a uniform rate, and the cutter C oscillates. (a) Using a scale of 1 in. = $\frac{1}{4}$ in. and considering the driver to be in its zero position when C is at its most clockwise position, divide the crank circle (A) into 24 equal portions. (*b*) Determine the corresponding positions of *B* and *C*. (*c*) Plot the displacement (arc distance in inches) of *C* as a function of the crank position. (*d*) On the basis of a crank speed of 8,600 rpm, calculate the time scale in units of seconds per inch.

2-6. The drag-link mechanism in Fig. P 4 is to be the basis for a quick-return mechanism with link 2, the driver. (*a*) Complete the mechanism by adding a $3\frac{1}{2}$-in. connecting rod and a slider. (*b*) Determine the length and the direction of the working stroke. (*c*) Determine the time ratio.

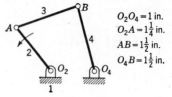

$O_2O_4 = 1$ in.
$O_2A = 1\frac{1}{4}$ in.
$AB = 1\frac{1}{2}$ in.
$O_4B = 1\frac{1}{2}$ in.

Fig. P 4

2-7. The requirement for rotation of both the driver and follower is that there be no dead points. Using the drag-link mechanism in Fig. P 4, show that this condition is met when $AB > O_2O_4 + O_4B - O_2A$, $AB > O_2O_4 + O_2A - O_4B$, and $AB < O_2A + O_4B - O_2O_4$.

Hint: Set up the relationships required for link 2 to "drag" link 4 through the line of centers (O_2-O_4) and for link 4 to drag link 2 through the line of centers.

2-8. Utilizing the criteria from Prob. 2-7 for rotation of both driver and follower, determine two combinations of dimensions for the driver and the connecting rod of the drag-link part of a quick-return mechanism with a center distance (O_2O_4 in Fig. P 4) of 3 in. and a working stroke of 8 in.

2-9. Figure P 5 shows side-rod drives as used on early electric locomotives *a* and on steam locomotives *b*. When the linkage on only one side of a locomotive is considered, both drives have the same problem in relation to dead points. However, the torque and force characteristics are quite different. For example, when starting up, the electric motor will be considered to exert a constant torque, whereas the force on the crosshead of the steam locomotive will be a constant value equal to the product of the steam pressure and the area of the piston.

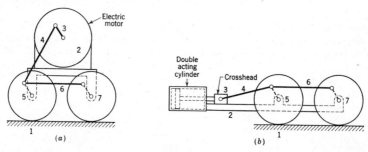

Fig. P 5

Make free-body diagrams as required. (*a*) Draw approximate curves of tractive effort, i.e., the force between the wheels and the rail, vs. time for one revolution of the wheels. (*b*) On the basis of the tractive-effort curves, which locomotive will be expected to give the smoother start? (*c*) Draw approximate curves of the force along connecting rods 4 vs. time for one revolution of the wheels. (*d*) What is your conclusion concerning the practicality of the electric-motor drive as shown? (*e*) How would you minimize the problem of dead points? (*f*) Should separate motors be used for each side of the electric locomotive, or should the cranks 3 be fixed to each end of a common motor shaft?   Why?   (g) Draw the approximate curves for tractive effort vs. time for one revolution of the wheels when both sides are being driven and each pair of wheels is fixed to a common axle.

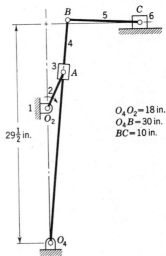

2-10. The fixed dimensions of a crank-shaper mechanism are given in Fig. P 6.   Crank 2 rotates clockwise at a uniform angular velocity. Determine (a) the direction of motion of the ram 6 during the working stroke; (b) both analytically and graphically, the length of link 2 (distance O_2A) required when the working stroke is 18 in.   [Use a scale of 1 in. = 6 in. (2 in. = 1 ft 0 in.)]; and (c) both analytically and graphically, the time ratio when the working stroke is 18 in.

$O_4O_2 = 18$ in.
$O_4B = 30$ in.
$BC = 10$ in.

$29\frac{1}{2}$ in.

Fig. P 6

2-11. Plot the displacement-time diagram for the motion of the ram 6 of the crank-shaper mechanism in Prob. 2-10.   Divide the crank circle into 24 equal portions, with the zero position when the ram is in its farthest left position.

NOTE: Save the displacement-time diagram for graphical differentiation in Prob. 4-101.

2-12. (a) Design a crank-shaper quick-return mechanism that will give a time ratio of 1.75:1 with a working stroke of 24 in.   Use a scale of 1 in. = 6 in. (2 in. = 1 ft 0 in.). (b) How would you provide for an adjustable length of working stroke?   (c) Will a change in stroke affect the time ratio?   Explain.

2-13. The working stroke is to be upward for ram 6 of the Whitworth quick-return mechanism in Fig. P 7.   Draw the mechanism to a scale of 1 in. = 4 in. and determine (a) the direction of rotation of crank 2, (b) the time ratio, and (c) the length of the working stroke.

$O_2O_4 = 3$ in.
$O_2A = 6$ in.
$O_4B = 5$ in.
$BC = 14$ in.

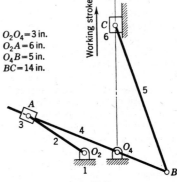

2-14. Plot the displacement-time diagram for the motion of ram 6 of the Whitworth quick-return mechanism in Fig. P 7.   Divide the crank circle into 24 equal portions, with the zero position when the ram is in its lowest position.

NOTE: Save the displacement-time diagram for graphical differentiation in Prob. 4-102.

Fig. P 7

2-15. (*a*) Design a Whitworth quick-return mechanism that will give a time ratio of 2.5:1 with a working stroke of 3 in.   Make the drawing full size.   (*b*) How would you provide for an adjustable length of working stroke?   (*c*) Will a change in stroke affect the time ratio?   Explain.

2-16. Design a Watt straight-line mechanism that will give a close approximation to a straight line over a distance of 3 in.

SUGGESTED PROCEDURE: Choose some reasonable dimensions such as 4 in. for *AB* + *CD* and 2 in. for *BC* in Fig. 2-27, plot the path of *p*, determine the length of the straight portion, and multiply all dimensions by the ratio of the desired length of 3 in. to the measured length from the trial design.

2-17. Design a pantograph linkage with a 3:1 ratio.

2-18. Design a pantograph linkage with a 1.75:1 ratio.

2-19. The Bacharach engine-indicator linkage in Fig. 2-30 has the dimensions given in Fig. P 8.   (*a*) Determine the usable length of travel of the stylus *P*.   (*b*) Determine the average multiplication ratio over the usable range of motion.   (*c*) Evaluate the linkage with respect to the degree to which the motion of *p* is reproduced at *P*.

SUGGESTED PROCEDURE: Draw the mechanism to as large a scale as convenient, the larger the scale the greater the accuracy.   A ten-times-size drawing, that is,

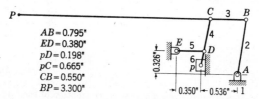

$$AB = 0.795''$$
$$ED = 0.380''$$
$$pD = 0.198''$$
$$pC = 0.665''$$
$$CB = 0.550''$$
$$BP = 3.300''$$

FIG. P 8. (*Courtesy Bacharach Industrial Instrument Company.*)

1 in. = 0.1 in., would be desirable, but with reasonable care a four-times-size drawing, that is, 1 in. = ¼ in., on a C-size (17- by 22-in.) sheet of drawing paper will give good results.   With exacting care, a double-size drawing on an 8½- by 11-in. sheet of paper will give results satisfactory for the purpose of this problem.   Let 5 denote the position of *p* when link 5 (*ED*) is horizontal.   Then move *p* both up and down in increments of 0.05 in. (actual motion), and locate the corresponding positions of *P*.

Draw a smooth curve through the points *P* and determine the length of the portion most closely approximating a straight line.   Draw a straight line parallel to the path of *p* through one end of the usable portion of the path of *P*.

Then plot a curve of the vertical displacement of *P* vs. the displacement of *p*; it will be desirable to use different scales for the displacements of *P* and *p*.   The average multiplication factor and the evaluation of the linkage accuracy may best be based upon this curve.   It is important to note that the evaluation will have little significance unless the drawing is done accurately.

2-20. An engine indicator is to be designed to give a multiplication ratio of 4:1 with a stylus travel of at least 2 in.   The 2-in. travel of the stylus is to extend from approximately ½ in. below to approximately 1½ in. above its position when the pencil arm is horizontal.   The design is to use, as far as practical, parts that are identical with those of the indicator in Figs. 2-30 and P 8.   This means that the perpendicular distance between the paths of *P* and *p* will be 2.655 in., the distance *PC* will be 2.750 in., and link 4 (*pDC*) will be the same as in Fig. P 8.

Using extra care, draw the mechanism twice size, i.e., a scale of 1 in. = ½ in., on an

8½- by 11-in. sheet of paper or four-times size on a C-size (17- by 22-in.) sheet of drawing paper.

After the design is completed, it will be worthwhile to check it (on a new sheet of paper) by making an evaluation of the linkage as outlined in Prob. 2-19.

Chapter 3. Motion in Machines

3-1. A conventional automobile weighing 4,200 lb is undergoing performance tests on a concrete highway. The coefficient of friction between the tires and the road is 0.67, and the automobile accelerates from rest to 60 mph in 13.0 sec. (a) What is the "average" value for the acceleration in feet per second per second and in g's, that is, the calculated acceleration divided by the acceleration of gravity (32.2 ft/sec²)? (b) On the basis of a uniform acceleration equal to the "average" value calculated in (a), how far does the car travel during the 13.0-sec period of acceleration? (c) Assuming that sufficient power is available, neglecting the rotational inertia of the wheels, and assuming that 60 per cent of the weight of the car is carried by the rear wheels during acceleration, what is the maximum possible acceleration? (d) How many seconds will be required to reach 60 mph from rest when the acceleration is that calculated in (c)?

3-2. The Safety Code for Elevators, Dumbwaiters, and Escalators[1] specifies that an elevator with a rated operating speed of 800 fpm must be provided with a safety brake that will be set by the tripping of a governor at a speed of 970 fpm. It also specifies that the elevator be uniformly retarded, with the stopping distance not less than 4 ft 6 in. and not greater than 12 ft 6 in. (a) What will be the value of the acceleration for the minimum and maximum stopping distances, in feet per second per second and in g's, that is, the acceleration divided by the acceleration of gravity (32.2 ft/sec²)? (b) How many seconds will elapse from the time the governor trips until the elevator cage comes to rest?

3-3. A direct-connected electric motor and pump reach rated running speed of 1,750 rpm in 0.3 sec after the switch is closed. (a) What is the average angular acceleration in rpm per second? (b) What is the average angular acceleration in radians per second per second?

3-4. What is the acceleration in radians per second per second of the armature of the motor in the Sunbeam Shavemaster (Fig. 2-4) if the normal running speed of 8,600 rpm is reached in 0.2 sec after the current is turned on?

3-5. An electric interurban train can accelerate to its top speed of 65 mph at the rate of 0.8 ft/sec², and its normal rate of deceleration is 2 ft/sec². What is the minimum running time between stations 3 miles apart?

3-6. Given the vectors in Fig. P 9, use a scale of 1 in. = 10 units and determine the following vectors: (a) **H** = **A** + **B**, (b) **I** = **A** − **B**, (c) **J** = **C** + **D** − **B**, (d) **K** = **G** − **E** + **F**, and (e) **L** = **E** − **A** − **D** + **F**.

3-7. Given the vectors in Fig. P 9, use a scale of 1 in. = 10 units and determine the following vectors: (a) **H** = **B** + **E**, (b) **I** = **B** − **E**, (c) **J** = **A** − **E** + **C**, (d) **K** = **F** + **G** + **D** − **C**, and (e) **L** = **A** − **G** + **D** − **E** + **B** − **C** + **F**.

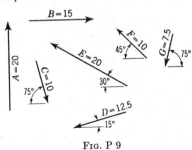

FIG. P 9

---

[1] ASA A17.1-1960, American Society of Mechanical Engineers, New York, 1960.

3-8. Given the velocity vectors in Fig. P 10, use a scale of 1 in. = 20 fpm and determine the following vectors: (a) $\mathbf{V}_D = \mathbf{V}_A + \mathbf{V}_B$, (b) $\mathbf{V}_{A/B}$, (c) $\mathbf{V}_{B/A}$, (d) $\mathbf{V}_E = \mathbf{V}_A + \mathbf{V}_C$, and (e) $\mathbf{V}_{B/C}$.

3-9. Given the velocity vectors in Fig. P 10, use a scale of 1 in. = 20 fpm and determine the following vectors: (a) $\mathbf{V}_D = \mathbf{V}_B + \mathbf{V}_C$, (b) $\mathbf{V}_{B/C}$, (c) $\mathbf{V}_{C/B}$, (d) $\mathbf{V}_{C/A}$, and (e) $\mathbf{V}_E = \mathbf{V}_A + \mathbf{V}_{B/C}$.

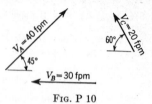

FIG. P 10

3-10. The rotating members in a machine have angular velocities $n_2 = 100$ rpm, cw (NOTE: We shall use n for revolutions per unit time and reserve ω for radians per unit time); $n_3 = 300$ rpm, ccw; $n_4 = 800$ rpm, ccw; and $n_5 = 400$ rpm, cw. Determine (a) $n_{3/2}$, (b) n_5/n_2, (c) $n_{2/5}$, (d) $n_{2/4}$, and (e) n_2/n_4.

3-11. Use the angular-velocity theorem to determine the ratio ω_2/ω_4 when crank 2 in Fig. P 1 is at an angle of 30° with the horizontal (O_2O_4).

3-12. By use of the angular-velocity theorem, determine the crank angle at which link 4 in Fig. P 1 will have a maximum angular velocity when crank 2 is rotating at a uniform speed.

3-13. Same as Prob. 3-11 except that crank 2 should be at an angle of 60° with the horizontal.

3-14. Draw the mechanism in Fig. P 1 full size. Let the zero position of the crank be that corresponding to the farthest-to-the-right position of follower 4. Divide the crank circle (ccw) into 12 equal portions. (a) Use the angular-velocity theorem to determine the ratio ω_2/ω_4 for each of the 12 positions of the crank. (b) Use the angular-velocity theorem to determine the ratio ω_4/ω_2 for each of the 12 positions of the crank. (c) Plot curves of the ratios of ω_2/ω_4 and of ω_4/ω_2 vs. crank position. (d) Considering crank 2 to be rotating at a uniform rate, which of the two curves in (c) better illustrates the operational characteristics of the mechanism?

3-15. (a) What is the significance when the line of transmission is parallel to the line of centers? (b) What is the significance when the line of transmission lies along the line of centers? (c) Where will the line of transmission cut the line of centers when the driving and driven links rotate in the same direction; in opposite directions?

3-16. Use the angular-velocity theorem to determine the ratio ω_4/ω_2 when crank 2 in Fig. P 2 makes an angle of 60° with the horizontal (O_2O_4).

3-17. Same as Prob. 3-16 except that crank 2 makes an angle of 120° with the horizontal.

3-18. Same as Prob. 3-16 except that crank 2 makes an angle of 180° with the horizontal.

3-19. Same as Prob. 3-12 except use the mechanism in Fig. P 2.

3-20. Draw the Sunbeam Shavemaster mechanism (Figs. 2-4 and P 3) to a scale of 1 in. = ½ in. Let the zero position of the crank be that corresponding to cutter C in its most clockwise position. Divide the crank circle into 12 equal portions (ccw). (a) Use the angular-velocity theorem to determine the ratio ω_4/ω_2 for each of the crank positions. (b) Plot the values of ω_4/ω_2 vs. the crank position. (c) Considering the crank to be rotating in a ccw direction at a uniform speed of 8,600 rpm, calculate the scales necessary to convert the curve of ω_4/ω_2 vs. crank position into a curve of the magnitude of the velocity of the cutter C in inches per second vs. time in seconds.

NOTE: Save the data for graphical differentiation in Prob. 4-100.

3-21. Use the angular-velocity theorem to determine the maximum and minimum values of the ratio ω_4/ω_2 for the drag-link mechanism in Fig. P 4.

3-22. Using the angular-velocity theorem, show that for any given angular positions of links 2 and 4 of the crank-shaper mechanism in Fig. P 6 the ratio ω_4/ω_2 is independent of the position of O_2 on the vertical line through O_4.

3-23. Use the angular-velocity theorem in showing that changing the length of stroke, i.e., changing the distance O_4B, of the Whitworth quick-return mechanism in Fig. P 7 has no effect on the ratio ω_4/ω_2.

3-24. Make sketches of the cam-follower mechanisms in Fig. P 11. (a) Use the angular-velocity theorem to determine the ratio ω_2/ω_3 for each mechanism. (b)

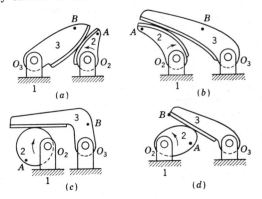

FIG. P 11

Assume a vector $1\frac{1}{2}$ in. long for the velocity of the point of contact on the driving member (P_2), and determine the velocity of sliding relative to the driven member for each mechanism.

3-25. The cam 2 in Fig. P 12 is directly connected to a 1,750-rpm $\frac{1}{4}$-hp electric motor. For the phase shown, the motor is delivering rated horsepower at rated speed.

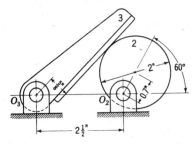

FIG. P 12

(a) Neglecting friction, what is the magnitude of the force being transmitted from the cam to the follower? (b) What is the ratio of the torque on the camshaft to that on the follower shaft, that is, T_2/T_3? (c) What is the ratio ω_2/ω_3? (d) What does the product $(T_2/T_3)(\omega_2/\omega_3)$ equal? What is the significance of this value? (e) Determine the velocity of sliding in feet per minute. (f) If the normal force is the value calculated in (a) and the coefficient of friction is 0.05, what is the instantaneous rate at which power is being lost because of friction (hp $= FV/33,000$, where F is in pounds and V is in feet per minute)?

3-26. Figure P 13 shows part of a cam-follower mechanism.   The cam is a circular disk with its geometric center on the line O_2A, and the follower is as shown.   (a) Determine the size and location of the cam if, in the phase shown, the ratio $\omega_2/\omega_3 = 4/1$.   (b) What will be the value of the ratio ω_2/ω_3 when the cam determined in (a) has rotated cw through an additional 30° angle?

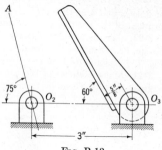

FIG. P 13

Chapter 4.   Velocity and Acceleration Analysis

NOTE: Many of the problems in this chapter involve drawing the mechanism and velocity and acceleration diagrams to scale.   It is recommended that the student make rough, order-of-magnitude sketches to assist him in locating the various diagrams, so that they do not interfere with one another on an 8½- by 11-in. sheet of paper, and in understanding the principles involved.

4-1. Use instant-center methods only.   Make a sketch of Fig. P 14a and assume a vector about 1½ in. long for the velocity of point A.   (a) Determine the vector for the velocity of point B.   (b) Write the equation for determining the magnitude of the angular velocity of link 3.   What is its direction?

4-2. Same as Prob. 4-1 except use Fig. P 14b.

4-3. Same as Prob. 4-1 except use Fig. P 14c.

4-4. Same as Prob. 4-1 except use Fig. P 14d and also determine the vector for the velocity of point C.

4-5. Same as Prob. 4-1 except use Fig. P 14e and determine the vector for the velocity of point C.

4-6. Same as Prob. 4-1 except use Fig. P 14f and determine the vector for the velocity of point D.

4-7. Same as Prob. 4-1 except use Fig. P 14g and determine the vector for the velocity of point C.

4-8. Same as Prob. 4-1 except use Fig. P 14h.

4-9. Same as Prob. 4-1 except use Fig. P 14i and determine the vectors for the velocity of points C and D.

4-10. Same as Prob. 4-1 except use Fig. P 14j and determine the vectors for the velocity of points B and C.

4-11. Same as Prob. 4-1 except use Fig. P 14k and determine the vector for the velocity of point C.

4-12. Same as Prob. 4-1 except use Fig. P 14l and determine the vector for the velocity of point D.

4-13. Same as Prob. 4-1 except use Fig. P 3 and determine the vector for the velocity of point C.

4-14. Same as Prob. 4-1 except use Fig. 2-6 and determine the vector for the velocity of point C.

4-15. Same as Prob. 4-1 except use Fig. P 6 and determine the vector for the velocity of point C.

4-16. Same as Prob. 4-1 except use Fig. P 7 and determine the vector for the velocity of point C.

4-17. Same as Prob. 4-1 except use Fig. P 11a.

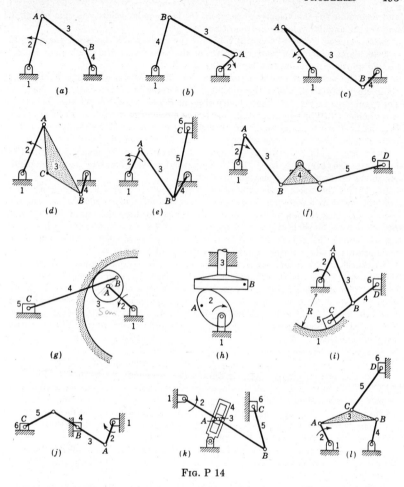

Fig. P 14

4-18. Same as Prob. 4-1 except use Fig. P 11*b*.

4-19. Same as Prob. 4-1 except use Fig. P 11*c*.

4-20. Same as Prob. 4-1 except use Fig. P 11*d*.

4-21. Use instant-center methods only.   Make a full-size drawing of the mechanism in Fig. P 1 for the phase in which crank 2 is at an angle of 60° with the horizontal. Crank 2 is rotating cw at 750 rpm.   Use a velocity scale of 1 in. = 50 in./sec and determine (*a*) the velocity of point *B* and (*b*) the angular velocity in rpm of link 4.

4-22. Same as Prob. 4-21 except solve for the phase when the crank angle is 150°.

4-23. Same as Prob. 4-21 except solve for the phase when the crank angle is 270° and also determine the magnitude in rpm and the direction of ω_3.

4-24. Use instant-center methods only.   Make a full-size drawing of the mechanism in Fig. P 2 for the phase in which crank 2 is at an angle of 0° with the horizontal. Crank 2 is rotating ccw at 200 rpm.   Use a velocity scale of 1 in. = 10 in./sec and determine (a) the velocity of point B, (b) the angular velocity in rpm of link 3, and (c) the angular velocity in rpm of link 4.

4-25. Same as Prob. 4-24 except solve for the phase when the crank angle is 90°.

4-26. Same as Prob. 4-24 except solve for the phase when the crank angle is 225° and do not determine the angular velocity of link 3.

4-27. The parallel-line construction (Sec. 4-9) offers a means of quickly determining complete information about the velocity characteristics of a four-bar mechanism. For example, if in Fig. 4-11 the velocity of point B were left as the distance Bb on the extension of $O_{41}B$ for a number of phases of the mechanism during one cycle of operation, a closed curve could be drawn through the points b, and the result would be a plot, or polar diagram, of the velocity of B as a function of the position of link 4.

Draw the O_2ABO_4 portion of the Sunbeam Shavemaster mechanism in Fig. P 3 four-times size (1 in. = ¼ in.). Let the zero position of the crank be that corresponding to the farthest-to-right position of point B.

(a) Divide the crank circle into 12 equal portions. Crank 2 is rotating ccw at a uniform speed of 8,600 rpm. Use a velocity scale of 1 in. = 200 in./sec and the parallel-line construction to make the polar diagrams for the velocities of points A and B. Note that the polar diagram for the velocity of A is a circle; also that, when moving to the left, the velocity of B (Bb) automatically lies radially outward and when moving to the right, radially inward, from the arc of travel of B.

(b) Since crank 2 is rotating at a constant speed, the crank angle is a direct function of time. Thus, the information expressed in the polar diagram in (a) may also be used to plot a velocity-time diagram for the velocities of points B and C. Use a velocity scale of 1 in. = 200 in./sec, let the 12 crank positions be represented by a line 6 in. long, and plot the velocity of point C as a function of the crank position. Let velocity to the left (ccw direction) be positive and velocity to the right (cw direction) be negative.

(c) Calculate the time scale in units of seconds per inch.

NOTE: Save this information for graphical differentiation in Prob. 4-100.

4-28. Use instant-center methods only. The drag-link mechanism in Fig. P 4 may be made into a drag-link quick-return mechanism by connecting a slider (at point C) moving along the line O_2O_4 to point B by means of a connecting rod. Use a connecting rod 3½ in. long, and make a full-size drawing of the mechanism for the phase when crank 2 is at an angle of 120°, measured ccw from its farthest-to-the-right position on the line of centers O_2O_4.

Crank 2 is rotating ccw at a uniform rate of 300 rpm. Use a velocity scale of 1 in. = 40 in./sec and determine (a) the velocity of point C and (b) the angular velocity of link 4.

4-29. Same as Prob. 4-28 except solve for the phase when the crank angle is 225°.

4-30. Same as Prob. 4-28 except solve for the phase when the crank angle is 345°.

4-31. Use instant-center methods only. Draw the crank-shaper mechanism in Fig. P 6 to a scale of 1 in. = 6 in. for the phase when crank 2 is at an angle of 30° with the vertical, measured in the direction of rotation from when point A is at its uppermost position. Crank 2 (O_2A) is 5½ in. long and is rotating cw at a uniform speed of 35 rpm. Use a velocity scale of 1 in. = 60 fpm and determine (a) the shaper-tool (point C) velocity and (b) the angular velocity of link 4.

4-32. Same as Prob. 4-31 except solve for the phase when the crank angle is 120°.

4-33. Use instant-center methods only. Draw the Whitworth quick-return mechanism in Fig. P 7 to a scale of 1 in. = 4 in. for the phase when crank 2 is at an angle of 120° (ccw) with the line O_2O_4. Crank 2 is rotating ccw at a uniform speed of 180 rpm. Use a velocity scale of 1 in. = 60 in./sec and determine (a) the velocity of point C and (b) the angular velocity of link 5.

4-34. Same as Prob. 4-33 except solve for the phase when the crank angle is 225°.

4-35. Same as Prob. 4-33 except solve for the phase when the crank angle is 330°.

4-36. Where is the instant center relative to the ground located for a wheel when (*a*) the wheel motion is pure rolling, (*b*) the wheel motion is pure sliding, and (*c*) the wheel motion is a combination of rolling and sliding? Assuming that in every case the wheel has the same angular velocity, use vectors to show for each case the velocity of the wheel at the point of contact, at the center, and at the highest point on its circumference.

4-37. Use relative-velocity methods only. Make a sketch of Fig. P 14*a* and assume a vector about 1½ in. long for the velocity of point *A*. (*a*) Draw the complete velocity polygon. (*b*) Place at point *B* the vector representing the velocity of point *B*. (*c*) Write the equation for determining the magnitude of the angular velocity of link 3. What is its direction?

4-38. Same as Prob. 4-37 except use Fig. P 14*b*.

4-39. Same as Prob. 4-37 except use Fig. P 14*c*.

4-40. Same as Prob. 4-37 except use Fig. P 14*d* and also determine the vector for the velocity of point *C*.

4-41. Same as Prob. 4-37 except use Fig. P 14*e* and determine the vector for the velocity of point *C*.

4-42. Same as Prob. 4-37 except use Fig. P 14*f* and determine the vector for the velocity of point *D*.

4-43. Same as Prob. 4-37 except use Fig. P 14*g* and determine the vector for the velocity of point *C*.

4-44. Same as Prob. 4-37 except use Fig. P 14*h* and also determine the vector for the velocity of sliding between links 2 and 3.

4-45. Same as Prob. 4-37 except use Fig. P 14*i* and determine the vectors for the velocity of points *C* and *D*.

4-46. Same as Prob. 4-37 except use Fig. P 14*j* and determine the vectors for the velocity of points *B* and *C*.

4-47. Same as Prob. 4-37 except use Fig. P 14*k* and determine the vectors for the velocity of point *C* and the velocity of sliding between links 3 and 4.

4-48. Same as Prob. 4-37 except use Fig. P 14*l* and determine the vector for the velocity of point *D*.

4-49. Same as Prob. 4-37 except use Fig. P 3 and determine the vector for the velocity of point *C*.

4-50. Same as Prob. 4-37 except use Fig. 2-6 and determine the vector for the velocity of point *C*.

4-51. Same as Prob. 4-37 except use Fig. P 6 and determine the vectors for the velocity of point *C* and the velocity of sliding between links 3 and 4.

4-52. Same as Prob. 4-37 except use Fig. P 7 and determine the vectors for the velocity of point *C* and the velocity of sliding between links 3 and 4.

4-53. Same as Prob. 4-37 except use Fig. P 11*a*.

4-54. Same as Prob. 4-37 except use Fig. P 11*b*.

4-55. Same as Prob. 4-37 except use Fig. P 11*c*.

4-56. Same as Prob. 4-37 except use Fig. P 11*d*.

4-57. Same as Prob. 4-21 except use relative-velocity instead of instant-center methods.

4-58. Same as Prob. 4-21 except use relative-velocity instead of instant-center methods and solve for the phase when the crank angle is 150°.

4-59. Same as Prob. 4-21 except use relative-velocity instead of instant-center methods and solve for the phase when the crank angle is 270°. Also determine the magnitude in rpm and the direction of ω_3.

4-60. Same as Prob. 4-24 except use relative-velocity instead of instant-center methods.

4-61. Same as Prob. 4-24 except use relative-velocity instead of instant-center methods and solve for the phase when the crank angle is 90°.

4-62. Same as Prob. 4-24 except use relative-velocity instead of instant-center methods and solve for the phase when the crank angle is 225°.

4-63. Use relative-velocity methods and draw the Sunbeam Shavemaster linkage in Fig. P 3 twice size (1 in. = ½ in.) in the phase when crank 2 is at an angle of 60° (ccw) with the line O_2O_4. The crank is rotating ccw at a uniform speed of 8,600 rpm. Use a scale of 1 in. = 100 in./sec and (a) draw the complete velocity polygon, (b) place at point C the vector for the velocity of C, and (c) calculate the magnitudes in rpm and specify the directions of ω_3 and ω_4.

4-64. Same as Prob. 4-63 except solve for the phase when the crank angle is 120°.

4-65. Same as Prob. 4-28 except use relative-velocity instead of instant-center methods.

4-66. Same as Prob. 4-28 except use relative-velocity instead of instant-center methods and solve for the phase when the crank angle is 225°.

4-67. Same as Prob. 4-28 except use relative-velocity instead of instant-center methods and solve for the phase when the crank angle is 345°.

4-68. Same as Prob. 4-31 except use relative-velocity instead of instant-center methods.

4-69. Same as Prob. 4-31 except use relative-velocity instead of instant-center methods and solve for the phase when the crank angle is 120°.

4-70. Same as Prob. 4-33 except use relative-velocity instead of instant-center methods.

4-71. Same as Prob. 4-33 except use relative-velocity instead of instant-center methods and solve for the phase when the crank angle is 225°.

4-72. Same as Prob. 4-33 except use relative-velocity instead of instant-center methods and solve for the phase when the crank angle is 330°.

4-73. Crank 2 of the mechanism in Fig. P 15 accelerates cw at a uniform rate from 0 to 1,800 rpm in 0.15 sec. The mechanism is shown in the phase corresponding to the time 0.05 sec after starting from rest. Draw the mechanism half size. Use a velocity scale of 1 in. = 50 in./sec and an acceleration scale of 1 in. = 2,000 in./sec². Determine (a) the acceleration of point B and (b) the angular acceleration of link 4.

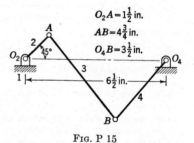

$O_2A = 1\frac{1}{2}$ in.

$AB = 4\frac{3}{4}$ in.

$O_4B = 3\frac{1}{2}$ in.

Fig. P 15

4-74. Crank 2 of the mechanism in Fig. P 15 rotates ccw at a uniform speed of 1,800 rpm. Draw the mechanism half size in the phase shown. Use a velocity scale of 1 in. = 10 ft/sec and an acceleration scale of 1 in. = 2,000 ft/sec². Determine (a) the acceleration of point B and (b) the angular acceleration of link 4.

4-75. Make a half-size drawing of the mechanism in Fig. P 15 for the phase when link 4 is at its extreme left (cw) position. Link 2 is rotating ccw at a uniform speed of 150 rpm. Use a velocity scale of 1 in. = 10 in./sec and an acceleration scale of 1 in. = 200 in./sec². Determine (a) the acceleration of point B and (b) the angular acceleration of links 3 and 4.

4-76. Crank 2 in Fig. P 16 is rotating cw at a uniform speed of 360 rpm. Draw the mechanism in the phase shown to a scale of 1 in. = 6 in. Use a velocity scale of 1 in. = 100 in./sec and an acceleration scale of 1 in. = 3,000 in./sec². Determine (a) the acceleration of point B and (b) the angular acceleration of link 3.

4-77. Same as Prob. 4-76 except solve for the phase when the crank angle is 90°.

4-78. Same as Prob. 4-76 except solve for the phase when the crank angle is 180°.

4-79. Same as Prob. 4-76 except solve for the phase when the slider 4 is at its uppermost position.

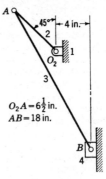

$O_2A = 6\frac{1}{2}$ in.
$AB = 18$ in.

FIG. P 16

4-80. Crank 2 of the mechanism in Fig. P 17 is rotating cw at a uniform speed of 85 rpm. Draw the mechanism in the phase shown to a scale of 1 in. = 1 ft. Use a velocity scale of 1 in. = 5 ft/sec and an acceleration scale of 1 in. = 30 ft/sec². Determine (a) the acceleration of point B, (b) the acceleration of point C, and (c) the angular acceleration of link 6.

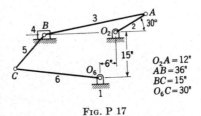

$O_2A = 12"$
$AB = 36"$
$BC = 15"$
$O_6C = 30"$

FIG. P 17

4-81. Same as Prob. 4-80 except solve for the phase when the crank angle < 180° and link 6 is horizontal.

4-82. Link 2 of the Watt straight-line mechanism in Fig. P 18 is rotating cw with a uniform angular velocity of 1.20 rps. Draw the mechanism half size. Use a velocity scale of 1 in. = 10 in./sec and an acceleration scale of 1 in. = 60 in./sec². Determine (a) the acceleration of point B and (b) the angular acceleration of link 3.

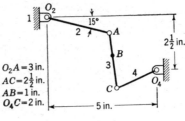

$O_2A = 3$ in.
$AC = 2\frac{1}{2}$ in.
$AB = 1$ in.
$O_4C = 2$ in.

FIG. P 18

4-83. Same as Prob. 4-82 except that link 2 has an instantaneous cw angular velocity of 1.20 rps and a ccw angular acceleration of 3 rps².

4-84. The direct-connected motor accelerates ccw crank 2 of the Sunbeam Shavemaster electric shaver in Figs. 2-4 and P 3 from at rest to 8,600 rpm in 0.2 sec. Use the dimensions in Fig. P 3 and draw the mechanism twice size in the phase when crank 2 is at an angle of 60° with the line O_2O_4. Use a velocity scale of 1 in. = 20 in./sec and an acceleration scale of 1 in. = 3,000 in./sec². For the time 0.05 sec after the current is turned on, determine (a) the acceleration of the cutter (point C), (b) the centrifugal force due to the cutter weight of 0.00093 lb, and (c) the angular acceleration of link 4.

4-85. Same as Prob. 4-84 except solve for the time 0.15 sec after the current has been turned on. Use a velocity scale of 1 in. = 60 in./sec and an acceleration scale of 1 in. = 30,000 in./sec².

4-86. Same as Prob. 4-84 except solve for the time 5 sec after the current has been turned on. Use a velocity scale of 1 in. = 80 in./sec and an acceleration scale of 1 in. = 50,000 in./sec².

4-87. Same as Prob. 4-84 except solve for the phase when links 2 and 3 are collinear (point B is at its farthest-to-the-right position).

4-88. Same as Prob. 4-84 except solve for the phase when links 2 and 3 are collinear (point B is at its farthest-to-the-right position) and the time is 10 sec after the current has been turned on. Use a velocity scale of 1 in. = 80 in./sec and an acceleration scale of 1 in. = 50,000 in./sec².

4-89. The steam locomotive in Fig. P 19 is moving to the right at a constant speed of 70 mph. Draw the mechanism in the phase shown to the scale of 1 in. = 3 ft 0 in. Use a velocity scale of 1 in. = 60 ft/sec and an acceleration scale of 1 in. = 2,000 ft/sec². Determine (a) the acceleration of point C, (b) the acceleration of point D, (c) the angular acceleration of link 3, and (d) the instantaneous center of curvature and the instantaneous radius of curvature of the cycloidal path of point C.

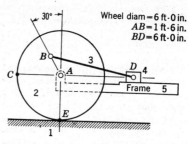

Wheel diam = 6 ft-0 in.
AB = 1 ft-6 in.
BD = 6 ft-0 in.

FIG. P 19

HINT: Utilize the fundamental principles for the acceleration of a point, as discussed in Sec. 3-4.

4-90. Link 2 of the crank-shaper mechanism in Fig. P 20 is rotating cw at a uniform speed of 120 rpm. Determine the magnitude and direction of the acceleration of the ram 6 when $\theta = 60°$. Use scales of 1 in. = 5 in., 1 in. = 3 fps, and 1 in. = 30 ft/sec².

4-91. Same as Prob. 4-90 except $\theta = 30°$.

4-92. Same as Prob. 4-90 except the slider is in its extreme position to the right.

4-93. Same as Prob. 4-90 except $\theta = 135°$ and use a scale of 1 in. = 50 ft/sec².

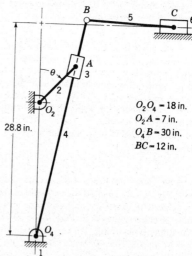

$O_2 O_4 = 18$ in.
$O_2 A = 7$ in.
$O_4 B = 30$ in.
$BC = 12$ in.

FIG. P 20

4-94. Link 2 of the Whitworth quick-return mechanism in Fig. P 7 is rotating ccw at a uniform speed of 600 rpm. Determine the acceleration of the slider 6 when link 2 is at an angle of 120° ccw from O_2O_4.   Use scales of 1 in. = 4 in., 1 in. = 200 in./sec, and 1 in. = 10,000 in./sec².

4-95. Same as Prob. 4-94 except link 2 is at an angle of 90° cw from O_2O_4.

4-96. Same as Prob. 4-94 except slider 6 is in its lowest position.

4-97. A recording speedometer was used in an acceleration test on an automobile with a new variation of torque-converter transmission.   The data are given in Table P 1 in the form of the velocity at 1-sec time intervals measured from the instant the car started moving.

TABLE P 1

Time, sec.....	0	1	2	3	4	5	6	7	8	9	10	11
Velocity, mph	0	8.9	17.6	26.0	33.7	40.2	45.7	50.3	54.3	57.9	61.2	64.3

(a) Plot the velocity-time curve for the test data.   Use a time scale of 1 in. = 3 sec and a velocity scale of 1 in. = 20 ft/sec.   (b) Graphically differentiate the velocity-time curve and plot the acceleration-time curve.   Use the same time scale as in (a) and choose the distance for the horizontal leg of the slope triangle (PB in Fig. 4-24) so that the vertical leg (BC in Fig. 4-24) will be the acceleration at a scale of 1 in. = 4 ft/sec².   (c) When does the maximum acceleration occur?   Is this reasonable? What is the limiting factor at this time?   (d) What is the acceleration when the automobile reaches 40 mph?   60 mph?   (e) What power is required, over and above that for overcoming friction, wind resistance, etc., to give the 4,200-lb automobile the acceleration in (d) at 60 mph?

HINT: hp = $FV/33,000$, where F is in pounds and V is in feet per minute.

4-98. One phase of the performance testing of an automobile with a new-design torque-converter transmission consisted of accelerating from rest to over 60 mph in the shortest possible time.   The record for the test consists of data, as given in Table P 2, in terms of the distance from the starting point at time intervals of 1 sec.

TABLE P 2

Time, sec......	0	1	2	3	4	5	6	7	8	9	10	11	12	13	14
Distance, ft....	0	4	18	42	75	116	165	221	284	353	428	509	595	686	786

(a) Plot the test data as a displacement-time curve.   Use a displacement scale of 1 in. = 100 ft and a time scale of 1 in. = 3 sec.   (b) Graphically differentiate the displacement-time curve and plot the velocity-time curve for the test data.   Use the same time scale as in (a) and choose the distance for the horizontal leg of the slope triangle (PB in Fig. 4-24) so that the vertical leg (BC in Fig. 4-24) becomes the velocity at a scale of 1 in. = 20 ft/sec.   (c) How many seconds elapsed before the automobile reached a speed of 60 mph?   (d) Graphically differentiate the velocity-time curve and plot the acceleration-time curve for the test data.   Use the same time scale as in (a) and (b) and choose the distance for the horizontal leg of the slope triangle so that the vertical leg becomes the acceleration at a scale of 1 in.. = 4 ft/sec².   (e) What is the value of acceleration at the instant the automobile reaches 60 mph?   What horsepower over and above that for overcoming friction, wind resistance, etc., is required to give a 4,200-lb automobile this acceleration at 60 mph?

HINT: hp = $FV/33,000$, where F is in pounds and V is in feet per minute.

4-99. A disk cam with a flat-face follower, similar to Fig. 5-1, has been used in a machine in which the cam 2 rotates at 200 rpm.   It is proposed to increase production by modifying the machine to permit operation at 450 rpm.   One of the critical points is that of the motion-time characteristics of the cam-follower mechanism, particularly the acceleration of the follower. The original design specifications are not available, and the only information on hand is that determined by the gage laboratory in terms of follower displacement as a function of the angular position of the cam. The displacement–cam-position data are given in Table P 3.

TABLE P 3

Cam position, deg.......	0	15	30	45	60	75	90	105	120	135	150	165
Displacement, in.........	0.000	0.016	0.064	0.144	0.257	0.402	0.562	0.723	0.867	0.980	1.060	1.109
Cam position, deg.......	180–	195	210	225	240	255	270	285	300	315	330	345
Displacement, in.........	1.125	1.125	1.125	1.125	1.125	1.082	0.960	0.778	0.562	0.347	0.165	0.043

(a) Use graphical differentiation to determine the velocity-time and acceleration-time diagrams for the motion of the follower. It is suggested that the cam-position scale (time axis) be 1 in. = 60° and that the displacement scale be 1 in. = ¼ in. It is also suggested that the horizontal legs of the slope triangles (PB in Fig. 4-24) be chosen so that the vertical leg (BC in Fig. 4-24) becomes in the first differentiation the velocity at a scale of 1 in. = 20 in./sec and in the second differentiation the acceleration at a scale of 1 in. = 1,000 in./sec². (b) What is the maximum acceleration of the follower in inches per second per second and in g's, that is, the ratio of the follower acceleration to the acceleration of gravity? (c) Will the follower be expected to remain in contact with the cam at all times without the application of an external force, such as by a compressed spring? Explain.

4-100. The information needed to make a velocity-time diagram for point C on the Sunbeam Shavemaster mechanism in Fig. P 3 was determined in Probs. 3-20 and/or 4-27. (a) Complete the calculations and plot the velocity-time diagram with the 12 crank positions equally spaced along a 6-in. base line and using a velocity scale of 1 in. = 100 in./sec. (b) Graphically differentiate the velocity-time curve and plot the acceleration-time curve. Use the same time scale as for the velocity-time diagram, and choose the distance for the horizontal leg of the slope triangle (PB in Fig. 4-24) so that the vertical leg (BC in Fig. 4-24) will be the acceleration at a scale of 1 in. = 100,000 in./sec².

4-101. The displacement of the ram (point C) of the crank-shaper mechanism in Fig. P 6 was plotted in Prob. 2-11 as a function of the position of crank 2. The crank is rotating at a uniform speed of 35 rpm. Use the same time scale for all diagrams. (a) Graphically differentiate the displacement-time curve to get the velocity-time curve. Choose the length of the horizontal leg of the slope triangle (PB in Fig. 4-24) so that the vertical leg (BC in Fig. 4-24) is the velocity at the scale of 1 in. = 20 in./sec. (b) Graphically differentiate the velocity-time curve and plot the acceleration-time curve. Choose the length of the horizontal leg of the slope triangle so that the vertical leg is the acceleration at a scale of 1 in. = 100 in./sec².

4-102. The displacement-time diagram for the slider (point C) of the Whitworth quick-return mechanism in Fig. P 7 was plotted in Prob. 2-14 with the time axis represented by 24 equal divisions of the crank circle. Crank 2 is rotating ccw at a uniform speed of 180 rpm. Use the same time scale for all diagrams. (a) Graphically differentiate the displacement-time curve and plot the velocity-time curve. Choose the length of the horizontal leg of the slope triangle (PB in Fig. 4-24) so that the vertical leg (BC in Fig. 4-24) is the velocity at a scale of 1 in. = 60 in./sec. (b) Graphically differentiate the velocity-time curve and plot the acceleration-time curve. Choose the length of the horizontal leg of the slope triangle so that the vertical leg is the acceleration at a scale of 1 in. = 2,000 in./sec²

4-103. Figure P 21 is an oscillographic record of data taken during the testing of a 330-kv 25,000-mva circuit breaker. The acceleration-time characteristics of the motion of the lift rod in the period immediately following the release of the trigger is of major interest. The displacement-time data for the lift rod during the first 0.05 sec after the trigger is released are given in Table P 4.

TABLE P 4

Time, sec	0	0.005	0.010	0.015	0.020	0.025	0.030	0.035	0.040	0.045	0.050
Displacement, in	0	0.07	0.25	0.58	1.19	1.97	2.88	3.80	4.72	5.64	6.54

(a) Plot the displacement-time curve using a time scale of 1 in. = 0.01 sec and a displacement scale of 1 in. = 2 in. (b) Graphically differentiate the displacement-time curve and plot the velocity-time curve. Choose the length of the horizontal leg of the slope triangle (PB in Fig. 4-24) so that the vertical leg (BC in Fig. 4-24) will be the velocity at a scale of 1 in. = 50 in./sec. (c) Graphically differentiate the velocity-time curve and plot the acceleration-time curve. Choose the length of the horizontal leg of the slope triangle so that the vertical leg will be the acceleration to a scale of 1 in. = 5,000 in./sec².

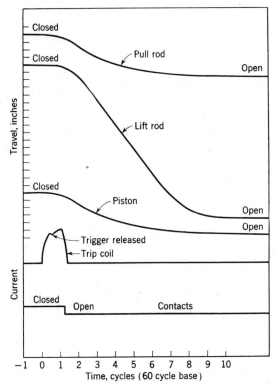

FIG. P 21. (*From R. C. Van Sickle, A Loading Fixture for Testing Power Circuit-breaker Mechanisms, Paper 54-SA-18, American Society of Mechanical Engineers, New York, 1954.*)

Chapter 5.   Cams

NOTE: It is suggested that the cam layouts in Probs. 5-9 to 5-20 be made full size if an $8\frac{1}{2}$- by 11-in. sheet of paper is used or twice size if a C-size (17-in. by 22-in.) sheet is used.

5-1. Draw the base curve for a cam follower with a total displacement of $1\frac{1}{2}$ in. and the following motion characteristics: rise with uniformly accelerated and decelerated motion during $\frac{8}{24}$ revolution of the cam, dwell for $\frac{6}{24}$ revolution, and return with uniformly accelerated and decelerated motion during the remainder of the revolution.

5-2. Draw a base curve for a cam follower with a total displacement of $1\frac{1}{2}$ in. and the following motion characteristics: rise with simple harmonic motion during $\frac{8}{24}$ revolution of the cam, dwell for $\frac{6}{24}$ revolution, and return with simple harmonic motion during the remainder of the revolution.

5-3. Draw the base curve for a cam follower with a total displacement of $1\frac{1}{2}$ in. and the following motion characteristics: rise with simple harmonic motion during $\frac{10}{24}$ revolution of the cam, dwell for $\frac{2}{24}$ revolution, and return with uniformly accelerated and decelerated motion during the remainder of the revolution.

5-4. Draw the base curve for a cam follower with a total displacement of $1\frac{1}{2}$ in. and the following motion characteristics: rise with a combination of uniformly accelerated, constant-velocity, and uniformly decelerated motion during $\frac{4}{24}$, $\frac{4}{24}$, and $\frac{4}{24}$ revolution, respectively, of the cam; dwell for $\frac{6}{24}$ revolution; and return with simple harmonic motion during the remainder of the revolution.

5-5. Draw the base curve for a cam follower with a total displacement of $1\frac{1}{2}$ in. and the following motion characteristics: rise to 1-in. displacement with uniformly accelerated and decelerated motion in $\frac{8}{24}$ of a revolution of the cam, dwell at 1-in. displacement for $\frac{3}{24}$ revolution, continue rise to $1\frac{1}{2}$-in. displacement in $\frac{6}{24}$ revolution, dwell for $\frac{1}{24}$ revolution, and return with uniformly accelerated and decelerated motion in the remainder of the revolution.

5-6. Draw the base curve for a follower with a total displacement of $1\frac{1}{2}$ in. and the following motion characteristics: dwell at zero displacement for $\frac{2}{24}$ revolution of the cam, rise to $\frac{3}{4}$-in. displacement with simple harmonic motion during $\frac{6}{24}$ revolution, dwell at $\frac{3}{4}$-in. displacement for $\frac{2}{24}$ revolution, continue rise to $1\frac{1}{2}$-in. displacement with simple harmonic motion during $\frac{6}{24}$ revolution, and return with uniformly accelerated and decelerated motion in the remainder of the revolution.

5-7. Draw the base curve for a follower with a total displacement of $1\frac{1}{2}$ in. and the following motion characteristics: rise with a combination of uniformly accelerated, constant-velocity, and uniformly decelerated motions during $\frac{4}{24}$, $\frac{3}{24}$, and $\frac{6}{24}$ revolution, respectively, of the cam; return to zero displacement with uniformly accelerated and decelerated motion in $\frac{10}{24}$ revolution; and dwell for the remainder of the revolution.

5-8. Draw the base curve for a follower with a total displacement of $1\frac{1}{2}$ in. and the following motion characteristics: rise with a combination of uniformly accelerated, constant-velocity, and uniformly decelerated motions such that the displacements are $\frac{1}{2}$ in., $\frac{1}{2}$ in., and $\frac{1}{2}$ in., respectively, during a total of $\frac{10}{24}$ revolution of the cam; dwell for $\frac{2}{24}$ revolution; return to zero displacement with simple harmonic motion in $\frac{8}{24}$ revolution; and dwell for the remainder of the revolution.

5-9. Draw the base curve for a follower with a total displacement of $1\frac{1}{2}$ in. and the following motion characteristics: rise with cycloidal motion during $\frac{8}{24}$ revolution of the cam, dwell for $\frac{6}{24}$ revolution, and return with cycloidal motion during the remainder of the revolution.

5-10. (a) Lay out the profile of a disk cam with a reciprocating flat-face follower.

The cam is to rotate in the cw (ccw) direction, the minimum distance from the cam axis of rotation to the follower face is to be 1 in., and the follower motion is to be that given in Prob. 5-1 (5-2, 5-3, 5-5, 5-6, 5-7, 5-8, 5-9).   (b) Use dotted lines to show the follower in its inverted position at which the maximum face length is required.   (c) Draw the follower in its zero position.   It is to be symmetrical about its axis, and the face is to extend $\frac{1}{8}$ in. beyond the maximum length required for proper action.

5-11. (a) Lay out the profile of a disk cam with a reciprocating flat-face follower. The flat face is to be at an angle of 60°, measured cw (ccw), with the follower axis; the cam is to rotate in a cw (ccw) direction; the minimum distance from the cam axis of rotation to the point of intersection of the follower face and the follower line of motion is to be $1\frac{1}{2}$ in.; and the follower motion is to be that given in Prob. 5-1 (5-2, 5-3, 5-5, 5-6, 5-7, 5-8, 5-9).   (b) Use dotted lines to show the follower in its inverted position at which the maximum face length is required.   (c) Draw the follower in its zero position.   The face is to extend both sides of the follower axis $\frac{1}{8}$ in. beyond the maximum length required for proper action.

5-12. It is proposed to design a disk cam with a flat-face follower to have the motion characteristics of Prob. 5-4 when the minimum distance from the cam axis of rotation to the follower face is $\frac{3}{4}$ in.   The cam is to rotate cw (ccw).   (a) Will the proposed design be satisfactory? Explain.   (b) What is the minimum value of the minimum distance (to the nearest $\frac{1}{8}$ in.) from the cam axis of rotation to the follower face that will permit the specified displacement-time characteristic of the follower?   (c) Lay out the cam profile on the basis of the dimension determined in (b).   (d) Draw the follower symmetrical about its line of motion, and make the face length extend $\frac{1}{8}$ in. beyond the maximum required for proper action.

SUGGESTED PROCEDURE: Study the base curve to determine whether the outward or the return stroke is more likely to be critical. If this is not apparent, it will be necessary to lay out the entire profile in order to determine whether or not the cam will be satisfactory and, if not, to determine the critical region.   After the critical part of the motion has been located, make twice-size (1 in. = $\frac{1}{2}$ in.) layouts of that portion of the cam profile, using minimum distances from the cam axis of rotation to the follower face of $\frac{7}{8}$ in., 1 in., $1\frac{1}{8}$ in., etc., until proper action is achieved.

5-13. (a) Lay out the profile of a disk cam with a radial roller follower.   The cam is to rotate in the cw (ccw) direction, the minimum distance from the cam axis of rotation to the center of the roller is to be 1 in., the roller diameter is to be $\frac{3}{4}$ (1, $1\frac{1}{2}$) in., and the follower motion is to be that given in Prob. 5-1 (5-2, 5-3, 5-4, 5-5, 5-6, 5-7, 5-8, 5-9).   (b) Draw the follower in its zero position.   (c) Determine the maximum pressure angle, and use dotted lines to show the follower in its inverted position at which the maximum pressure angle occurs.

5-14. Same as Prob. 5-13 except design a face cam instead of a simple disk cam.

5-15. (a) Lay out the profile of a disk cam with an offset roller follower.   The vertical follower line of motion is to be offset $\frac{1}{2}$ in. to the right (left) of a vertical line through the cam axis of rotation, the minimum distance from the cam axis of rotation to the center of the roller is to be $1\frac{1}{4}$ in., the roller diameter is to be $\frac{3}{4}$ (1, $1\frac{1}{2}$) in., the cam is to rotate cw (ccw), and the follower motion is to be that given in Prob. 5-1 (5-2, 5-3, 5-4, 5-5, 5-6, 5-7, 5-8, 5-9).   (b) Draw the follower in its zero position.   (c) Determine the maximum pressure angle, and use dotted lines to show the follower in its inverted position at which the maximum pressure angle occurs.

5-16. (a) Lay out the profile of a disk cam with an oscillating flat-face follower. The follower is shown in its zero position in Fig. P 22.   The cam is to rotate in the cw (ccw) direction, and the follower is to oscillate through a 20° angle.   The motion is to be that given in Prob. 5-1 except that the linear displacement of $1\frac{1}{2}$ in. now becomes an angular displacement in degrees.   In the interest of accuracy and convenience, it is desirable to utilize the relationship between arc distances and angles in designing the cam. In this case the 20° angle may best be considered as an arc distance of 1½ in. at a radius of 4.30 in. (NOTE: The designation of angles by arc distances permits also the direct use of the base curves in Probs. 5-2, 5-3, 5-4, 5-5, 5-6, 5-7, 5-8, and 5-9 to describe the motion-time characteristics of the oscillating flat-face follower.) (b) Draw the follower in its zero position, making the length of the face extend ⅛ in. beyond the maximum lengths required for correct action.

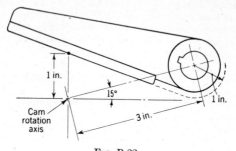

FIG. P 22

5-17. (a) Lay out the profile of a disk cam with an oscillating roller follower. The follower is shown in its zero position in Fig. P 23. The cam is to rotate in the cw (ccw) direction, the roller diameter is to be ¾ (1, 1½) in., and the follower oscillation is to be defined in terms of the displacement of the roller center along its arc of travel. The follower is to have the displacement-time characteristics given in Prob. 5-1 (5-2, 5-3, 5-4, 5-5, 5-6, 5-7, 5-8, 5-9). (b) Design the follower arm so that there will be at least ⅛-in. clearance between the cam and the arm. Use dotted lines to show the follower in its inverted position when clearance is minimum. (c) Draw the follower at its zero position.

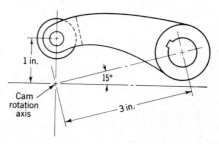

FIG. P 23

5-18. (a) Lay out the profile of a positive-return single-disk cam with a reciprocating flat-face yoke follower (constant-breadth cam). The cam is to rotate cw (ccw), the minimum distance from the cam axis of rotation to the follower face is to be 1 in., and the follower motion is to be the applicable part of that given in Prob. 5-1 (5-2, 5-3, 5-4, 5-8, 5-9). (b) Draw the follower in its zero position. There is to be at least ⅛-in. clearance between the cam and the sides of the yoke.

5-19. (a) Lay out the profile of a positive-return single-disk cam with a reciprocating roller follower (constant-diameter cam). The cam is to rotate cw (ccw), the minimum distance from the cam axis of rotation to the roller centers is to be 1 in., the roller diameters are ¾ (1, 1½) in., and the follower motion is to be the applicable part of that given in Prob. 5-1 (5-2, 5-3, 5-4, 5-8, 5-9). (*b*) Draw the follower in its zero position. There is to be at least ⅛-in. clearance between the cam and the sides of the yoke. (*c*) Determine the maximum pressure angle. At what position(s) does it occur?

5-20. (*a*) Lay out the profiles of the disks for a double-disk positive-return cam with a reciprocating flat-face yoke follower. The cam is to rotate cw (ccw), the minimum distance from the cam axis of rotation to the follower faces is to be 1 in., and the follower motion is to be that given in Prob. 5-1 (5-2, 5-3, 5-4, 5-5, 5-6, 5-7, 5-8, 5-9). (*b*) Draw the follower in its zero position. There is to be at least ⅛-in. clearance between the cam and all nondriven surfaces on the follower.

5-21. (*a*) Lay out the profiles of the disks for a double-disk positive-return cam with a reciprocating roller follower. The cam is to rotate cw (ccw), the minimum distance from the cam axis of rotation to the roller centers is to be 1 in., the roller diameters are ¾ (1, 1½) in., and the follower motion is to be that given in Prob. 5-1 (5-2, 5-3, 5-4, 5-5, 5-6, 5-7, 5-8, 5-9). (*b*) Draw the follower in its zero position. There is to be at least ⅛-in. clearance between the cam and all nondriven surfaces on the follower. (*c*) Determine the maximum pressure angle. At what position does it occur?

5-22. Use instant centers to prove that the minimum length of face for a reciprocating flat-face follower is independent of the cam size and is dependent only upon the ratio of the maximum follower velocity to the angular velocity of the cam.

Chapter 6.   Fundamentals of Dimension Determination

6-1. The link in Fig. P 24 is to be part of a machine used for pouring concrete in highway construction. Under normal use, the tension load P on the rod varies between 4,000 and 6,000 lb several times each hour. The rod and rod ends are forged as an integral part from hot-rolled 1020 steel. The only operations subsequent to forging are the machining, drilling, and reaming required at the ends. The pins are to be made from cold-drawn 1020 steel. Specify (*a*) the rod diameter d_1; (*b*) the pin diameter d_2, assuming no bending of the pin; (*c*) the eye diameter D; and (*d*) the width of the rod end w, assuming that $w = 4a$.

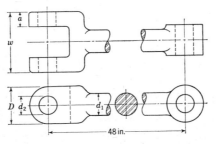

FIG. P 24

6-2. The link in Fig. P 24 is to operate under loading that varies from 1,000 to 3,500 lb once every 5 sec. The machine is used indoors, and reliability is a major consideration. Determine the rod diameter d_1 on each of the following bases: (*a*) The rod is made from hot-rolled 1020 steel, and the surface is to be "as received" from the mill; (*b*) the rod is made from hot-rolled 1020 steel, and the rod is ground to diameter after the ends are formed; (c) the rod is ground to size from 1040 steel which has been quenched in oil and tempered at 1100°F after the ends are formed.

6-3. The Safety Code for Elevators, Dumbwaiters, and Escalators[1] states that the factor of safety for all parts of the driving machine is to be 8 if the elongation is >14 per cent in 2 in. and 10 if the elongation is <14 per cent in 2 in. Assuming that, as in most codes, the design is to be made using equations from elementary strength of materials, for example, $s = P/A$, $s = Mc/I$, etc., and the ultimate strength of the materials, what are more realistic values of the factor of safety for parts designed in accordance with the code when (a) the part is made from hot-rolled 1020 steel and carries a static load and (b) the part has a maximum thickness of 2 in.; the surface is machined; the stress-concentration factor $K_f = 1.80$; the material is 1040 steel, quenched in oil from 1575°F and tempered at 1000°F; and the part carries a load that reverses its direction many times a second?

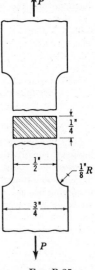

6-4. The link in Fig. P 25 was designed to carry a dead-weight load of 8,000 lb. The link is machined from 3140 steel that has been quenched in oil and tempered at 1000°F. What is the factor of safety?

6-5. The link in Fig. P 25 is machined from hot-rolled 1020 steel. What is the factor of safety when the link is carrying a load that varies repeatedly between 1,000 and 3,000 lb?

6-6. A link, with the same relative proportions as in Fig. P 25, is to carry a load that varies repeatedly between 5,000 and 12,000 lb. The link is to be machined from hot-rolled 1030 steel. What width should be specified for a factor of safety of 2½?

Fig. P 25

6-7. Same as Prob. 6-6 except the link is to be made from 3140 steel and it will be ground all over after the rough-machined part has been quenched in oil and tempered at 1100°F.

6-8. A link with the same relative proportions as in Fig. P 25 is to carry a load that varies repeatedly between 5,000 and 12,000 lb. Assuming that the yield strength of the material is 85,000 psi and that fatigue tests on machined specimens result in a plot identical with that in Fig. 6-1, what width is required if not more than 2.5 per cent of the parts are to fail in 200,000 cycles of load application?

Note: The relation of the yield strength to fatigue failures in a finite life is even less well understood than for infinite life as discussed in Sec. 6-5. For the purpose of this problem it is necessary to assume that the yield strength value is statistically compatible with the fatigue test data.

6-9. Same as Prob. 6-8 except not more than 0.15 per cent of the parts are to fail in 700,000 cycles.

[1] ASA A17.1-1960, American Society of Mechanical Engineers, New York, 1960.

6-10. Same as Prob. 6-8 except not more than 2.5 per cent of the parts are to fail in 100 million cycles.

6-11. The handbooks give empirical formulas for calculation of the diameter of shafts for use in applications where the horsepower and speed are known but it is difficult to calculate the bending moments. Examples are head shafts and line shafts with several bearings and pulleys. One such formula for a head shaft made from cold-rolled shafting is

$$d = \sqrt[3]{\frac{100 \times hp}{n}}$$

where hp = horsepower transmitted and n = speed in rpm. (a) Assuming that the material is the equivalent of 1020 steel, determine the apparent factor of safety based on torsional stress only. (b) Does the value seem reasonable? Explain.

6-12. A 15-hp motor-generator set is being designed for a portable test stand for use in the aircraft industry. The motor and generator are to run at 3,500 rpm (3,600 nominal). To facilitate servicing of the set, the motor and generator will be made as separate units and will be connected by means of a flexible coupling (Chap. 11) that will not transmit a bending moment from one shaft to the other. The motor breakdown torque is 200 per cent of rated torque and the electrical control circuit will not permit the generator output to exceed 110 per cent of rated capacity. The generator dimensions given in Fig. P 26 have been determined largely by the electrical requirements. Determine the diameter d required for the generator shaft if it is ground all over and the material is 1040 steel, quenched in oil and tempered at 900°F.

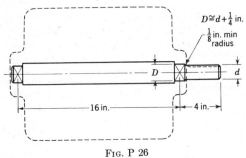

Fig. P 26

6-13. The shaft in Fig. P 27 rotates but carries no torque. The material is 3140 steel, quenched in oil and tempered at 1000°F, and the shaft is ground after heat-treating. What is the maximum load P that the shaft can carry with a factor of safety of 3?

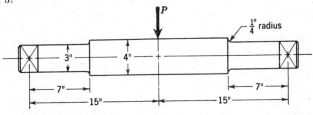

Fig. P 27

6-14. The NEMA standards state that, when a frame No. 326U motor is used with a flat-belt drive to transmit 20 hp at 1,200 rpm, the pulley should have a diameter of at least 7 in.   The actual belt forces will depend upon a number of factors (Chap. 14), but a typical application under these conditions would have a resultant force perpendicular to the motor-shaft axis of about 1,200 lb.   The dimensions U and $N - W$ of the shaft in Fig. P 28 are the same for all NEMA frame No. 326U motors, while the other dimensions vary somewhat from manufacturer to manufacturer.   (a) Assuming that the belt forces act in a plane at the mid-point of the $N - W$ portion of the shaft and that the shaft is machined from hot-rolled 1040 steel, what is the factor of safety?   (b) What is the factor of safety if the axial position of the pulley is such that the plane of the belt forces is at the end of the shaft extension?

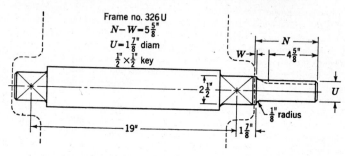

Frame no. 326U
$N - W = 5\frac{5}{8}"$
$U = 1\frac{7}{8}"$ diam
$\frac{1}{2}" \times \frac{1}{2}"$ key

FIG. P 28

6-15. The NEMA standards state that, when a frame No. 326U motor is used with a chain drive (Chap. 15) to transmit 30 hp at 1,800 rpm, the sprocket pitch diameter should be at least 3.67 in.   Under these conditions, the chain force is approximately 570 lb.   The dimensions U and $N - W$ of the shaft in Fig. P 28 are the same for all NEMA frame No. 326U motors, while the other dimensions vary somewhat from manufacturer to manufacturer.   (a) Assuming that the chain force acts in a plane at the mid-point of the $N - W$ portion of the shaft and that the shaft is machined from hot-rolled 1040 steel, what is the factor of safety?   (b) What is the factor of safety if the axial position of the sprocket is such that the chain force acts at the end of the shaft extension?   (c) For what reasons is it not desirable to use sprockets with pitch diameters less than the specified minimum?

6-16. The NEMA standards as given in Appendix G permit the use of frame No. 326U for motors transmitting 30 hp at 1,800 rpm, 20 hp at 1,200 rpm, and 15 hp at 900 rpm.   The dimensions of a typical shaft for a frame No. 326U motor are shown in Fig. P 28.   (a) In terms of factor of safety, are the power-speed combinations consistent?   (b) The NEMA standards include motor nominal speeds of 720, 600, 514, and 450 rpm in addition to those listed in Appendix G.   What will be the maximum standard horsepower rating for a frame No. 326U motor at these additional speeds?

6-17. The railroad freight-car axle in Fig. P 29 is being designed to carry a rated load of 24,000 lb.   The axle is to be turned from a normalized 1040 steel forging, the journal (diameter d_1) is to be ground, and the wheels are to be press-fitted on the axle. If $d_2/d_1 = 1.25$ and $d_3/d_2 = 1.1$, what are the minimum dimensions you would recommend for diameters d_1 and d_2?

6-18. Most railroad cars, particularly freight and Pullman cars, are used on a large number of different railroads and may require servicing while away from the owning railroad.   As a result, the American Association of Railroads (AAR) has issued standards covering the dimensions, materials, etc., for many of the important parts of the cars.   For example, a freight-car axle designed to carry a load of 42,000 lb

(21,000 lb per wheel) is shown in Fig. P 29, where $d_1 = 5\frac{1}{2}$ in., $d_2 = 6\frac{5}{8}$ in., $d_3 = 7$ in., $d_4 = 6\frac{3}{4}$ in., and $d_5 = 5\frac{7}{8}$ in. If the axle is turned from a normalized 1040 steel forging, the journal (diameter d_1) is ground, and the wheels are press-fitted on the axle, what is the factor of safety under normal operating conditions?

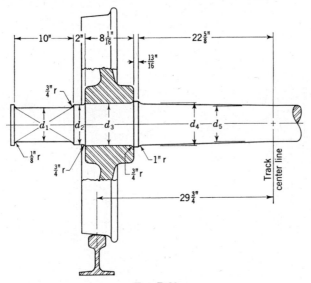

Fig. P 29

6-19. What is the per cent decrease in strength if a solid steel shaft is replaced by a hollow shaft made from the same material and with the same outside diameter but with $d_i = 0.6d$? What is the per cent decrease in weight?

6-20. The offset bell crank in Fig. P 30 is to be part of the clutch linkage on a heavy-duty road grader. The tubular part K is supported by a solid steel pin through the bushings so that the tube carries only a torque load. Assuming that the tubing is welded, low-carbon steel (equivalent to hot-rolled 1020) and that 0.01 in. must be machined from the inside surface so that the bushings will seat properly, what do you recommend for the outside diameter (OD) of the tubing if the inside diameter (ID) is to be $0.8 \times$ OD and the force P on the rod A is 100 lb?

6-21. What dimensions do you recommend for the section E-E of the offset bell crank in Fig. P 30 when the lever is machined from hot-rolled 1030 steel plate, the load P varies slowly and infrequently between 500 and 1,500 lb, the thickness of the plate is to be one-third the depth of the section, the radius of curvature r is to be three-fourths the depth of the section, and human life will be endangered if the part fails?

6-22. Same as Prob. 6-21 except the load varies fifteen times each minute.

6-23. The load P on rod A in Fig. P 30 varies several times each minute between 1,500 lb to the left and 500 lb to the right. The bell crank is part of a feeding device, and failure would result in shutting down an assembly line involving 35 persons and a number of expensive machines. Assuming that pins D and F are to be ground from 1030 steel rod that has been quenched in water and tempered at 1000°F, what diameter do you recommend if the pins are to be identical (to minimize the number of spare parts)?

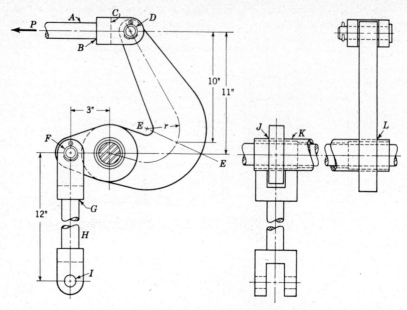

Fig. P 30

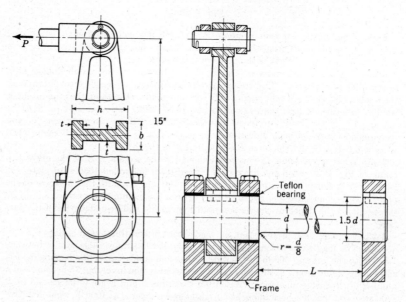

Fig. P 31

6-24. Figure P 31 shows a proposed scheme for a torsional spring. The lever is to rotate through an angle of 30° (±15° from the position shown) when the load P varies in service between 1,000 and 3,000 lb. The crank, or lever, is to be forged from 1020 steel, and only the bearing surfaces will be machined. What do you recommend for the value of h at a distance of 10 in. from the center of the small pin if $b = h/2$, $t = h/4$, the load varies several times each minute, and it would be inconvenient but not dangerous if the part fails?

6-25. What do you recommend for the diameter d and length l of the torsion spring in Prob. 6-24 if the rod is to be ground all over and the material is 3140 steel, quenched in oil and tempered at 1000°F?

NOTE: The shear modulus of elasticity of steel is approximately 11.5×10^6 psi.

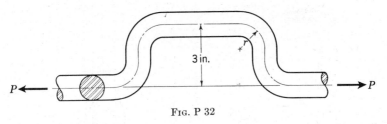

FIG. P 32

6-26. A control rod (Fig. P 32) is to be used under an essentially static tension load of 400 lb. The rod must be bent as shown to clear other parts of the machine. The rod is to be made from cold-drawn 1020 steel rod, and the radius of curvature r of the rod centerline is to be not less than the rod diameter d. What is the minimum diameter you would recommend for the rod?

NOTE: Include the stresses due to both the tension load and the bending moment

6-27. The lever in Fig. P 33 is to be heat-treated after it is forged. The material will be 3140 steel, and it will be quenched in oil and then tempered at 1200°F. The relative proportions of the I sections will correspond to those given in Fig. 6-20. The radius of curvature of the neutral axis of the I section is to be at least 0.75 times the total depth of the section. Determine the required dimensions of the lever sections A-A, B-B, and C-C.

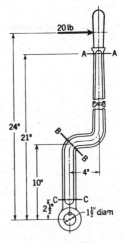

FIG. P 33

6-28. The shop has expressed an urgent need for a 10-ton-capacity press for use in an assembly operation of an experimental production run of a new product. Since immediate delivery of a suitable press cannot be secured, it is decided to build one from materials and parts on hand. The proposed design in Fig. P 34 utilizes a hydraulic cylinder and a welded frame made largely from parts that are flame-cut from steel plate. The main part of the frame (the C-shaped side plates) consists of two identical pieces of plate. A relief valve will be used in the hydraulic circuit to prevent the pressure exceeding the value which gives a force of 10 tons, and the material is to be hot-rolled 1020 steel plate. What is the minimum thickness t you would recommend for the side plates?

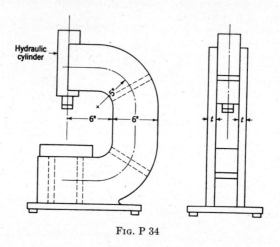

Fig. P 34

6-29. The link in Fig. P 24 is to be made from hot-rolled 1020 steel and must carry a 20,000-lb load in compression. What is the minimum dimension you would recommend for d_1?

6-30. Same as Prob. 6-29 except that the rod is to have a rectangular cross section with proportions such that the link has equal strengths, as a column, in the planes of and perpendicular to the pin axes.

6-31. The load P on rod A of the offset bell crank in Fig. P 30 varies infrequently between zero and 400 lb. What diameter do you recommend for rod H if the rod and the ends (clevises) are machined integrally from a 1030 steel upset forging and only minor inconvenience will result from a failure of the part?

6-32. The rod H and rod ends (clevises) in Fig. P 30 are to be machined integrally from a 1030 steel upset forging. The load P varies many times each minute between zero and 400 lb. Assuming that the transition between the rod and rod ends is equivalent to a two-to-one increase in diameter and that the radius at the transition is 0.2 times the rod diameter, what do you recommend for the rod diameter if only minor inconvenience will result from a failure of the part?

6-33. The load P on rod A of the offset bell crank in Fig. P 30 varies infrequently between 1,000 and 3,000 lb. What diameter do you recommend for rod H if it is made from cold-drawn 1030 steel and only minor inconvenience will result from a failure of the part?

6-34. An experimental investigation of the stresses introduced in truck-tire rims under different combinations of radial and axial loads—simulating conditions on straight roads and in turning corners—necessitated the design and construction of a special testing machine. The major consideration was to design the loading and supporting structure so that the axial and radial loads would be, at least for practical purposes, independent of each other. The scheme selected for applying the radial load is shown in Fig. P 35. The platform corresponds to the road surface and is maintained horizontal at all times by the use of a spherical linkage consisting of parallel linkages in two perpendicular planes. Since the angle through which the links move is small, the radial load F_R is carried, essentially, by the hydraulic jack only, and the axial load F_A is carried, essentially, by the links only. Thus, the pressure in the hydraulic-jack cylinder may be related directly to the radial load on the tire.

The links were made from $1\frac{1}{4}$-in. standard black pipe (OD = 1.660 in., ID = 1.380 in.), with spherical rod ends (Sec. 18-19) attached to each end of the pipe. (a) Assuming that the material properties of the pipe approximate those of hot-rolled

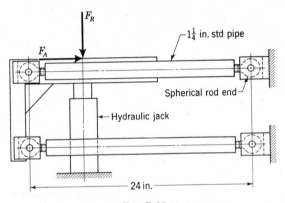

FIG. P 35

1015 steel, what is the maximum load the links may carry in compression? (b) What is the maximum load you would recommend that the links carry in an experimental apparatus such as this?

6-35. A loading device similar to that shown in Fig. P 35 and discussed in Prob. 6-34 is being designed for use under conditions such that the maximum compression load on the links will be 25,000 lb. If the links are to be made from cold-drawn 1020 steel tubing with a wall thickness equal to 0.1 of the outside diameter, what diameter of tubing would you recommend in this application?

6-36. The Traxcavator (Fig. P 36) is a tractor-excavator used in earth-moving applications as a bulldozer, grader, loader, and carrier. The bucket and linkage positions during digging and grading operations correspond closely to those shown in solid lines, and the positions when dumping are shown in dotted lines. The maximum load expected to be carried by each lift cylinder and piston rod is 37,500 lb in compression when the bucket is in the position shown by the solid lines. In this case, the unsupported length of the piston rod in the cylinder, i.e., between the piston and the cylinder head, is 34 in. However, when lifting a normal load of $1\frac{1}{4}$ yd³ of gravel and dirt (3,600 lb) to the height necessary for dumping the load into a truck, each lift cylinder and piston rod is subjected to a compressive load of 17,000 lb.

What minimum diameter would you recommend for a solid piston rod on the basis of using a low-carbon steel with properties similar to hot-rolled 1020 steel?

NOTE: A more rigorous analysis would consider the rod and cylinder as a stepped column. However, the cylinder is expected to be considerably stiffer than the rod, and a conservative, i.e., on-the-safe-side, answer will be reached if the rod itself is considered to be 76⅛ in. long.

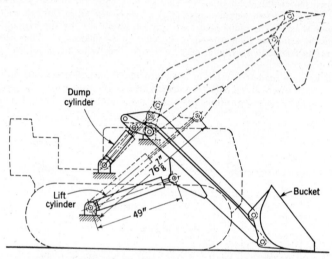

FIG. P 36. (*Courtesy Caterpillar Tractor Company.*)

6-37. The dump cylinder piston rod of the Traxcavator, shown in Fig. P 36 and discussed in Prob. 6-36, carries a negligible load when lifting and dumping but must carry as much as 56,700 lb in tension when "bulldozing" and 56,700 lb in compression when "backdozing." In both cases, the unsupported length of piston rod in the cylinder, i.e., between the piston and the cylinder head, is 13 in. What minimum diameter would you recommend for a solid piston rod on the basis of using a low-carbon steel with properties similar to hot-rolled 1020 steel?

Chapter 7.   Castings and Weldments

7-1. A strut is to be made by welding ends, machined out of bar stock, to a ¾-in.-diameter hot-rolled 1020 steel rod, as shown in Fig. P 37. The welds will be made by the manual metallic-arc process with coated electrodes. What should be the weld

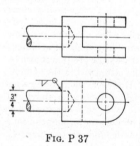

FIG. P 37

dimension if the joint is to have a strength equal to that of the rod when used under a steady tension load?

7-2. The bracket in Fig. P 38 is to be welded to the wall of a steel casting. The maximum value of the load P is to be 8,500 lb. The welds will be made with coated electrodes. Specify the weld dimensions required if (*a*) the load is static and (*b*) the load varies repeatedly from zero to 8,500 lb.

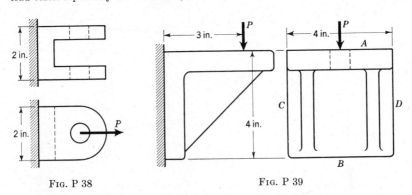

FIG. P 38   FIG. P 39

7-3. The cast-steel bracket in Fig. P 39 is to be welded to a large steel beam and is to carry a dead-weight load of 18,000 lb. What size of welds should be specified if coated-electrode metallic-arc fillet welds are used along the top and bottom edges (at A and B) of the bracket?

7-4. Calculate the allowable load P on the bracket in Fig. P 39 if the load is reversed and $\frac{1}{2}$-in. fillet welds (coated electrode) are used (*a*) at A and B only and (*b*) at C and D only. (*c*) What is the relative efficiency of the weldment in (*b*) as compared with that in (*a*)?

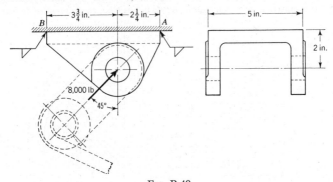

FIG. P 40

7-5. The bracket in Fig. P 40 is a support for one end of a leaf spring of a special-purpose trailer. The dotted links rotate as the curvature of the spring changes with a change in load. The maximum force on the bracket is 8,000 lb when the side links are at an angle of 45° with the vertical, as shown in the figure. What are your recommendations for welds at A and B?

7-6. The brake lever in Fig. P 41 is to be a weldment.   What are your recommendations for the welds?

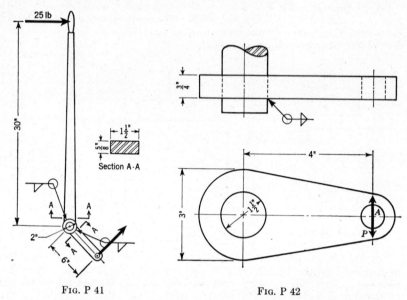

FIG. P 41   FIG. P 42

7-7. A hydraulic cylinder acting at A on the lever is used to apply a reversing torque of 1,000 lb-ft on the shaft in Fig. P 42.   What are your recommendations for the weld?

7-8. The welded gear blank in Fig. P 43 is to carry a steady torque of 1,200 lb-ft. What are your recommendations for the welds at A and B?   Specify intermittent welds if the required size for continuous welds is less than recommended in Table 7-5.

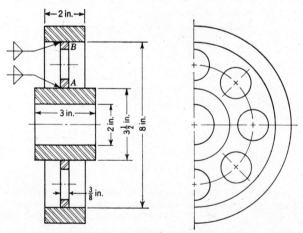

FIG. P 43

7-9. The welded gear blank in Fig. P 43 is to carry a torque that varies between 200 and 1,200 lb-ft fifteen times each minute. What are your recommendations for the welds at A and B?

7-10. The welded gear blank in Fig. P 43 is to carry a reversing torque of $\pm 1,200$ ft-lb. What are your recommendations for the welds at A and B if the torque reverses every 2 sec?

7-11. The load P on rod A of the offset bell crank in Fig. P 30 varies frequently between 1,000 and 3,000 lb. The outside diameter of the tube K is 3 in. What are your recommendations for the welds at J and L?

7-12. The load P on rod A of the offset bell crank in Fig. P 30 varies infrequently between zero and 4,000 lb. The outside diameter of the tube K is 3 in. What are your recommendations for the welds at J and L?

Chapter 8.   Mechanical Fasteners

8-1. (a) Assuming that a 1-in. bolt and nut with Unified coarse threads are made from the same material, what is the height of nut required for the threads to be as strong in shear as the bolt is in tension under static loads? (b) What is your conclusion about the manner in which failure of a standard bolt and nut will occur?

8-2. Calculate the depth of Unified coarse threads required in tapped holes if the threads in the tapped hole are to have the same factor of safety as the bolt has in tension when the load is static, if the bolt is made from cold-drawn 1020 steel, and if the holes are tapped in (a) cold-drawn 1020 steel, (b) ASTM 30 cast iron, and (c) 113 aluminum alloy, as cast.

8-3. (a) Show that the energy stored in a member elastically deformed in tension is

$$U = \frac{s^2 V}{2E} \quad \text{in.-lb}$$

where s = tensile stress, psi

V = volume of material under stress, in.^{3}

E = modulus of elasticity of the material, psi

(b) Determine the ratio of the maximum energy absorptions of the 3-in. portion of the bolt in Fig. P 44a to that in Fig. P 44b when both are made from the same material and have the same factor of safety under dynamic loading.

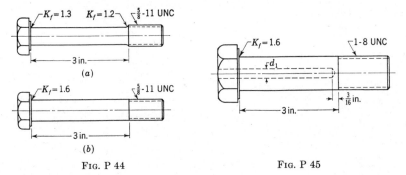

FIG. P 44   FIG. P 45

8-4. (a) What is the maximum diameter d_1 for a drilled hole if the bolt in Fig. P 45 is to have the same factor of safety under dynamic loads with and without the hole? (b) Using the relationship for energy absorption in tension given in (a) of Prob. 8-3, calculate the ratio of the energy absorption in the 3-in. portion of the bolt with the hole to that of the bolt without the hole when both bolts are made from the same material and have the same factor of safety under dynamic loads.

8-5. Figure P 46 shows the construction details for the ends of the tubular links used in a special testing machine (see Prob. 6-34 and Fig. P 35). Steel plugs are welded into the ends of the pipe and are then drilled and tapped for the rod ends. Precise adjustment of the length of the link is provided for by using LH and RH threads at opposite ends to give a turn-buckle effect. The rod ends are made from heat-treated alloy steel (equivalent of 3140, oil-quenched and tempered at 1000°F), and the plugs are made from hot-rolled 1020 steel. What do you rec-

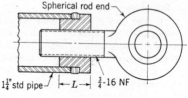

Spherical rod end

$1\frac{1}{4}"$ std pipe $\mid\leftarrow L\rightarrow\mid$ $\frac{3}{4}$-16 NF

Fig. P 46

ommend for the length L of the plug if the maximum static tension or compression load on the link is to be 11,000 lb?

8-6. The load P on rod A of the offset bell crank in Fig. P 30 varies infrequently between 1,000 and 3,000 lb. It is proposed to make the rod A out of cold-drawn 1020 steel and connect it to the clevises by using screw threads. What diameter do you recommend for the rod if UNF threads are used and failure would endanger human life?

8-7. Same as Prob. 8-6 except the load varies ten times each minute.

8-8. A 5-in.-diameter hydraulic cylinder is being designed for use in an operation in which the pressure varies from 0 to 1,500 psi every 10 sec. It is proposed that eight UNC bolts be used to join the head to the cylinder (Fig. 8-15). The bolts will be made from cold-drawn 1030 steel, and the threads will be die-cut. What size of bolts would you recommend if (*a*) a compressible gasket is used and the joint is made "steam-tight" and (*b*) an O ring is used to provide sealing while the cylinder head makes metal-to-metal contact with the cylinder flange? (*c*) What should be the mini-mum initial tightening loads on the bolts in (*a*) and (*b*)?

8-9. The cylinder and cylinder head in Fig. P 47 are made from ASTM 30 cast iron, and the bolts are made from cold-drawn 1020 steel. Assuming that the entire bolt deflection occurs over the length L, neglecting the decrease in bolt section area due to the threads, and assuming that the cylinder flange and head are uniformly compressed over the area of contact, what is the ratio of the rigidity of the joint to that of the bolts?

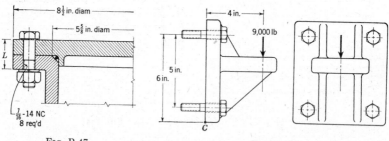

Fig. P 47   Fig. P 48

8-10. The bracket in Fig. P 48 is to be bolted to the flange of a beam by four identical UNC bolts made from cold-drawn 1020 steel. Assuming that the bolts are accurately fitted in line-reamed holes so that each bolt carries one-fourth of the load in shear, and that the bracket does not bend but tends to tip about point C, what size of bolts do you recommend for this application?

SUGGESTED PROCEDURE: Set up equations for the tensile stress and the shear stress in terms of force on and stress area of the bolts. Use the maximum-shear-stress theory [Eq. (6-32)] to determine the maximum shear stress which may then be related to the appropriate property of the material in calculating the required area. Note that the stress area in shear is actually that of the full cross section of the bolts and not the stress area of the threads. Thus, the answer given by this procedure will be conservative, and an additional check using actual diameters may be worthwhile if the required area in the first trial is just slightly greater than the stress area for one of the standards bolts.

8-11. What is the relationship between the length of a standard square key, whose width is one-fourth the shaft diameter, and the shaft diameter if both the key and shaft are made from the same material and have the same strength under static torsional loads?

8-12. The breakdown torque of a 15-hp Design B 1,800-rpm three-phase induction motor is given as 200 per cent of rated torque. The motor-shaft diameter is $1\frac{5}{8}$ in., and the $4\frac{1}{2}$-in.-long keyway is machined for a $\frac{3}{8}$- by $\frac{3}{8}$-in. square key. Assuming that a cold-drawn 1020 steel key is used, (a) what will be the factor of safety for the key based upon the breakdown torque of the motor? (b) Is this a reasonable value? (c) What is the minimum length of key you would recommend for use with this motor?

8-13. The welded gear blank in Fig. P 43 will transmit a steady torque of 1,200 ft-lb to the shaft. (a) What is the minimum length of key you would recommend if it is to be made from square cold-drawn 1020 steel key stock? (b) Under what conditions is it desirable for the key to be shorter than the hub?

8-14. The shaft for a belt-driven centrifugal blower was designed in Example 6-1. The steady torque on the shaft was 2,100 lb-in., and the diameter required for the shaft at the pulley was found to be $1\frac{9}{16}$ in. If the key is to be made from cold-drawn 1020 steel, what minimum length do you recommend if the key is (a) square and (b) flat?

8-15. The rod (with length L) in Fig. P 31 is a torsion spring and is designed to carry the torque due to the load P. What dimensions do you recommend for the keys if they are made from cold-drawn 1020 key stock, the diameter D is 3 in., the load varies infrequently in service between 1,000 and 3,000 lb, and a failure would endanger human life?

8-16. Same as Prob. 8-15 except the load varies many times each minute.

8-17. Feather keys and splines for use when there must be sliding while the torque is being transmitted are often designed on the basis of an allowable pressure of 1,000 psi between the key and the keyway. (a) Derive an equation for the shaft diameter as a function of a constant and the torque when two diametrically opposite square feather keys are used, the allowable contact pressure is 1,000 psi, and the hub length is one and one-half times the hub bore. (b) Using the expression derived in (a), specify the shaft diameter and hub length required if a gear (similar to Fig. P 43) is to transmit a torque of 400 lb-ft in a machine-tool application where the gear remains in a given axial position while the shaft moves axially as the cutting tool is fed into the work. (c) Assuming a coefficient of friction of 0.08 between the keys and keyways for the shaft in (b), what is the magnitude of the force required to move the shaft axially while transmitting the torque of 400 lb-ft?

Chapter 9. Springs

9-1. Design a steel helical compression spring with squared and ground ends for use under average service conditions when the maximum working load is 40 lb, the maximum working deflection is $\frac{1}{2}$ in., and the solid deflection is $\frac{9}{16}$ in.

9-2. The degree of agreement between calculated and experimentally determined values for springs is dependent upon the accuracy of the values used for wire diameter, mean coil diameter, and modulus of elasticity.   This is particularly important in applications where the tolerance on load-deflection characteristics must be small, i.e., when the permissible variation from a nominal value must be small, because the cost of a spring increases rapidly with decreasing tolerances.   A reasonable tolerance for a commercial spring wire with a diameter of 0.0800 in. (W & M No. 14) is ±0.002 in., and a reasonable tolerance on the mean coil diameter for a spring with a spring index of 8 and a wire diameter of 0.0800 in. is ±0.0155 in.

What is the ratio of the maximum spring rate to the minimum spring rate when (a) the mean coil diameter is its nominal value (0.6400 in.) and the wire diameter is within its tolerance; (b) the wire diameter is its nominal value and the mean coil diameter is within its tolerance; (c) both the wire and mean coil diameters are within their tolerances?

9-3. If the spring rate must be within ±3 per cent of the value calculated on the basis of nominal values of wire and mean coil diameters, what is the permissible per cent variation in (a) the wire diameter when the mean coil diameter is its nominal value and (b) the mean coil diameter when the wire diameter is its nominal value?

9-4. Derive an equation for the resultant spring rate k (P/δ) when three springs with spring rates k_1, k_2, and k_3 are used (a) in series, as in Fig. P 49a, and (b) in parallel, as in Fig. P 49b.

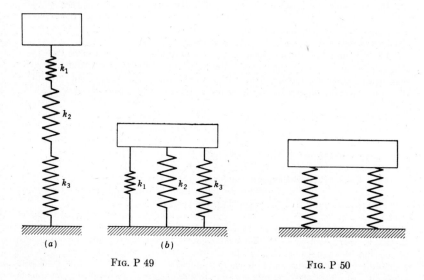

Fig. P 49   Fig. P 50

9-5. What is the resulting spring rate if a compression spring with a spring rate k (P/δ) is cut into two equal parts and the parts are then used in parallel, as in Fig. P 50?

9-6. The allowable stresses given in Table 9-1 for light service are to be used with loads that are essentially static, whereas those for severe service are to be used with loads that vary more than one million times between some maximum and a value one-half or less of the maximum.   Assuming that the light-service stresses are based on applying a factor of safety to the torsional yield strength and that the same factor of safety applies to the severe-service stresses when $s_{max} = 2s_{min}$, what value of allowable stress can be used in designing a ½-in.-diameter torsion bar that will be made of the same material and will have the same factor of safety while being subjected to a torque that reverses completely more than one million times?

9-7. An automobile engine is being designed for operation at speeds up to 6,000 rpm. The overhead valves are to open ½ in., and the valve-spring forces are to be 60 lb when the valve is closed and 150 lb when the valve is open. The valve mechanism will be similar in principle of operation to that shown in Fig. 2-9. Design a helical compression spring with squared and ground ends for this application.

9-8. Design a steel helical compression spring with squared and ground ends for use in a recoil system where it must absorb 100 ft-lb of energy with a maximum force of 600 lb. The recoil system will be used frequently.

9-9. Design a steel helical compression spring with squared and ground ends for use under severe-service conditions when the maximum force is to be 1,000 lb, the spring rate 750 lb/in., and the outside diameter cannot exceed 2¼ in.

9-10. An automotive-type clutch, as shown in Fig. 12-2, utilizes friction in transmitting a torque from one member to another. The normal force on the friction surfaces is supplied by helical compression springs so that the clutch is always engaged unless an external force is applied to compress further the spring and release the clutch. The specifications for a heavy-duty truck clutch are that the normal force must not be less than 2,020 lb when the clutch facing has worn 0.125 in. and the normal force should not be greater than 2,470 lb when the clutch is new. A total of 15 springs will be used to supply the normal force.

Design steel helical compression springs with squared and ground ends for this application.

9-11. The spring used in the engine indicator in Fig. 2-30 is unusual because it is wound with a lead twice the pitch so that, in effect, it is the same as two identical springs in parallel. Figure P 51 shows how the wire crosses over at the top of the spring so that both coils are one continuous length of wire.

Design a spring of this type for use in an indicator when the stylus P position is to represent the pressure on the piston p to a scale of 1 in. = 100 psi. The total displacement of P is to be 2 in.; the linkage multiplication ratio is 6:1, i.e., the displacement of P is six times the displacement of p; the compression of the spring is equal to the travel of the piston; and the piston area is 0.500 in.² Consider the top of the spring as having a plain end while the bottom is squared and ground.

FIG. P 51

9-12. The Safety Code for Elevators, Dumbwaiters, and Escalators[1] states that elevators with car speeds equal to or less than 200 fpm may use spring buffers (shock absorbers), while those with higher speeds must use oil buffers. The code further specifies that the stroke of the springs is to be 4 in. for car speeds of 151 to 200 fpm and that the total weight (car plus rated load) is to be more than one-third and less than one-half the sum of load ratings of all spring car buffers installed.

Assuming that the total weight is 8,000 lb, the car speed is 200 fpm, and eight buffer springs are to be used, (a) design steel helical compression buffer springs on the basis of the total weight equal to 0.4 of the sum of the load ratings of the springs when the spring deflection is 4 in. (b) What will be the maximum deflection if the loaded car

[1] ASA A17.1-1960, American Society of Mechanical Engineers. New York, 1960.

fails to stop at the bottom landing and hits the buffers designed in (a) when moving at 200 fpm? (c) What will be the maximum acceleration experienced by the car and occupants during the impact in (b)?

9-13. A spring coupling, similar to that shown in Fig. 11-14, is being designed as an integral part of a drive gear for an electric locomotive. The springs are to be installed under initial compression so that there is no angular deflection of the coupling, i.e., additional compression of the springs, until the torque exceeds 3,500 lb-ft and the angular deflection is to be 2° when the coupling is carrying its rated torque of 14,000 lb-ft. Stops are to be provided to prevent motion, without the spring being compressed solid, when the torque exceeds 125 per cent of rated torque.

Space limitations require that the springs must be located outside a 13-in.-diameter circle and inside a 30-in.-diameter circle. Service conditions will be average, but the springs will be shot-peened after forming and heat-treating, and the allowable stresses for light service in Table 9-1 may be used in the design.

Preliminary calculations have shown that it is impossible to design the coupling using four springs and a spring index of 8. In fact, it has been found that the coupling will probably require at least eight springs, i.e., in effect, two couplings in parallel, and that the spring index will have to be in the order of 4.

Design the eight springs with squared and ground ends for use in the coupling when the centerlines of the springs are tangent to a circle with a diameter of 19 in., as shown in Fig. P 52.

NOTE: If time is available, it will be worthwhile to make additional trials with different numbers of springs and values of spring index to determine the range of combinations with which a satisfactory design may be achieved.

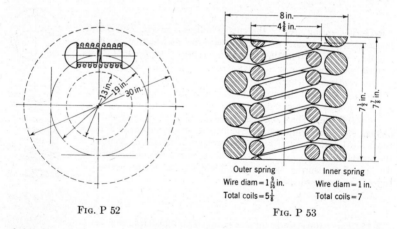

FIG. P 52

8 in.
4⅝ in.

7½ in.
7⅞ in.

Outer spring
Wire diam = 1 9/16 in.
Total coils = 5⅛

Inner spring
Wire diam = 1 in.
Total coils = 7

FIG. P 53

9-14. One of the concentric springs in a pedestal-type truck for a 100,000-lb-capacity railroad freight car has the free dimensions shown in Fig. P 53. The length of both the inner and outer springs is 6⅝ in. when carrying the rated load. (a) What is the rated load for the concentric spring? What per cent of the rated load is carried by the outer spring? By the inner spring? (b) What is the design stress for the outer spring? For the inner spring? (c) Discuss the significance of the values in (b) in relation to the allowable stresses in Table 9-1.

9-15. The optimum design for concentric springs is obtained when the material in each spring is stressed to the same relative level, i.e., the stress and wire diameters are related according to service conditions.   In general, it is not practical to make a direct solution, and the wire diameters and load distribution must be found by a trial-and-error solution.   However, in the case of two concentric springs with the same value of allowable stress, the same spring index, and zero clearance between the springs, the ratio of the allowable load carried by the outer spring to that carried by the inner spring may be shown to be a function of the spring index only.

(*a*) For two concentric springs meeting the conditions specified above, determine the equation for P_1/P_2 as a function of C, where P_1 = allowable load for the outer spring, P_2 = allowable load for the inner spring, and C = the common spring index. (*b*) Using the equation from (*a*), calculate the ratio P_1/P_2 for a concentric spring with a spring index of 8.   (*c*) What is the effect on the ratio P_1/P_2 of providing clearance between the springs?

9-16. The valves of a heavy-duty diesel engine, similar to Fig. 2-9, are to be opened 0.8 in. by the valve mechanism.   The concentric valve springs are to exert a force of about 130 lb when the valves are closed and about 280 lb when the valves are open.

(*a*) Design the springs with squared and ground ends on the basis of both springs having the same length at all times and with the outer spring carrying approximately 65 per cent of the load.   Provide at least $\frac{1}{8}$-in. diametral clearance between the springs.   (*b*) What will be the valve-spring forces when the valves are closed and opened if the initial compression of the outer spring remains as in (*a*) while the inner spring is initially compressed an additional $\frac{1}{4}$ in. to permit the use of a combination guide and seat, as in Fig. 2-9?

9-17. Design a helical tension spring which is to be used to maintain an approximately uniform force as a part moves $\frac{3}{4}$ in.   The force is to be at least 18 lb and not to exceed 20 lb.   Service conditions may be considered to be average.   Space limitations are not critical, but it is desirable that the spring be wound with the maximum possible initial tension.   The spring is to have full round hook ends with the relative proportions indicated in Fig. P 54.

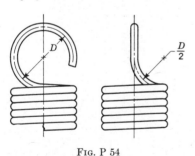

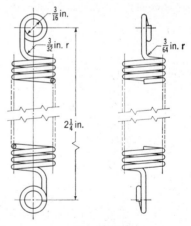

FIG. P 54   FIG. P 55

9-18. The governor element of the self-lowering screw jack in Fig. 10-6 consists of a centrifugal brake (Fig. P 76) in which the friction resistance to lowering is a function of centrifugal force which, in turn, is dependent upon the square of the speed of the rotating element.   Tension springs are used to keep the shoes from contacting the drum until a predetermined speed has been reached.   Design a tension spring for a similar application in which the force in each spring is to be 1.2 lb when the spring is extended as shown in Fig. P 55.

Chapter 10.   Translation of Power Screws

10-1. A $1\frac{1}{2}$-in. steel screw and a bronze nut are carrying a load of 6,000 lb.   The screw has a single, modified square thread with three threads per inch.   If the speed of rotation of the screw is 550 rpm, the operating conditions are excellent, and the screw is supported on a ball thrust bearing, determine (a) the starting torque, (b) the efficiency, and (c) the horsepower required to drive the screw at its rated speed of 550 rpm.

10-2. Same as Prob. 10-1 except that a hardened-steel thrust collar running on a bronze washer is substituted for the ball thrust bearing.   Assume d_c is $1\frac{1}{2}$ in.

10-3. What is the per cent (a) decrease in starting torque and (b) increase in running efficiency if a ball thrust bearing is substituted for a soft-steel thrust collar on a cast-iron surface when the screw has a single two-threads-per-inch modified square thread with a diameter of 2 in. and the mean diameter of the thrust collar is 3 in.?

10-4. Considering the efficiency of the screw with a true square thread as 100, what are the relative efficiencies of screws with (a) modified square and (b) general-purpose Acme threads when all three are single five-threads-per-inch 1-in. steel screws and bronze nuts manufactured and used under average conditions with ball thrust bearings?

10-5. Same as Prob. 10-4 except use double-thread screws instead of single-thread screws.

10-6. Figure P 56 shows a proposed design for an inexpensive automobile-bumper jack with a rated capacity of 1,500 lb.   (a) What should be the dimension l for the hand lever if the maximum operating force is to be 30 lb?   (b) What will be the maximum force required to lower the load if the lever in (a) is used?   (c) What will be the maximum operating force if the lever in (a) is used and a ball thrust bearing is substituted for the bronze thrust washer?

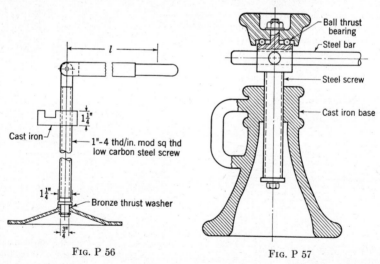

FIG. P 56   FIG. P 57

10-7. Bell-bottom screw jacks, as in Fig. P 57, are used as all-purpose jacks, e.g., in the construction industry, in moving and installing machinery, and in the maintenance of heavy equipment.   Since use is relatively infrequent and wear is not too critical a problem, jacks of this type are usually designed on the basis of an allowable bearing pressure that is between two and three times the suggested values in Table 10-3.

(a) Design a screw with a modified square thread for use in a 20-ton-capacity bell-bottom jack with a maximum raise of 5 in. (b) What must be the length of the steel bar if the maximum operating force is not to exceed 30 lb under rated load? (c) What will be the force required to lower the jack under rated load when the bar in (b) is used? (d) What diameter do you recommend for the bar if it is made from cold-drawn 1040 steel?

10-8. Same as Prob. 10-7 except that the jack capacity is to be 12 tons instead of 20 tons.

10-9. Certain control surfaces of an airplane must be adjustable to permit compensation for changes in weight, etc., that will affect the flight path of the plane when the controls are in their neutral positions. In some cases, on very large planes, the nose-up or nose-down trim is accomplished by rotating the entire horizontal tail surface. One company has used a servo-controlled jackscrew driven by a hydraulic motor, such as in Fig. 17-28, in this application.

Assuming that in a new design the force required is 75 tons, that the total distance of travel is ±6 in. from neutral, that the speed of operation is slow, and that the number of cycles of use under maximum load are so few that the allowable pressures in Table 10-3 may be safely tripled, (a) what is your recommendation for a steel screw and a bronze nut with modified square threads? (b) What will be the maximum torque required to rotate the nut if it is supported on roller thrust bearings?

10-10. A small screw press, as in Fig. P 58, is being designed with a capacity of 5,000 lb. (a) What are your recommendations for a steel screw and bronze nut with modified square threads? (b) What will be the handwheel diameter D if a bronze thrust washer is used and the operator is not to have to exert over 15-lb force with each hand when using the press at rated capacity?

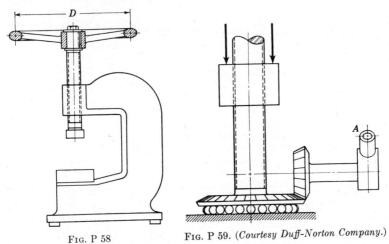

Fig. P 58   Fig. P 59. (*Courtesy Duff-Norton Company.*)

10-11. The essential parts of the raising mechanism of the 35-ton-capacity Duff-Norton governor-controlled jack (Fig. 10-6) are shown in Fig. P 59. The operating lever is 60 in. long and fits into the socket at A. The operating force is transmitted through a ratchet and becomes a torque on the bevel-pinion shaft (e.g., in Fig. 12-17, arm 3 corresponds to the operating lever, and ratchet wheel 2 corresponds to the bevel-pinion shaft). The torque on the bevel-pinion shaft is multiplied by the gear-tooth ratio 25/12 and is transmitted to the screw. The heat-treated steel screw has double, left-hand, modified square threads in which the height of thread h is $0.3p$ instead of $0.5p$ as shown in Fig. 10-1b. The thread pitch p is 0.625 in., the screw outside diameter is 1.980 in., and the bronze-nut length is 5 in. The base is sealed to prevent the entry of dirt and to prevent the loss of lubricant.

(a) Neglecting the efficiency of the bevel-gear train, what force will the operator have to apply to raise the rated load of 35 tons? (b) If the bevel-gear-train efficiency is 97 per cent, what force will the operator have to apply to raise the rated load of 35 tons? (c) What torque will be required on the screw to keep a 35-ton load from lowering itself? (d) What torque will be required on the screw if a 35-ton load is to lower itself at a constant speed? (e) What is the design value of bearing pressure between the screw and nut threads? Compare this value with the recommendations in Table 10-3, and discuss the significance of and reasons for the difference.

10-12. Someone has proposed using a translation screw in a servo positioning device. The device must be self-locking, and the efficiency must be as high as possible. Assuming that a single-thread screw with modified square threads is to be used, the pitch is to be 1 in., and the steel screw and bronze nut will be of the highest quality and used under ideal conditions with ball thrust bearings, what will be (a) the minimum permissible pitch diameter and (b) the operating efficiency, using a screw with the pitch diameter calculated in (a)?

Chapter 11.   Couplings

11-1. Flanged shaft couplings are usually designed with a strength equal to the static torsional strength of the shaft. (a) On this basis, what should be the dimensions l, t, and b for the coupling in Fig. P 60 if the shaft and bolts are made from hot-rolled 1020 steel, the coupling halves are made from ASTM 30 cast iron, and the bolts are fitted in line-reamed holes? (Recall from Sec. 6-4 that, for static loads, a cast-iron part is usually designed with a factor of safety twice that for a similar steel part.) (b) What will be the factor of safety for the coupling if used on a 60-hp 1,750-rpm three-phase electric motor that has a breakdown torque of 225 per cent of rated torque?

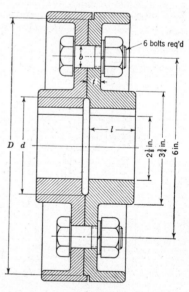

Fig. P 60

11-2. What size of bolts will be required if the torque is transmitted from one half of the flanged shaft coupling in Fig. P 60 to the other by means of friction between the contact surfaces and if the torque capacity of the coupling is to equal the static strength of the shaft?   Assume that the bolts and shaft will be made from hot-rolled 1020 steel, that a torque wrench will be used to tighten the bolts, and that the tightening torque may vary ±30 per cent from the desired value.

11-3. Assuming that the torque-speed characteristics in Fig. 11-16 apply to a hydraulic coupling driven by a 100-hp electric motor, (a) determine the output speed and coupling efficiency when the load torque is 5,000 lb-in. (b) At what rate in Btu per minute must energy be dissipated under the operating conditions in (a)? (c) What is the motor speed at the instant the load starts moving if the load breakaway torque is 120 per cent of normal operating torque?   At 180 per cent of normal operating torque?   (d) How is the excess of motor torque over load torque utilized in the drive?

11-4. A machine requires an input torque of 2,100 lb-in. when running at its normal input-shaft speed of approximately 1,800 rpm.   However, if the machine has to start in one particular phase, the breakaway torque is 4,400 lb-in.   (a) On the basis of the electric-motor and hydraulic-coupling torque-speed characteristics in Fig. 11-16, what is the minimum standard motor horsepower that may be used in this application?   (b) For the motor selected in (a), what will be the normal input-shaft speed of the machine?   (c) For the motor selected in (a), what will be the coupling efficiency when the machine is running at normal speed?

11-5. A diesel-engine-driven rock crusher is provided with an extra-heavy flywheel which will, for short periods of time, supply extra energy required when a larger than normal piece of rock is put in the crusher.   The disadvantage is that, if a piece of iron or steel is accidentally dropped between the jaws, the crusher will jam and the torque supplied by the flywheel during the rapid (almost instantaneous) deceleration will be great enough to damage the gears in the reduction unit.   Since this situation will occur only infrequently, if ever, it is proposed to install a shear pin, as shown in Fig. P 61, to prevent the transmission of torque between the flywheel and the shaft in excess of about 75,000 lb-in.

Assuming that the shear pin is made from a low-notch-sensitivity steel with an ultimate shear strength of 40,000 psi and that the necked-down portions of the pin are 4½ in. apart, what is the maximum diameter that you would recommend for the necked-down section?

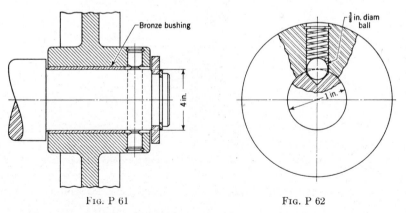

FIG. P 61   FIG. P 62

11-6. If friction is neglected, what must be the spring force on the ball in the ball-detent overload release coupling in Fig. P 62 if the ball is to start to move when the torque reaches 10 lb-in.?   Assume a 90° detent included angle.

11-7. A slip clutch, similar to that shown in Fig. 11-29, has two friction planes with OD = 8.00 in. and ID = 5.00 in.   Twelve helical compression springs are used to supply the normal force on the friction surfaces, each spring has a spring rate of 400 lb/in., and the adjusting nut has 16 threads per inch.

(*a*) How many revolutions of the adjusting nut are required to set the clutch to slip at its maximum rated torque of 300 lb-ft when the coefficient of friction is 0.35? (*b*) If the clutch had been adjusted previously to transmit a torque of 300 lb-ft with a coefficient of friction of 0.35, how many revolutions must the adjusting nut be backed off if the clutch is to be reset to transmit a torque of 225 lb-ft?

11-8. Slip couplings are often used to supply a constant torque under changing speed.   In such applications, the rate of heat dissipation becomes the critical factor. For example, a standard Hilliard slip coupling, as in Fig. 11-29, that has a torque rating of 1,058 lb-ft will, when the ambient temperature is 70°F, dissipate heat at a continuous rate of 0.99 hp when the input speed is 300 rpm and 2.00 hp when the input speed is 1,400 rpm.

(*a*) What is the permissible per cent slip for continuous use under rated torque with an input speed of 300 rpm?   (*b*) What is the permissible per cent slip for continuous use under 25 per cent of rated torque with an input speed of 1,400 rpm?

11-9. The coupling in Fig. P 63 is a single unit combining the features of a roller-chain flexible coupling and a slip coupling.   On the basis of an average pressure (P/A) of 50 psi on the friction surfaces and a coefficient of friction of 0.25 (*a*) what will be the dimensions D and d for a coupling with two friction planes and $d/D = 0.7$ if the clutch is to slip when the torque exceeds 44 lb-ft?   (*b*) What will be the force on the Belleville spring under the conditions in (*a*)?

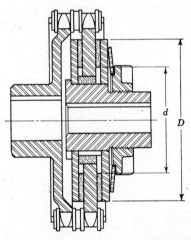

Fig. P 63. (*Courtesy Morse Chain Company, Division of the Borg-Warner Corporation.*)

Chapter 12.   Clutches and Other Intermittent-motion Mechanisms

12-1. Several square-jaw clutches are required for an application in which the shaft diameter is 2 in. and the maximum torque carried by the engaged clutch is to be 7,500 lb-in.   What are the minimum values you would recommend for the outside diameter and length of the jaws, that is, D and l in Fig. P 64, if the parts are to be machined from hot-rolled 1020 steel bar stock and if the force is considered to act at the mean radius of the jaws?

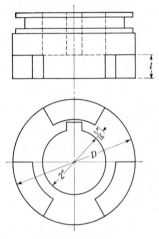

FIG. P 64

12-2. A complete line of square-jaw clutches is being designed for distribution by equipment-supply houses. Since the conditions of use can be neither specified nor accurately foreseen, the clutches will be designed to carry a torque equal to the torsional strength of a solid cold-drawn 1020 steel shaft.

What are your recommendations for the outside diameter D and the length of jaw l for the 2-in.-size clutch shown in Fig. P 64 if the parts are to be machined from steel castings with properties similar to those of hot-rolled 1020 steel and if the force is considered to act at the mean radius of the jaws?

12-3. What is the minimum value for the ratio d/D (Fig. 12-3) at which the capacity of a plate clutch will decrease not more than 10 per cent during the initial-wear period?

12-4. A clutch, similar to that in Fig. 12-2, is being designed for heavy-duty service in a truck with an engine that is rated at 185 hp at 3,600 rpm and delivers a maximum torque of 340 lb-ft at 1,800 rpm. The clutch is to have a capacity 25 per cent in excess of the maximum engine torque when the clutch-plate friction material has worn a total of $\frac{1}{8}$ in. Past experience has indicated that in similar applications a reasonable clutch life will be obtained if the average pressure does not exceed about 27 psi and that it is practical to design compression springs for the clutch when the ratio of the new clutch (after initial wear) capacity to the worn clutch ($\frac{1}{8}$-in. wear) capacity is in the order of 1.15:1 or greater.

(a) Under the conditions given above, what are your recommendations for the inner and outer diameters of the friction material on the driven disk? (b) If 15 springs are used, what are your specifications for the spring rate and the initial deflection of each?

12-5. The driven disk of an automobile clutch, similar to that shown in Fig. 12-2, has friction surfaces with 6-in. inner and 9-in. outer diameters. Nine springs, each compressed $\frac{1}{4}$ in. to give a force of 140 lb when the clutch is new, are used to provide the normal force during engagement. When the clutch is used with an automobile engine having the characteristics shown in Fig. P 65, (*a*) what is the factor of safety for the new clutch (after initial wear) and (*b*) how much wear of the friction material can take place before the clutch will slip?

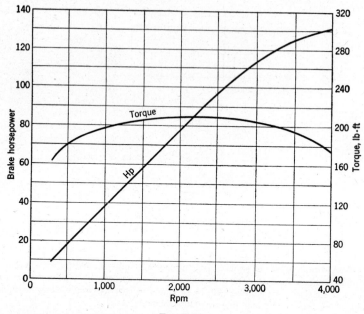

Fig. P 65

12-6. Figure P 66 shows a plate clutch designed so that the user may install a V-belt sheave having the diameter required in a particular situation. Assuming that a composition friction material with an allowable pressure of 25 psi is to be used, determine the outer diameter *D* of the friction surfaces if the clutch is to transmit 6 hp at 1,200 rpm.

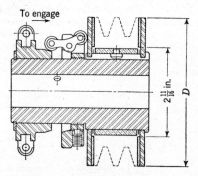

Fig. P 66. (*Courtesy Carlyle Johnson Machine Company.*)

12-7. What maximum torque capacity would you specify for the clutch in Fig. 12-5 if the inner and outer friction-surface diameters are $1\frac{3}{4}$ and $2\frac{1}{2}$ in., respectively, and the clutch is used (*a*) dry and (*b*) in an oil atmosphere?

12-8. How many metal plates do you recommend for a dry clutch that is to transmit 12 hp at 450 rpm, when the outside diameter of the friction surfaces cannot exceed $5\frac{1}{2}$ in.?

12-9. Hardened-steel clutch plates with inner and outer friction-surface diameters of 2 and 3 in., respectively, are available in your plant. You are asked to design a clutch utilizing these plates that will transmit $5\frac{1}{2}$ hp at 825 rpm in an oil atmosphere. Specifically, (*a*) how many plates will be required and (*b*) what magnitude of axial force will be required during engagement?

12-10. The SynchroGear clutch in Fig. P 67 and the synchromesh transmission in Fig. 12-8 are similar in operation in that both utilize a relatively low-capacity friction

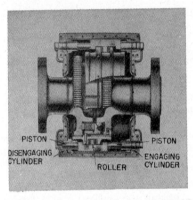

FIG. P 67. (*Courtesy Mechanical Power Transmission Division, Zurn Industries, Inc.*)

clutch to synchronize the rotative speeds of the input and output shafts so that the internal and external teeth of a high-capacity spline clutch may engage smoothly. The SynchroGear clutch is hydraulically operated; the top half of the figure shows the clutch disengaged, and the bottom half shows the clutch engaged. The major differences between the two units are that the SynchroGear clutch uses a multiple-disk instead of a conical friction clutch and it is used in applications requiring a greater capacity. One of its major fields of application is in the engaging and disengaging of dual power plants and in reversing gears of marine propulsion drives. In these applications, the comparatively large inertia of the rotating members, e.g., steam or gas turbines and propellers, and the effect of the power absorbed by the propeller when the ship is coasting through the water require the use of a high-capacity friction clutch for quick synchronization. However, even in this case, the friction clutch may be required to transmit only a few per cent of the full-power torque that the spline clutch must transmit.

A dual-power-plant propulsion system for an atomic-powered submarine is being designed for optimum part-load efficiency. A preliminary study of the inertia and power loads has indicated that two clutches, similar to that shown in Fig. P 67, will be required, each capable of transmitting a torque of 87,500 lb-ft when fully engaged and 2,000 lb-ft during synchronization. Space limitations are such that the friction-surface inner diameter cannot be less than 10 in. and the outer diameter not greater than 12 in.   If the clutch is to have metal plates and operate with a copious supply of cooling oil, (a) what is the minimum number of plates you would recommend and (b) what magnitude of axial force will be required during the synchronizing operation?

12-11. An air-operated clutch, similar to that shown in Fig. 12-6, has fourteen 5-in.-wide friction shoes in contact with a 24-in.-diameter drum.   The shoes are short enough (Sec. 13-1) that the resultant normal and tangential (friction) forces may be considered to act at the mid-point of the contact between each shoe and the drum. Assuming that the coefficient of friction is 0.30, that the rubber-tube width is such that the pressure between the shoes and the drum is equal to the air pressure, and that each shoe and its portion of the rubber tube weighs 2.4 lb with the radius to its center of gravity equal to the drum radius, (a) what is the clutch torque capacity at 75 psi air pressure and zero rpm, (b) what is the clutch torque capacity at 75 psi air pressure and 1,000 rpm, and (c) what is the horsepower capacity at 100 psi air pressure and 1,000 rpm?

12-12. An air-operated clutch, similar to that shown in Fig. 12-6, is to be designed to have a torque capacity of 50,000 lb-in. when the air pressure is 75 psi.   Twelve friction shoes, made from a composition material which has a coefficient of friction of 0.30 and an allowable pressure of 75 psi, are to be used, and the effective width of the rubber tube is to be approximately equal to the width of the shoes so that, at zero speed, the air pressure will be equal to the pressure between the friction shoes and the drum.   The width of the shoes is to be one-fourth the drum diameter, and the shoes are short enough that the resultant normal and tangential (friction) forces may be considered to act at the mid-point of the contact (Sec. 13-1) between each shoe and the drum.

(a) What will be the dimensions for the drum diameter and friction-shoe width? (b) What will be the decrease in torque capacity at 1,000 rpm if each shoe and its portion of the rubber tube weighs 1.8 lb and the radius to its center of gravity is equal to the drum radius?   (c) What must be the air pressure if the clutch capacity at 1,000 rpm is to be 50,000 lb-in?

12-13. The exact analysis of the application of a flywheel-clutch combination, similar to that shown in Fig. 12-7, to a punch press is quite complicated.   However,

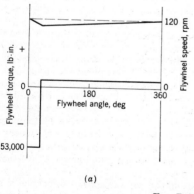

(a)

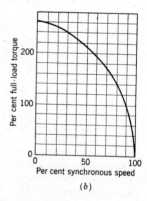

(b)

FIG. P 68

in many cases, sufficient information may be gained by an approximate analysis which considers that the punching energy comes entirely from the flywheel slowing down at a constant rate and that, after the punching operation is completed, the motor exerts a constant torque on the flywheel until the flywheel once again reaches normal running speed.

Specify (a) the maximum energy available, (b) the angle through which the flywheel rotates during the punching operation, and (c) the minimum required rated horsepower of a 1,200-rpm synchronous-speed Design D electric motor for a punch press when the flywheel $WR^2 = 1,600$ lb-ft², the flywheel normal running speed is 120 rpm, the flywheel speed is to decrease only 10 per cent during the punching operation, the flywheel speed and torque vary with the angle of rotation as shown in Fig. P 68a, the motor torque-speed curve is as shown in Fig. P 68b, and the clutch torque capacity is 53,000 lb-in.

NOTE: Recall from dynamics that the mass moment of inertia of a rotating body $I = Mk^2$ (in this case, $Mk^2 = WR^2/g$); that the kinetic energy of a rotating body $U = I\omega^2/2$; and that a change in kinetic energy of a rotating body may be expressed also as $\Delta U = T\theta$, where θ is in radians.

12-14. If a torque of 20 lb-in. is required to accelerate the gears and clutch plate at a rate sufficient to ensure smooth engagement of the spline clutch in a synchromesh transmission, as in Fig. 12-8, what is the minimum value of axial force that the ball detent must be designed to transmit if the mean diameter of the hardened-steel conical friction surfaces is 2 in. and the cone-face angle is 15°?

12-15. The following information is known about a centrifugal clutch (Fig. 12-10):

	Outer shoes	Inner shoes
Number of shoes........................	6	4
Weight of each shoe, oz..................	17	22
Friction-drum diameter, in...............	8	5½
Coefficient of friction....................	0.35	0.35
Radius to center of gravity of shoes, in......	3½	2¼

Assuming that the normal and tangential (friction) forces act at the center of the contact between each shoe and its drum and that the coupling is to be used with a 1,150-rpm electric motor, determine (a) the torque supplied during starting by the outer shoes, (b) the total horsepower capacity of the coupling when the load has come up to running speed, and (c) the per cent overload capacity at running speed if the outer shoes are designed to transmit rated motor power.

12-16. A centrifugal clutch is required for a centrifuge drive. The centrifuge is to rotate at 7,500 rpm, and preliminary calculations have indicated that the rated torque of a 5-hp 1,750-rpm electric motor will bring the centrifuge up to running speed in 2 min 45 sec. One proposal is to use a single ring of six rotating shoes, similar to the driver half of the coupling shown in Fig. 12-10.

Assuming that the coefficient of friction is 0.35, that the friction material has an allowable pressure of 25 psi, that the shoe width is to be one-fourth the drum diameter, that the separator ribs between the shoes will take up 15 per cent of the drum area, that the radius to the center of gravity of the shoes is 85 per cent of the drum radius, and that the normal and tangential (friction) forces act at the center of the contact between each shoe and the drum, determine (a) the drum diameter and width of the shoes, (b) the weight of each shoe, and (c) the temperature rise in the drum during each starting operation if the drum is made of cast iron with a wall thickness of 5/16 in. and if all the heat generated is considered to be stored in the cylindrical portion of the drum. [Density of cast iron is about 0.260 lb/in.³; specific heat of cast iron is about 0.12 Btu/(lb)(°F).]

12-17. A spring-controlled centrifugal clutch has a single rotating ring of shoes that are spring-loaded, as shown in Fig. P 69, so that they do not make contact with the drum until the inner member reaches a predetermined speed.   The drum diameter is 16 in., each of the eight shoes weighs 2.5 lb, the radius to the centers of gravity of the shoes is $7\frac{1}{4}$ in., and the spring force at the instant of contact is 84 lb on each shoe.   (a) What is the speed at which the shoes first contact the drum?   (b) What is the horsepower capacity of the coupling at 1,200 rpm?

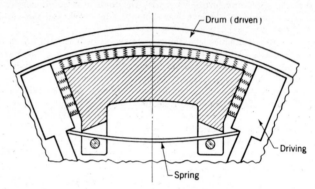

FIG. P 69. (*Courtesy O. S. Walker Company, Inc.*)

12-18. A small manufacturing plant has a 1,200-rpm 200-kw generator that may be driven by either a steam turbine or a diesel engine.   As shown in Fig. P 70, both power units are to be connected to the generator by means of spring-controlled centrifugal clutches to permit either unit to be at rest or warming up, at an idling speed, while the other is driving.   Figure P 69 shows the details of a spring-loaded shoe.

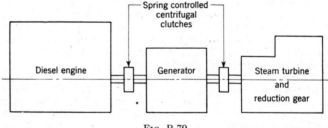

FIG. P 70

Assuming that the coefficient of friction is 0.35, that the allowable pressure for the friction material is 25 psi, that the shoe width is 30 per cent of the drum diameter, that the separator ribs between the shoes will take up 15 per cent of the drum area, that the radius to the center of gravity of the shoes is 85 per cent of the drum radius, and that the normal and tangential (friction) forces act at the center of the contact between each shoe and the drum, determine (a) the required weight of each of the eight shoes and (b) the spring force required on each shoe at the instant the shoes contact the drum if the clutches are to meet the following specifications: (1) The shoes are to contact the drum at 500 rpm and (2) the capacity at 1,200 rpm is to be 140 per cent of the generator rating (1 kw = 1.341 hp).

12-19. Lay out a Geneva-wheel mechanism in which the driven member will rotate ¼ revolution during each revolution of the driving member. The distance between the center of the driving and driven shafts is to be 3 in., and the driving-pin diameter is to be ½ in. Provide for positive locking in place of the driven member during the interval when the driving pin is not in the slot. Also provide at least ⅟₃₂-in. clearance between the driving pin and the bottom of the slot.

12-20. Determine the angular acceleration of the Geneva wheel designed in Prob. 12-19 at the instant a pin enters a slot when the driving member is rotating at the uniform speed of 200 rpm. What is the significance of your answer in relation to the use of the Geneva wheel at very high speeds? Use scales of 1 in. = 1 in., 1 in. = 20 in./sec, and 1 in. = 200 in./sec².

12-21. Same as Prob. 12-19 except that the driven member is to rotate ⅓ revolution during each revolution of the driving member.

12-22. Same as Prob. 12-20 except apply to the Geneva wheel designed in Prob. 12-21.

12-23. Same as Prob. 12-19 except that the driven member is to rotate ⅕ revolution during each revolution of the driving member.

12-24. Same as Prob. 12-20 except apply to the Geneva wheel designed in Prob. 12-23.

12-25. Same as Prob. 12-19 except that the driven member is to rotate ⅙ revolution during each revolution of the driving member.

12-26. Same as Prob. 12-20 except apply to the Geneva wheel designed in Prob. 12-25.

Chapter 13. Brakes

13-1. Determine the value of the operating force P required for the brake in Fig. P 71 for the cases when the lever pivot is at A, at B, and at C if the braking torque is 105 lb-in. and $\mu = 0.35$.

13-2. Where along the vertical line through A-B-C must the lever pivot be located for the brake in Fig. P 71 to be just self-locking when $\mu = 0.35$?

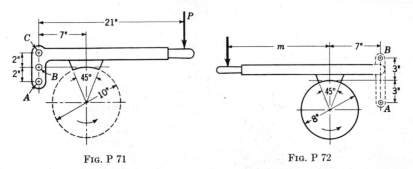

Fig. P 71   Fig. P 72

13-3. The brake in Fig. P 72 is to be self-energizing and is to have a braking torque of 16 lb-ft. Assuming that the coefficient of friction is 0.35, determine (a) which point, A or B, should be the brake-lever pivot point; (b) the required value of m; and (c) the horsepower absorbed if the drum is rotating at 105 rpm.

13-4. Some references suggest that brakes be designed on the basis of horsepower per square inch of lining. The text suggests that $pV = 30,000$ is a reasonable basis for design of brakes to be used under continuous application of load and poor dissipation of heat.   Assuming that the friction material is to be asbestos with a coefficient of friction of 0.35, (*a*) what is the equivalent rating in terms of horsepower per square inch?   (*b*) How many square inches of lining will be required for shoe brakes to meet the conditions of Example 13-1, which are to lower a load of 3 tons at the rate of 369 fpm?

13-5. The spring-operated brake in Fig. P 73 is to be used intermittently to exert a braking torque of 1,000 lb-ft on a drum rotating at 200 rpm.   Assuming that the shoes are lined with an asbestos friction material, (*a*) determine the required spring force.   (*b*) If the shoes are to be identical, what do you recommend as their minimum width?   (*c*) For the direction of rotation shown, which shoe will have the greater rate of wear and what will be the ratio of the rates of wear of the two shoes?

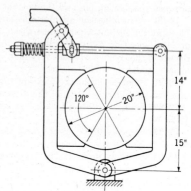

Fig. P 73

13-6. The spring-set solenoid-released brake in Fig. P 74 is part of a geared elevator machine (similar to Fig. 13-9 except that in the geared machine there is a worm-gear

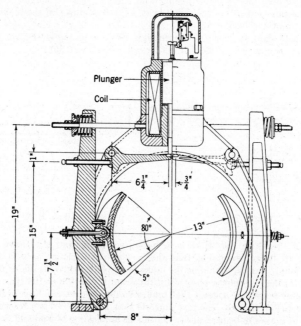

Fig. P 74. (*Courtesy Westinghouse Electric Corporation.*)

reduction unit between the motor and the cable sheave) and has a rated torque capacity of 170 lb-ft. If it is assumed that the coefficient of friction is 0.27 and that the springs are compressed equally so that, when releasing the brake, the load on the plunger will be along its axis of translation and wear of the plunger guide will be minimized, what are the required magnitudes of (a) the spring forces for setting the brake and (b) the plunger force for releasing the brake?

13-7. Same as Prob. 13-6 except that, instead of the springs being compressed equally, they are compressed so that each shoe exerts one-half the total torque when the brake drum rotates in a cw direction. Also determine the couple acting on the plunger guide.

13-8. A spring-set hydraulically released brake is to be designed to have a torque capacity of 400 lb-ft under almost continuous duty when the brake drum is rotating at 400 rpm in either direction. The brake lining is to be an asbestos composition, the coefficient of friction will be 0.35, the width of the brake shoes is to be one-third the diameter of the drum, and the remaining proportions are to be as given in Fig. P 75.

Determine (a) the required brake-drum diameter and width of the brake shoes, (b) the spring force required to set the brake, and (c) the hydraulic pressure required to release the brake if the hydraulic-cylinder internal diameter is 2 in.

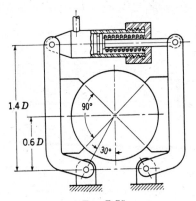

Fig. P 75

13-9. The centrifugal and thumbscrew brakes used to control the descent of a self-lowering jack, similar to that in Fig. 10-6, are shown in Fig. P 76. The jack

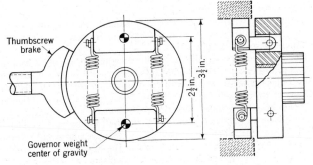

Fig. P 76. (*Courtesy Duff-Norton Company.*)

capacity is to be 25 tons, and the torque on the screw required to keep the rated load from lowering itself has been calculated to be 5,400 lb-in. The double-thread screw has a pitch of 0.625 in. The gear train is to consist of four pairs of gears with numbers of teeth such that the governor speed of rotation is 94.9 times that of the screw. Assuming that both brakes are short-shoe brakes, that the centrifugal weights are made of bronze, that the coefficient of friction between the bronze and the cast-iron stationary drum will not be less than 0.15, that the center of gravity of the shoes will be located as shown in Fig. P 76, that the thumbscrew-brake shoe will be lined with an asbestos composition with a coefficient of friction of 0.30 between it and the steel rotating member, and that the efficiency is 97 per cent for each of the four steps in the gear train, determine (a) the thumbscrew force required to hold the rated load in position, (b) the required weight of each governor weight if the lowering speed is not to exceed ½ in./sec when the thumbscrew brake is completely released and the rated load is on the jack (a force of 1.2 lb is required to stretch each spring when the weights make initial contact with the drum), and (c) the maximum lowering speed when the governor weights determined in (b) are used and the load on the jack is 10 tons.

13-10. Assuming that the coefficient of friction is 0.35 and the angle of wrap is 180° for the band brake in Fig. P 77, determine (a) the location of the lever pivot

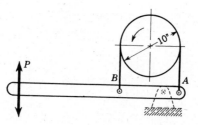

FIG. P 77

if the brake is just at the point of self-locking, (b) the region along AB in which the lever pivot may be located if the brake is to be self-energizing but always under control of the operator, (c) the region along AB in which location of the lever pivot would make the brake self-locking, and (d) the region along AB in which the lever pivot should be located if the brake is to be non-self-energizing.

13-11. Derive the general equation for the ratio of the torque capacity with cw drum rotation to that with ccw drum rotation for the brake in Fig. 13-14 when the operating force is the same in both cases.

13-12. Compare a pivoted-shoe brake with a band brake of equal torque capacity with respect to (a) efficiency in use of brake lining or friction material and (b) sensitivity to changes in coefficient of friction.

13-13. The air-operated brake in Fig. P 78 is designed for use in oil-field operations and is to have a torque capacity of 18,000 lb-ft. The brake band consists of a number of 9-in.-wide blocks of asbestos composition riveted to a cold-rolled 1040 steel band. The coefficient of friction is 0.35. (a) Which direction of drum rotation should be specified so that the air-cylinder operating force will be a minimum? (b) For the direction of rotation in (a), what diameter is required for the air cylinder if the air pressure is 150 psi? (c) What is the minimum thickness you would recommend for the steel band? (d) What are the magnitude and the direction of the resultant force on the lever pivot pin? (e) At what value of coefficient of friction will the brake become self-locking? (*f*) What will be the torque capacity of the brake if it is accidentally used when the drum rotates in the direction opposite to that specified in (*a*)?

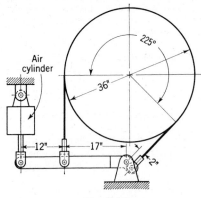

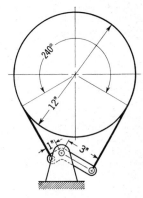

FIG. P 78   FIG. P 79

13-14. For the backstop in Fig. P 79, determine (*a*) the direction of rotation for which it is self-locking, (*b*) the minimum coefficient of friction with which it will perform satisfactorily, and (*c*) the rated torque capacity if the coefficient of friction is 0.35 and the band tension force is not to exceed 2,000 lb.

13-15. A belt-type coal conveyor is being designed for a power plant. The rated capacity is to be 740 tons/hr, the lift height is 95 ft, the belt speed is to be 300 fpm, the belt-drive pulley diameter is to be 30 in., and the conveyor slope angle is 20° with the horizontal. Neglecting the friction losses in the belt and conveyor, (*a*) determine the horsepower required at rated capacity and (*b*) design a backstop that will prevent reverse motion of the conveyor if power is turned off while the conveyor is loaded to capacity. The backstop is to have a 240° angle of wrap, as shown in Fig. P 80; the coefficient of friction will normally be 0.30, but the backstop is to hold the load even if the coefficient of friction is as low as 0.15; the maximum pressure on the band lining under normal conditions is to be 100 psi; and the band width is to be one-third of the drum diameter. Determine (1) the drum diameter and band width, (2) the dimensions *a* and *b* if the shorter is to be 3 in., and (3) the location of the lever pivot.

FIG. P 80

13-16. A brake for a brake motor, similar to that shown in Fig. 13-15, has asbestos-composition friction surfaces with inner and outer diameters of $4\frac{1}{2}$ and 7 in., respectively. Assuming that the springs are set for a total force of 70 lb, that the coefficient of friction is 0.35, and that the brake is mounted on a 1,750-rpm motor, determine (*a*) the stopping time if the equivalent total WR^2 ($I = WR^2/g$) of the motor and machine is 1.78 lb-ft² and (*b*) the energy in Btu absorbed by the brake during each operation.

13-17. One basis for the design of a brake for a brake motor, as shown in Fig. 13-15, is to have the brake capacity exceed the maximum motor torque. Thus, if the current to the brake coils is accidentally interrupted, the motor will be stopped, the current to the motor will be cut off by its thermal protection device, and neither the motor nor the brake will be damaged. This also provides for a rate of deceleration somewhat greater than the rate of acceleration.

(a) What outer diameter would you recommend for the friction surface of a brake with the same number of plates as in Fig. 13-15 for use with a 5-hp 1,750-rpm Design B squirrel-cage motor which has a breakdown torque of 225 per cent of full-load torque if the maximum magnetic force available is 90 lb and the coefficient of friction is between 0.3 and 0.4? (b) If the friction material is an asbestos composition, what is your conclusion about the expected life of the brake?

Chapter 14. Belt and Rope Drives

14-1. What is the maximum belt velocity at which the centrifugal-force term may be neglected without introducing an error greater than 10 per cent in the calculation of the capacity of a flat-leather-belt drive?

14-2. Derive the equation for the optimum belt velocity of a flat-leather-belt drive as a function of the allowable stress s_1.

14-3. A 12-in.-wide light double leather belt is used on 18-in.-diameter cast-iron pulleys rotating at 850 rpm. (a) What horsepower rating would you recommend for this drive? (b) What is the minimum value of initial tension for the drive capacity in (a)?

14-4. A machine requiring 50 hp at about 350 rpm is to be driven by a 1,150-rpm electric motor by use of a flat leather belt. If it is assumed that the center distance will be 5 ft and that both pulleys will be cast iron, what are your recommendations for (a) pulley diameters, (b) combinations of widths and thicknesses of belting, (c) length of the belt, and (d) minimum initial tension?

14-5. A 7½-hp 1,750-rpm electric motor is to drive the input shaft of a small hammer mill at 900 rpm. Space is at a premium, and one proposal is to use a short-center flat-belt drive with a center distance equal to the diameter of the larger pulley. Remaining within recommendations in the text, determine the combination of pulley materials, pulley diameters, belt thickness, and belt width that will result in the most compact drive.

14-6. A continuous-belt conveyor is being designed to carry stone out of a quarry to the crusher, as shown in Fig. P 81a. The 42-in.-wide rubber-impregnated cord

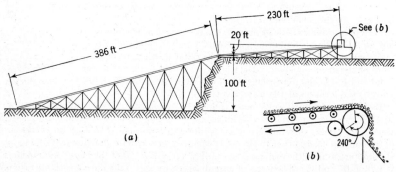

Fig. P 81

belt has an allowable tensile strength of 70 lb/in. of width per ply. The conveyor velocity is to be 300 fpm, and its capacity is to be 950 tons/hr.

The horsepower required is made up of three parts: (1) that required to move the empty belt horizontally, which for a 42-in.-wide belt is approximately $0.00005VL$, where V is the velocity in feet per minute and L is the length of the conveyor; (2) that required to move the load horizontally, which is approximately $(0.4 + 0.00325L)$ (tons per hour)/100;* and (3) that required to lift the load.

The steel conveyor-drive pulley will be at the upper end of the conveyor and will have an angle of wrap of 240°, as shown at b. The drive-pulley diameter is to be not less than 7 in. per ply of the belt. (a) What is the total horsepower required for the conveyor? (b) What is the minimum ply rating you would recommend for the belt? (c) What is the minimum value of initial belt tension?

14-7. In Eq. (14-35) the factor of safety is applied to the service load. Although not so conveniently handled, another possible approach would be to apply the factor of safety to the life of the wire rope. Assuming that a 6 × 19 wire rope had been selected on the basis of a factor of safety of 2 in Eq. (14-35) for a life of 150,000 bends, what will be the factor of safety of this rope in terms of life? HINT: The new factor of safety will be defined as the ratio of the expected life when carrying the service load to 150,000.

14-8. The four cables that lift the dipper of a large power shovel (Fig. P 82) must carry a maximum load of 180 tons when moving overburden in a strip mine. The shovel has a cycle time of 45 sec (fill dipper, rotate 90° and dump, and return). (a) Considering this service to be the equivalent of "cranes and derricks," specify the diameter of a 6 × 19 regular-lay plow-steel wire rope on the basis of the factors of safety in Table 14-6. (b) How many hours of continuous use will be expected before rope failure if the sheaves have the recommended minimum diameters? (c) The rope will be inspected daily and will be replaced well before wear has progressed to the point where breaking is imminent. If the rope is to be replaced when three-fourths of the expected life has been used up, it is reasonable to consider that, in terms of life, the factor of

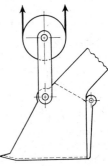

FIG. P 82

safety is 1.33. To what value of factor of safety, in terms of load, does this correspond? (d) What will be the per cent increase in expected life for the rope if the average recommended sheave diameter is used instead of the minimum?

14-9. The car speed for a freight elevator is to be 300 fpm. Each of the 6 × 19 traction-steel wire ropes must carry a maximum load of 3,000 lb. (a) According to the code, what is the minimum diameter of rope that may be used? (b) If the sheaves have the average recommended diameters, what number of cycles will correspond to one-half of the expected life of the rope?

14-10. A 2-ton electric hoist (Fig. P 83) is being designed for sale with a "guaranteed life of five years' service" on the basis of its being used three times each hour, 16 hours each day, and five days each week. However, it will be impossible to prevent the user from slightly overloading the crane, and it will be difficult to prove, at a later date, that the crane was misused. Also, the design data are not sufficiently accurate to permit designing for an exact number of cycles. Therefore it is proposed that the rope be designed to carry 150 per cent of rated load for a life 25 per cent greater than guaranteed.

(a) What material and diameter of regular-lay rope do you recommend if the hoist is to be as light and compact as possible? (b) For the rope recommended in (a), what is the nominal factor of safety in terms of ultimate strength?

---

* L. S. Marks (ed.), revised by T. Baumeister, "Mechanical Engineers' Handbook," 6th ed., p. 10-67, McGraw-Hill Book Company, Inc., New York, 1958.

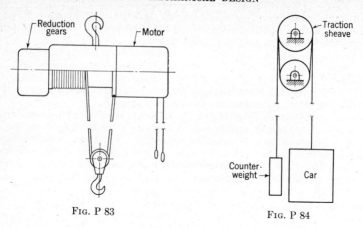

FIG. P 83   FIG. P 84

14-11. Traction-type elevators (Fig. P 84) using variable-voltage gearless elevator machines (Fig. 13-9) are used in almost all cases where the car speed exceeds 300 fpm. Each of the four ropes passes over the traction sheave twice, and the motor torque is transmitted to the ropes by friction, as in a belt drive.

If it is assumed that the car in Fig. P 84 is to accelerate in 3 sec to its running speed of 1,000 fpm, that the load capacity is to be 4 tons, and that the coefficient of friction between the lubricated ropes and the traction sheave is 0.1, (a) what must be the weights of the empty car and the counterweight to ensure that the loaded car can be raised and the empty car will provide enough tension in the rope so that the counterweight can be raised and, thus, the empty car lowered? (b) If friction losses are ignored, what is the maximum horsepower required for the drive? (c) What size of 6 × 19 traction-steel rope will be required by the code? (d) The minimum diameter of suspension sheaves permitted by the code for passenger elevators is $40d_r$. If this diameter is used, how many round trips of the fully loaded car can be expected before the ropes fail?

14-12. The car spotter in Fig. P 85 is basically a motor-driven sheave. The operator can "spot" a railroad car into position for loading or unloading by pulling on the loose end of the cable that has been attached to the car and then wrapped around the capstan (sheave) one or more times. When in use, the capstan rotates at a constant speed. The desired operating specifications are: a car speed of 45 fpm, a starting pull of 10,000 lb, and a running pull of 9,000 lb.

Assuming that the full-load speed of the electric motor is 1,750 rpm and the coefficient of friction between the wire rope and the capstan is 0.1, what are your recommendations for (a) type and size of rope, (b) diameter of capstan, (c) number of turns

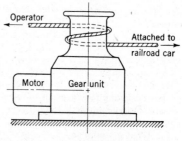

FIG. P 85

the rope must be wrapped around the capstan if the pull by the operator is not to exceed 40 lb, (d) horsepower of the motor, and (e) gear-reduction ratio between the motor and capstan? (f) What per cent of the total power will be supplied by the operator?

Chapter 15. Chain Drives

15-1. Roller chains find extensive use as tension links in applications such as in the lifting mechanism of fork-lift trucks, shown schematically in Fig. P 86. The chains carry only the weight of the platform and load plus a small additional load due to guide-roller friction. If the truck is to be rated to lift 1½ tons to a height of 8 ft, what do you recommend for (a) the pitch of the roller chains and (b) the stroke of the hydraulic cylinder?

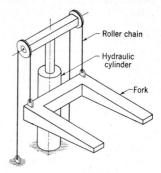

Roller chain

Hydraulic cylinder

Fork

Fig. P 86

Chapter 16.   Gears

16-1. Two involute gears, 2 and 3, with base-circle radii of 1.7321 and 5.1962 in. and addendum-circle diameters of 5.500 and 12.750 in., respectively, are meshed together on a center distance of 8.0000 in.   Gear 2 is driving and rotating ccw.

Make a full-size layout of a portion of the two gears showing the tooth profiles at the first and last points of contact. Indicate (a) the base circles, (b) the addendum circles, (c) the pitch circles, (d) the line of action, (e) the pressure angle, (f) the angular-velocity ratio ω_2/ω_3, (g) the pitch radii, (h) the angles of approach, (i) the angles of recess, (j) the arcs of action, (k) the addenda, and (l) the maximum possible addenda without involute interference.

16-2. Assuming that the gears in Prob. 16-1 are transmitting 20 hp and that the pinion is driving at 1,750 rpm, determine (a) the transmitted force F_t, (b) the normal force F_n, and (c) the maximum velocity of sliding in feet per minute.

16-3. An involute gear 2 with base-, pitch-, and addendum-circle diameters of 7.2505, 8.0000, and 9.500 in., respectively, is rotating ccw and is driving an involute rack 3 (a gear with an infinite pitch diameter) that has an addendum of 0.500 in.

Make a full-size layout of a portion of the gear and rack showing the tooth profiles at the first and last points of contact. Indicate (a) the gear base circle, (b) the gear pitch circle, (c) the gear addendum circle, (d) the rack pitch line, (e) the rack addendum line, (f) the line of action, (g) the pressure angle, (h) the angle of approach for the gear, (i) the angle of recess for the gear, (j) the arcs of action, and (k) the maximum possible addenda without involute interference.

16-4. If the gear in Prob. 16-3 is rotating instantaneously at 240 rpm and the force, resisting motion of the rack, is 2,900 lb, what will be (a) the torque on the gear, (b) the normal force F_n, and (c) the maximum velocity of sliding in feet per minute?

16-5. A 20-tooth pinion with 3-pitch 20° full-depth teeth is driving a 35-tooth gear. Calculate (a) the pitch diameters, (b) the base-circle diameters, (c) the addenda, (d) the dedenda, (e) the circular pitch, and (f) the base pitch.

Make a full-size layout of a portion of the gears showing a pinion tooth in contact with a gear tooth at the pitch point.   Indicate (g) the pitch circles, (h) the base circles, (i) the addendum circles, (j) the dedendum circles, (k) the directions of rotation of the gears when the pinion is driving, (l) the pressure angle, (m) the line of action, (n) the first and last points of contact, (o) the angles of approach and recess for the pinion, (p) the arc of action for the pinion, (*q*) the length of contact, and (*r*) the maximum possible addendum for the gear without involute interference. (*s*) Determine the value of the contact ratio.

16-6. A 14-tooth gear with 2-pitch 20° full-depth teeth is driving a rack. Calculate (*a*) the pitch-circle diameter of the gear, (*b*) the base-circle diameter of the gear, (*c*) the addenda, (*d*) the dedenda, (*e*) the circular pitch, and (*f*) the base pitch of the gear.

Make a full-size layout of a portion of the gears showing a gear tooth in contact with a rack tooth at the pitch point. Indicate (*g*) the pitch circle of the gear, (*h*) the pitch line of the rack, (*i*) the base circle of the gear, (*j*) the addendum circle of the gear, (*k*) the addendum line of the rack, (*l*) the dedendum circle of the gear, (*m*) the dedendum line of the rack, (*n*) the direction of motions of the gear and rack, (*o*) the pressure angle, (*p*) the line of action, and (*q*) the part of the rack tooth that would have to be removed in order for there to be no involute interference. (*r*) Determine the value of contact ratio if the rack tooth height is decreased as in (*q*).

NOTE: If the gear teeth are generated with a rack-type cutter, the actual contact ratio will be somewhat less than that determined here because of the loss of part of the involute near the base circle, as shown in Fig. 16-12.

16-7. One manufacturer of gears gives the capacity of its stock gears in terms of the horsepower and load which a 25-tooth gear of the set can transmit at a given pitch surface velocity. For example, a 25-tooth cast-iron gear with hobbed 6-pitch 20° full-depth teeth and a face width of 2 in. has a rated capacity of 11.00 hp at 1,200 fpm. Assuming that values of $K_L = K_O = 1.0$ and $K_R = 2$ apply to the catalogue ratings, (*a*) what value of allowable stress was used in the beam-strength equation in determining the rating? (*b*) How does the wear-strength capacity compare with the beam-strength capacity for a pair of 25-tooth gears from the set? (*c*) What will be the horsepower capacity at 850 rpm of a 60-tooth gear from the set?

16-8. A spur-gear speed reducer has been brought to your office by the shop foreman. The name plate (and rating) has been lost. His question is, "What is the horsepower capacity of the unit when used with a 1,750-rpm motor?" Upon examination, you have determined that the 8-pitch 20° full-depth teeth were shaved after hobbing, the 24-tooth pinion was through-hardened to 400 Bhn, the 75-tooth gear was through-hardened to 300 Bhn, and the face width is 1½ in.

What is your answer to the foreman's question?

16-9. Three identical gears, shown in Fig. P 87, are to be used to give the same speed and direction of rotation to two shafts in a slow-speed positioning device. The torque capacity is to be 2,100 lb-in. Space is limited, and it is proposed to use the smallest number of 20° full-depth teeth that may be hobbed without undercutting. The material is to be 1040 steel quenched and tempered to 300 Bhn. If the device is used so infrequently (less than 1,000 cycles total) that wear is not a consideration, (*a*) what do you recommend for the pitch of the gears and (*b*) what will be the center distance *l* between the input and output shafts?

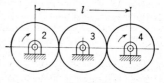

FIG. P 87

16-10. Same as Prob. 16-9 except that the use is frequent (100,000 cycles total) and wear must be considered. Also, the number of teeth may be increased, if desired, provided that the beam strength does not become less than the wear strength.

16-11. An auxilary drive requires the transmission of 20 hp between a shaft rotating at 1,200 rpm and one rotating at 750 rpm. The shafts are 8 in. apart and are parallel. The smoothest possible operation, the maximum reliability, and a long service life are essential. What is the minimum face width you would recommend in this application if the spur gears are made from steel heat-treated to a hardness of 300 Bhn?

16-12. The shaft keyed to gear 4 of the hydraulic cylinder-rack-gear drive in Fig. P 88 must oscillate through about 1 revolution while carrying a steady torque of 6,000 lb-ft. The maximum instantaneous speed of rotation of gear 4 will be 10 rpm. Reasonably high reliability is desired, and the service life required is 10,000 cycles.

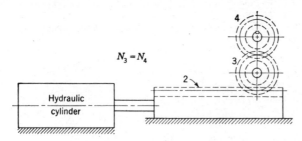

FIG. P 88

What pitch and face width do you recommend if the teeth are 20° full-depth and are hobbed from steel heat-treated to 250 Bhn?

16-13. Twenty-five horsepower must be transmitted from one shaft rotating at 3,600 rpm to another shaft rotating at 1,200 rpm when the center distance is required to be 6 in. The power source is a 3-phase synchronous motor, and the load is uniform. If it is assumed that 20° full-depth teeth will be used, (a) what quality of gears do you recommend, (b) what do you recommend for the minimum properties of the gear materials, and (c) what pitch and numbers of teeth do you recommend?

16-14. Same as Prob. 16-13 except that the gears are to be as light and quiet-running as possible with spur gears and the material will be 3140 steel oil-quenched and tempered at 1000°F. (a) What quality of gears do you recommend? (b) What is the minimum possible face width? (c) What pitch and numbers of teeth do you recommend for the gears?

16-15. Design the gears required for the first reduction in the hoist reduction unit in Example 16-1 if a 900-rpm crane and hoist motor is used instead of the 1,800-rpm motor originally specified.

16-16. Design the gears required for the first reduction in the hoist reduction unit in Example 16-1 if a triple reduction is used instead of the double reduction originally specified.

16-17. Design the gears required for the first reduction in the hoist reduction unit in Example 16-1 if the pinion is heat-treated to 350 Bhn and the gear is made of steel heat-treated to 300 Bhn.

16-18. Design the gears required for the first reduction in the hoist reduction unit in Example 16-1 if a 900-rpm crane and hoist motor is used, the pinion is heat-treated to 350 Bhn, and the gear is made of steel heat-treated to 300 Bhn.

16-19. Design the gears required for the first reduction in the hoist reduction unit in Example 16-1 if double-helical gears with 30° helix angles are used, the pinion is heat-treated to 350 Bhn, and the gear is made of steel heat-treated to 300 Bhn.

16-20. Problem 13-9 was concerned with the design of the centrifugal brake for a governor-controlled jack similar to that shown in Fig. 10-6. The torque on the screw required to prevent self-lowering at a speed in excess of $\frac{1}{2}$ in./sec was given as 5,400 lb-in.   The step-up gear train consists of a pair of bevel gears and three pairs of spur gears, as shown in Fig. P 89.   The double-thread screw has a pitch of 0.625 in., $N_2 = 25$, $N_3 = 12$, $N_4 = 50$, and $N_5 = 15$.

A reasonable service life for the jack is estimated to be 4,000 in. of travel under rated load.   If it is assumed that the efficiency of each step in the gear train is 97 per cent, what (a) pitch, (b) pitch diameters, and (c) face width do you recommend for gears 4 and 5 if they are to have 20° full-depth teeth cut on a gear shaper from a steel heat-treated to 250 Bhn?

Note: The 15-tooth pinion 5 may be machined without undercutting by use of a 15-tooth pinion shaper cutter.

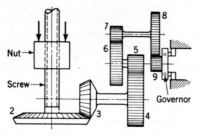

Fig. P 89. (*Courtesy Duff-Norton Company.*)

Fig. P 90

16-21. A diesel locomotive designed for high-speed freight service is to have four driving motors with a total of 1,500 hp.   A nose-suspension mounting will be used for the motors, as shown in Fig. P 90, in which part of the weight is carried on the truck frame and part on the axle.   The distance l between the motor-shaft and axle centers should be between 17 and 20 in. in order to permit the use of existing motors and trucks.   The maximum tractive effort, i.e., the tangential force between the wheels and the rail, for the locomotive is to be 55,000 lb when starting, and the maximum tractive effort when full power may be used is to be 35,100 lb at 16 mph.   These operating conditions require a gear ratio as close to 16/61 as is practical.

What (a) pitch, (b) numbers of teeth, (c) diameters, and (d) face width do you recommend for the gears if the 20° full-depth teeth are to be shaved after heat-treating to give a 350-Bhn pinion and a 300-Bhn gear?

16-22. The control mechanism of a missile requires a gear reduction with a ratio of 4.5:1 that can carry a torque of 500 lb-in. on the low-speed shaft for a few (less than 1,000) cycles at speeds near zero.   The power source may be considered uniform, and the load as moderate shock.   The gears must be as light as possible, and 99 per cent reliability is considered sufficient.   Although the cycles are few, it is essential that the accuracy not be impaired by too rapid wear.   The bearing mounting is to be as simple as possible, and it is desirable to keep friction losses in the bearings to a minimum value.

Design the gears for this application.

16-23. A portable generator is being designed for use by spies who have parachuted into enemy territory.   It is proposed to use a solid-fuel 5-hp gas turbine running at 14,000 rpm to drive a generator at 5,000 rpm.   The fuel charge will last only 10 min, and the entire unit will then be discarded.   Weight must be kept to an absolute minimum, but good reliability and a simple design are essential.

Design the gears for this application.

16-24. A portable generator is being designed for use by troops fighting guerrillas in the mountains and jungles. It is proposed to use a 10-hp 2-cycle gasoline engine running at 6,000 rpm to drive a generator at 3,600 rpm. A life of 50 hr with 99 per cent reliability is considered sufficient.

Design the gears for this application.

16-25. A rock crusher is being designed. It is proposed to use a 300-hp diesel engine running at 950 rpm to drive the crusher mechanism (a toggle mechanism, as shown in Fig. 2-17) at the rate of 60 cpm. A double-reduction gear train is to be used, and ratios of 4.5:1 and 3.5:1 have been suggested in the interest of equalizing the diameters of the gears.

Design the gears for the first reduction of 4.5:1.

16-26. Same as Prob. 16-25 except design the gears for the second reduction of 3.5:1.

16-27. The analysis of a proposed design for a disk cam with a $1\frac{1}{4}$-in.-diameter radial roller follower indicates that the critical stress will be at one point where the cam radius of curvature is 0.563 in. and the normal force is 540 lb. Since gear teeth are just special cams, the allowable contact pressures q_o in Table 16-7 and Eqs. (16-28) and (16-29) may be used to determine the required length of contact between the cam and follower. If, in this case, the roller has a hardness of R_c 58 (600 Bhn), what length of roller will be required if (a) the cam is made from ASTM 30 cast iron and (b) the cam is made from carburized steel with a surface hardness of R_c 58 (600 Bhn)?

16-28. The analysis of a proposed design for a disk cam with a mushroom follower indicates that the critical stress will be at one point where the cam radius of curvature is 0.375 in. and the normal force is 290 lb. Since gear teeth are just special cams, the allowable contact pressures q_o in Table 16-7 and Eqs. (16-28) and (16-29) may be used to determine the required length of contact between the cam and follower. If, in this case, the follower-surface hardness is R_c 58 (600 Bhn), what do you recommend for the width of the cam if (a) it is made from ASTM 30 cast iron and (b) it is made from carburized steel with a surface hardness of R_c 58 (600 Bhn)?

16-29. Compare the horsepower capacity of a 24-tooth 10-pitch 20° full-depth spur pinion driving a 48-tooth gear with that of a similar pinion driving a 48-tooth internal gear when all gears are cut on a gear shaper from 250-Bhn steel and the pinions are rotating at 1,500 rpm.

16-30. An 18-tooth 12-pitch 20° full-depth spur pinion is driving a 36-tooth gear. A change in the function of the machine requires the replacement of these gears with a pair having 20 and 36 teeth, respectively. Since this cannot be done with standard spur gears, it becomes necessary either to use nonstandard spur gears or to replace the spur gears with helical gears. For the latter method, specify (a) the pitch of the standard 20° full-depth hob that will be used to cut the gears, (b) the helix angle, and (c) the outside diameters of the gear blanks.

16-31. Designate the larger gear on the countershaft in Fig. 16-23 as 3 and the smaller as 4. Derive an equation for the required helix angle on gear 4 as a function of the pitch diameters and the helix angle on gear 3 if there is to be no net axial force on the countershaft.

16-32. The helical-gear speed reducer in Fig. P 91 is rated to transmit 20 hp at an input speed of 1,750 rpm. The helix angle is 15°, and the 22-tooth pinion and the 54-tooth gear were hobbed with an 8-pitch 20° full-depth hob. Determine (a) the pitch diameters of the gears, (b) the outside diameters of the gears, (c) the axial or thrust force on the output shaft, and (d) the resultant forces acting in the planes perpendicular to the shaft axis at bearings A and B. (e) What will happen to the forces on the bearings if the direction of rotation is reversed?

NOTE: The worst reactions on each bearing, from (d) and (e), would be combined with the forces due to the belt, chain, or gear on the overhung part of the output shaft to permit the calculation of shaft bending moments in determining the required diameter of the shaft and to permit the selection of the proper bearings to support the shaft.

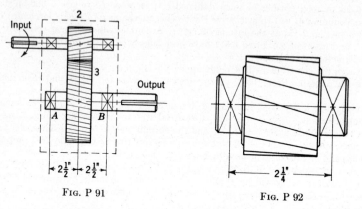

FIG. P 91

FIG. P 92

16-33. Each planet pinion (Fig. P 92) in a planetary gear train transmits a force of 2,450 lb. The 14-tooth pinions have a 15° helix angle, and the teeth were hobbed by use of a 7-pitch 20° full-depth hob. What are the resultant bearing reactions?

16-34. Figure P 93 shows a 15-tooth 4-pitch straight bevel pinion 2 driving a 34-tooth bevel gear 3. Calculate (a) the pitch diameters, (b) the pitch angles, (c) the equivalent pitch radii, and (d) the equivalent numbers of teeth.

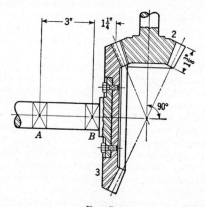

FIG. P 93

16-35. The 15-tooth 4-pitch 20° involute straight bevel pinion 2 in Fig. P 93 is driving a 34-tooth bevel gear 3. Assuming that the torque on the bevel pinion is 6,700 lb-in., determine (a) the tooth force components on gear 3 and (b) the resultant forces acting in planes perpendicular to the shaft axis at bearings A and B.

Note: These forces must be known in order to design the shaft and to select proper bearings to support the shaft.

16-36. Complete the sketches of the crossed helical gears in Fig. P 94, indicating (*a*) the helix angles, (*b*) the hands of the gears, and (*c*) the directions of rotation.

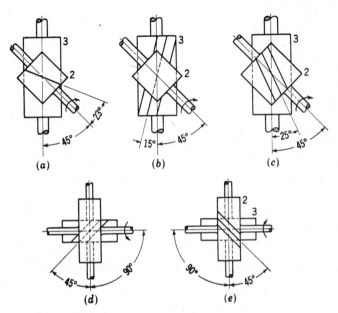

Fig. P 94

16-37. A RH 30°-helix-angle 18-tooth pinion 2 is driving a 28-tooth gear 3 when the angle between the shafts is 90°. The pinion teeth are hobbed by use of a 10-pitch 20° full-depth hob, and the pinion is rotating at 2,500 rpm.

(*a*) What hob must be used to cut the teeth on gear 3? Why? (*b*) What is the helix angle on gear 3? (*c*) Calculate the pitch diameters. (*d*) Calculate the distance between the shafts. (*e*) Determine the velocity of sliding in feet per minute.

16-38. It is desired to connect two perpendicular, nonintersecting shafts that are 4.000 in. apart by use of standard 20° full-depth helical gears, and the angular-velocity ratio is to be 23:19. (*a*) What do you recommend for the pitch of the hob? (*b*) What helix angles do you recommend?

ONE SUGGESTED PROCEDURE: Assume 19 and 23 teeth for the gears, assume helix angles of 45°, and solve for the required normal pitch. Round off the pitch to the closest numerically larger standard pitch, e.g., round off 8.34 to 9, and then use a trial-and-error solution to determine the combination(s) of helix angles that will give the desired center distance. NOTE: There will be no solution if the pitch is rounded off to the closest numerically smaller pitch.

16-39. Same as Prob. 16-38 except that the angular-velocity ratio is to be 24:19.

16-40. A 38-tooth 8-pitch 20° full-depth spur gear is to be finish-machined by the rotary shaving process utilizing a 15°-helix-angle cutter with 64 teeth. At what speed must the cutter rotate if the cutting speed is to be 200 fpm?

16-41. The hobbing of gears is discussed in Sec. 16-17.   It is pointed out that the angular velocity ratio is simply the inverse of the ratio of the numbers of teeth when a spur gear is being generated and that, when a helical gear is being cut, the ratio must be modified to take into account the fact that the hob is being fed along the axis of the blank rather than along the helical path of the teeth.

A 10-pitch single-tooth right-hand hob with a lead angle of 2°30′ is available. What must the angular velocity ratio be to generate (a) a 30-tooth spur gear and (b) a 30-tooth right-hand helical gear with a 15° helix angle when the axial feed of the hob is 0.030 in. per revolution of the gear blank?

NOTE: The ratio for (b) will most likely be unobtainable, and thus, the feed will have to be modified before the gear can be cut.

16-42. Same as Prob. 16-41 except to generate a left-hand helical gear.

16-43. A series of worm and worm-gear combinations is being designed for sale directly to machinery manufacturers.   One of the major considerations is the choice of dimensions so that the center distances may be easily calculated for any combination of worms and worm gears.   One series is to consist of 4-pitch (in plane of rotation) bronze worm gears and a single-thread hardened-steel worm with a pitch diameter of 3.0000 in.   The hob will be identical, except for clearance, to the worm, and the tooth proportions will correspond to the 20° full-depth involute system.

Determine for the pair consisting of the worm and a 48-tooth worm gear (a) the center distance, (b) the axial pitch of the worm, (c) the lead angle of the worm, and (d) the outside diameter of the worm.

If the worm and worm gear are mounted on rolling-contact bearings, the worm speed is 1,150 rpm, the torque on the worm-gear shaft is 750 lb-ft, and the coefficient of friction is 0.04, what will be (e) the required input horsepower, (f) the required heat-dissipation rate in Btu per minute when the unit is in continuous use, and (g) the axial force on the worm shaft?

16-44. Same as Prob. 16-43 except that the worm is to have a quadruple thread.

Chapter 17.   Gear Trains, Variable-speed Drives, and Differential Mechanisms

17-1. The input-shaft speed is 500 rpm in the direction indicated.   What are the output-shaft speed and direction of rotation for the gear train in Fig. 95?

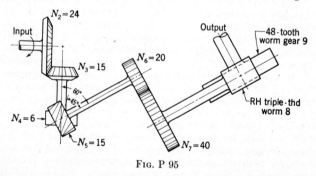

FIG. P 95

17-2. Figure P 96 shows a gear train for the power-feed drive of a production drill press.   The clutch, similar to Fig. 12-5, is set so that it will slip and thus protect the other members of the train if the drill hits a hard spot or otherwise jams.   A spring (not shown) returns the quill rack to its uppermost position each time the clutch is disengaged.   The numbers of teeth are chosen for the 12-pitch change gears 6 and 7 to give the feed per revolution of the drill required by the given combination of drill size and material being drilled.

If it is assumed that a feed of about 0.007 in./revolution of the drill is required and that the drill must rotate in the direction indicated, (a) what numbers of teeth should be specified for gears 6 and 7 and (b) what should be the hand of the worm 4?

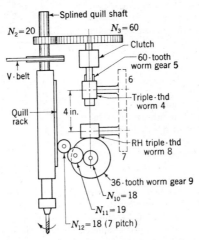

Fig. P 96

17-3. The gear train for the governor-controlled self-lowering jack in Fig. 10-6 is shown in Fig. P 89. The screw has a double LH thread with a pitch of 0.625 in., and the numbers of teeth on the gears are $N_2 = 25$, $N_3 = 12$, $N_4 = 50$, $N_5 = 15$, $N_6 = 48$, $N_7 = 15$, $N_8 = 49$, and $N_9 = 14$. What are the speed and the direction of rotation of the governor when the nut is descending at the rate of $\frac{1}{2}$ in./sec?

17-4. The speedometer drive in most automobiles is taken from the transmission output shaft, as shown in Fig. P 97. Since it is customary to furnish a number of

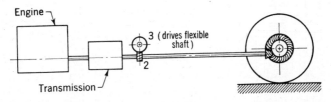

Fig. P 97

different rear-axle gear ratios to suit the customers' requirements, it becomes necessary to provide either a number of speedometer designs or a number of different gear ratios for the flexible-shaft drive. From manufacturing and service viewpoints, it is preferable to use the different gear ratios, and, in general, it is not possible to get the exact desired ratio because of the necessity of using whole numbers of teeth.

Assuming that a car with 7.10 × 15 tires, which make 745 revolutions/mile, is offered with rear-axle ratios of 4.3:1, 4.1:1, 3.9:1, 3.74:1, and 3.42:1, that the odometer reads correctly at 1,001 revolutions/mile, and that a 7-tooth pinion 2 will be used in each case, determine and tabulate for each rear-axle ratio (a) the optimum number of teeth for gear 3 and (b) the per cent error in the speedometer and odometer readings.

17-5. In Sec. 16-6 the dynamic factor K_V was shown to be a function of the pitch velocity of the gears. For planetary gear trains, it is no longer so simple, because the rotation of the carrier introduces a velocity that must also be considered. Actually, the velocity of interest is the velocity at the point of tangency of the pitch circles relative to the velocity of the coincident point on the carrier.

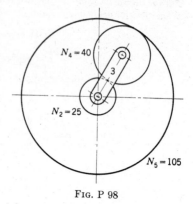

FIG. P 98

The gears in the train in Fig. P 98 have standard 20° 6-pitch full-depth teeth. Gear 2 is rotating cw at a speed of 1,750 rpm, and gear 5 is fixed. What value of velocity should be used in calculating K_V for (a) contact between gears 2 and 4 and (b) contact between gears 4 and 5?

17-6. Determine the speed and the direction of rotation of the planet carrier 3 if, in Fig. P 98, gear 5 is fixed and gear 2 is driving cw at 300 rpm.

17-7. Determine the speed and the direction of rotation of gear 5 if, in Fig. P 98, gear 2 is fixed and the planet carrier 3 is driving ccw at 550 rpm.

17-8. Determine the speed and the direction of rotation of gear 5 if, in Fig. P 98, gear 2 is driving ccw at 200 rpm and the planet carrier 3 is driving cw at 300 rpm.

17-9. Determine the speed and the direction of rotation of the planet carrier 3 if, in Fig. P 98, gear 2 is driving cw at 500 rpm and gear 5 is driving cw at 100 rpm.

17-10. Calculate the gear ratios for the Ford Model T transmission in Fig. 17-13.

17-11. Calculate the gear ratios for the Buick Dynaflow transmission in Fig. 17-14.

17-12. Figure P 99 shows the compound planetary gear train used in an aircraft servo. Determine the angular-velocity ratio ω_2/ω_7.

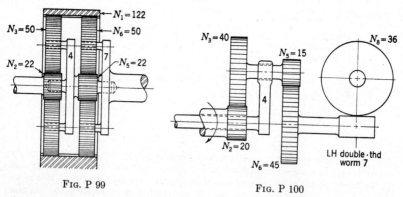

FIG. P 99   FIG. P 100

17-13. In the gear train in Fig. P 100, worm gear 8 may be driven either by holding the carrier 4 fixed and rotating gear 2 or by holding gear 2 fixed and rotating the carrier 4. (a) Determine the angular-velocity ratio ω_2/ω_8 and the direction of rotation of the worm gear if the carrier is fixed and gear 2 is rotating in the direction indicated. (b) Determine the angular-velocity ratio ω_4/ω_8 and the direction of rotation of the worm gear if gear 2 is fixed and the carrier is rotating in the direction indicated.

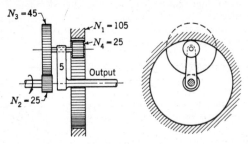

FIG. P 101

17-14. The input shaft 2 of the gear train in Fig. P 101 is rotating at the speed of 2,000 rpm. What is the speed of the output shaft and what is its direction?

17-15. Very large gear ratios can be obtained using a planetary gear train with nonstandard spur or standard helical gears.

(a) Show that for the gear train in Fig. P 102 the maximum value for ω_2/ω_5 is 15.5:1 when the center distance for the gears is fixed at 3.00 in. and standard 10-pitch spur gears are used.

(b) If nonstandard spur or standard helical gears (Prob. 16-30) are used, the pitches (in the plane of rotation), and thus the numbers of teeth, can be varied without changing the center distance. What is the maximum ratio for the gear train under these conditions if the numbers of teeth do not vary more than ±1 from 30?

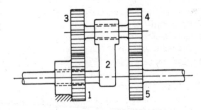

FIG. P 102

17-16. Same as Prob. 17-15 except standard 32-pitch spur gears are used in (a) and the numbers of teeth cannot vary more than ±1 from 96.

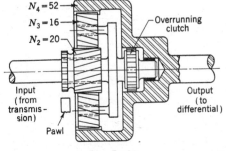

FIG. P 103

17-17. Figure P 103 shows part of an overdrive unit used in some automobiles to decrease the over-all reduction ratio from the engine to the wheels. The overdrive operates when the pawl engages the toothed wheel and locks the sun gear 2 to the frame. When the sun gear is not held stationary, the drive is through the overrunning clutch and the unit is "freewheeling," i.e., the output shaft is free to run faster than the input shaft; thus the engine has no braking effect when the driver takes his foot off the gas. A clutch (not shown) is provided so that the input and output shafts may be locked together whenever the overdrive or freewheeling features are not desired or may be dangerous, as in hilly or mountainous country.

For the unit shown, what is the velocity ratio when it is in overdrive?

17-18. The spur-gear differential in Fig. P 104 is similar in construction to that in Fig. 17-16. However, the differential is now being used to control the phase of the output shaft relative to the input shaft. The pinions have 18 teeth, and the gears have 60 teeth.

Determine (a) the hand of the single-thread worm and (b) the number of teeth on the worm gear that will result in a 5°-angular relative advance, i.e., in the direction of rotation, of the output shaft for one cw revolution of the handwheel.

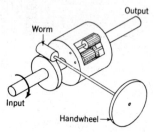

Fᴵɢ. P 104

17-19. A belt-type variable-speed transmission similar to that shown in Fig. 17-19 is designed to give a ratio of maximum output speed to minimum output speed of 6:1. If the sheaves can be adjusted equally from their neutral (no-reduction) positions and the motor speed is 1,750 rpm, what are the maximum and minimum output speeds of the transmission?

17-20. Figure P 105 shows the skeleton outline of one of the linkages used in a Zero-Max transmission (Fig. 17-21). Assuming that four linkages are used, with the cranks equally spaced, (a) draw one of the linkages with O_4 set for the maximum output speed, (b) specify the quarter of a crank revolution during which this linkage will be driving, and (c) use any of the methods discussed in Chaps. 3 and 4 to determine the values and then plot the curve of the angular-velocity ratio ω_6/ω_2 as a function of the crank angle during the quarter revolution this linkage is driving.

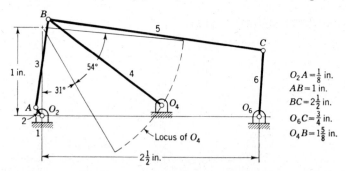

$O_2A = \frac{1}{8}$ in.
$AB = 1$ in.
$BC = 2\frac{1}{2}$ in.
$O_6C = \frac{3}{4}$ in.
$O_4B = 1\frac{5}{8}$ in.

Fᴵɢ. P 105. (*Courtesy Revco Incorporated.*)

Be sure to include the maximum and minimum points. (d) What is the per cent variation of the output speed, considering the maximum as 100 per cent? (e) What will be the effect of load inertia on the output speed variation? Explain. (f) What will be the effect of reversing the direction of rotation of the input shaft?

17-21. Figure P 106 shows the basic parts of one model of a Graham variable-speed transmission.   The inside diameter of reaction ring 1 is 4½ in., the maximum and minimum effective diameters of the tapered roller 3 are 1.8 and 1.2 in., respectively, ring gear 5 has 52 teeth, and pinion 4 has 14 teeth.   Through what range of reduction ratios will the unit operate?

17-22. Figure P 106 shows the basic parts of one model of a Graham variable-speed transmission.   If the inside diameter of the reaction ring is 4½ in., ring gear 5 has 57 teeth, and pinion 4 has 19 teeth, (*a*) what are the maximum and minimum effective diameters of the tapered roller if the output speed range is from one-fifth through zero to minus one-fifth of the input speed?   (*b*) Describe, with respect to the large-diameter end of the tapered roller, the location of the reaction ring when the output speed is zero.

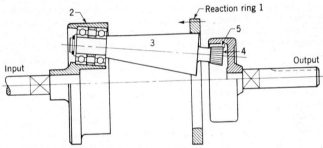

Fig. P 106. (*Courtesy Graham Transmissions, Inc.*)

17-23. The drive requirements for a high-speed packaging machine are that a certain shaft speed must be set within 0.1 rpm of any specified speed between 450 and 470 rpm.   This would be almost impossible to accomplish with a d-c motor and difficult with any standard variable-speed drive, particularly with changing loads. It is proposed to use two 1,800-rpm synchronous motors in combination with gear reducers, a variable-speed drive, and a differential as shown in Fig. P 107.

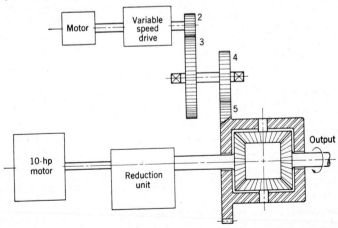

Fig. P 107

The variable-speed drive unit is similar to that shown in Fig. P 106 and will give output speeds varying continuously between $-1/100$ to $+1/3$ of the input speed. (a) In what directions should the motor shafts rotate? (b) Assuming that $\omega_2/\omega_3 = \omega_4/\omega_5$, what should these ratios be to permit the maximum sensitivity in adjustment of the output speed? (c) Assuming an over-all efficiency of 70 per cent for the variable-speed drive and the gear trains, what horsepower rating would be required for the small motor?

17-24. The differential adjustment for a Fenwal Thermoswitch (Fig. 17-34) is composed of special screws with 42 and 38 threads per inch. If an axial adjustment of 0.00235 in. results in a change of 100°F in the temperature at which the contacts open, what is the change in temperature setting given by one revolution of the adjusting screw?

17-25. A Thermoswitch, similar to that in Fig. 17-34, is being designed for an application in which one revolution of the adjusting screw is to change the temperature setting by about 50°F. If an axial adjustment of 0.00235 in. results in a temperature-setting change of 100°F and the outside screw has 42 threads per inch (special), (a) how many threads per inch should the inside screw have and (b) what is the closest approach to 50°F/revolution possible with a whole number of threads per inch on the inside screw?

17-26. Equations (17-35) and (17-36) for the operating force and mechanical advantage, respectively, of a differential chain hoist neglected the effect of friction of the sprocket bearings. If rolling-contact bearings are used in both sprockets, the friction may be neglected without introducing serious error. However, in such a case a brake or some other means of locking the compound sprocket against rotation will be required to prevent the load from falling when the operating force is removed. In the interest of economy and simplification, most differential chain hoists use simple bushings, or no bushings at all, as bearings and depend upon the bearing friction to hold the load in place when the operator releases his hold on the chain.

For example, consider a ½-ton-capacity differential chain hoist, designed on the basis of Eq. (17-35), to be operated by a force of 50 lb. (a) What is the required smaller pitch diameter of the compound sprocket if the larger pitch diameter is 5.00 in.? (b) What is the required pitch diameter of the load sprocket if the two load-carrying chains are to be parallel? (c) If it is assumed that both sprockets from (a) and (b) rotate on 1-in.-diameter pins, what is the minimum value of coefficient of friction required if the pin friction alone is to hold the load in an elevated position? (d) If the coefficient of friction between the sprockets and pins is 0.15, what will be the operating force required at rated capacity?

Chapter 18.   Bearings and Lubrication

18-1. A journal bearing has been operating satisfactorily for many years under the following conditions:

Load, 3,000 lb
Speed, 2,000 rpm
Journal diameter, 3 in.
Bearing length, 3 in.
Diametral clearance, 0.006 in.
Oil, SAE 10 at 160°F (temperature maintained by circulating the lubricating oil through a cooler)

It is proposed to use the machine under identical conditions as before except that the journal speed is to be 1,000 rpm. (a) Do you recommend using the bearing at the new operating speed? Justify your conclusion. (b) If your answer to (a) is yes, determine the horsepower loss in the bearing.   (c) If your answer to (a) is no, suggest a change that may be made easily and prove that your suggestion will work.

18-2. What is the minimum speed of operation you would recommend for a hydrodynamic journal bearing operating under the following conditions:

Load, 15,000 lb
Journal diameter, 4.5 in.
Bearing length, 6 in.
Clearance ratio, 0.002 in./in.
Oil, SAE 10 at 180°F

18-3. A 1,000-kw generator is to be driven at 3,600 rpm by a steam turbine. The two 4-in.-diameter by 5-in.-long journal bearings of the generator are required to carry only the 3,500-lb weight of the rotor.   Assuming that the diametral clearance will be 0.006 in., the oil will be SAE 10, and the temperature of the oil leaving the bearing will be maintained at 140° by circulating the oil through a heat exchanger, determine (a) the operating minimum oil-film thickness, (b) the power loss in the bearings, (c) the oil flow in gallons per hour (gph) required if all the heat is removed by the oil and the oil temperature rise when passing through the bearings is to be 10°F [specific gravity of lubricating oils at temperature $t°F$ is approximately $0.91 - 0.000365(t - 60)$], (d) the speed at which the minimum thickness of the oil film is 0.0001 in., and (e) the minimum length of bearing that would give a satisfactory oil-film minimum thickness.

18-4. A 4,500-hp gas turbine is being designed for stationary service in driving a compressor used in the transmission of natural gas through pipe lines.   Preliminary calculations have indicated that the rotating parts of the turbine (both compressor and turbine stages) will weigh about 15,000 lb and that the two bearings will have journal diameters of about 4½ in.   The turbine speed will be 10,000 rpm.

Assuming that the turbine bearings will carry only the equally distributed weight of the turbine rotor and that an oil-circulating system with a heat exchanger will be used to maintain the temperature of the oil leaving the bearings at 160°F, (a) select an oil and recommend a diametral clearance and a length for the bearing, (b) prove that your choice and recommendations in (a) will result in satisfactory operation, (c) determine the power loss in the turbine bearings, and (d) determine the oil flow in gallons per hour (gph) required if all the heat is removed by the oil and the oil temperature rise through the bearings is to be 10°F [specific gravity of lubricating oils at temperature $t°F$ is approximately $0.91 - 0.000365(t - 60)$].

18-5. A journal bearing with a diameter of 8 in., a length of 6 in., and a diametral clearance ratio of 0.0015 in./in. is to carry a load of 4,150 lb when the journal speed is 150 rpm.   If possible, the bearing is to operate in 100° ambient temperature without external cooling with a maximum oil temperature of 200°F.   If external cooling is required, it is to be as little as possible to minimize the required oil flow rate and heat-exchanger size.

(a) What SAE oil do you recommend for this application?   (b) Will the bearing operate without external cooling?   (c) If your answer to (b) is yes, determine the operating oil temperature.   (d) If your answer to (b) is no, determine the oil flow in pounds per minute required to carry away the excess of heat generated over heat dissipated when the oil temperature rises from 190 to 200°F when passing through the bearing.

18-6. It is proposed that a journal bearing operate under the following conditions:

Oil, SAE 20
Load, 1,000 lb

Speed, 370 rpm
Journal diameter, $2\frac{1}{2}$ in.
Bearing length, $2\frac{1}{2}$ in.
Diametral clearance, 0.005 in.
Ambient temperature, 80°F

(a) What will be the oil equilibrium temperature? (b) Will the bearing perform satisfactorily? Justify your conclusion. (c) If your answer to (b) is no, what change or changes would you suggest?

18-7. For completely reversed loading, the performance of a squeeze-film bearing can be related in terms of maximum and minimum values of eccentricity ratio to a modified version of the load number by simply using n in cycles per minute rather than revolutions per minute. Starting with Eq. (18-14) and letting $p = -p_0 \sin 2\pi nt$, show that the relationship is

$$\frac{1}{(1 - \epsilon_{max})^{0.28}} - \frac{1}{(1 + \epsilon_{max})^{0.28}} = \frac{N_L}{52.7}$$

18-8. A furnace is being designed for annealing castings on a production-line basis. Small four-wheeled trucks loaded with castings are to be run through the furnace at a speed of 3 fpm. The truck-axle bearings are somewhat shielded from the heat when in the furnace, but they may still reach a temperature of 350°F. Assuming that the total weight of the truck and castings will be 3,800 lb, the wheels will be 12 in. in diameter, and the journal diameters will be 2 in., (a) recommend a suitable bearing material for this application, (b) determine the minimum possible length for the bearings, and (c) recommend a length for the bearings.

18-9. If it is assumed that a porous bronze bearing material can be used at values of $pV = 35,000$, what is the minimum journal diameter that can be used if the bearing length is equal to the bearing diameter, the load is 80 lb, and the journal speed is 1,375 rpm?

18-10. A roller bearing is rated by one manufacturer to have an average life of 1,000 hr when carrying a load of 9,000 lb at 1,800 rpm. (a) What is the 90 per cent life expectancy for these bearings when carrying 9,000 lb at 1,800 rpm? (b) What is the rated load capacity of this bearing for an average life of 50,000 hr at 1,800 rpm?

18-11. (a) What is the per cent decrease in expected life of a group of ball bearings when used under a load 10 per cent greater than rated load? (b) What is the per cent increase in expected life of a group of ball bearings when used under a load 10 per cent less than rated load?

18-12. A ball bearing is to be selected to carry a load of 130 lb when running continuously at 550 rpm for three years without requiring replacement of more than 10 per cent of the bearings. What catalogue rating should be specified if the bearing manufacturer rates his bearings on the basis of (a) 90 per cent life expectancy of 1,000,000 revolutions, (b) an average life of 10,000 hr at 900 rpm, and (c) an average life of 3,800 hr at 1,000 rpm?

18-13. Ball bearings were selected on the basis of having only 10 per cent failures in 10,000 hr when carrying a nominal load of 2,000 lb. However, service reports indicate that 10 per cent of the bearings have actually failed by the end of 4,000 hr of operation.

What value of nominal load should be used in selecting new bearings to meet the original specifications?

18-14. A particular size 1,750-rpm electric motor has been redesigned, and the rating has been increased by 50 per cent. In the past, the life of the average bearing has been found to be 9,000 hr.

Assuming that the bearing loads are directly proportional to the motor torque, by what factor should the design load on the old bearings be multiplied to ensure that no more than 10 per cent of the bearings in the new-model motor will fail in 10,000 hr?

18-15. Ball bearings are to be selected for an application in which the shaft rotates at 2,100 rpm and the radial load variation during each revolution may be approximated by 2,000 lb for 90°, 6,000 lb for 260°, and 15,000 lb for the remainder of the revolution.

What basic dynamic load rating should be specified to ensure not more than 10 per cent failures in 500 hr?

18-16. Same as Prob. 18-15 except roller bearings are to be selected.

18-17. A cam shaft rotating at 350 rpm is to be supported between ball bearings. The radial load on the cam will vary sinusoidally each revolution between a minimum load of 200 lb and a maximum load of 750 lb.

What basic dynamic load rating should be specified to ensure not more than 10 per cent failures in 12,000 hr of operation?

Chapter 19.   Vibrations in Machinery

19-1. An elevator is supported by four $\frac{5}{8}$-in. 6 × 19 traction-steel wire ropes. If the total weight of the car and passengers is 8,000 lb, the distance from the traction sheave (Fig. P 84) to the car at the basement landing is 480 ft, and the distance from the traction sheave to the car at the top-floor landing is 15 ft, determine (a) the static deflection of the cable and (b) the natural frequency in cycles per second of the vertical motion of the car when the car is at the basement landing and at the top-floor landing.

HINT: Utilize the definition of modulus of elasticity and the wire-rope data in Table 14-5.

19-2. What must be the value of the radius of gyration about the Z-Z axis if the system in Fig. 19-10 has the same natural frequency for oscillation about the Z-Z axis as it does for vibration in the vertical direction?

19-3. If the elasticity of the tires is neglected, the automobile suspension system may be represented by Fig. 19-9. The most important modes of vibration are the vertical translatory motion and the oscillatory motion about the Z-Z axis. If the automobile weighs 4,000 lb, the spring rate for each spring is 200 lb/in., the wheelbase ($2a$ in Fig. 19-9) is 122 in., the radius of gyration of the body about the Z-Z axis is 50 in., and, as an approximation, $b = 0$, what are the two important natural frequencies in cycles per second?

19-4. The a-c–d-c motor-generator set for an elevator drive in an apartment building rotates at 1,750 rpm. Even though the rotating parts are dynamically balanced to a high degree, the vibration due to the remaining unbalance is disturbing to those residing in the apartments on the floor below. If the total weight of the motor-generator set and base is 2,500 lb, (a) what spring rate would you recommend for the suspension system? (b) What materials or types of isolators would be satisfactory in this case?

19-5. A reciprocating compressor is belt-driven at 750 rpm by a 1,750-rpm motor. The total weight of the compressor, motor, and base is 435 lb. (a) What spring rate would you recommend for a suspension system that will transmit only 10 per cent of the compressor inertia forces to the floor? (b) What materials or types of isolators would be satisfactory in this case? (c) For the spring rate in (a), what per cent of the force due to motor unbalance will be transmitted to the floor?

19-6. An 80-lb electronic instrument is to be attached to the deck of a ship. Under certain conditions the deck has been found to vibrate with an amplitude of 0.006 in. at a frequency of 40 cps.

What spring rate is required for each of the four isolators if the amplitude of vibration of the instrument is to be no greater than 0.001 in.?

19-7. An attempt is being made to obtain precise measurements with electronic apparatus in a laboratory located over a subway. The vibration of the building structure excited by the subway trains has seriously affected the stability, and thus the accuracy, of the equipment. Measurements made with a vibration meter and analyzer have indicated that the disturbance has an amplitude of 0.000250 in. at a frequency of 17.3 cps. Further investigations, made by use of a vibrating table in another laboratory, have shown that the apparatus is unaffected by vibrations at a frequency of 17.3 cps, provided that the amplitude does not exceed 0.000028 in.

(*a*) What spring rate would you recommend for a suspension system for this 50-lb piece of equipment? (*b*) What materials or types of isolators would be satisfactory?

19-8. A 1,150-rpm electric motor drives the propeller in a small wind tunnel, as shown in Fig. P 108. The mass moments of inertia about the axis of rotation of the motor rotor and the propeller are 8.10 lb-in.-sec² and 293 lb-in.-sec², respectively. What will be the minimum diameter of a steel shaft if the critical frequency for torsional vibrations is to be at least 40 per cent above the normal running speed?

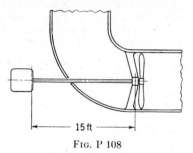

FIG. P 108

19-9. A pilot-plant model of a new machine for making photographic film is being constructed. A ¾-hp single-phase 1,750-rpm electric motor is to drive the casting roll through a reduction gear. It is important to isolate as much of the 120-cps motor-torque variation as possible to ensure a uniform film thickness. It is proposed to use a coupling, similar to that in Fig. 11-11a, that operates with a twist angle of 15° under rated torque. The mass moment of inertia of the motor rotor about its axis of rotation is 0.056 lb-in.-sec², and the equivalent mass moment of inertia of the casting roll and gear train is 0.28 lb-in.-sec². If the shafts are considered to be infinitely rigid relative to the coupling, what per cent of the motor-torque variation will be transmitted to the casting rolls?

19-10. A 20-hp steam turbine (Fig. P 109) is being designed for operation at 12,000 rpm. The turbine wheel weighs 11.2 lb. Strength considerations require the diameter of the solid steel shaft to be at least ⅜ in. Determine (*a*) the shaft diameter required if the critical speed for shaft whirl is 40 per cent above the normal running speed and (*b*) whether or not the ⅜-in. shaft diameter can be used. (*c*) What would you recommend for the shaft diameter of this turbine? Justify your recommendation.

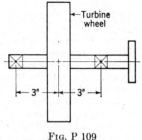

FIG. P 109

19-11. The critical frequency of shaft whirl for a shaft without a concentrated mass may be found by considering the shaft to have a concentrated mass, whose weight is 0.486 times the weight of the shaft, located at the center of the bearing span. Using this relationship, determine the required diameter of a solid steel shaft if the critical whirling frequency of the wind-tunnel drive shaft in Fig. P 108 is to be at least 40 per cent above the normal running speed of 1,150 rpm. (Density of steel = 0.283 lb/in.³)

INDEX